General Electronics Circuits

SECOND EDITION

J.J. DeFrance

Professor Emeritus
Department of Electrical Technology
New York City Community College
The City University of New York

HOLT, RINEHART AND WINSTON
NEW YORK

Library of Congress Cataloging in Publication Data

DeFrance, Joseph J
    General electronics circuits.

    Includes index.
    1. Electronic circuits. I. Title.
TK7867.D43   1976   621.3815'3   75-25718
ISBN 0-03-015481-2

PRINTED IN THE UNITED STATES OF AMERICA

6 7 8 9   038   9 8 7 6 5 4 3 2 1

Contents

7

Additional Power Supply Circuits   171

8

The Decibel   186

12

Bipolar Transistor
Equivalent Circuits   287

13

Distortion in R-C Amplifiers   303

17

Audio Devices   454

18

Special Audio Circuits   488

19

Linear Integrated Circuits 511

Appendix A

The Cathode-Ray Tube 539

Appendix B

The Cathode-Ray Oscilloscope 554

Appendix C

Current and Voltage
Letter Symbols 580

Preface

PREFACE TO THE SECOND EDITION

Three factors have made this second edition necessary. The first, and major, reason should be obvious. Since the first edition was written (back in the dark ages of 1962), major advances have been made in the semiconductor industry. As a result, solid-state devices have replaced vacuum tubes in most applications—especially at low power levels. New designs do not use tubes; and so in this edition, semiconductors are given priority treatment. In this respect, this new edition is long overdue. Yet, even now, vacuum-tube treatment cannot be eliminated completely. Some equipment is still being manufactured with tubes. (For example, some current models of TV receivers are only partially "transistorized.") In addition, many products with vacuum tubes are in use and require maintenance. Finally, at very high power levels, the vacuum tube is still the only suitable active device. Therefore, it is felt that the study of vacuum-tube characteristics—although not nearly as important as in the earlier days—cannot be dropped altogether.

The treatment in this edition is primarily, and heavily, solid state. Where pertinent, a brief presentation is also made of the older vacuum-tube circuits. In addition, Chapter 4 is devoted exclusively to vacuum tube fundamentals.

A second reason for this revision is to change the prior background requirements. The first edition had as a prerequisite "a good foundation in . . . vacuum-tube and semiconductor characteristics." This was not practical, and so, in accordance with reader comments, the fundamentals of vacuum tubes and semiconductors are added here.

Finally, the third change is to use the international system (SI) of units and abbreviations for all electrical quantities.

PREFACE TO THE FIRST EDITION

This text, which is intended for use in a first course on electronics circuitry, is aimed primarily at the level of the engineering technician. Prerequisite to understanding the subject matter in this volume is a background in the principles of algebra, vector algebra, and basic trigonometry, as well as a good foundation in direct-current fundamentals, alternating-current fundamentals, and vacuum-tube and semiconductor characteristics.

Whereas the above prerequisites are the "tool" subjects, this volume begins the application of these tools to actual practical circuits. The circuits are developed, their action is analyzed, and the function of the various components is explained. Typical component values are generally given so as to acquaint the student with practical circuitry. The circuits discussed in this volume may be found in the simplest home radio or in the most complex automation control equipment. Since these circuits are common to all areas of the electronics field, this volume has been titled *General Electronics Circuits*. After completion of this common-core subject matter, the student is prepared to advance into more specialized subject areas such as Communications, Industrial Electronics, or Computers.

Although some texts start with an integrated treatment of vacuum-tube and transistor circuitry, such a treatment (logical to an experienced person or advanced student) is often confusing for a newcomer in the field of electronic circuitry. The earlier chapters of this text were therefore written to present vacuum-tube circuits first, followed by their transistor counterparts. Then, as the student becomes technically more mature, the treatment in the later chapters is integrated.

In a direct personal approach, the author has used the same conversational style that has proved successful in previous texts. The emphasis is on practical concepts without sacrifice of technical depth or accuracy. The *programmed machine-question technique* used for review learning is discussed under *Instructional Notes*.

ACKNOWLEDGMENTS

The author is indebted to the following industrial organizations for their cooperation in furnishing photographs, tables, characteristics, and technical information on their products: The Astatic Corporation; CBS, Inc.–Hytron; Allan B. Dumont Labs Inc.; Electrovoice, Inc.; Fairchild Semiconductor; Freed Transformer Co.; General Electric Company; International Electronic Research Corp.; International Rectifier Corporation; James B. Lansing Sound, Inc.; Jensen Manufacturing Company; Jerrold Electronics Corp.; P. R. Mallory & Co.; Merit Coil & Transformer Corp.; Motorola Semiconductors; National Semiconductor Corp.; Radio Corporation of America; Radio Receptor Co., Inc.; Raytheon Manufacturing Co.; Shure Brothers, Incorporated; Sola Electric Company; Sylvania Electric Products, Inc.; Tektronix Inc.; Texas Instruments, Inc.; The Turner Company; United Transformer Co.; University Loudspeakers, Inc.; Westinghouse Electric Corp.

Garden City, New York   J. J. DeFrance

Instructional Notes

Many studies of the psychology of learning have shown that effective learning must involve active participation by the learner, and his correct responses must be "rewarded." Teaching machines developed in keeping with these basic principles have been very successful. In general, these machines give factual information; ask questions (in small steps) based on this information; elicit some form of response; and finally give or confirm the correct answer.

However, machines have serious limitations. One type, although it allows the student to make any answer, merely states the correct response and continues. Another type restricts the student to a choice of one out of only four answers. If a wrong response is selected, the machine indicates why it is wrong and then allows the student to make another response. When he selects the correct response, the program advances to the next step.

In a classroom situation, a live instructor can combine the better features of each type of machine. Not only can he allow a student complete freedom of response, but he can also modify his teaching "program" instantaneously to fit any response.

This text was written with such a teaching technique in mind. In the body of the text, factual information is given and circuit operation is described in more detail than usual, so that the instructor will not have to spend valuable class time lecturing at length to supplement missing items or skimpy treatment. Instead, the lesson time can be spent using a question-discussion-guidance technique, with heavy emphasis on student participation. To help implement this type of lesson, the author has incorporated many review questions at the end of each chapter. These questions follow the text sequence and represent small "bits" of each topic, much in the manner of a teaching machine. Sufficient class time should also be allowed for a satisfying analysis of all problems assigned for homework. If additional time is available, it can be well spent in enriching and motivating each lesson from the instructor's own practical experience.

This system has been tested by the author with several classes, with very gratifying results. Not only were class averages raised, but even more important, the students actually enjoyed these lessons and came to class better prepared to join in the general discussion.

Throughout this text, the term *current flow* is used synonymously with *electron flow,* the direction being from negative to positive. If reference to the opposite direction of flow is intended, it is clearly indicated by using the term *conventional current flow*.

1

Semiconductors—Diodes

The tiny transistor emerged into public view when the Bell Telephone Laboratories announced the development of the point-contact transistor in June of 1948. Much excitement was created because this speck of solid matter could control the flow of electrons in much the same manner as an electron tube. Voltage, current, and power amplification could all be achieved with this simple device consisting of a small crystal of the element germanium and two pointed wires making contact to it. The transistor required no heating power, and it operated effectively on only a few volts of power supply. The space and weight advantages of this midget device were tremendous. However, these early transistors had their drawbacks. They produced high noise levels; they had serious frequency limitations; and they were unreliable, as it was impossible to produce two units with identical characteristics. Would the transistor replace the electron tube, or would it remain a laboratory curiosity?

Research continued. Manufacturing techniques were improved, and within a few years many of the limitations of the early transistor were sufficiently overcome. Meanwhile a new type, the *junction* transistor, was developed. By 1953 commercial applications of transistorized equipment began to appear on the market. Their small size and low power requirements made transistors ideal for such uses as hearing aids and miniature portable radios. The transistor was here to stay.

SEMICONDUCTOR FUNDAMENTALS

Before we can discuss the characteristics of a transistor it is necessary to study the fundamentals that make its operation possible. We have already mentioned that transistors are made from a small crystal of germanium. Other materials, such as silicon, can be used. What is it about these materials that leads to devices which function like electron tubes? Electrically, these materials are classified as *semiconductors;* that is, they are neither good conductors like copper nor good insulators like glass. Germanium, for example, offers a resistivity of 60 Ω per cubic centimeter as compared to 1.7×10^{-6} Ω for copper and 9×10^{13} Ω for glass. You might wonder, if germanium is neither a good conductor nor a good insulator, why not classify it as a resistance material? To answer this let us first check the resistance of nichrome, one of the highest resistance materials. The resistance of nichrome is approximately 60 times that of copper, or 1×10^{-4} $(1.7 \times 10^{-6} \times 60)$ Ω per cubic centimeter. Compare this value with germanium and you will see that as a conductor germanium is 600,000 times poorer. In other words, with regard to the general classifications as insulators, semiconductors, and conductors, resistance materials fall under the category of conductors. Germanium and silicon are not the only semiconductor materials. Other materials in this class are elements such as selenium, sulfur, cesium, boron, and most oxides and carbides.

It is interesting to note that some peculiar properties of semiconductors were discovered many years ago. For example, around 1880 it was found that current would flow only in one direction through a junction of selenium and a conductor. Yet the commercial application of this discovery was not made until 1930 when the selenium rectifier was developed. Even more interesting is the example of the crystal detector used in the early 1900s when radio was in its infancy. It consisted of a lump of semiconductor crystal such as lead sulfide (galena), silicon, or silicon carbide with a pointed wire, the *catswhisker,* bearing on it under a slight spring pressure. This combination also had the property of allowing current flow in only one direction. Unfortunately this early crystal detector had an annoying defect. The crystal had sensitive spots that were destroyed by overload or could be readily lost by a slight displacement of the catswhisker. The early radio operators were always fiddling with this contact trying to find a more sensitive spot. So, with the invention of the vacuum-tube diode, the crystal detector became obsolete. Then, as electronic applications went to higher and higher frequencies, the performance of the vacuum-tube diode was impaired. Research on crystals as diodes was resumed, and during World War II silicon and germanium crystal diodes were in extensive use as mixers and detectors in radar and other UHF equipment.

VALENCE ELECTRONS.   As a first step in explaining semiconductor action let us review the structure of the atom. Any student of chemistry, physics, or electricity is probably familiar with the theory of the planetary structure of the atom—a compact nucleus in the center and electrons revolving around it. We are primarily interested in the electrons. The chemical and electrical characteristics of each element are closely related to the quantity and arrangement of these electrons.

In the Periodic Table of elements,* all the elements are given an *atomic weight* and an *atomic number*. The atomic weight corresponds to the number of neutrons and protons in the nucleus, and the atomic number represents the number of electrons revolving around the nucleus. An electron as it revolves around the nucleus may be considered as having a definite orbit. However, unlike the motion of the planets in our solar system, the plane of the electron orbit keeps shifting, so that in time the electron will have traversed the surface of a sphere, or shell. This is similar to the way a ball of cotton is made. As you wind each turn at a different angle from the previous turn, the end result is a ball rather than flat loops. Now to complete the picture for *any one electron,* also imagine that the radius of motion changes slightly from time to time so that instead of a thin-line orbit we have a band and eventually a thick shell as the total surface traversed. The distance of this "pathway" from the nucleus depends on the *energy level* of the electron. Such a pathway is common to all electrons having this energy level.

The electrons of an atom do not all travel at the same energy level. Some are at a second and higher energy level, others at a third and still higher energy level, etc. There is a definite pattern to their arrangement as the atomic number and atomic weight of the element increase. The maximum number of shells for any known element is seven. (The shells are labeled alphabetically K through Q starting with the innermost shell.) Each shell has a definite maximum limit as to the number of electron orbits it can have. This limit is given by

$$\text{Maximum orbital paths} = 2n^2 \qquad \text{(1-1)}$$

where n is the shell number. For example, the L or second shell has a maximum limit of 8 (i.e., 2×2^2) electron orbits; the third shell 18 (i.e., 2×3^2); and so on.

However, there are two other limitations. The outermost shell is filled when it reaches 8 electrons; and the next-to-last shell cannot contain more than 18 regardless of its quota. Some examples follow:

1. Argon (Ar), atomic number 18, has an electron grouping of 2-8-8.

*Consult any general chemistry text.

The last shell is filled. (Yet, as the third shell, its maximum quota is 18.)

2. Potassium (K), atomic number 19, has four shells, with an electron grouping of 2-8-8-1.

3. Iron (Fe), atomic number 26, has an electron grouping of 2-8-14-2. Notice that since the third shell is no longer the last shell, it can build up toward its quota of 18.

4. Cesium (Cs), atomic number 55, has an electron grouping of 2-8-18-18-8-1. The last electron cannot go into shell 4, because it would then be the next-to-last shell and its maximum is 18; nor can this electron go into shell 5, because that would exceed the last-shell limitation of 8.

The electrical (and chemical) characteristics of each element depend on the number (and arrangement) of the electrons at the outermost level. These electrons are known as *valence electrons*. Table 1-1 shows the distribution of electrons for several elements that we will discuss later.

Table 1-1
DISTRIBUTION OF ELECTRONS AT VARIOUS ENERGY LEVELS

ELEMENT	ATOMIC NUMBER	FIRST LEVEL	SECOND LEVEL	THIRD LEVEL	FOURTH LEVEL
Boron	5	2	3*		
Carbon	6	2	4*		
Silicon	14	2	8	4*	
Copper	29	2	8	18	1*
Gallium	31	2	8	18	3*
Germanium	32	2	8	18	4*
Arsenic	33	2	8	18	5*
Selenium	34	2	8	18	6*

*Valence electrons.

SEMICONDUCTOR CRYSTAL LATTICE.   From Table 1-1, we see that copper has only one valence electron, whereas germanium has four and selenium six. Yet copper is a good conductor while the other elements are not. To explain this we must now analyze the interatomic structure of each type of element. Any element is held together by the bonding action of the valence electrons. In copper and other conductors, the valence electrons comprising these bonds are not specifically associated with any particular atom but exist as a communal group. They are therefore relatively free to move, and as one electron moves away another moves in to maintain the bonding structure. That is why these electrons are often referred to as *free electrons*.

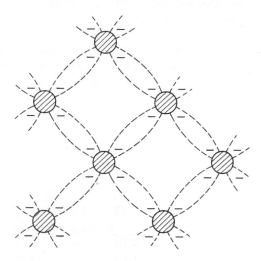

Figure 1-1   Pure germanium crystal lattice structure.

With semiconductors we have a somewhat different interatomic structure. Each valence electron of one atom forms a direct bond with a valence electron from an adjacent atom. These bonds are called *covalent pair bonds,* and the individual electrons are not free to move away from their associated atom. Figure 1-1 shows this type of structure diagrammatically. In this diagram the circle represents the nucleus *and all the electrons in the inner shells.* Four valence electrons are shown around each circle. The dashed lines between circles represent the covalent pair bond between atoms. Although the diagram is for a germanium crystal, it can just as well represent the crystal structure for carbon or silicon since each of these also has four valence electrons. Unfortunately, Fig. 1-1 has a drawback which cannot be avoided. The crystal structure should be shown in three dimensions, with some of the atoms coming up from the paper and others down. With this revised picture in mind, we can then visualize the selenium crystal structure (valence 6), with two more valence electrons making covalent pair bonds with two other adjacent atoms.

Based on this crystal lattice structure with covalent pair bonds, a pure semiconductor should have no conductivity whatever, and therefore theoretically it should be an insulator. There are two factors that create the semiconductor characteristics. Impurities in the crystal structure will produce an excess or deficiency of valence electrons, or absorption of energy from light or heat can break the covalent bonds. In the case of germanium, thermal agitation even at room temperature will create one free electron for every 10 million germanium atoms. This may seem a ridiculously small value, but when you consider the number of atoms even in a tiny bit of matter, germanium is not an insulator.

DONORS AND ACCEPTORS.   Conduction caused by impurities is the basic principle underlying semiconductor diodes and transistors. In the preparation of germanium for use in transistors, it is first refined to a purity of not more than one foreign atom for each 100 million (10^8) germanium atoms. Then carefully controlled impurities are introduced in a ratio of one-millionth of a gram of impurity to 1600 grams of crystal. Still using germanium as an example, let us see the effect of adding impurities. Since germanium has four valence electrons, atoms having five valence electrons (such as arsenic, antimony, or phosphorous) are often used as impurities. Four of their five electrons will enter the interatomic bonds of the crystal lattice structure. The fifth electron will be bound only by the relatively weaker electric charge on the nucleus. This is more in the nature of the communal bond found in conductors. These extra electrons are therefore sufficiently free to wander through the crystal lattice structure and conduct electricity. This effect is shown in Fig. 1-2. A semiconductor having such excess electrons is known as *N type* because the conduction is by negative carriers. The impurities used to produce this type of semiconductor are called *donors* since they provide the excess or conduction electrons.

Semiconductor properties can also be obtained by using, as impurities, elements having one less valence electron. With reference to germanium, suitable impurities would be boron, aluminum, gallium, or indium, each

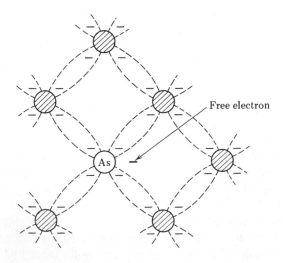

Figure 1-2   Electron released by donor (arsenic) impurity.

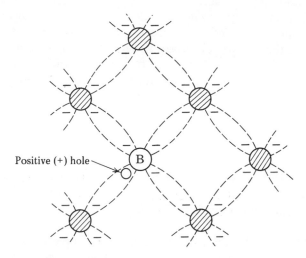

Figure 1-3   Hole produced by acceptor (boron) impurity.

of which has only three valence electrons. As an impurity atom enters into the crystal lattice structure, there will be a deficiency of one electron, leaving a vacancy or *hole* in the covalent pair bonds. This is shown in Fig. 1-3. Since the hole is an absence of an electron, it acts as a positive charge. An electron from a nearby bond can jump into this hole, leaving a hole behind. Still another electron can be attracted into the newly formed hole, creating a new hole. This process can continue so that the hole wanders throughout the crystal lattice structure and in so doing conducts electricity. A semiconductor having such properties is known as *P type* because conduction by holes is similar to conduction by positive carriers. The impurities used to produce this type of semiconductor are called *acceptors* because the holes they produce can accept electrons.

ELECTRIC CHARGE OF *N*- AND *P*-TYPE SEMICONDUCTORS.   So far we have seen that in an *N*-type semiconductor conduction is due to excess electrons, while in a *P*-type semiconductor conduction is by holes or positive carriers and is due to a deficiency of electrons. This might lead you to conclude that the *N*-type material has a negative charge and the *P*-type material a positive charge. Such a conclusion is fallacious, erroneous—in fact it is absolutely wrong! Let us analyze the situation more carefully. We have said that the *N*-type semiconductor has excess electrons. True—but these extra electrons were supplied by atoms of the donor impurity, and each atom of the donor impurity is electrically balanced. The number of negative electron charges equals the number of positive charges in the nucleus; the total charge is zero. The pure semiconductor is also electrically neutral, and so the *N-type semiconductor is still electrically neutral*. When the impurity atom is added, the term *excess electron* refers to an excess with regard to the number of electrons needed to fill the covalent crystal lattice structure. The extra electron is a free electron and increases the conductivity of the semiconductor. Remember that free electrons in a copper wire do not give the wire a negative charge.

The situation with regard to *P*-type semiconductors is quite similar. The crystal lattice structure has a deficiency in electrons—or holes. Conduction is by means of these holes. Acceptor impurity atoms created these holes because they were shy one electron (compared to the semiconductor atom), but remember that each acceptor atom also has one less positive charge in its nucleus and is electrically neutral. So—what is the charge of a piece of *P*-type semiconductor? Zero!

MAJORITY AND MINORITY CARRIERS.   The preceding picture of *N*-type and *P*-type semiconductors is oversimplified. It may create the impression that in an *N*-type semiconductor there are no holes or positive current carriers and also that there are no free electrons or negative current carriers in a *P* type. This is not the case. You will recall that even in a pure semiconductor heat or light energy can break the covalent bonds, releasing some electrons and creating holes. This effect also happens when a semiconductor has been doped with acceptor or donor impurities. Therefore *both* holes and free electrons are present regardless of the type of impurity added. In an *N*-type semiconductor some holes and electrons are present due to the breaking of covalent bonds, and in addition a very much greater number of electrons are produced by the donor impurity. Therefore electrons are considered as the *majority carriers* and the holes are *minority carriers*. Using a similar line of reasoning, it follows that in a *P*-type semiconductor the holes form the majority carriers and electrons are the minority carriers.

P-N JUNCTION.   If a piece of *P*-type semiconductor makes intimate contact with another piece of *N* type, a *P-N* junction is formed. Such a junction has the characteristic that it will allow relatively easy passage to electric current flow in one direction but offers high opposition to current flow in the opposite direction. This is the "peculiar" property of semiconductors mentioned earlier in this chapter. To explain this property of the semiconductor junction, first let us recall that the current carriers in semiconductors are the holes (positive carriers) in the *P* region and excess electrons (negative carriers) in the *N* region. Normally, these current carriers are distributed more or less evenly throughout each piece of semiconductor material. At a *P-N* junction, however, free electrons from the

N side will diffuse into the P side to fill holes in the P material—adjacent to the junction. Similarly, hole carriers from the P side will diffuse into the N material and recombine with free electrons. This interchange of holes and electrons wipes out the current carriers—in and around the junction—forming a *depletion region*. The N side (near the junction), having lost electrons and gained holes, acquires a positive charge. Conversely, the P side has gained electrons and lost holes. It becomes negatively charged. Obviously, a difference of potential, or barrier voltage, is created *at the junction*. (Each potential drops to zero, as we penetrate back into the respective N and P materials.) The creation of this barrier voltage prevents (repels) other carriers from diffusing across the junction. The magnitude of the barrier voltage depends on the semiconductor material used and on the concentration of the donor and acceptor impurities added. In typical germanium junctions this voltage is approximately 0.2 to 0.3 V; in silicon units it is approximately 0.6 to 0.7 V.

So far, we have been considering only the action of the majority carriers. To complete the picture, let us also recall that because of thermal activity, even at room temperature, covalent bonds are broken, releasing some electrons and creating holes—on each side of the junction. Now such holes on the N side (minority carriers) are repelled by the positive charge on this side of the junction and are attracted by the negative charge on the P side. These minority carrier holes will drift to the P side. Similarly, the minority-carrier electrons on the P side of the junction, under the influence of the barrier voltage, will drift across the junction to the N side. A dynamic equilibrium condition is reached wherein the *drift* currents of the minority carriers equal the *diffusion* currents of majority carriers. Actually, then, the net current crossing the junction is zero, but minority and majority carriers are crossing the junction back and forth, maintaining this equilibrium condition.

Now let us apply a direct voltage across a *P-N* junction, with a polarity such as to make the N region negative and the P region positive. (This voltage must be greater than the barrier potential of the junction. Also notice that the polarity of the supply voltage is opposite to the barrier potential.) Such a condition is shown in Fig. 1-4a. The negative potential applied to the N region will repel the excess or free electrons toward the junction and through the junction. Simultaneously, with the positive side of the supply source connected to the P region, holes will be injected* into the P material and repelled to and through the junction into the N

---

*It should be realized that current flow is always a result of electron motion. Holes do not flow; holes do not move. Phrases such as "hole-current flow" or "holes are injected" are used to distinguish between the electron flow in an N-type semiconductor as compared to the electron flow in a P-type material. Texts dealing with the physics of semiconductor materials show that the free electrons in an N-type semiconductor are raised to the conduction band

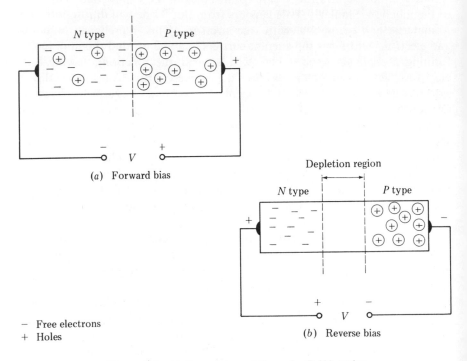

(a) Forward bias

Depletion region

(b) Reverse bias

− Free electrons
+ Holes

Figure 1-4 Unilateral current flow of a *P-N* junction.

side. The depletion region no longer exists, and an easy flow of current (or a low-resistance path) is established for the entire circuit. When such a supply voltage is applied, the junction is said to be *forward-biased*.

In Fig. 1-4*b*, the power supply polarity has been reversed so as to apply a negative potential to the *P* region and a positive potential to the *N* region. This time the excess electrons (the majority carriers in the *N* side) will be attracted toward the positive terminal of the power supply and away from the junction. In addition, the hole carriers in the *P* region will be attracted toward the negative terminal of the power source and again away from the junction. Notice that the area adjacent to the junction is devoid of current carriers (holes or electrons), thereby becoming an insulator. The

and that the current flow (electrons) is in the conduction band. On the other hand, in a *P*-type semiconductor there are no free electrons. Instead there are missing electrons—or holes—in the valence band. As an electron from an adjacent covalent bond jumps over to fill such a hole, it leaves a hole behind. Notice, again, that the *electrons* are traveling, not the holes. However, the electron travel in this case is always within the valence band. To avoid confusion between the free-electron travel in the conduction band and this electron travel in the valence band, this latter current is called a hole current flowing in the opposite direction.

depletion region at the junction has been widened. Current cannot flow. A *P-N* junction so connected is said to be *reverse-biased.* So far, we have been considering only the majority carriers. Now let us add the effect of this reverse bias on the minority carriers. The holes created in the *N* region by the breaking of covalent bonds will be repelled toward and through the junction; and the electrons, or minority carriers, in the *P* region will also be repelled to and through the junction. As a result, a small current will flow—even with reverse bias—and the resistance to current flow will be high, but not infinite. This current is often small enough (especially with silicon semiconductors) that it can be neglected.

SEMICONDUCTOR DIODES

We have seen that a semiconductor *P-N* junction has the property of allowing easy current flow in one direction and practically no current flow if the applied voltage is reversed. This *unilateral* current flow is similar to the action that takes place in a vacuum-tube diode. It is therefore not surprising to find that semiconductor diodes have practically replaced vacuum-tube diodes in new-equipment designs. Solid-state diodes are available commercially in a variety of sizes and shapes depending on their power rating, voltage rating, and current-carrying ability. A few typical units are shown in Fig. 1-5. Figure 1-5a shows a general-purpose *signal* diode.* It is a low-power (250 mW) device that can be used in receiver circuits as a mixer or a demodulator (detector) or in computer circuits as a switch. Its physical size is only 0.26 in. long and 0.105 in. in diameter. Only slightly larger is the *rectifier*† diode shown in Fig. 1-5b. This type can withstand peak reverse voltages up to 1000 V and can provide an average current of up to 1.5 A. Figure 1-5c and *d* shows two more rectifier diodes, both with peak reverse voltage ratings up to 600 V. The "top-hat" unit (DO-1 case) in Fig. 1-5c has a current rating of 5 A, while the stud unit has an average current rating of up to 40 A. The next two diodes are high-current, high-power devices. The model in Fig. 1-5e has peak reverse voltage ratings of up to 2000 V and average forward current ratings of up to 280 A. According to the manufacturer, the unit shown in Fig. 1-5f is the highest current rectifier available (1500 A average forward current and up to 1000 V peak inverse voltage). The last three types are intended for mounting on a metal surface or *heat sink* for better heat dissipation.

Regardless of their type or shape, all solid-state diodes have two terminals, or leads. The one attached to the *P* side is the *anode* lead; the

---

*Notice that it looks very much like—and could readily be mistaken for—a ⅛ W resistor.

†Rectifiers are used to supply dc power from the ac power line. These circuits will be discussed in a later chapter.

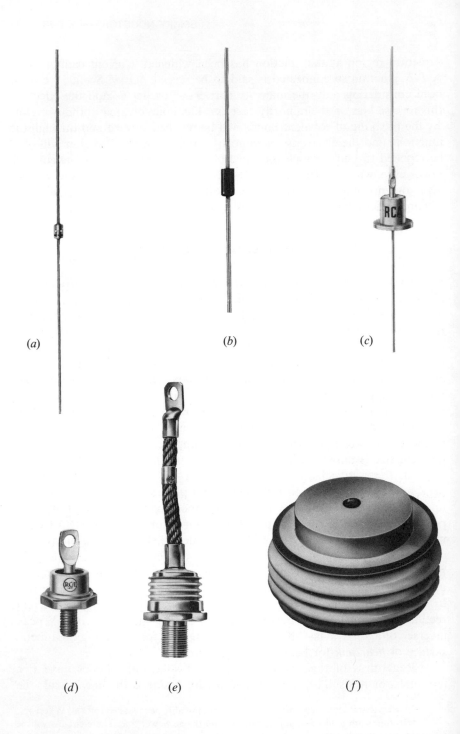

(a)

(b)

(c)

(d)

(e)

(f)

Figure 1-5 Typical solid-state diodes (*General Electric, RCA, Westinghouse*).

one attached to the N side is the *cathode* lead. Various techniques are used to distinguish the two terminals or leads. In Fig. 1-5a, the band at one end is used to indicate the cathode or N side. Some manufacturers go further and use the standard color-code bands on the cathode end not only to identify the cathode but also to specify the diode type number. For example, the 1N163 has brown, blue, and orange bands to indicate 1, 6, and 3, respectively. Another technique is used in Fig. 1-5b. Here the rounded end is the cathode side. In stud units, the stud is generally the cathode. However, these units are also available as "reverse polarity" units with the stud as the anode. Such reverse polarity types are designated with an R suffix after the type number. In some cases manufacturers will imprint a + on the cathode end, or they will imprint the schematic symbol of the diode directly on the case.

SCHEMATIC SYMBOL.   In circuit diagrams, semiconductor diodes are represented by the schematic symbol shown in Fig. 1-6, where the bar

N-type material —— ——P-type material

Figure 1-6   Schematic symbol for solid-state diode. (*Note:* The circle around the diode is often omitted.)

of the symbol is the cathode and the arrowhead is the anode. Since electron flow is from the N side (cathode or bar) to the P side (anode or arrow), it should be carefully noted that electron flow is *opposite* to the arrowhead direction.*

DIODE CHARACTERISTIC CURVE.   The characteristic curve for a diode is a plot of diode current versus diode voltage. Data for the curve is obtained by varying the voltage applied across the diode and noting the amount of current flow. Figure 1-7a shows a suitable circuit. The voltage is set at selected values by varying potentiometer P.† Resistor R is a current-limiting resistor, chosen so that with the potentiometer at maximum, the current will not exceed the maximum current rating of the diode. Since a semiconductor has some conduction in both directions, it is necessary to take data with both forward and reverse supply-voltage polarities.

A typical diode characteristic curve is shown in Fig. 1-7b. Notice that when the diode is biased in the forward direction—forward bias—a rela-

---

*The unilateral current flow action in semiconductor materials was discovered as early as 1883. At that time the electron theory had not been developed and a conventional current flow was in use. Naturally, the symbol was drawn with the arrowhead to indicate the direction of easier *conventional* current flow.

† The potentiometer can be omitted if a variable-voltage power supply is available.

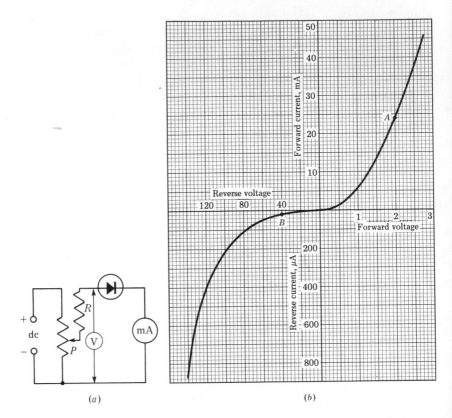

Figure 1-7   Circuit (a) and static characteristics (b) of a germanium diode.

tively high forward current flows, even with low applied voltages. Obviously, the forward resistance is low. With reversed polarity of applied voltage— reverse bias—only a small current flows (due to minority carriers) and in a reverse direction. Now the resistance is high. It is therefore common practice, when plotting diode characteristic curves, to employ different scale magnitudes for each direction. Notice in Fig. 1-7b that the forward current is in milliamperes while the reverse current is in *microamperes.* Notice also the difference in the voltage scales. (With silicon diodes, reverse conduction is of the order of 1000 times smaller.)

DIODE TESTING.   The difference in the forward and reverse resistance of a diode makes possible a simple quality check by using an ohmmeter. If we connect an ohmmeter across a diode, first with one polarity and then with the other, one reading should be appreciably higher than the other. (The high resistance is obtained when the ohmmeter biases the diode in the reverse or *back* direction; the low resistance is obtained with forward or *front* bias.) The ratio of these two resistance readings is known as the *back-to-front* ratio. For signal diodes, this can be quite high (in the hundreds); for power-rectifier diodes, it may be as low as 10:1. However, if the resistance is high with both ohmmeter polarities, the diode is open. Conversely, a low resistance with both ohmmeter connections indicates that the diode is shorted.

DIODE RESISTANCE.   In a dc resistive circuit we know that the current flowing through the circuit is proportional to the applied voltage—the proportionality factor being the resistance of the circuit. If we plot current versus voltage, the result will be a straight line since the resistance of a given circuit is constant. The slope of this line will be a measure of the resistance; i.e., the steeper the slope, the lower the resistance and vice versa. This relationship is shown in Fig. 1-8 for three resistance values. The circuit resistance for the middle curve is not marked. How can we find this resistance value? Let us see the method from an illustrative example.

EXAMPLE 1-1

Find the resistance represented by the slope of the middle curve in Fig. 1-8.

Solution

1. Select any point such as P on the curve.
2. Read the current (2.0 A) and the voltage (40 V) corresponding to this point.
3. Apply Ohm's law:

$$R = \frac{V}{I} = \frac{40}{2} = 20 \ \Omega$$

How does this review of dc apply to a diode? We saw that the current

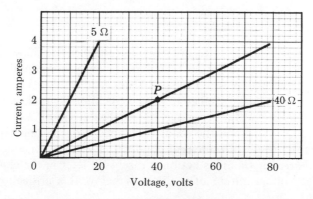

Figure 1-8   Effect of resistance value on slope of V-I curve.

flowing through the diode varies with the applied voltage. In this respect the diode action is similar to that of the dc circuit, and the diode characteristic curve must represent the internal resistance of the unit. But a typical diode characteristic is not a straight line. (See Fig. 1-7.) This must mean that the current is not directly proportional to the voltage—or that the proportionality factor, the diode resistance, is not a constant. If the diode is operated at a condition corresponding to point A of Fig. 1-7, the slope of the curve is steep and the diode resistance is low. At point B, however, the slope of the curve approaches the horizontal—the diode resistance is very much higher. It is therefore important, when evaluating diode resistance, that this quantity be determined at the actual operating point.

A diode characteristic can be used (in the same manner as shown in Example 1-1) to evaluate the resistance of the diode. First we must know (or select) the operating point. The ratio of the total diode voltage to the total diode current—at that point—is known as the *dc diode resistance* R_D. To illustrate this technique, the diode characteristic of Fig. 1-7b is repeated here as Fig. 1-9.

EXAMPLE 1-2

Using the characteristic curve of Fig. 1-9, evaluate:

1. The dc resistance of this diode if operated at point A
2. The dc resistance at point B

Solution

1. To find the dc resistance at point A (a forward resistance),

$$R_{DF} = \frac{V_D}{I_D} = \frac{2.0}{24 \times 10^{-3}} = 83.3 \ \Omega$$

2. To find the dc resistance at point B (the reverse resistance R_{DR}),

$$R_{DR} = \frac{V_D}{I_D} = \frac{40}{20 \times 10^{-6}} = 2.0 \ \text{M}\Omega$$

In the diode characteristics of Figs. 1-7 and 1-9, the slope of the curve for negative bias voltages of 40 V or less is nearly horizontal. The diode resistance in this region is very high. This is desirable. We have a high back-to-front ratio, and the efficiency of rectification is good. Now examine the curve at reverse voltages above 100 V. The slope of the curve is approaching the vertical, and the diode resistance is decreasing rapidly. This condition is the result of increased conduction in the semiconductor material *due to breakdown*. (This action is discussed later in the chapter.) The peak inverse voltage rating of a diode is below this critical value by a safe margin.

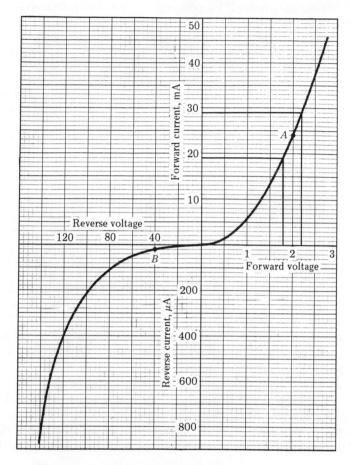

Figure 1-9   Static characteristics of a germanium diode.

Since diodes are generally used with ac or varying potentials, an important quantity is their ac (dynamic, small-signal, or incremental) resistance r_d. This value is obtained from the ratio of a *small change* in the diode voltage to the resulting change in diode current, or

$$r_d = \frac{\Delta V_D}{\Delta I_D}$$

(1-2)

The symbol Δ (delta) is used to denote a small change.

Due to the curvature of the diode characteristic, the ac resistance will vary depending on the operating point. Consequently, it is important that the small increment of voltage or current be taken halfway on each side of the selected operating point. This procedure is used in the following example and is shown in Fig. 1-9.

EXAMPLE 1-3

Find the ac resistance for the diode of Fig. 1-9, using point A as the operating point.

Solution

1. At point A, the diode static voltage is 2.0 V. Let us take increments of ± 0.2 V, so that $\Delta V = 0.4$ V.
2. The corresponding current values are:
   (a) At 1.8 V, $I = 19$ mA.
   (b) At 2.2 V, $I = 29$ mA.
   (c) ΔI (from 19 to 29) $= 10$ mA.

3. $r_d = \dfrac{\Delta V_D}{\Delta I_D} = \dfrac{0.4}{10 \times 10^{-3}} = 40\ \Omega$

DIODE EQUIVALENT CIRCUIT.   If we examine the forward portion of the characteristic curve in Fig. 1-9 carefully, we see that the current is negligible until the forward voltage is increased to about 0.2 to 0.3 V.* This point in the curve is variously known as the *cut-in, offset, knee,* or *threshold* voltage. We will refer to it as the threshold voltage (V_y).† In silicon diodes this threshold voltage is about 0.6 to 0.7 V.

In an ideal diode the threshold voltage would be zero, the forward resistance also zero, and the reverse resistance infinite. The diode, then, is acting as a switch. When forward-biased, the switch is closed; when reverse-biased, the switch is open. Although such an ideal diode does not exist, this simplified *equivalent circuit* of a diode can be used with good accuracy if:

1. The circuit voltage is high compared to the threshold voltage (for example, 30 V compared to 0.3 V).
2. The circuit resistance is high compared to the forward diode resistance.
3. The circuit resistance is low compared to the diode reverse resistance.

Many rectifier circuits meet these three requirements, and this ideal-diode equivalent circuit can be used. (See Fig. 1-10a.) However, if the source voltage is low, the threshold voltage cannot be neglected. This gives the equivalent circuit of Fig. 1-10b. Finally, if neither the threshold voltage nor

---

*You will recognize this as the barrier potential for a germanium P-N junction.
†This subscript is the lowercase Greek letter gamma.

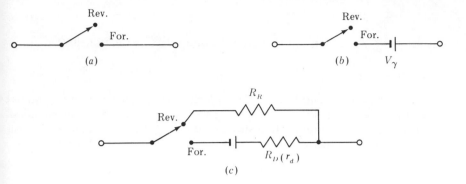

Figure 1-10   Diode equivalent circuits.

the diode resistance can be neglected, we must use Fig. 1-10c to represent the diode action.

The ac resistance of a diode can also be evaluated without the use of the characteristic curve. Theoretically, the lower-current region is an exponential curve and follows the equation*

$$I = I_o(\varepsilon^{V/\eta V_T} - 1) \tag{1-3}$$

Solving for $\Delta V/\Delta I$ at room temperature (25°C) yields

$$\frac{\Delta V}{\Delta I} = \frac{26}{I}$$

where I is in milliamperes. But this is the equation for ac resistance. At low-current values (in the curved region), this resistance is due to the junction action and can be called the junction resistance (r_j). Therefore

$$r_j = \frac{26}{I} \tag{1-4}$$

where I is the average (dc) value of the diode current in milliamperes.

Unfortunately, there is wide variation in individual diode characteristics, so that this ideal equation does not fit all diodes. (In practice, junction resistance values may range from $20/I$ to $50/I$.)

At higher-current values, the diode resistance is essentially due to the bulk resistance (R_B) of the material. This value can be calculated from data generally available in manufacturers' catalogs and manuals, if we realize that the forward voltage drop (V_F) across a diode must be the sum of the

*J. Millman and C. Halkias, *Integrated Electronics,* New York: McGraw-Hill Book Company, 1972, chap. 3.

barrier voltage and the IR drop due to the bulk resistance, or

$$V_F = V_\gamma + IR_B \qquad (1\text{-}5)$$

An example will readily show how we can apply this relationship.

EXAMPLE 1-4

In a transistor manual the 1N4785 is listed as a germanium diffused-junction diode that is used in 18-kV TV horizontal deflection systems. Under characteristics, the manual states: "Forward voltage drop, static $(I_F = 7\ \text{A})\ldots V_F \ldots 0.77\ \text{V}$." Find the bulk resistance of this diode.

Solution

1. For a germanium diode, $V_\gamma \cong 0.3\ \text{V}$.

2. $V_F = V_\gamma + IR_B$

$$R_B = \frac{V_F - V_\gamma}{I} = \frac{0.77 - 0.3}{7} \cong 0.07\ \Omega$$

DIODE SPECIFICATION DATA.   Each diode has certain specific electrical ratings and characteristics that define the type. Such data is listed by the manufacturers in their manuals and data sheets. A typical data sheet starts with an introductory paragraph stating the broad capabilities and intended service of the device. This is followed by a listing of *absolute maximum ratings,* which gives the limiting conditions of usage beyond which catastrophic failure will occur. (These may include electrical, mechanical, and thermal limitations.) Ratings are determined by the manufacturer on the basis of extensive reliability tests. Equipment using the device should be so designed that, even under the worst probable operating conditions, no specified maximum rating will ever be exceeded. Typical electrical ratings include:

1. *Maximum peak reverse voltage.*   This is the highest reverse voltage that can be applied before breakdown occurs. It can be given as a repetitive value, $V_{RM(rep)}$,* or as a transient or nonrepetitive value, $V_{RM(nonrep)}$.

2. *Forward current ratings.*   These ratings may include:
   (a) *Maximum-allowable average forward current* $I_{F(av)}$
   (b) *Maximum-allowable repetitive peak forward current* $I_{FM(rep)}$
   (c) *Maximum-allowable peak surge current* $I_{FM(surge)}$

Ratings are generally followed by *characteristics.* As distinguished from a rating, a characteristic is a measurable property of the device.

---

*This value is sometimes abbreviated PRV or may be called *peak inverse voltage* (abbreviated PIV).

For example, the voltage drop across the diode when current is flowing is a characteristic. This may be specified as a peak, dc, or average value (V_{FM}, V_F, and $V_{F(av)}$, respectively). Another commonly listed characteristic is the reverse current that flows under given conditions—maximum peak reverse current I_{RM}; maximum dc reverse current I_R; and the average value $I_{R(av)}$. Characteristics data is often given in curves in addition to specific values.

ZENER DIODES.   A typical characteristic curve for a reverse-biased silicon diode is shown in Fig. 1-11a. As mentioned earlier, the current flow with reverse bias is extremely low (of the order of nanoamperes) until the breakdown point is reached. At this voltage a very sharp increase in current occurs. Two phenomena contribute to this sharp rise in current—the *avalanche* effect and the *zener* effect.

In the first case, when a sufficiently high reverse voltage is applied across a diode, the relatively few free electrons (the minority carriers in the P side) are accelerated to very high velocities. In their path across the junction they strike neutral atoms with such force as to knock some valence electrons out of their covalent bonds. These newly freed electrons in turn are accelerated, strike other atoms, and release still more electrons. The action is cumulative—hence the term *avalanche* breakdown.

The second effect is also caused by high voltage. However, it does not depend on the presence of minority carriers. Instead, the voltage gradient across the depletion region can become high enough to tear electrons out of the covalent bonds of the atoms in the region. The value of reverse voltage at which this breakdown occurs is called the *zener voltage* V_Z.

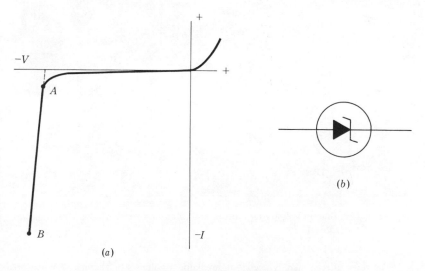

Figure 1-11   Reverse-bias characteristic of silicon diode.

Notice in Fig. 1-11*a* that the voltage drop across the diode in the region *A* to *B* is practically constant over a wide range of current. A diode operated in this region can serve as a *voltage regulator* (see Chapter 6) or as a *voltage-reference* source. Operation in this breakdown region will not damage the diode as long as the power dissipation rating of the device is not exceeded. Diodes specifically made for such service are called *zener diodes*—regardless of whether the actual breakdown is a zener or an avalanche effect.* The schematic symbol shown in Fig. 1-11*b* is used to distinguish a zener diode from a standard diode. The breakdown voltage of a zener diode depends on the resistivity of the silicon material used, and it can be controlled to a great degree in the manufacturing process. It is therefore possible to manufacture these units for almost any desired voltage. Zener diodes are available with V_Z ratings from as low as 2.0 V to over 200 V. These units are also available in wide power (and current) ratings, from less than 1 W (or a few milliamperes) to as much as 50 W (and as high as 10 A). An idea of their size can be obtained from Fig. 1-12, which

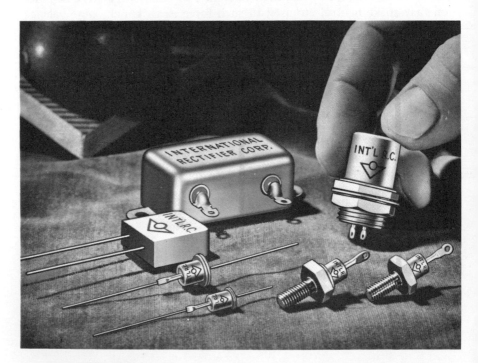

Figure 1-12   Silicon voltage regulators and reference zener diodes (*International Rectifier Corporation*).

*The zener effect generally occurs at breakdown voltages of about 6 V. When breakdown occurs at higher voltages, it is generally caused by avalanche action.

shows an assortment of units of various power ratings. Because of their excellent stability, temperature-compensated units (reference diodes) are used to develop reference voltages for measurements or control equipment. Such applications are found in recorders, telemetering, guided missiles, and a variety of automatic machinery.

OTHER SEMICONDUCTOR DIODES.   The preceding discussion is basic and applies to all semiconductor diodes in general. There are, in addition, several special-purpose diodes and earlier-vintage diodes that deserve further mention:

1. *Tunnel diodes.*   If the impurity density in the P and N regions is increased, the junction depletion region becomes narrower and narrower. With sufficient increase in these densities, it becomes possible for electrons to "tunnel" through from the valence to the conduction band and overcome the barrier potential at the junction. Because of this quantum-mechanical action, such diodes are called *tunnel diodes.* The characteristic curve of a typical tunnel diode and its schematic symbol are shown in Fig. 1-13. Notice the points marked A and B. These are known as the *peak* and *valley,* respectively. A high ratio of peak-to-valley current is desirable. Notice also that the slope of the characteristic between points A and B is opposite to the general diode *I-V* curve slope. This represents a *negative-resistance characteristic.* By operating the diode in this region, it is possible to use this type of diode as an amplifier or as an oscillator.

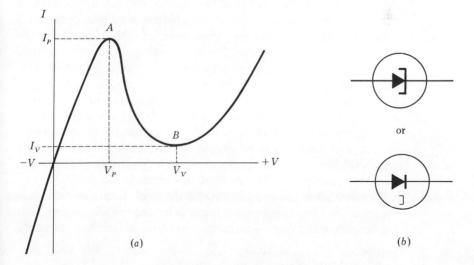

Figure 1-13   Tunnel diode.

2. *Varactor diodes.* We saw earlier that a *P-N* junction unbiased, or with reverse bias, has a depletion region. This is a region of no conductance. To either side we have conducting regions. But this is the basic construction for a capacitor—so a reverse-biased diode can be used as a variable capacitor. Its capacitance varies with the width of the depletion region or inversely with the value of the reverse-bias voltage. A diode specifically designed to make use of this feature is known as a *varactor diode* (coined from *variable reactor*). They are also more generally known as voltage-variable-capacitance (VVC) diodes. Typical symbols for these diodes are shown in Fig. 1-14. VVC diodes are used in frequency-control cir-

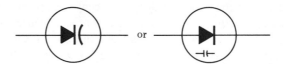

Figure 1-14   Schematic symbols for varactor diodes.

cuits, in electronic tuning circuits for radio and TV receivers, and as harmonic generators in RF frequency multipliers.*

3. *Photodiodes.* In a photodiode, the *P-N* junction is embedded in a clear plastic so that light can fall on the junction. The light energy will break covalent bonds in the junction area, producing electron-hole pairs and developing a voltage across the diode. The magnitude of this voltage varies with the light intensity. These diodes find wide application in counters, card readers, alarm systems, and in a variety of light-controlled circuits. A striking application of this photovoltaic principle is the solar battery used to supply power for the electrical equipment in space vehicles. These have generally been low-power applications. In 1973, however, scientists from the University of Arizona proposed construction of a 1000 MW generating plant in the Mojave Desert, using solar cells.

4. *Light-emitting diodes (LEDs).* When a junction is forward-biased, electrons from the *N* region are repelled through the junction, where they may recombine with excess holes. By proper design such recombinations can be encouraged. Some semiconductor materials release energy in the form of light (photons) when recombinations take place. (Notice that this effect is the reverse of the action in a photodiode.) Because of their solid-state construction (and the resulting ruggedness) these light sources are finding wide application as the display element or readout devices in computers,

*J. J. DeFrance, *Communications Electronics Circuits* (2nd ed.), New York: Holt, Rinehart and Winston, 1972, chaps. 4, 8, 9.

digital measuring instruments, watches, and a variety of warning or fault indicators.

5. *Earlier semiconductor diodes.* Solid-state diodes have been in use since about 1920. They were used mainly as rectifiers at power-line frequency and were generally known as *dry rectifiers* or *metallic rectifiers*. The first of these devices was the copper oxide rectifier. A few years later came the magnesium–copper sulfide rectifier and then (around 1938) the selenium rectifier. The first two, because of high leakage or reverse current and because of low voltage rating (6 V or lower), have since become obsolete. Selenium units are still in use, but they are being superseded by silicon units. A single selenium cell has a relatively low voltage rating (26 to 52 V), but by stacking cells in series, units are available for voltages up to 500,000 V. Their current rating depends on the area of the cell. Rectifiers are available for currents ranging from a few microamperes to over 100,000 A.

REVIEW QUESTIONS

1. **(a)** Name two commonly used semiconductor materials. **(b)** How does the electrical conductivity of these materials compare with that of copper or rubber?

2. **(a)** What is a valence electron? **(b)** How many valence electrons do each of the following materials have: germanium, silicon, boron, arsenic? **(c)** In a semiconductor material, is there any relation between the valence electrons of adjacent atoms? **(d)** What is the effect of such bonds with regard to the electrical conductivity of the material?

3. Refer to Fig. 1-1: **(a)** What do the hatched circles represent? **(b)** What do the four minus signs around each circle represent? **(c)** What do the broken-line arcs connecting the circles represent? **(d)** Can light or heat energy affect these bonds? Explain. **(e)** Explain why a semiconductor is not a perfect insulator at room temperature.

4. **(a)** Why are impurities generally added to a pure (intrinsic) semiconductor material? **(b)** How does a "donor" impurity atom differ in atomic structure from a pure silicon atom? **(c)** Name an opposite type of impurity atom, and tell how it differs from a silicon atom.

5. Refer to Fig. 1-2: **(a)** What does the circle marked "As" represent? **(b)** Is this a donor or acceptor impurity? How can you tell? **(c)** Account for the symbol marked "free electron." **(d)** What is the effect of adding donor impurities to a pure semiconductor material? **(e)** What is a material doped with donor atoms called, as distinguished from a pure semiconductor?

6. Refer to Fig. 1-3: **(a)** What does the circle marked "B" represent? **(b)** How many valence electrons does this material have? **(c)** Account for the symbol identified as a hole. **(d)** Can such holes affect the conducting qualities of a material? **(e)** Explain *hole current*. **(f)** What is the type of impurity used here called? **(g)** What is a semiconductor doped with such impurity atoms called?

7. **(a)** What is the electric charge on a piece of pure (intrinsic) semiconductor material? **(b)** What is the charge on a piece of N-type semiconductor? **(c)** Since an N type has "excess" electrons, why is the charge still zero? **(d)** What is the charge on a piece of P-type semiconductor? Explain.

8. **(a)** At room temperature, are there any holes in a piece of N-type material? Explain. **(b)** Which would be more abundant, the holes or the electrons? Why? **(c)** What are the holes in such a material called?

9. In a piece of P-type semiconductor, which are the majority carriers and which are the minority carriers?

10. **(a)** What is a P-N junction? **(b)** How does such a junction affect the electrical conductivity?

11. With reference to a P-N junction, **(a)** what is the *depletion region*? **(b)** Where is it found? **(c)** What is the *barrier voltage*? **(d)** What is the value of this voltage for a germanium junction? For a silicon junction?

12. Refer to Fig. 1-4: **(a)** What do the minus symbols represent? **(b)** What do the plus symbols represent? **(c)** Why are there any holes at all in the N side? **(d)** In the wiring, how does diagram (a) differ from diagram (b)? **(e)** From observation of the battery polarity, state a simple way of distinguishing forward from reverse bias. **(f)** Why isn't a depletion region shown in diagram (a)? **(g)** What is the effect of forward bias on the conduction through the junction? **(h)** If the applied voltage in diagram (a) were 0.15 V, would this still apply? Explain. **(i)** Account for the wide depletion region in diagram (b). **(j)** What is the effect of reverse bias on conduction? **(k)** Will the current flow with reverse bias be absolutely zero? Explain.

13. Refer to Fig. 1-5: **(a)** What is the main physical difference between these diodes? **(b)** How does this affect their ratings?

14. **(a)** What is a *heat sink*? **(b)** Why are they used with semiconductors?

15. **(a)** In a semiconductor diode, to which side of the P-N junction is the anode lead attached? **(b)** What is the lead to the other side called?

16. **(a)** In Fig. 1-5a, which end of the diode is the cathode? How can you tell? **(b)** Repeat for the diode in Fig. 1-5b.

17. Refer to Fig. 1-6: **(a)** If this diode is forward-biased, in which direction will the current (electrons) flow? **(b)** Why is the arrowhead symbol opposite to the direction of true current flow? **(c)** For forward bias, what is the correct polarity for the left-side terminal compared to the right side? **(d)** Which is the anode and which is the cathode?

18. Refer to Fig. 1-7a: **(a)** What is this circuit used for? **(b)** What is the function of the potentiometer? **(c)** Is a potentiometer always necessary? Explain. **(d)** Why is resistor R used? **(e)** Can the complete characteristic curve as shown in diagram (b) be obtained with the circuit as shown? Explain.

19. Refer to Fig. 1-7: **(a)** Since the curves in the first and third quadrant look very much alike, does this mean that the conduction in the forward and reverse directions is similar? Explain. **(b)** What is the current at 2 V forward voltage? **(c)** What is the current at 2 V reverse voltage? **(d)** Why are different scales used for the reverse and forward directions?

20. **(a)** The resistance of a diode is measured with an ohmmeter. The reading is 200,000 Ω no matter which way the ohmmeter leads are connected across the diode.

Is this normal? Explain. **(b)** What does the term *back-to-front ratio* signify? **(c)** Is a ratio of 3:1 considered satisfactory?

21. Refer to Fig. 1-8: **(a)** What does each of the three "curves" represent? **(b)** What is the relation between the slope of these curves and the resistance value? **(c)** What is the significance of the fact that these "curves" are straight lines?

22. **(a)** What letter symbol is used to represent the dc resistance of a diode? **(b)** Repeat for the ac resistance. **(c)** Give three other names for the ac resistance. **(d)** Give the letter symbol for the reverse resistance.

23. Refer to Fig. 1-9: **(a)** What does a steep slope signify with respect to resistance? **(b)** What does an almost horizontal slope signify? **(c)** Is the resistance in the forward direction constant? Explain. **(d)** When discussing diode resistance, why is it necessary to specify the operating point? **(e)** Is the reverse resistance high or low at approximately 40 V reverse voltage? Explain. **(f)** Why does the reverse resistance decrease rapidly at higher reverse voltages? **(g)** At approximately what forward voltage does the diode begin to conduct? **(h)** Give three names used to designate this voltage. **(i)** Give the letter symbol for this threshold voltage. **(j)** Account for the lack of conduction for applied voltages below this threshold value.

24. Refer to Fig. 1-10: **(a)** Which of these diagrams applies to an "ideal" diode? **(b)** What is the resistance of this diode in the reverse direction? **(c)** In the forward direction? **(d)** Give three conditions under which this simple equivalent circuit can be used with good accuracy. **(e)** In diagram (b), what does the battery symbol represent? **(f)** When must this threshold voltage be considered? **(g)** When must the forward resistance be considered? **(h)** When must the reverse resistance be considered?

25. **(a)** What is the cause for the high diode resistance at the low-current (curved) region? **(b)** Give the letter symbol for this quantity. **(c)** In Equation 1-4, what does the symbol I stand for? **(d)** Is this an ac or a dc value? **(e)** Is this current value in amperes, in milliamperes, or in microamperes? **(f)** Does this theoretical equation apply accurately to all diodes? Explain.

26. **(a)** At high-current values, what is the diode resistance due to? **(b)** Give the letter symbol for this quantity. **(c)** Can we obtain this resistance value from a manufacturer's data sheet? Explain.

27. **(a)** In a manufacturer's data sheet, what do *absolute maximum ratings* signify? **(b)** How are these values determined? **(c)** What does the *maximum peak reverse voltage* rating signify? **(d)** Give its letter symbol. **(e)** Give two other terms (and their symbols) for this voltage rating. **(f)** Give forward current ratings and their letter symbols.

28. **(a)** How does a *characteristic* differ from a rating? **(b)** What does $V_F = 0.77$ V at $I_F = 7$ A signify? **(c)** Is data on characteristics given in specific values as in part (b) or in curves?

29. Refer to Fig. 1-11: **(a)** How can you tell that this characteristic curve is for reverse bias? **(b)** Why is there a sharp rise in current at a voltage value corresponding (approximately) to V_Z? **(c)** Name two effects responsible for this breakdown. **(d)** Briefly explain each of these effects. **(e)** What does the symbol V_Z stand for? **(f)** Can a diode be operated safely in the breakdown region (A-B)? **(g)** What is a diode designed to operate in this region called? **(h)** Give an application for such a diode.

30. What is the significance of the bent ends on the bar of the diode symbol in Fig. 1-11*b*?

31. **(a)** What is a *reference diode*? **(b)** What is it used for?

32.(a) In manufacturing, how does a tunnel diode differ from general-purpose diodes? **(b)** Give an application for this device. **(c)** How does the schematic symbol for a tunnel diode differ from other diodes?

33. Refer to Fig. 1-13*a*: **(a)** What are points *A* and *B* respectively known as? **(b)** What does V_P represent? **(c)** What does I_V represent? **(d)** To get amplifier action, in what region of its characteristic is this diode operated? **(e)** What special property does operation in this region produce?

34. **(a)** With reference to diodes, what does VVC stand for? **(b)** Give another name for such diodes. **(c)** Is such a diode operated with forward or with reverse bias? **(d)** How does the capacitance vary with the applied voltage? **(e)** How does the schematic symbol for such a diode differ from other diodes? **(f)** Give an application for this diode.

35. **(a)** How does a photodiode differ in construction from other diodes? **(b)** What is the effect produced when light energy is applied to the junction area? **(c)** What is this effect called? **(d)** Give two applications for these diodes.

36. What is a solar cell or solar battery?

37. **(a)** With reference to diodes, what does LED stand for? **(b)** What is the basic principle of these devices? **(c)** Give two applications for these diodes.

38. **(a)** What does the term *dry rectifier* refer to? **(b)** Name three such specific rectifiers. **(c)** What is the general disadvantage of these units? **(d)** Which of these units is still in use to any extent?

PROBLEMS AND DIAGRAMS

1. Selenium has an atomic number of 34. **(a)** How many electrons does it have? **(b)** How many shells does it have? **(c)** What is the number of electrons in each shell? **(d)** How many valence electrons does it have? **(e)** Is the last shell filled? Explain.

2. Repeat Problem 1 for zinc (atomic number 30).

3. **(a)** Find the dc resistance of the diode of Fig. 1-7 at 1 V of forward bias. **(b)** Find its ac resistance.

4. Repeat Problem 3 for a forward bias of 2.5 V.

5. Find the reverse resistance of the preceding diode at **(a)** 24 V of reverse bias and **(b)** 90 V of reverse bias.

6.(a) Draw the equivalent diagram of a diode, assuming that no parameter value of the diode can be neglected. Label the parameters. **(b)** In an ideal diode, what are the values of R_R, R_d, r_d, and V_γ?

7. Find the junction resistance of a diode (at room temperature) if the operating point is fixed so that the dc value of the diode current is 20 μA, 2 mA, 6 mA.

8. Find the bulk resistance of the following silicon diodes: **(a)** 1N659 with $I_F = 6$ mA at $V_F = 1$ V; **(b)** 1N4380 with $I_F = 570$ mA at $V_F = 1.4$ V; **(c)** S508 with $I_F = 500$ mA at $V_F = 1$ V.

9. Draw the schematic symbol for **(a)** a zener diode, **(b)** a tunnel diode, **(c)** a varactor diode.

2

Bipolar Junction
Transistors

Although diodes, whether of the vacuum-tube type or semiconductor type, are suitable for many electronic purposes (except for the tunnel diode), they cannot perform the essential function of amplification. This great advancement in electronics was not achieved until 1907 when Dr. Lee De Forest invented the vacuum-tube triode. By introducing a grid between the cathode and plate (anode) of a diode and applying battery power (dc) between plate and cathode, he found that a small signal applied between grid and cathode could produce enlarged replicas of the original signal in the circuit between plate and cathode. Power supplied by the battery was converted to additional signal power. The result—amplification!

The characteristic curves for vacuum-tube diodes and semiconductor diodes are very similar. Would it not be possible to add a "third electrode" to a crystal diode and get amplification? Research on this idea led to the invention of the *point-contact transistor* in 1948 by Dr. J. Bardeen and Dr. W. Brattain of the Bell Telephone Laboratories. Unfortunately, the point-contact transistor had several serious shortcomings. Among the more important of these limitations were that they produced high noise levels and their use was restricted to very low power applications of around 100 mW. The low power-handling capacity of the point-contact transistor is directly attributable to its point contacts—particularly the collector point. Even a few milliamperes of current represents extremely high current densities. Considerable heat may be generated in the contact and surrounding germanium. If this heat is not conducted away from the point of contact, the resulting rise in temperature will cause breakdown of the covalent electron bonds. The resistance of the germanium will decrease and the current will increase, causing further rise in temperature. The effect is cumulative and the transistor is permanently damaged.

Research to reduce these limitations led to the development of *P-N* junctions and the bipolar junction transistor.* This new addition to the semiconductor family was in many respects superior to the point-contact transistor. Junction transistors were found to produce much lower noise levels, were more stable in operation, had higher power gain, higher power-handling capacity, and greater efficiency. Point-contact transistors are no longer in use.

CONSTRUCTION.   Bipolar junction transistors basically consist of three layers of alternate type (*P* or *N*) semiconductor material. This results in two possible transistors: the *NPN* or the *PNP*. Both types are in use. (In some circuitry, the use of a dual combination of one of each type—complementary symmetry—is a decided advantage.) Junction transistors can also be classified with regard to the method of fabrication:†

1. The earliest units were *grown-junction* transistors, formed by successively adding *P*-type and *N*-type impurities while growing the crystal from its melt. This technique has since been abandoned.
2. Chronologically, the *alloy-junction* method came next. This starts with a semiconductor wafer of either *P*-type or *N*-type material. Then a dot of opposite-type impurity is alloyed, or fused, to each side of the wafer. However, in spite of improvements in the alloy-junction techniques (except for high-power germanium units), this type of construction is also being abandoned.
3. The *diffused-junction* process was a major step forward in the art. It provided better manufacturing controls, raised the high-frequency limits by several orders of magnitude, and facilitated the use of silicon (in place of germanium), thereby improving temperature and power-output ratings. It also led to the development of integrated circuits. Basically, a transistor of this type consists of a semiconductor wafer that has been subjected to gaseous diffusion of both *P*-type and *N*-type impurities to form two *P-N* junctions in the original material. There are many variations to this technique, leading to transistor types such as *single-, double-,* and *triple-diffused, mesa, planar,* and *hometaxial.* A combination of growing and diffusing is used to produce the *epitaxial* transistor.

*The term *bipolar* is used because conduction involves both majority and minority current carriers.

†For details on fabrication techniques, see Texas Instruments, Inc., *Transistor Circuit Design,* New York: McGraw-Hill Book Company, 1963, chap. 1.

Figure 2-1   Comparative size of early junction transistors (vintage 1958) (*Raytheon*).

Among the advantages of the early transistors (over tubes) were their smaller size (which can be seen from Fig. 2-1), their theoretical long life (there is no cathode to wear out), and the facts that they did not require heating power and could operate with very low direct-voltage sources (3 to 12 V). (Vacuum tubes, on the other hand, require at least 90 V and preferably closer to 250 V.) All these factors were a great asset to portable and mobile equipment. For example, a portable radio could now be supplied from one battery no larger than the conventional type D flashlight cell. Mobile equipment could now be operated directly from the vehicle's storage battery. With transistors, there is no need for expensive, space-consuming ac-to-dc converters or motor-generator sets.

Unfortunately, these very advantages also created limitations. Small units could not dissipate much heat—the power output limitation was severe. Low power-supply voltage also meant low output voltage. So these units could not be used in power amplifier stages—and especially not in the horizontal output circuits of TV receivers, where high output voltages are needed to deflect the beam of the cathode-ray tube. Furthermore, receivers using these transistors could not be operated directly from the house ac supply (nominally 120 V) using a simple rectifier circuit: they could not withstand such high voltages. Instead, the power supply required a step-down transformer; otherwise a heavy power loss, using series voltage-dropping resistors, was necessary.

Improved manufacturing processes—especially the epitaxial technique —produced junctions that can operate at several hundred volts without

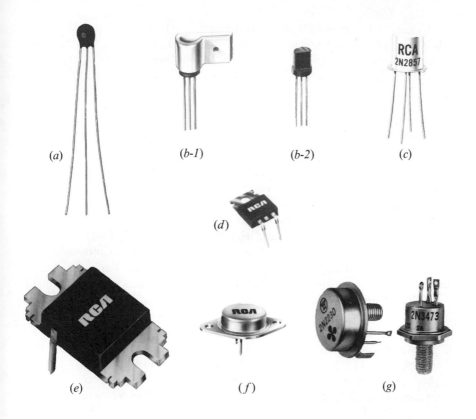

Figure 2-2 Commonly used bipolar junction transistors (*General Electric, Westinghouse, RCA*).

breaking down. Increased size (and use of heat sinks) gave transistors higher power-handling ability. Modern transistors are no longer necessarily small, low-voltage devices. A variety of bipolar junction transistors are shown in Fig. 2-2. Figure 2-2a shows a small-signal, low-power, plastic-encapsulated transistor. The body of this unit is less than $\frac{1}{8}$ in. in diameter. Figure 2-2b shows a larger, plastic-cased unit (approximately $\frac{1}{4}$ in. in base diameter) with and without a heat sink. Its power rating is 360 mW without the heat sink and 560 mW with it. The metal-encased transistor shown in Fig. 2-2c is another low-power device (200 mW). Its case is less than $\frac{1}{4}$ in. in diameter. The "versawatt" unit of Fig. 2-2d is larger. The plastic body is approximately $\frac{3}{8} \times \frac{1}{2}$ in. and has a power rating of 36 W. Notice the heavy metal lug at the rear, which must be mounted to a suitable heat sink. The plastic unit shown in Fig. 2-2e (body approximately $\frac{1}{2} \times \frac{1}{2}$ in.) can dissipate 100 W. The flat unit in the TO-3 case shown in Fig. 2-2f can dissipate 150 W, while the stud-mounting units shown in Fig. 2-2g have ratings as high as 300 W.

TRANSISTOR ELEMENTS AND DIAGRAMS.   It was mentioned earlier that a bipolar junction transistor consists of three layers of alternate types of semiconductor materials. This results in two possible units—the *NPN* and the *PNP*—shown diagrammatically in Fig. 2-3. (The schematic symbol for each type is also shown.) The three layers, or elements, are known as the *emitter,* the *base,* and the *collector,* with the base serving as the meat of this transistor sandwich. Notice that this sandwich construction creates two *P-N* junctions—one between emitter and base, and the other between collector and base. For proper transistor action, *the emitter-to-base junction must be biased in the forward direction, while the collector-to-base junction must be biased in the reverse direction.* The term *bias* is used to indicate the dc potentials applied to the elements and the resulting direct *current* flow.

Now let us examine the schematic symbols. Notice that the only distinction between the *NPN* and the *PNP* types is the direction of the arrowhead on the emitter. If we consider this arrowhead as indicating the direction of current flow in the emitter-base circuit, notice that it is a carry-over from the old metallic rectifier symbol and denotes the direction of conventional current flow. Electron flow is then opposite to the arrowhead symbol.

In specific circuit applications, the transistor symbol may often be seen rotated to simplify interconnection with other circuit components. This, however, does not alter the basic symbol. At times, modified versions of this representation may be seen. For example, the circle around the elements may be omitted or the base may be shown as a thinner or heavier line.

TRANSISTOR CONFIGURATIONS.   The schematics drawn in Fig. 2-3 lead to circuits wherein the base is common both to the input side (usually shown at the left) and to the output side (on the right). Such circuits are known as *common-base* circuits. Since the base (or common element) is often at ground potential, this configuration is also called a *grounded-base*

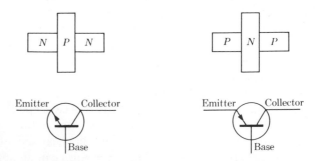

Figure 2-3 *NPN* and *PNP* transistors—basic construction and schematic symbols.

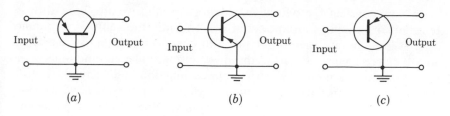

Figure 2-4   Basic transistor configurations.

connection. A simplified schematic of the grounded-base or common-base connection (without power supply and associated circuitry) is shown in Fig. 2-4a. Transistors can also be connected with either of the other two elements as the common or grounded element. This results in the *common-emitter* connection shown in Fig. 2-4b and the *common-collector* connection shown in Fig. 2-4c. In each case, the output and input would be connected to the points indicated. The choice as to which connection to use is often dictated by the impedances and gain requirements of the circuit. The input impedance Z_i* in any of these circuits is the impedance as seen looking into the input circuit, *with the output circuit connected to its load.* Variation of the load value can make a drastic change in the input impedance because there is appreciable feedback in a transistor. (Internal feedback is covered in Chapter 12 in the discussion of equivalent circuits.)

The input impedance for the common-base circuit is very low, ranging from approximately 20 to 200 Ω, the impedance increasing with higher values of load. This low resistance should be apparent from two aspects. First, the emitter-to-base junction (which is the major component of this resistance) is forward-biased and therefore has a low resistance. Second, the input current is the emitter current I_E, which is fairly large. Consequently, this resistance is low. In the grounded-emitter configuration, the emitter-to-base junction is again forward-biased and has a low resistance. This time, however, the input current is the base current I_B, which is much lower than I_E. Obviously the input impedance is appreciably higher—but still low. Typical values range from approximately 300 to 2000 Ω, the values decreasing with increase in load resistance value.

In considering the input impedance for the common-collector circuit, notice that the input circuit now contains the base-to-collector junction. This junction is reverse-biased. Obviously, then, the input impedance is, in comparison, very high. It ranges from approximately 3000 Ω to 1 MΩ. This wide variation is caused by interaction from the output circuit, the input impedance increasing with increase in load resistance value.

*This parameter is also referred to by the symbol h and appropriate subscripts.

Now consider the output impedances. Again, we are faced with interaction from input to output, and consequently the value of impedance as seen looking into the output side will depend on the resistance of the signal source connected across the input circuit. The common-base configuration now contains the reverse-biased junction and therefore has the highest output impedance. Its value ranges from approximately 100,000 Ω to 1 MΩ, increasing with the source resistance value. On the opposite end, the common-collector circuit has the lowest output impedance, ranging from below 50 Ω to approximately 5000 Ω. Again, the output impedance increases with increase in the source resistance value. The common-emitter circuit has an intermediate value of output impedance (between 20,000 and 100,000 Ω), and the impedance *drops* with increase in source resistance. A summary of these impedance values, together with the typical current gains for each configuration, is shown in Table 2-1.

Table 2-1
COMPARISON OF MAJOR TRANSISTOR CIRCUIT VALUES

QUANTITY	COMMON BASE	COMMON EMITTER	COMMON COLLECTOR
Input impedance	30 to 200 Ω	300 to 2000 Ω	3000 Ω to 1 MΩ
Output impedance	100 kΩ to 1 MΩ	20 to 100 kΩ	35 Ω to 5 kΩ
Current gain	0.85 to 0.999	20 to 600	20 to 600

TRANSISTOR CURRENTS—*NPN*.   Figure 2-5 shows a transistor connected in what is known as the common-base (or grounded-base) connection. The base element is common to both the input and the output sides of the transistor. In this preliminary circuit the emitter has deliberately been left open. Notice that the collector-base circuit is reverse-biased, as it should be. Because of this reverse bias the majority carriers (holes in the base and electrons in the collector) are pulled away from the junction, leaving the junction area depleted of current carriers. Current flow due to majority carriers cannot exist. However, even at normal ambient temperatures some

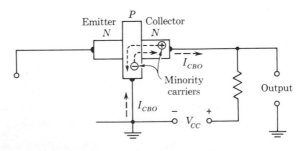

Figure 2-5   Transistor leakage current (I_{CBO}).

covalent bonds are broken and electron-hole pairs are formed in both the base and the collector. The electrons so freed in the base region and the holes so created in the collector are minority carriers. As far as these minority carriers are concerned, the bias is in the forward direction. Therefore a minority current flows (Fig. 2-5). For simplicity, only these minority carriers are shown. This current is considered a *cutoff current* or *leakage current* and is generally quite low, only a few microamperes. By symbol, it is properly referred to as I_{CBO}.* Quite often the reference to the base B is omitted and its symbol is reduced to I_{co}.

Now let us complete the emitter circuit (Fig. 2-6). Notice that the emitter-base junction is forward-biased. Because of this forward bias, it might at first be assumed that, as in a diode, relatively high current would flow from the supply, through the emitter, through the P-N junction, through the base, and back to the power supply. But this ignores the action of the collector and its potential. So, let us see the combined action:

1. The emitter-base junction is biased in the forward direction—low resistance to current flow.
2. Electrons leave the power supply (V_{EE}) and enter the emitter (current I_E). Since this is an N-type semiconductor, electrons are the current carriers.
3. Electrons move through the emitter region and across the P-N junction into the base area (still I_E).
4. Here, in the base, since it is a P-type semiconductor, normal conduction is by holes. Due to the field created by the power supply (V_{EE}), a hole is injected into the base region, at the positive terminal of the supply, as an electron is pulled out of its covalent bond. An electron from an adjoining atom fills this hole and a hole "moves" further in. Somewhere one of the electrons that crossed the P-N junction from the emitter fills this hole. This constitutes

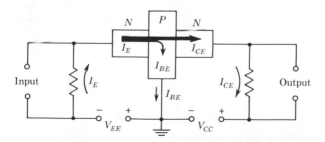

Figure 2-6   Current distribution—*NPN* transistor, common-base connection.

* This subscript signifies collector-to-base current, with the third element (the emitter) open.

the base current I_B. *Were it not for the collector action, all the emitter current electrons would end up as base current I_B.*

5. However, once electrons cross the *P-N* junction from the emitter side into the base area, they feel the attractive force due to the higher positive potential of the collector. So, instead of combining with holes to create base current, most of these electrons flow on through the base-to-collector junction, through the collector as collector current I_{CE} to the positive side of the supply voltage V_{CC}, and back to supply source V_{EE} to complete the circuit. In this way, the collector current rises from the few microamperes of leakage to a much higher value. The greater the emitter current I_E, the greater the collector current I_{CE}. This action is enhanced by deliberately doping the emitter much more heavily than the base, so that there are many more electron current carriers in the emitter than there are hole carriers in the base. Also, the base region is made very thin. Therefore most of the emitter current goes through the base region and through the base-collector junction. Only a very few emitter electrons (usually less than 5 percent) recombine with base holes and become base current. This effect is shown graphically by the current arrow in Fig. 2-6. Algebraically the current relationships are given by

$$I_E = I_{CE} + I_{BE} \qquad (2\text{-}1)$$

Now if we consider the cutoff or leakage current I_{CBO}, the following current relations can be added:

$$I_C = I_{CE} + I_{CBO} \qquad (2\text{-}2)$$

and

$$I_B = I_{BE} - I_{CBO} \qquad (2\text{-}3)$$

Quite often, since the cutoff current is small compared to the other values, I_{CBO} is neglected and I_C and I_B are considered the same as I_{CE} and I_{BE}, respectively. Although this leakage current can (at high temperatures) affect the static or dc bias conditions, it does not enter the picture with regard to ac signal changes. For ac operation, it is therefore quite correct to say that

$$I_e = I_c + I_b \qquad (2\text{-}4)$$

So far we have been discussing the current relations for a common-base circuit. These same current relations also apply for the common-emitter or common-collector circuits. This is shown in Fig. 2-7 for the common-emitter configuration. Notice that this time the cutoff current is I_{CEO} and represents the current that flows between emitter and collector when the input circuit (base) is open. Notice also that in normal operation

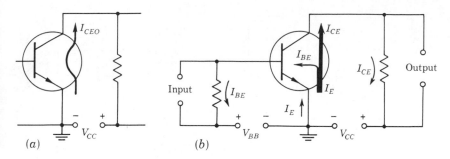

Figure 2-7   Current relations—*NPN* transistor, common-emitter connection.

(Fig. 2-7*b*) the emitter current I_E divides into two components I_{BE} and I_{CE}. Again the base current is very small, with most of the emitter current passing through the base into the collector.

TRANSISTOR LETTER SYMBOLS.   Notice in Equations 2-1 to 2-4 that the first three equations have capital-letter subscripts, whereas the last one —for ac operation—has lowercase subscripts. This notation is in accordance with the IEEE standards. Some clarification is in order:

1. Capital I for current (or V for voltage) is used to represent a *fixed* or constant value, such as dc, rms, or maximum current (or voltage)

2. If the quantity is an *instantaneous* value that changes with time, the lowercase i is used for current (or lowercase v for voltage).

3. Capital-letter subscripts indicate that the value is either a *dc component* or the *total* value (dc + ac)

4. Lowercase subscripts are used to represent only the *ac (varying) component* by itself.

5. The letter used as subscript tells us to which element the current (or voltage) applies.

For example, in Fig. 2-6 there are no ac inputs; therefore, all currents are dc. Hence, capital I and V with capital-letter subscripts are used. Notice also that the supply voltages (dc) use the doubled capital-letter subscript for the element being supplied. On the other hand, in Equation 2-4, the subscripts are lowercase letters; therefore, these values refer to the ac components of current. (The accompanying text mentions "for ac operation.") Since capital I is used, these are fixed values of the ac components and represent the rms values. A more complete description of letter symbols, with a supporting diagram, is given in Appendix C.

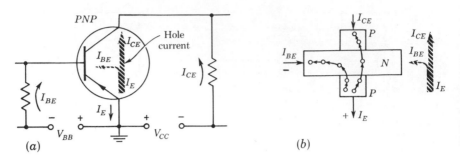

Figure 2-8   Schematic and pictorial view showing current flow in a *PNP* transistor.

CURRENT FLOW—*PNP* TRANSISTOR.   The current relations for a *PNP* transistor are identical to those already discussed for the *NPN* type; that is, $I_E = I_{BE} + I_{CE}$ for the dc or static conditions and $I_e = I_b + I_c$ for the ac or signal conditions. There are, however, two differences—the type of current flow within the transistor and the direction of current flow in the external circuit. Both these differences are pictured in Fig. 2-8 for a common-emitter connection. The same differences would hold true for the common-base or common-collector connections. But before we study this diagram, remember that for transistor operation—any type and any connection—the emitter-base junction must be forward-biased and the collector-base junction reverse-biased. Now let us see how this applies to Fig. 2-8.

Since the transistor is of the *PNP* type, the emitter must be made positive with respect to base and the collector made negative (reverse bias) compared to base. From Fig. 2-8a, it might appear as if the collector and base were at the same potential. However, the proper difference in potential is readily established by using a higher potential for the voltage supply V_{CC} and/or by proper selection of the resistors. (This matter is covered in detail in a later chapter.) Comparing this diagram with Fig. 2-7b, notice that the supply voltages and the external currents in the corresponding leads are reversed. This should be obvious: one is a *PNP* transistor, the other an *NPN* unit. However, notice that the currents within the *PNP* transistor (Fig. 2-8) are shown as flowing in the same direction as for the *NPN* of Fig. 2-7. The currents shown in each case are majority currents. The emitter of the *PNP* transistor has holes as the majority current carriers. The positive polarity applied to the emitter drives these holes toward the emitter-base junction and across the junction. They now become minority carriers in the *N*-type base and diffuse toward the collector. A few combine with base electrons and allow power-supply electrons to enter the base, but most of the emitter holes pass through to the collector and are attracted toward the negative collector terminal where they are neutralized by power-supply electrons entering at the collector terminal. Meanwhile, to counter-act for the loss of hole carriers in the emitter (as they pass through to the base region), electrons simultaneously leave the emitter at the opposite end and flow to the power supply. This creates *new* holes in the emitter. These new holes are then also driven toward the junction, and the process con-tinues. Hole currents within the *PNP* transistor have the same direction of flow as the electron current flow in the *NPN* type.

CURRENT GAIN—COMMON BASE.   From the study of current rela-tions in a common-base circuit (Fig. 2-6 and Equation 2-1), we saw that

$$I_E = I_{CE} + I_{BE}$$

If we neglect the relatively small value of cutoff current I_{CBO}, this equation can be rewritten as

$$I_E = I_C + I_B \qquad (2\text{-}5)$$

Now remember that transistors are manufactured so as to minimize recom-binations of the emitter majority-current carriers with the base majority-current carriers. Therefore, the ratio of I_C to I_E can be considered as a measure of how close we approach this goal. This ratio is known as the *current-gain* factor of the transistor and is represented by the Greek letter alpha (α). Since the current values (Equation 2-5) are dc, it is specifically called the dc current gain, α_{dc}. (A ratio of $1:1$ would represent an ideal transistor.) Expressed algebraically,

$$\alpha_{dc} = \frac{I_C}{I_E} \qquad (2\text{-}6)$$

Practical values for current gain α in commercial transistors range from 0.90 to as high as 0.999.

We are generally more interested in the ac operation of the transistor. The current relations for this were given by Equation 2-4 as

$$I_e = I_c + I_b$$

Since these ac components of current will cause corresponding changes (above and below) the original static current values, it follows that

$$\Delta I_E = \Delta I_C + \Delta I_B \qquad (2\text{-}7)$$

This leads directly into an evaluation for the ac or *small-signal current gain* (α_{ac}) as the ratio of a small change in collector current (ΔI_C) to a small change in emitter current (ΔI_E). However, to prevent the collector voltage from also affecting the collector current, the collector-base voltage (V_{CB}) must be held constant:

$$\alpha_{ac} = \frac{\Delta I_C}{\Delta I_E} = \frac{I_c}{I_e} \qquad (V_{CB} \text{ constant}) \qquad (2\text{-}8)$$

At times the current-gain symbol (α)* is used without subscript. The context in which it is used will generally clarify whether the dc or ac alpha is intended.

VOLTAGE AND POWER GAIN.   At first thought one might question why we use this gadget at all, if it provides no gain. In spite of a current loss (gain less than 1), however, the common-base circuit will produce a *voltage and a power gain.* This is so because the input resistance is much lower than the output resistance. Since the emitter-base junction is forward-biased, the input resistance is quite low (similar to the junction resistance of a diode), generally less than 100 Ω. On the other hand, the output resistance—because of the reverse-bias junction—can be over 100,000 Ω. Then, since the external circuit resistances should match the transistor input and output resistances, the voltage gain could be

$$A_v = \frac{V_o}{V_i} = \frac{I_c R_o}{I_e R_i} = \alpha \frac{R_o}{R_i}$$

and for a current gain of $\alpha = 0.95$,

$$A_v = 0.95 \times \frac{100,000}{100} = 950$$

Similarly, the power gain would be

$$A_p = \frac{P_o}{P_i} = \frac{I_c^2 R_o}{I_e^2 R_i} = \alpha^2 \frac{R_o}{R_i}$$

$$= (0.95)^2 \times \frac{100,000}{100} = 903$$

CURRENT GAIN—COMMON EMITTER.   With this configuration, the current on the output side (see Fig. 2-7) is I_{CE} and the current on the input side is I_{BE}. Again, neglecting the relatively small value of the cutoff current, these currents can be simplified to I_C and I_B. The current gain is now taken as the ratio of these two currents, and since these are dc values, this is the dc current gain. It is represented by the Greek letter beta (β):[†]

$$\beta_{dc} = \frac{I_C}{I_E} \tag{2-9}$$

---

*These current-gain factors are also expressed in *hybrid parameters* as the *forward current transfer ratio*—h_{FB} for the dc value and h_{fb} for the ac value. Hybrid parameters are discussed in detail in Chapter 12.

[†]Using hybrid-parameter notation, the symbols h_{FE} and h_{fe} are used to represent the dc and ac forward current transfer ratios, respectively. The first subscript (capital or lowercase) distinguishes the dc from the ac value; the second subscript (E or e) signifies the common-emitter connection.

The ac current gain can be found by using small changes in the dc values or from the ac components of these currents when a small-signal ac input is applied. As with the alpha evaluation, the collector potential must be kept constant:

$$\beta_{ac} = \frac{\Delta I_C}{\Delta I_E} = \frac{I_c}{I_e} \qquad (V_{CE} \text{ constant}) \qquad \textbf{(2-10)}$$

Since the current values used in calculating α and β are interrelated, it follows that these current-gain factors are also interrelated. We can derive this relation by using simple algebra:

$$\beta = \frac{\Delta I_C}{\Delta I_B}$$

and since $\Delta I_B = \Delta I_E - \Delta I_C$,

$$\beta = \frac{\Delta I_C}{\Delta I_E - \Delta I_C}$$

Dividing numerator and denominator by ΔI_E, we get

$$\beta = \frac{\Delta I_C / \Delta I_E}{(\Delta I_E - \Delta I_C)/\Delta I_E} = \frac{\alpha}{1 - \alpha} \qquad \textbf{(2-11)}$$

What is the significance of this relationship? We already know that the range of alpha in commercial transistors is from 0.90 to 0.999. Using a common value of 0.95, the base-current amplification factor would be $0.95/(1 - 0.95) = 19$. As alpha approaches unity, beta approaches infinity. Common values for beta range from around 20 to 600.

CUTOFF (LEAKAGE) CURRENTS.   In discussing the currents in a common-base circuit, we saw (Fig. 2-5) that a base-collector current flows even when the emitter current is zero. This reverse current was called a leakage or cutoff current (I_{CBO}) and was a flow of *minority-current carriers*. These minority carriers are a result of heat energy breaking the covalent bonds. Obviously, then, cutoff current is temperature-dependent, increasing with rise in temperature. In germanium transistors I_{CBO} doubles for every $10°C$ rise in temperature. In silicon transistors it doubles for every $6°C$ rise.

As you will learn later, leakage current is undesirable. If it increases too much, it can cause serious distortion in the circuit response and can even destroy the transistor unless suitable precautions are taken. Consequently, germanium transistors might seem preferable to silicon units because of the slower rise in leakage current with temperature. However, silicon transistors have a much lower original (room temperature) value of leakage current, so that in spite of their higher rise with temperature, they still have a lower final I_{CBO}, even at elevated operating temperatures. This can be seen from the following example.

EXAMPLE 2-1

Calculate the cutoff current in the following two cases, if the temperature rises from normal room temperature (25°C) to 85°C:

1. Germanium transistor 2N2635 with $I_{CBO} = 5 \,\mu\text{A}$ at 25°C
2. Silicon transistor 2N2484 with $I_{CBO} = 10$ nA at 25°C

Solution

In either case, the total change in temperature is 25°C to 85°C, or

$$\Delta T = 85 - 25 = 60°\text{C}$$

1. For germanium, I_{CBO} doubles every 10°C. Therefore, it doubles $\frac{60}{10} = 6$ times.

$$\text{New } I_{CBO} = 5 \,\mu\text{A} \times 2^6 = 5 \times 64 = 320 \,\mu\text{A}$$

2. For silicon, I_{CBO} doubles every 6°C. Therefore, it doubles $\frac{60}{6} = 10$ times.

$$\text{New } I_{CBO} = 10 \text{ nA} \times 2^{10} = 10 \times 1024 = 10{,}240 \text{ nA} = 10.24 \,\mu\text{A}$$

Notice that with silicon, in spite of the leakage current doubling 10 times, the final leakage current (at 85°C) is far less than in the germanium unit. When operation at high temperature may be a factor, silicon units are definitely preferred. One such application is in the automotive industry, where electronic control circuits are mounted in the hot engine-compartment area. In fact, silicon is now used in about 90 percent of all semiconductor applications.

Using Fig. 2-7, we also discussed the current relations in a common-emitter circuit. The cutoff or leakage current is now labeled I_{CEO}. Notice that it represents the current that flows between emitter and collector *when the base connection is open*. To analyze this in greater detail, let us redraw Fig. 2-7a in pictorial form (see Fig. 2-9). As before, the transistor is an *NPN* type. If we look at the potential gradient along the path from emitter to collector, we see that this single power supply will both forward-bias the emitter-base junction and reverse-bias the base-collector junction. The dashed arrow shows the "normal" I_{CBO} in the reversed-biased base-collector

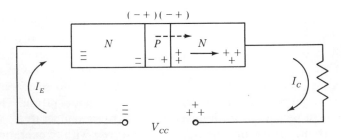

Figure 2-9   Cutoff current conditions—*NPN* transistor.

junction. The minority carriers in an *NPN* unit are electrons in the base (and holes in the collector) liberated when heat energy breaks the covalent bonds. These carriers will go through the junction because they are aided by the reverse bias. (Remember that this bias is "reverse" only to the majority carriers.) These minority carriers continue through the *N*-type collector, through the resistor, to the positive side of the power supply. However, this current I_{CBO} cannot return directly to the base to complete its circuit because the base is open. Therefore it must flow to the emitter, and it becomes emitter current I_{BE}. In turn, current I_{BE} does not go out through the base, as shown in Fig. 2-7b—the base is open. Instead, it continues through the base to the collector, and by transistor amplifier action it becomes βI_{BE}. This current is shown in the collector as the solid arrow. The total collector current is then

$$I_C = I_{CBO} + \beta I_{BE} \tag{2-12}$$

In this special case (base open), I_C is I_{CEO} and I_{BE} is I_{CBO}, so that

$$I_{CEO} = I_{CBO} + \beta I_{CBO}$$
$$= I_{CBO} (1 + \beta) \tag{2-13a}$$

When β is large compared to unity (as it usually is), this equation simplifies to

$$I_{CEO} \cong \beta I_{CBO} \tag{2-13b}$$

Obviously, the collector cutoff current I_{CEO} for the common-emitter circuit is much higher than the cutoff current I_{CBO} for the common-base circuit, depending on the current gain β. Consequently, special precautions may be necessary to minimize temperature effects in common-emitter (CE) circuits. With silicon transistors, however, improved technology has reduced leakage current to such low values that it is generally neglected.

STATIC CHARACTERISTIC CURVES.   When studying diodes, we saw how static characteristic curves could be obtained. Similarly, *families* of voltage-versus-current curves can be secured for transistors. These curves can be quite useful in designing transistor circuits or in evaluating their performance. Since current flows in both the input and the output sides of a transistor, curves should be taken for both sides. A suitable circuit is shown in Fig. 2-10a. The supply sources V_{EE} and V_{CC} should be variable.

A typical *input* family of curves for a common-base (CB) circuit is shown in Fig. 2-10b. The procedure for obtaining the curve data is quite simple:

1. Set the collector supply voltage to the desired V_{CB} value.
2. Raise the emitter supply voltage, in steps, from zero to the desired maximum value.

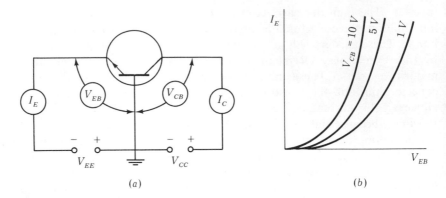

Figure 2-10 Input static characteristics of an *NPN* transistor (common base connection).

3. Record the emitter current at each step.
4. Repeat the process for as many V_{CB} (fixed) values as desired.

Note that the shape of these curves is very similar to the typical diode characteristic curve, but notice also that there is interaction between the output and input sides. A change in the collector-base voltage affects the input characteristic. This can be explained as follows. Increasing the reverse V_{CB} voltage widens the depletion region at the collector-base junction. The majority carriers on the base side (in this case holes, because the base is *P* type) are repelled away from the collector side and toward the emitter side. This increases the conductivity of the emitter-base path, and the *I-V* curve rises more sharply.

A similar set of input characteristic curves can be obtained with the transistor connected in the common-emitter (CE) configuration. The data gathering procedure is identical, but the data nomenclature is changed to fit the CE connection. The vertical axis current is now base current (I_B), the independent variable is the base-emitter voltage (V_{BE}), and the constant collector voltage is V_{CE}.

A more valuable and commonly used set of curves is the *output* or collector family of characteristic curves. The test setup is the same as before. (See Fig. 2-10 for an *NPN* transistor.) Now, however, the emitter-base voltage (V_{EB}) is held constant while the collector-base voltage is varied in steps over the desired range. At each step, the collector current is recorded.

Figure 2-11 shows the collector characteristic curves for a *PNP* transistor in the CB connection. Several peculiarities may be seen in this diagram:

1. The collector voltages are marked negative. This indicates that the collector potential is negative compared to the base, which is normal for any *PNP* transistor. (Had the transistor been of the *NPN* type, collector voltages would have been positive and no polarity

mark would have been shown.) This technique is not universally used, however, and so an unmarked polarity could be either positive or negative.

2. Notice that the collector currents are also marked negative. This indicates that the current is a reverse current flowing in the direction of high junction resistance.

3. Notice (at the left) that collector voltages are also shown with positive polarity. This biases the *collector-base* junction in the forward direction. The *reverse* current drops rapidly to zero. If the collector voltage had been increased in this positive (forward) direction, the forward (positive) collector current would increase so rapidly, due to the much lower forward resistance, that the transistor would be instantly damaged.

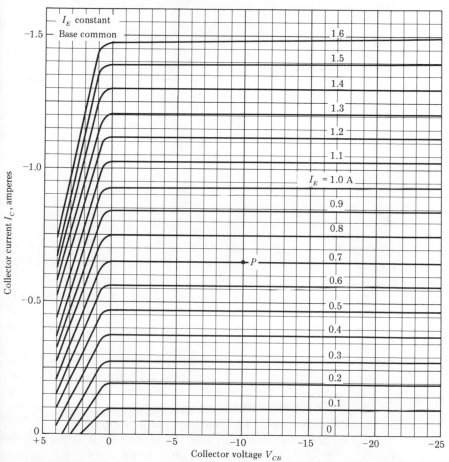

Figure 2-11 Output static characteristic curves for a bipolar junction transistor (common-base connection).

This third point brings up a precaution that must be observed when connecting power sources to transistors. If the collector voltage is accidentally reversed, the resulting high currents will ruin the transistor. (Contrast this with vacuum tube practice. If the plate voltage is reversed, nothing will happen—because current cannot flow from plate to cathode!)

Collector family curves can be used to evaluate the current gain of a transistor. An example will make this clear.

EXAMPLE 2-2

Find the ac current gain for the transistor of Fig. 2-11 at an operating point of $V_{CB} = 10$ V and $I_E = 0.7$ A.

Solution

This is a common-base configuration, and we can find α_{ac} from Equation 2-8:

$$\alpha_{ac} = \frac{\Delta I_C}{\Delta I_E} \quad (V_{CB} \text{ constant})$$

1. Locate the operating point $P V_{CB} = 10$ V; $I_E = 0.7$ A.
2. For a small change in I_E, we swing 0.1 A *above and below* point P, along the vertical line $V_{CB} = 10$ V. The corresponding I_C values are

$$\text{At } I_E = 0.8 \text{ A, } I_C = 0.75 \text{ A}$$
$$\text{At } I_E = 0.6 \text{ A, } I_C = 0.56 \text{ A}$$

3. $\qquad\qquad \Delta I_E = 0.8 - 0.6 = 0.2$ A

4. $\qquad\qquad \Delta I_C = 0.75 - 0.56 = 0.19$ A

5. $\qquad\qquad \alpha_{ac} = \dfrac{\Delta I_C}{\Delta I_E} = \dfrac{0.19}{0.2} = 0.95$

Although theoretically we can use these curves to find the current gain, in actual practice the accuracy of such a determination is not very good. This is because each characteristic curve is almost horizontal, and changes in I_C cannot be taken off accurately. Gain values are therefore generally found by direct measurement or are calculated from experimental data. However, static curves still have value in circuit design, as you will see in Chapter 11.

Figure 2-12 shows the static characteristic curves for the preceding *PNP* transistor in the CE connection. Again the collector voltage is varied, and the effect on the collector current is noted. This time, the constant value is the base current (I_B). These curves could be used to evaluate the beta gain of the transistor (see Problem 16), but again more accurate results are obtained by direct measurement. As before, these curves are of great value in design and evaluation, especially for power amplifier circuits.

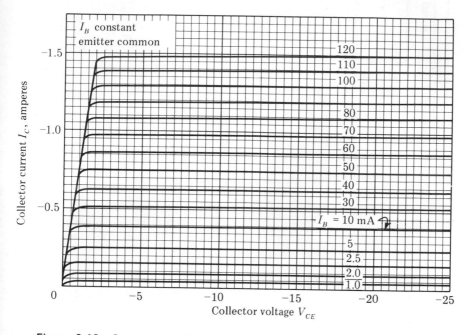

Figure 2-12   Output static characteristic curves for a bipolar junction transistor (common-emitter connection).

MANUFACTURERS' DATA SHEETS.   Manufacturers' data sheets (or manuals) are normally used when selecting transistors suitable for some specific application. The information is generally presented in the following format. At the top, with the identifying number, there is a brief summary of the general characteristics—for example, the device material (silicon or germanium; the type (*NPN* or *PNP*); the fabrication process (epitaxial, planar, etc.); and the typical application or intended service. This section may also include a few of the important electrical features of the device. This general information is then followed by absolute maximum ratings. Here are listed the maximum values of voltages, currents, power, and temperature the device may be subjected to. Next are the electrical characteristics or parameters, together with the test conditions under which they were obtained. The data is always given in chart form and is often supplemented with graphs to show the variation in parameter values under various operating conditions.

A typical data sheet is shown in Fig. 2-13. (The graphs have been omitted to conserve space.) Notice the wide variation (minimum to maximum) in the current-gain factor. This is an unfortunate situation. Two transistors with the same designation—made by the same manufacturer,

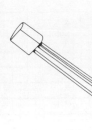

PNP	2N5447
	2N5448

Silicon
Transistors

ELECTRONIC
INNOVATIONS
IN ACTION
SEMICONDUCTORS

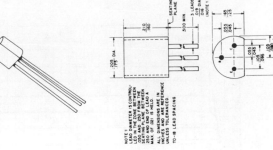

NOTE:
LEAD DIAMETER IS CONTROLLED IN THE ZONE BETWEEN .050 AND .250 OF THE SEATING PLANE. BETWEEN .250 AND END OF LEAD A MAX. OF .021 IS HELD.

ALL DIMENSIONS ARE IN INCHES AND ARE REFERENCE UNLESS TOLERANCED

TO-18 LEAD SPACING

The General Electric 2N5447 and 2N5448 are silicon, PNP planar, epitaxial, passivated transistors, designed for general audio frequency applications and linear amplifiers. For complimentary NPN types see 2N5449, 2N5450 and 2N5451 specification. Voltage and current valves for PNP are negative, observe proper bias polarity.

absolute maximum ratings: $(T_A = 25°C$ unless otherwise specified)

		2N5447	2N5448	
Voltages				
Collector to Emitter	V_{CEO}	25	30	Volts
Collector to Base	V_{CBO}	40	50	Volts
Emitter to Base	V_{EBO}	5	5	Volts
Current				
Collector	I_C	← 200 →		mA
Dissipation				
Total Power $T_A \leqslant 25°C$	P_T	← 360 →		Watts
Total Power $T_C \leqslant 25°C$	P_T	← 500 →		Watts
Derate Factor $T_A > 25°C$		← 2.88 →		mW/°C
Derate Factor $T_C > 25°C$		← 4 →		mW/°C
Temperature				
Storage	$T_{STG.}$	-65 to +150		°C
Lead (1/16'' ± 1/32'' from case for 10 sec.)	T_L	+260		°C

electrical characteristics: $(T_A = 25°C$ unless otherwise specified)

STATIC CHARACTERISTICS

	Symbol	2N5447 Min.	2N5447 Max.	2N5448 Min.	2N5448 Max.	Units
Collector-emitter breakdown voltage ($I_C = 10mA$, $I_B = 0$)	$V_{(BR)CEO}$*	25		30		Volts
Collector-base breakdown voltage ($I_C = 100uA$, $I_E = 0$)	$V_{(BR)CBO}$	40		50		Volts
Emitter-base breakdown voltage ($I_E = 100uA$, $I_C = 0$)	$V_{(BR)EBO}$	5		5		Volts
Collector cutoff current ($V_{CB} = 20V$, $I_E = 0$)	I_{CBO}		100		100	nA
Emitter-base reverse current ($V_{EB} = 3V$, $I_C = 0$)	I_{EBO}		100		100	nA
Forward current transfer ratio ($V_{CE} = 5V$, $I_C = 50mA$)	h_{FE}*	60	300	30	150	
Collector-emitter saturation voltage ($I_C = 50mA$, $I_B = 5mA$)	$V_{CE(sat)}$*		.25		.25	Volts
Base-emitter voltage ($V_{CE} = 5V$, $I_C = 50mA$)	$V_{BE(on)}$*	.6	1	.6	1	Volts

DYNAMIC CHARACTERISTICS

	Symbol	2N5447 Min.	2N5447 Max.	2N5448 Min.	2N5448 Max.	Units
Collector-base capacitance ($V_{CB} = 10V$, $I_E = 0$, $f = 1MHz$)	C_{cb}		12		12	pf
Forward current transfer ratio ($V_{CE} = 5V$, $I_C = 50mA$, $f = 20MHz$)	h_{FE}	5		5		

*Pulse Conditions: Pulse Width $\leq$ 300 μs and duty cycle $\leq$ 2%

GENERAL ELECTRIC

Figure 2-13 Typical data sheet (*General Electric*).

from the same batch, and operating under identical conditions—may have as much as a 10 to 1 variation. This makes exact calculation using typical parameter values rather meaningless. (Exact calculations can be made, however, if the parameters for a particular unit are measured experimentally, although the calculations would no longer apply if the unit was replaced.)

Abbreviated data on many transistors can also be obtained in chart form for quick reference. Transistors are grouped under common application headings. Such a chart is shown in Table 2-2.

REVIEW QUESTIONS

1. Name two types of bipolar junction transistors, based on the layers of semiconductor materials used.

2. Name three techniques used to manufacture these junction transistors.

3. Name three varieties of diffused-junction transistors.

4. State three advantages of transistors over tubes.

5. A bipolar transistor will operate satisfactorily with power-supply voltages as low as 6 V. A tube requires at least 90 V. **(a)** Explain where and why this is an advantage for the transistor. **(b)** Is the reverse ever true? Explain.

6. Refer to Fig. 2-2: **(a)** What is the major difference in operating characteristics between the units in diagrams (a) and (d)? **(b)** What is the physical difference between the two units in diagram (b)? **(c)** What is the metal lug on one of these called? **(d)** Explain the action and effect of this lug. **(e)** What is the purpose of the stud construction on the units of diagram (g)?

7. Refer to the schematic symbols in Fig. 2-3: **(a)** How can you distinguish the emitter from the collector? **(b)** How can you distinguish a *PNP* from an *NPN* unit? **(c)** What is the direction of electron flow compared to the emitter arrowhead?

8. Refer to Fig. 2-4: **(a)** Which of these diagrams represents the common-emitter configuration? How can you tell? **(b)** Give another name for this connection. **(c)** What is diagram (a) called? Give two names. **(d)** Give two names for the circuit in diagram (c).

9. Refer to Fig. 2-5: **(a)** Is the collector-base junction forward-biased or reverse-biased? How can you tell? **(b)** Is this the correct polarity? **(c)** What do the circles marked plus and minus represent? **(d)** How can both of them be called minority carriers? **(e)** What is the significance of the subscript CBO in the current symbol I_{CBO}? **(f)** Give two names for this current.

10. Refer to Fig. 2-6: **(a)** Is the emitter-base junction forward-biased or reverse-biased? How can you tell? **(b)** Is this the proper polarity? **(c)** Approximately what percentage of current I_E flows out through the base as current I_{BE}? **(d)** Give two reasons for this low value. **(e)** Where does the rest of current I_E go? **(f)** Since the collector-base junction is reverse-biased, why does most of the emitter current go through this junction and into the collector? **(g)** Is there any other component to the collector current besides the I_{CE} shown? **(h)** Give an equation for the total collector current. **(i)** It is permissible to consider I_{CE} and I_{BE} the same as I_C and I_B, respectively? Explain.

Table 2-2
TRANSISTOR SPECIFICATIONS

NPN Transistors

LOW LEVEL AMPS

Type No.	Case Style	V_CBO (V) Min	V_CEO (V) Min	V_EBO (V) Min	I_CBO (nA) Max	@ V_CB (V)	hFE / hFE (1kc)* Min	Max	@ I_C (mA)	& V_CE (V)	V_CE(sat) (V) Max	V_BE(sat) (V) Min	Max	@ I_C (mA)	C_ob (pF) Max	f_T (MHz) Min	Max	@ I_C (mA)	t_off (ns) Max	NF (dB) Max	Test Condition	Process No.
2N760	TO-18	45	45	8	200	30	76	333*	1	5	1	0.6	1.1	10	8	50						07
2N760A	TO-18	60	60	8	100	30	76	333*	1	5	1	0.9	1.1	10	8	50						07
JAN2N760A	TO-18	75	60	8	10	30	76	333*	1	5	1	0.6	1.1	10	6	60				24	1	07
2N929	TO-18	45	45	4	10	45	40 / 60	120 / 350	0.01 µA / 0.5 / 10	5 / 5 / 5	1	0.6	1.1	10	8	30	180	0.5		4	5	07
JAN2N929	TO-18	60	45	6	10	45	40 / 60	120 / 350	0.01 µA / 0.5 / 10	5 / 5 / 5	1	0.6	1.1	10	8	45	180	0.5		5 / 3 / 3	3 / 4 / 2	07
JANTX2N929	TO-18	65	45	6	10	45	40 / 60	120 / 350	0.01 / 0.5 / 10	5 / 5 / 5	1	0.6		10	8	45	180	0.5		5 / 3 / 3	3 / 4 / 2	07
JAN2N930	TO-18	60	45	6	10	45	100 / 150	300 / 600	0.01 / 0.5 / 10	5 / 5 / 5	1	0.6	1	10	8	45	180	0.5		5 / 3 / 3	3 / 4 / 2	07
JANTX2N930	TO-18	60	45	6	10	45	100 / 150	300 / 600	0.01 / 0.5 / 10	5 / 5 / 5	1	0.6	1	10	8	45	180	0.5		5 / 3 / 3	3 / 4 / 2	07
JANTXV2N930	TO-18	60	45	6	10	45	100 / 150	300 / 600	0.01 / 0.5 / 10	5 / 5 / 5	1	0.6	1	10	8	45	180	0.5		5 / 3 / 3	3 / 4 / 2	07
2N930A	TO-18	60	45	6	2	45	60 / 100 / 150	300 / 600	0.001 / 0.01 / 0.5 / 10	5 / 5 / 5 / 5	0.5	0.7	0.9	10	6	45	180	0.5		3	2	07
2N981	TO-18	80	80	8	1.0 µA	30	36	100*	1	5	3			10	5	45						07

Test Conditions:

1. I_C = 1.0 mA, V_CB = 5V, R_G = 500Ω, f = 1 kHz
2. I_C = 10 µA, V_CE = 5V, R_G = 10 kΩ, f = 10 kHz
3. I_C = 10 µA, V_CE = 5V, R_G = 10 kΩ, f = 100 Hz
4. I_C = 10 µA, V_CE = 5V, R_G = 10 kΩ, f = 1 kHz
5. I_C = 10 µA, V_CE = 5V, R_G = 10 kΩ, BW = 15.7 kHz
6. I_C = 5 µA, V_CE = 5V, R_G = 50 kΩ, f = 1 kHz
7. I_C = 5µA, V_CE = 5V, R_G = 50 kΩ, f = 10 kHz
8. V_CE = 5V, I_C = 100µA, R_G = 10kΩ, W.B.
9. V_CE = 5V, I_C = 30µA, R_G = 100kΩ, f = 1kHz
10. I_C = 20 µA, V_CE = 5V, R_S = 22 KΩ, W.B.
11. I_C = 20 µA, V_CE = 5V, R_S = 10 KΩ, f = 1 kHz
12. I_C = 100 µA, V_CE = 5V, R_G = 5 KΩ, W.B.
13. I_C = 100 µA, V_CE = 4.5V, R_G = 5 KΩ, W.B.

11. Refer to Fig. 2-7: **(a)** In diagram (*a*), what does I_{CEO} represent? **(b)** In diagram (*b*), why is the left-hand supply voltage polarity opposite to that shown in Fig. 2-6? **(c)** Is the emitter-base junction still forward-biased? **(d)** Notice that the base and collector potentials are both positive (with respect to ground). Is the collector-base junction still reverse-biased? Explain. **(e)** How do the current relations in this circuit compare with those in Fig. 2-6?

12. What letter symbols are used to represent each of the following: **(a)** a fixed value of current or voltage; **(b)** an instantaneous value of voltage or current; **(c)** a dc value of voltage or current; **(d)** an rms value of voltage or current?

13. Distinguish between each of the following symbols: **(a)** I_E and I_e; **(b)** i_b and i_B; **(c)** V_{CE} and V_{CC}; **(d)** V_{be} and v_{be}; **(e)** V_{ce}, V_{cem}, and v_{ce}.

14. Refer to Fig. 2-8*b*: **(a)** What type of transistor is this? **(b)** What do the arrows inside the body of this transistor represent? **(c)** Why is the direction of current flow inside the transistor opposite to the direction of flow shown outside the transistor?

15. Refer to Figs. 2-8*a* and 2-7*b*: **(a)** Which of these diagrams uses the common-emitter configuration? **(b)** In Fig. 2-7*b*, is the emitter-base junction forward-biased or reverse-biased? Explain. **(c)** Compare the polarity of the supply voltages V_{BB} in Figs. 2-7 and 2-8. **(d)** Is the base-emitter junction in Fig. 2-8 also forward-biased? Explain. **(e)** Notice the V_{CC} polarities in Figs. 2-7 and 2-8. Which one produces forward bias? Which one produces reverse bias? Explain.

16. What is the general rule for biasing a transistor with respect to **(a)** the emitter-base junction? **(b)** The collector-base junction?

17. (a) What symbol is used to represent the current-gain factor for a transistor connected in common base? **(b)** What is an ideal value for this parameter? **(c)** What is a typical commercial value? **(d)** Give the equation for calculating the dc current gain. **(e)** Repeat part (d) for the ac current gain. **(f)** Give another name for this current-gain factor. **(g)** Give another symbol for the dc parameter. **(h)** Give another symbol for the ac parameter.

18. Since the current gain in the common-base connection is less than 1, can voltage gain be obtained with this configuration? Explain.

19. (a) What symbol is used to represent the current gain of a transistor in the common-emitter configuration? **(b)** Give the equation for the dc current gain. **(c)** Repeat part (b) for the ac current gain. **(d)** Give another symbol used for the dc gain; give another for the ac gain. **(e)** Can this current gain exceed unity? **(f)** Give a typical value (or range of values) for β. **(g)** Give an equation showing the relation between α and β.

20. (a) Why is there some collector current flow in a common-base circuit when the emitter circuit is open? **(b)** Give two names for this current. **(c)** Give the symbol for this current. **(d)** Is such a current flow desirable? Explain. **(e)** How is the magnitude of this current affected by temperature? **(f)** Which has the lower initial (room temperature) leakage current, a germanium transistor or a silicon transistor? **(g)** What is the rate of rise (current versus temperature) for each material? **(h)** Which type of material (germanium or silicon) is preferable if high operating temperatures are expected? Why?

21. (a) What is the symbol for leakage current in a common-emitter circuit? **(b)** What happens to the value of leakage current when a transistor connection is changed from common base to common emitter?

22. Refer to Fig. 2-10: **(a)** What is the circuit in diagram (*a*) used for? **(b)** Are the supply sources V_{EE} and V_{CC} fixed or variable? Explain your choice. **(c)** In diagram (*b*), how is the curve marked $V_{CB} = 10$ V obtained? **(d)** How is the curve marked $V_{CB} = 1$ V obtained? **(e)** Explain why a change in V_{CB} affects these curves.

23. Refer to Fig. 2-10*a*: **(a)** What change in circuitry would be needed to take an output or collector characteristic curve? **(b)** What current or voltage is held constant when taking such a curve? **(c)** Explain the procedure for taking this curve.

24. Refer to Fig. 2-11: **(a)** From the curves, how can you tell whether this diagram is for a CB or a CE connection? **(b)** From the scale values, how can you tell whether this diagram is for a *PNP* or an *NPN* unit? **(c)** What would happen to the collector current if the collector voltage was increased to $+25$ V to match the negative values?

25. In Fig. 2-12, why are I_B values held constant, instead of I_E as in Fig. 2-11?

26. Refer to Fig. 2-13: **(a)** Why is the V_{EBO} rating so much lower than the V_{CBO} rating? **(b)** What is the maximum current these units can carry without damage? **(c)** Which transistor has the higher dc current gain? How much higher? **(d)** How much variation may be found in the current-gain factor of units of the same type? **(e)** Can either of these devices be operated at a collector supply voltage V_{CC} of 75 V? Explain.

27. If high current gain is the important criterion, which transistor at what operating point is the most suitable choice (use Table 2-1)? Explain your choice.

PROBLEMS AND DIAGRAMS

1. (a) Draw the schematic symbol for an *NPN* germanium transistor. Label the elements. **(b)** Repeat part (a) for an *NPN* diffused-junction transistor.

2. Draw a basic schematic for an *NPN* transistor in a common-base connection biased with dc supply sources, and indicate the three element currents.

3. Repeat Problem 2 for a *PNP* transistor in the common-emitter connection.

4. A germanium transistor (2N130) has a collector cutoff current I_{CBO} of 60 μA at 25°C. Calculate its cutoff current at 75°C.

5. The 2N3709 silicon transistor has a cutoff current I_{CBO} of 100 nA at 25°C. Calculate its cutoff current at 97°C.

6. Another silicon transistor (2N3242A) has an I_{CBO} of 10 nA at 25°C. Calculate its cutoff current at 115°C.

7. The 2N3242A in Problem 6 has a common-emitter current gain β of 150. Calculate the common-emitter leakage current at 25°C and at 115°C.

8. A transistor data sheet lists α as 0.95 but does not list the current gain for the common-emitter configuration. Calculate this value for the same operating point.

9. Repeat Problem 8 for an α of 0.985.

10. A transistor with $\alpha = 0.985$ has a cutoff current I_{CBO} of 60 μA. What would the cutoff current be in the common-emitter configuration?

11. A transistor with $\alpha = 0.988$ has a collector current I_{CBO} of 50 μA. Find its cutoff current I_{CEO}.

12. The 2N130 of Problem 4 has a common-base current gain α of 0.975. Calculate its leakage current when used in the common-emitter configuration at 25°C and at 80°C.

13. Using the grounded-base characteristic curves for the 2N130 of Fig. 2-14, calculate the ac current gain α for this transistor for an operating point of $V_{CB} = 8$ V and an emitter current of 2.0 mA.

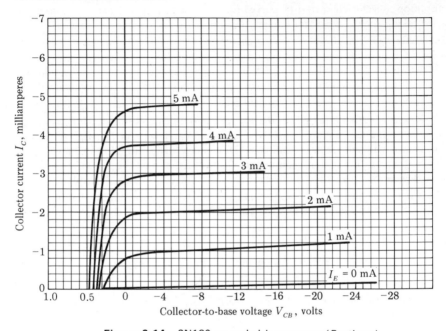

Figure 2-14 2N130 grounded-base curves (*Raytheon*).

14. Repeat Problem 13 for an operating point of $V_{CB} = 16$ V and $I_E = 1.0$ mA.

15. Repeat Problem 13 for an operating point of $V_{CB} = 4.0$ V and $I_E = 4.0$ mA.

16. Using the grounded-emitter curves for the 2N130 (Fig. 2-15), calculate the current gain β for this transistor for an operating point of $V_{CE} = 8$ V and $I_B = 150$ μA.

17. Repeat Problem 16 for an operating point of $V_{CE} = 15$ V and $I_B = 100$ μA.

18. Repeat Problem 16 for an operating point of $V_{CE} = 4.0$ V and $I_B = 350$ μA.

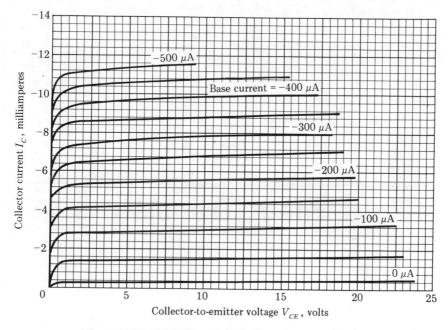

Figure 2-15   2N130 grounded-emitter curves (*Raytheon*).

3

Field-Effect Transistors

Field-effect transistor action was first presented in the 1920s—long before the advent of bipolar junction devices—but because of limitations in manufacturing know-how, production was not pursued. Later, in 1949, a workable field-effect device was proposed, but the timing was off. Bipolar junction transistor technology was now in full swing, and again the field-effect transistor (FET) was sidetracked. Then, in the early 1960s, further improvements in manufacturing processes made the FET a practical reality and extensive research on these devices began.

The name *field-effect transistor* comes from the fact that the current between two of the elements in this device is controlled by the electric field created by the third element (the *gate*). There are two basic types of field-effect units: the *junction field-effect transistor* (JFET or just FET) and the *metal-oxide-semiconductor transistor* (MOSFET). The latter type, because of its construction, is also known as an *insulated-gate transistor* (IGFET).

THE JUNCTION FIELD-EFFECT TRANSISTOR

Junction field-effect transistors are available in two types—*N-channel* or *P-channel* devices. The current carriers are electrons in the *N*-channel unit; in the *P*-channel unit, the carriers are holes. In either case, the terminal into which the current is "injected" is called the *source,* while the terminal from which the current is "extracted" is called the *drain*. As mentioned before, current flow is controlled by the action of the third element, the *gate*. Obviously, the polarities of the voltages applied to an *N*-channel unit are opposite to the polarities required for a *P*-channel device. The action is otherwise quite similar. We will use an *N*-channel device to explain this action.

FIELD-EFFECT TRANSISTOR ACTION.   Figure 3-1 shows, in simplified form, the construction of an *N*-channel JFET. It consists essentially of a bar of *N*-type silicon with heavily doped *P*-type regions on either side. These regions are the gates. *P-N* junctions are formed between the gates

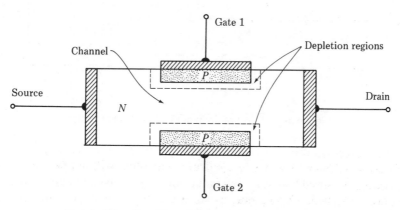

Figure 3-1   Simplified diagram of an *N*-channel JFET.

and the bar. Notice the depletion regions surrounding the junctions and the channel formed between the source and the drain. If we make the gates negative compared to the channel, we have reverse-biased junctions. The depletion region will widen, extending deeper into the channel. Since the channel is narrower, its resistance is increased.

Now let us apply a variable-voltage supply V_{DD} between source and drain, as shown in Fig. 3-2. Current (electrons) will flow from the grounded side of the power supply to the source, through the *N*-channel bar to the drain, and back to the positive side of the power supply. Neglecting the very low resistance of the wiring, the voltage drop in the channel is equal to the supply voltage V_{DD}, with the drain end positive compared to the source. (Since there is no other resistance in the circuit, the drain-source voltage V_{DS} is equal to the supply voltage V_{DD}.) Also, the potential all along the *N*-channel bar is increasing from zero to the full V_{DD} positive value as we approach the drain end.

Now notice that the gates are tied to the source, so that the full length of these gates is at ground potential. Consequently, the *P-N* junctions are

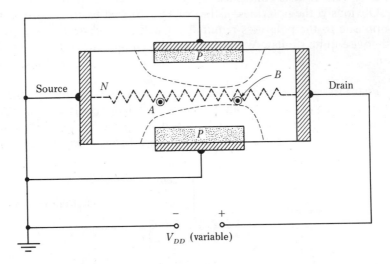

Figure 3-2   Effect of drain current on channel width.

reverse-biased, and this reverse bias is higher at the drain end. Notice how the depletion region is wider at the drain end. What will happen if the supply voltage is varied? At very low values of V_{DD}, the effect is negligible and the channel is wide throughout, as shown in Fig. 3-1. As we increase V_{DD} (and V_{DS}), the drain current I_D increases and the effect shown in Fig. 3-2 begins. As we continue to increase the supply voltage, the drain current increases further and the widening of the depletion region increases. Finally, at some value of voltage (and current I_D) the depletion regions *at the drain end* will just meet, pinching off the channel. This value of the drain-source voltage V_{DS} is called the *pinch-off voltage* V_P. Increasing the drain-source voltage above this value results in little or no further increase in drain current. This is similar to the saturation effect in bipolar transistors and in tubes.

It should be noted that when pinch-off occurs—due to increase in voltage V_{DS}—*current flow is not cut off:* it levels off. Remember that the widening of the depletion region is caused by the increase in reverse bias, which in turn is caused by the voltage drop $(I_D R)$ in the semiconductor material. Any tendency for the current I_D to decrease would reduce this reverse bias, increase the channel width, and improve conduction. Obviously, the channel cannot close completely. Instead, as V_{DS} is increased above V_P, the almost-touching depletion region that started at the drain end of the transistor begins to elongate toward the source end. The channel resistance increases, thereby offsetting the increase in voltage V_{DS}, and the current flow remains essentially constant.

STATIC CHARACTERISTICS.   In the preceding description of the field-effect action, the gate was tied to the source. In other words, the applied gate-source voltage V_{GS} is zero. Now, using a circuit similar to Fig. 2-10 for the bipolar transistor,* we can apply variable voltages to both the gate and the drain (compared to source). A family of drain current–drain voltage curves can be obtained for various fixed values of gate voltage V_{GS}. Such a set of curves is shown in Fig. 3-3. Notice that when the gate voltage is applied, the gate is made negative compared to source. In an N-channel FET this is reverse bias for the junction, making the depletion region extend

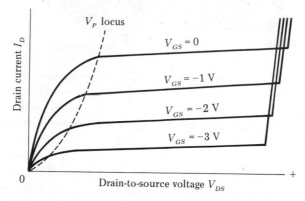

Figure 3-3   Output characteristic curves—N-channel JFET.

deeper into the channel and reducing conduction. Notice that as this bias (V_{GS}) is increased, the pinch-off effect occurs sooner—at lower values of drain voltage V_{DS} and at lower values of drain current I_D. (The dotted line shows the locus of the pinch-off point V_P with variation in gate voltage.) Two factors are now contributing to channel depletion: the IR drop in the channel due to I_D (as before) and the applied reverse gate voltage V_{GS}. Since at zero gate bias pinch-off occurs when $V_{DS} = V_P$, now the relationship is given by

For pinch-off: $$V_{DS} + V_{GS} = V_P \qquad (3\text{-}1)$$

where voltages V_{DS} and V_{GS} are added numerically without regard to their respective polarities. From this equation it follows that as the gate voltage is increased, pinch-off occurs at lower values of drain voltage; and since V_{DS} is lower, I_D will be lower. Finally, if the gate voltage is made equal to or higher than V_P (in magnitude), then pinch-off occurs at zero drain voltage and no current can flow. In this case, pinch-off is also cutoff.

*Drawing such a circuit is left as an exercise for the student. See Problem 1 at the end of this chapter.

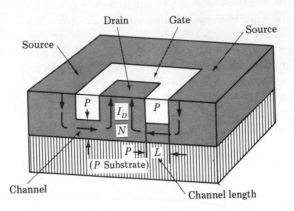

Figure 3-4 Single-sided geometry JFET (*Motorola Semiconductors*).

CONSTRUCTION AND SYMBOL. In actual practice, diffusion of gates to both sides of the semiconductor material is costly. Instead, a single-sided process is used. This is shown in Fig. 3-4. For an *N*-channel FET, fabrication starts with a *P*-type substrate. Onto this a layer of *N*-type material is grown epitaxially to act as the *N* channel. Then, still from the same side, a *P*-type gate is diffused. Metal contacts are added to complete the structure. The substrate in this diagram functions as gate 2 of Fig. 3-1. Usually these gates are connected together internally. (Sometimes the connection from the substrate is brought out separately, producing a tetrode device.)

P-channel junction FETs are made by starting with an *N*-type substrate, growing onto this a *P*-type channel, and diffusing into the channel an *N*-type gate. All that has been discussed above regarding *N*-channel FETs applies equally well to *P*-channel devices. Remember, however, that the drain-source and the gate-source voltages will be opposite in polarity compared to *N*-channel devices.

The electrical symbols used to represent JFETs are shown in Fig. 3-5.

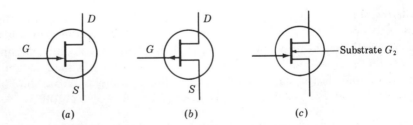

Figure 3-5 Schematic symbols for JFETs.

Figure 3-5*a* is for an *N*-channel FET. (Notice that the gate, being of *P*-type material, uses the same arrowhead direction as for the *P*-type emitter in a *PNP* transistor.) In Fig. 3-5*b* the gate arrowhead points out (*N*-type gate). This is the schematic for a *P*-channel FET. When the substrate is not internally tied to gate 1, this is shown by the schematic as in Fig. 3-5*c*.

MOSFETS OR IGFETS

The second major category of field-effect transistors is the *metal-oxide-semiconductor transistor* (MOSFET) or *insulated-gate transistor* (IGFET). As the latter name implies, the gate is electrically insulated from the channel material. No *P-N* junction exists. Current control is now dependent on the capacitive effect (dielectric field) between the gate and channel materials (as the "plates") separated by an insulating oxide layer. As with JFETs, these transistors are also available as *N*-channel or *P*-channel devices. In addition, depending on the polarity of the gate-bias voltage with which

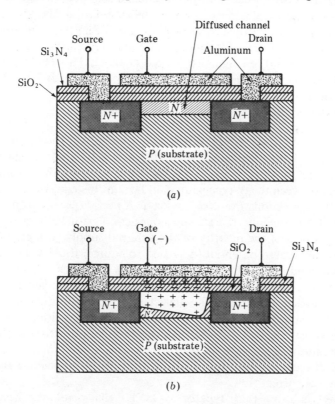

(a)

(b)

Figure 3-6   *N*-channel depletion-mode MOSFET (*Motorola Semiconductors*).

they are designed to operate, they are further classified as *depletion-mode* or *enhancement-mode* MOSFETs.

DEPLETION-MODE MOSFET.   The construction and basic action in a depletion-mode MOSFET can be explained with the aid of Fig. 3-6. An N-channel device is used. Fabrication starts with a P-type substrate. Into this are diffused two separate, heavily doped, N-type regions to act as source and drain; between them a moderately doped N-channel is produced. The surface is then covered with an oxide layer (SiO_2) to insulate the channel from the gate. The oxide is in turn covered with a nitride layer that has high resistance to contamination from outside effects. Holes are cut through the oxide and nitride layers for metallic (ohmic) contacts to the source and drain regions. A third metallic layer is deposited, centrally, over the insulation covering the channel region. This is the gate.

If an external power source is connected between source and drain, making the drain positive, considerable current will flow between source and drain. This is so because a full conductive channel is available between these elements. Now, if a negative potential is applied to the gate, by capacitive action through the dielectric (insulating oxide layer) it will attract holes from the P-type substrate into the channel and repel and/or neutralize the electron carriers in the originally N-type channel. This effect is shown in Fig. 3-6b. Notice that the channel is being depleted of carriers. Drain current will decrease. (This final effect is the same as that noted for the JFET. In other words, the JFET operates in the depletion mode.)

ENHANCEMENT-MODE MOSFET.   Figure 3-7a shows the construction of an N-channel enhancement-mode MOSFET. Notice that this diagram is almost the same as Fig. 3-6a. This time, however, there is no diffused N-channel between the N-type source and drain. Instead, we now have two back-to-back junctions (an N-P and a P-N) in series between these two elements. Consequently, there is no conduction path between source and drain, regardless of the polarity of the external supply voltage. Obviously, with the gate at zero potential (gate tied to source) the device is in cutoff. To obtain conduction, the gate must be biased with the same polarity as the drain—i.e., positive for an N-channel FET. The gate acts as the upper plate of a capacitor. Through the dielectric, electrons (the minority carriers in the P-type substrate) are attracted into the channel region, directly below the positive gate. (See Fig. 3-7b.) As the positive potential on the gate is increased, more electrons are attracted into this region, increasing the conductivity of the now N-channel path. In other words, a positive gate potential enhances current flow.

There is also a third type of MOSFET that combines these two ac-

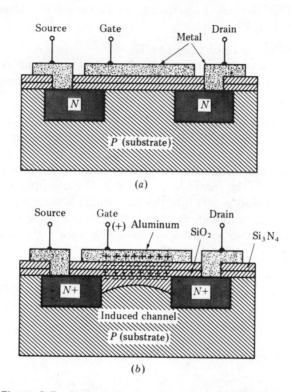

Figure 3-7   *N*-channel enhancement-mode MOSFET.

tions. In other words, it can be operated in the depletion and/or the en-hancement mode. For an *N*-channel unit, the channel material has a light *N*-type doping so that a moderate drain current will flow even at zero gate bias. Then positive gate biases will produce full (maximum rated) conduc-tion, whereas negative gate voltages will drive the unit into cutoff.

The three modes of operation are classified by industry as:

Depletion mode only: type A
Depletion/enhancement mode: type B
Enhancement mode only: type C

The electrical characteristics of these FETs are covered in a later section. Although the discussion has been on the *N*-channel MOSFET, all that has been said applies equally well to *P*-channel units. Remember, however, that the gate-source and drain-source voltages must be opposite in polarity to those used above (with *N*-type units).

MOSFET SCHEMATIC SYMBOLS. A variety of schematic symbols are used for MOSFETs to:

1. Distinguish them from JFETs.
2. Indicate whether they are N-channel or P-channel devices.
3. Distinguish between a depletion-mode and an enhancement-mode unit.

These symbols are shown in Fig. 3-8. Comparing these schematics with the JFET diagrams (Fig. 3-5), notice that the gate symbol for the MOSFETs does not touch the channel bar. Instead, a capacitive action is implied by the symbol used. To distinguish an N channel from a P channel, the arrowhead direction on the substrate is reversed. For example, in Fig. 3-8a the substrate arrowhead points into the bar, which is the symbol for a P-type element. Since the substrate is P type, the channel is N type. Conversely, the reversed substrate arrowhead in Fig. 3-8b is for a P-channel MOSFET.

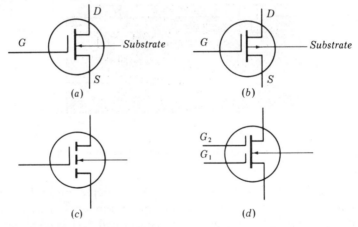

Figure 3-8 MOSFET schematic symbols.

Now compare Fig. 3-8a and c. Notice that the channel bar in (c) is a broken line, indicating that there is no immediate conducting path between source and drain. This symbol represents the enhancement-mode FET. Figure 3-8d shows a *dual-gate* MOSFET. (This type should not be confused with units using the substrate as a second gate.) These two gates are effectively in series along the channel, and the drain current is a function of the voltage applied to each gate. Dual-gate MOSFETs are useful when two input signals are to be mixed, and they are especially desirable as RF and IF amplifiers, with the second gate facilitating automatic gain control.*

*J. J. DeFrance, *Communications Electronics Circuits* (2nd ed.), New York: Holt, Rinehart and Winston, 1972, chap. 8.

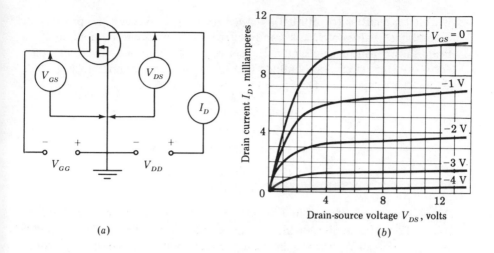

Figure 3-9 Test circuit and output characteristics for an *N*-channel depletion-mode MOSFET.

STATIC OUTPUT CHARACTERISTICS.   Figure 3-9 shows a circuit for obtaining the output characteristic curves of an *N*-channel depletion-mode MOSFET. Both power supplies, V_{GG} and V_{DD}, should be variable. For an output characteristic (I_D versus V_{DS}), the procedure has been discussed before: V_{GS} is held constant, V_{DD} is varied in steps, and the I_D and V_{DS} values are recorded for each step. The process is repeated for several values of V_{GS} to get a family of curves. Such a family of curves is shown in Fig. 3-9*b* for an *N*-channel depletion-mode MOSFET.

Remember that we also have *P*-channel and enhancement-mode MOSFETs. Is this circuit suitable for these devices? Are the supply polarities still correct as shown? Since in all cases we want data for I_D versus V_{DS}, the circuitry is satisfactory. However, the power supply polarities may need to be reversed. Let us consider some specifics. With a *P*-channel FET—of any type—the drain must be negative compared to source, so the V_{DD} polarity must be reversed. Now consider type C (enhancement) FETs. When V_{GS} is zero, the device is cut off. To make it conduct, the gate-bias voltage must be increased—in the same polarity as the drain voltage (i.e., positive for *N* channels and negative for *P* channels). Therefore, the V_{GG} polarity must be changed accordingly.

STATIC TRANSFER CHARACTERISTIC.   Quite often a more useful curve is obtained by holding the drain potential constant and varying the gate-source voltage. The data then shows how the drain current I_D varies with gate-source voltage V_{GS} for a constant drain voltage V_{DS}. Such a curve is called a *transfer* (or mutual) *characteristic.* By repeating the process for several fixed values of V_{DS}, a family of transfer curves can be obtained. If a drain-family (output characteristic) set of curves is already available, data for a transfer set can be obtained from it. For example, refer to Fig. 3-9*b*. Along the vertical line for $V_{DS} = 12$ V, we can pick off

$$
\begin{aligned}
I_D &= 10 \quad \text{mA at } V_{GS} = 0 \text{ V} \\
&= 6.6 \text{ mA at } V_{GS} = -1.0 \text{ V} \\
&= 3.6 \text{ mA at } V_{GS} = -2.0 \text{ V} \\
&= 1.4 \text{ mA at } V_{GS} = -3.0 \text{ V} \\
&= 0.2 \text{ mA at } V_{GS} = -4.0 \text{ V}
\end{aligned}
$$

These points are plotted in Fig. 3-10 to give a transfer characteristic for $V_{DS} = 12$ V. In similar fashion, transfer characteristics can be obtained for $V_{DS} = 8$ V, 4 V, or any other value. Figure 3-10 also shows—by broken line—the transfer curve for $V_{DS} = 4$ V. Notice that in spite of the great change in drain-source voltage, the two transfer curves are almost identical. The more nearly horizontal the output characteristics (beyond the pinch-off region), the more nearly identical the transfer curves will be. Therefore,

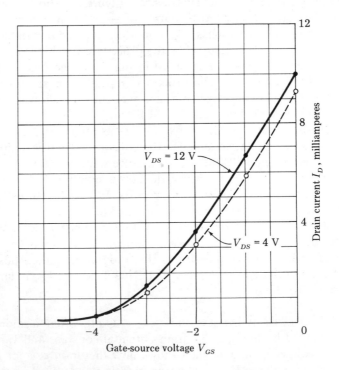

Figure 3-10   Transfer characteristic.

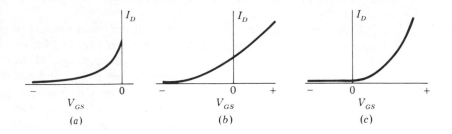

Figure 3-11   Transfer curves for types A, B, and C FETs.

to a close approximation, one transfer curve is all that is needed to represent the characteristics of a specific FET. (This is also true for JFETs.) Manufacturers' data sheets often give two transfer curves for each device. Because of manufacturing tolerances, there can be appreciable differences between two FETs of the same type. The two transfer curves then show the minimum and maximum possible characteristics for that specific model. Transfer characteristics readily show the differences between type A (depletion mode), type B (depletion/enhancement mode), and type C (enhancement mode only) FETs. Comparative curves are shown in Fig. 3-11.

ELECTRICAL RATINGS AND PARAMETERS.   Data sheets for FETs give quite a number of ratings and parameters. The more important of these will be reviewed briefly now:

1. *Breakdown voltages* are generally abbreviated BV (or V_{BR}), followed by three more subscripts. The first two subscripts show between what two elements the breakdown voltage applies. The third specifies the condition of the third element: S for shorted, O for open, and X for some specified voltage. For example:

   BV_{DGO} = breakdown voltage between drain and gate, with the source open-circuited

   BV_{DSX} = breakdown voltage between drain and source, with V_{GS} at some specified value

   BV_{DSS} = as above, but with $V_{GS} = 0$ (gate shorted to source)

2. $V_{GS(\text{off})}$ is the gate-source cutoff voltage and represents the gate voltage necessary to reduce I_D to a very low level. (For type A and type B FETs this current is less than 1 percent of the saturated value at $V_{GS} = 0$.) $V_{GS(\text{off})}$ is numerically equal to the pinch-off voltage V_P.

3. $V_{GS(\text{th})}$ applies to enhancement-mode FETs and is the gate thresh-

old voltage, or minimum voltage at which the FET starts to conduct. '

4. I_{GSS} is the gate-source leakage current (with the drain shorted to source). This is an extremely low value of current (generally less than 1 nA) and accounts for the very high input resistance of the FET. (Gate leakage current may also be specified as I_{GDO} or I_{GSO}.) *Note: This low leakage, or extremely high resistance, does produce a problem. It is possible for the gate to pick up and build up high static charges—high enough to break down the insulation between gate and channel, ruining the device. To prevent this, manufacturers twist the element leads together (providing a discharge path). Before untwisting these leads, it is recommended that a fine wire be wrapped around the terminals until the FET is soldered into place or mounted in its socket. The circuitry will then provide protection. New devices are often manufactured with built-in zener diodes to protect against high voltage buildup.*

5. I_{DSS} is the drain-source current when $V_{GS} = 0$. In a depletion-mode FET, this value is close to the maximum current rating. In B-type devices, it is approximately halfway between cutoff and saturation; in C-type devices, it is a very low value in the cutoff region.

6. $I_{D(off)}$ is the drain current when V_{GS} is beyond the cutoff value.

7. $I_{D(on)}$ is used with enhancement-mode (and also with B-type) FETs and represents full-conduction drain current, close to the maximum current rating.

8. y_{fs} and g_{fs} are key dynamic parameters for all FETs and are very useful in the design of amplifier circuits. Parameter y_{fs} is the *forward transadmittance:* it shows the relationship between input signal voltage and output signal current. Theoretically, the y_{fs} value for any transistor can be evaluated directly from its output characteristic curves as

$$y_{fs} = \frac{\Delta I_D}{\Delta V_{GS}} \quad \text{(with } V_{DS} \text{ held constant)} \tag{3-2}$$

In practice, such evaluations would not be accurate because the I_D–V_{DS} curves are almost horizontal. Being an admittance quantity, y_{fs} is of the form $\dot{y} = \dot{g} + jb$ and has a conductance (or "real") component and a susceptance (capacitance) component. The capacitance effects are quite small, however, and can be neglected up to about 100 MHz. It is therefore quite common, at lower frequencies, to see y_{fs} used interchangeably with g_{fs}. Parameter g_{fs} [sometimes referred to as $Re(y_{fs})$] is called the *transconductance.* Both quantities (y_{fs} and g_{fs}) are measured in millimho or micromho units. In general, the higher the y_{fs} or g_{fs} value, the better the device.

9. y_{os}, g_{os}, and $r_d - y_{os}$ is the *output admittance*. By equation,

$$y_{os} = \frac{\Delta I_D}{\Delta V_{DS}} \quad \text{(with } V_{GS} \text{ held constant)} \tag{3-3}$$

As before, this parameter could be evaluated from the I_D–V_{DS} curves, but the results would not be very accurate. Like y_{fs}, y_{os} is a complex quantity that contains a conductance component g_{os} and a susceptive (capacitive) component jb_{os}. At the lower frequencies, the susceptive component is negligible and y_{os} and g_{os} are considered equivalent. Some data sheets use still another variation: instead of specifying the output admittance or conductance, they show a *drain resistance* r_d, where

$$r_d = \frac{1}{g_{os}} \cong \frac{1}{y_{os}} \tag{3-4}$$

10. μ—In vacuum tubes, this parameter is the *amplification factor* and is equal to $g_m \times r_p$. The same relationship can be applied to a FET by using g_{fs} for g_m and r_d for r_p. The amplification does not appear on FET data sheets, however, and is not generally used.

11. C_{iss} is the small-signal, common-source, short-circuit *input capacitance*. It is the major component of the input admittance y_{is}—the real component, g_{is}, approaches zero (infinite resistance).

12. C_{rss} is the small-signal, common-source, short-circuit *transfer capacitance* and represents the drain-to-gate capacitance. It takes the place of the more general term y_{rs}, since this admittance has no real (conductance) component.

13. C_{oss} is the small-signal, common-source, short-circuit *output capacitance* and is approximately equal to C_{rss} in value.

PARAMETER VARIATIONS AND INTERRELATIONS.   The numerical parameter values given in data sheets apply only at some specified condition of current, voltage, temperature, and frequency. Most values vary appreciably with one or more of these factors. Therefore, full specifications include curves to show how the parameters vary with these factors. On the other hand, the more common abbreviated data sheets do not include curves and may also omit some of the parameters listed above. Knowledge of some parameter interrelations* can help in such cases:

*L. J. Sevin, *Field-Effect Transistors*, New York: McGraw-Hill Book Company, 1965; *JFET Applications and Specifications*, Teledyne Semiconductor, 1972.

For A-type and B-type FETs:

$$I_D = I_{DSS}\left(1 - \frac{V_{GS}}{V_{GS(off)}}\right)^2 \qquad (3\text{-}5)$$

For C-type FETs:

$$I_D = K(V_{GS} - V_{GS(th)})^2 \qquad (3\text{-}6)$$

$$V_P = V_{GS(off)} = \frac{2I_{DSS}}{y_{fso}} \qquad (3\text{-}7)$$

where $y_{fso} = y_{fs}$ at $V_{GS} = 0$.

$$y_{fs} = y_{fso}\left(1 - \frac{V_{GS}}{V_{GS(off)}}\right) \qquad (3\text{-}8)$$

$$y_{fs} = y_{fso}\sqrt{\frac{I_D}{I_{DSS}}} \qquad (3\text{-}9)$$

Equation 3-5 can be used to obtain the transfer characteristic when it (or the drain-family curves) is not given. All we need is I_{DSS} and $V_{GS(off)}$ or its equivalent V_P. An example will illustrate this.

EXAMPLE 3-1

The 2N5033, a Teledyne P-channel JFET, has $I_{DSS} = 0.3$ mA (min) and 3.5 mA (max) and $V_{GS(off)} = 0.3$ V (min) and 2.5 V (max). (All data is for $V_{DS} = -10$ V.) Calculate data for the maximum-condition transfer curve, and plot the curve.

Solution

$$I_D = I_{DSS}\left(1 - \frac{V_{GS}}{V_{GS(off)}}\right)^2$$

1. At $V_{GS} = 0$, $I_D = I_{DSS} = 3.5$ mA
2. At $V_{GS} = 0.5$ V,

$$I_D = 3.5\left(1 - \frac{0.5}{2.5}\right)^2 = 3.5(0.8)^2 = 2.24 \text{ mA}$$

3. At $V_{GS} = 1.0$ V,

$$I_D = 3.5\left(1 - \frac{1.0}{2.5}\right)^2 = 3.5(0.6)^2 = 1.26 \text{ mA}$$

4. At $V_{GS} = 1.5$ V,

$$I_D = 3.5\left(1 - \frac{1.5}{2.5}\right)^2 = 3.5(0.4)^2 = 0.56 \text{ mA}$$

5. At $V_{GS} = 2.0$ V,

$$I_D = 3.5\left(1 - \frac{2.0}{2.5}\right)^2 = 3.5(0.2)^2 = 0.14 \text{ mA}$$

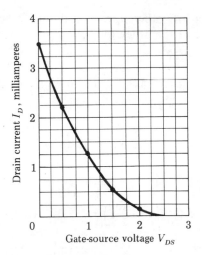

Figure 3-12

6. These points are plotted in Fig. 3-12.

In Example 3-1, the transfer curve is for $V_{DS} = -10$ V. However, it should be remembered that V_{DS} has little effect on the current values (see Fig. 3-10) and therefore this curve can be used, with little error, for any other drain-source voltage.

Equation 3-6 can be used in similar manner to obtain the transfer characteristic for enhancement-mode FETs. Notice, however, that the value of K is not specified. (It varies with each FET.) Furthermore, this value is not given in the data sheets, but it can be calculated from other given data. An example or two will make this clear.

EXAMPLE 3-2

The M164, a Siliconix enhancement-mode MOSFET, has the following "typical" parameter values: $V_{GS(th)} = 3.2$ V and $I_D = 15$ mA at $V_{GS} = 10$ V. Using Equation 3-6, evaluate the transfer constant K for this device.

Solution

From Equation 3-6,

$$K = \frac{I_D}{(V_{GS} - V_{GS(off)})^2} = \frac{15 \times 10^{-3}}{(10 - 3.2)^2} = 3.2 \times 10^{-4}$$

The transadmittance value y_{fs} varies with the operating point. Yet in charts only one value is given (usually y_{fso}). Equation 3-8 or 3-9 can then be used to find y_{fs} for any other operating point. Depending on the information available, one or the other may be more convenient.

EXAMPLE 3-3

The JFET of Example 3-1 has a y_{fs} of 5.0 mmho (max) at $V_{DS} = -10$ V and $V_{GS} = 0$. Find the value of this parameter at $V_{GS} = 1.0$ V.

Solution

Since we have the cutoff value of V_{GS}, Equation 3-9 is suitable:

$$y_{fs} = y_{fso}\left(1 - \frac{V_{GS}}{V_{GS(off)}}\right)$$

$$= 5 \times 10^{-3}\left(1 - \frac{1.0}{2.5}\right) = 5 \times 10^{-3}(0.6) = 3.0 \text{ mmho}$$

REVIEW QUESTIONS

1. (a) Name two major classifications of field-effect transistors. (b) Give the abbreviations for each. (c) Name the three elements of a transistor. (d) State briefly the function of each element.

2. (a) To which of the major classifications of FET does the N-channel device belong? Explain. (b) What are the current carriers in this type? (c) Are holes used as current carriers in FETs? Explain.

3. Refer to Fig. 3-1: (a) What makes this a junction FET? (b) Is there a junction between source and channel or between drain and channel? Explain.

4. Refer to Fig. 3-2: (a) What is the significance of the broken-line resistor symbol? (b) Will any current flow through this circuit? If so, trace the path. (c) When there is no current flow through the channel, what is the gate-source voltage? Why? (d) With the supply voltage set at 6 V, what is the potential of point A in the channel one-third of the way from source to drain? Explain. (e) Is this a forward or a reverse bias? (f) What is the potential at point B, two-thirds of the way from source to drain? (g) Are there any depletion regions in this diagram? (h) Account for their greater penetration at the drain end. (i) What happens to the channel width and resistance as the supply voltage is increased?

5. (a) What does the symbol V_P stand for? (b) In Fig. 3-2, if the supply voltage is set to this value, does any current flow? (c) What happens to this current if V_{DD} is increased above the pinch-off value? (d) What could this effect be called?

6. Refer to Fig. 3-3: (a) From the diagram, how can you tell that this is an N-channel device? (b) Why does the drain current remain almost constant when V_{DS} is increased to the right of the broken-line region? (c) At any given V_{DS} value, why is the drain current higher for $V_{GS} = -1$ V than for $V_{GS} = -3$ V? (d) Why does the locus of the V_P curve approach zero as V_{GS} is made more negative?

7. Refer to Fig. 3-5: (a) Which of these diagrams represents an N-channel FET? (b) How do we distinguish a P-channel FET from an N-channel FET?

8. Refer to Fig. 3-6a: **(a)** What is the gate-source voltage in this diagram? **(b)** Is there a conducting path from source to drain? If so, describe it. **(c)** Is this a full conductive channel or not? Explain.

9. Refer to Fig. 3-6b: **(a)** What are the plus symbols located in what was the N-channel of diagram (a)? **(b)** Account for their presence. **(c)** What happens to conduction as the gate-source voltage is made more negative?

10. Refer to Fig. 3-7: **(a)** In diagram (a), what is the gate-source voltage? **(b)** Is there a conducting channel? Explain. **(c)** In diagram (b), what is the gate polarity with respect to source? **(d)** What are the minus symbols located in the substrate region directly below the gate? **(e)** Account for their presence. **(f)** What happens to conduction as the gate is made positive with respect to source?

11. In Fig. 3-7, conduction is obtained by making the gate positive. Is this true for all enhancement-mode MOSFETs? Explain.

12. **(a)** Give the mode of operation for a type-A MOSFET. **(b)** Repeat for a type B. **(c)** Repeat for a type C. **(d)** To which class does a JFET belong?

13. Refer to Fig. 3-8: **(a)** Which of these are P-channel MOSFETs? **(b)** Which are enhancement-mode MOSFETs? **(c)** Describe the device shown in diagram (d).

14. Refer to Fig. 3-9a: **(a)** What type of MOSFET is this? **(b)** How can you tell whether it is an N-channel or P-channel device? **(c)** Describe the procedure for getting an $I_D - V_{DS}$ curve.

15. Refer to Fig. 3-9b: **(a)** What is the drain current if the gate bias is -2 V and the drain-source voltage is 8 V? **(b)** Repeat for $V_{GS} = -1$ and $V_{DS} = 6.5$ V. **(c)** Repeat for $V_{GS} = -2.5$ V and $V_{DS} = 9.0$ V. **(d)** What gate bias would be needed for a drain current of 6 mA at a drain-source voltage of 5.0 V? **(e)** Repeat part (d) for $I_D = 5$ mA and $V_{DS} = 7.0$ V.

16. **(a)** What are the coordinates of any transfer characteristic curve? **(b)** How can such a curve be obtained by using the circuit of Fig. 3-9a?

17. Refer to Fig. 3-10: **(a)** How are the points for these curves obtained? **(b)** Why is there so little difference between the two curves?

18. Which of the curves in Fig. 3-11 applies to which mode of operation?

19. What do the following abbreviations mean: BV_{DSS}, $V_{GS(off)}$, I_{DSS}, BV_{DGO}, I_{GSS}, $I_{D(off)}$, $V_{GS(th)}$, $I_{D(on)}$?

20. **(a)** What is y_{fs}? **(b)** How does it differ from g_{fs}? **(c)** When can they be considered equivalent?

21. **(a)** Distinguish between y_{os} and g_{os}. **(b)** When can they be considered equivalent? **(c)** How do they compare with r_d?

22. **(a)** Name three capacitance values inherent in any FET. **(b)** Give the abbreviation for each.

23. **(a)** Are FET parameters fixed values? **(b)** Name three operating conditions that can cause a change in these values. **(c)** State a technique used to show parameter values for any operating condition.

24. **(a)** Equations 3-5 and 3-6 show how the drain current varies with what other factor? **(b)** Do these equations apply for V_{DS} above V_P, below V_P, or both?

25. In Equation 3-7, what is y_{fso}?

26. In Equation 3-6, how is the K value obtained for a given MOSFET?

27. (a) In Example 3-3, is y_{fso} given? Explain. **(b)** This is the same FET that was used in Example 3-1. Explain how Equation 3-9 could have been used to solve this problem.

PROBLEMS AND DIAGRAMS

1. Draw a circuit diagram suitable for obtaining data for the output characteristic curves (I_D versus V_{DS}) for an N-channel JFET.

2. Draw the schematic symbol for a P-channel FET. Repeat for an N-channel device.

3. (a) Draw the schematic symbol for a P-channel enhancement-mode MOSFET. **(b)** Repeat for a dual-gate depletion-mode unit. **(c)** Repeat for an N-channel, single-gate, depletion-mode MOSFET.

4. Draw a circuit diagram suitable for obtaining data for the output characteristic curves of a P-channel enhancement-mode MOSFET.

5. Using the output characteristics from Fig. 3-9, tabulate data for a transfer curve for **(a)** $V_{DS} = 9.0$ V and **(b)** $V_{DS} = 6.0$ V.

6. Calculate the forward transadmittance value for the FET in Fig. 3-9b at an operating point of $V_{DS} = 8.0$ V and $V_{GS} = -1.5$ V. (Take V_{GS} increments of 0.5 V above and below the operating point.)

7. Repeat Problem 6 for an operating point of $V_{DS} = 6.0$ V and $V_{GS} = 2.0$ V. (Use a ± 1.0 V swing.)

8. Calculate the output admittance and drain resistance of the FET in Fig. 3-9b for an operating point of $V_{DS} = 8.0$ V and $V_{GS} = -1.0$ V.

9. Calculate the output admittance and drain resistance of the FET in Problem 7.

10. A 2N4869 JFET has $V_P = -1.8$ V (min) and -5 V (max); $I_{DSS} = 2.5$ mA (min) and 7.5 mA (max); $y_{fs} = 1300$ μmho (min) and 4000 μmho (max) at $V_{GS} = 0$, all at $V_{DS} = 20$ V. Tabulate data for the maximum-condition transfer curve.

11. Repeat Problem 10 for the minimum-condition transfer curve.

12. A 3N128 depletion-mode MOSFET has $I_{DSS} = 5$ to 25 mA; $V_{GS(off)} = -3$ to -8 V; and $g_{fs} = 5000$ to 12,000 μmho at $V_{GS} = 0$. Obtain data and plot curves for maximum-condition and minimum-condition transfer curves.

13. An M114 P-channel enhancement-mode MOSFET has typical values of $V_{GS(th)} = -2.9$ V; $I_D = 20$ mA at $V_{GS} = -10$ V; and g_{fs} (at $V_{GS} = -10$ V) $= 4000$ μmho. Find the transfer constant K for this transistor.

14. Using the FET from Problem 13, obtain data and plot the transfer characteristic.

15. Using the FET from Problem 10, find **(a)** the minimum and maximum forward transfer admittance at $V_{GS} = -1.0$ V. **(b)** Repeat for $I_D = 2.0$ mA.

16. Using the FET from Problem 12, find **(a)** the minimum and maximum forward transconductance at $V_{GS} = -2.0$ V. **(b)** Repeat for $I_D = 4.0$ mA.

17. Using the FET from Problem 13, find **(a)** the forward transconductance value g_{fso}. ($V_{GS(th)}$ is the equivalent of $V_{GS(off)}$ in a depletion-mode FET.) **(b)** Find the transconductance at $V_{GS} = -6$ V.

4

Active Devices—
Vacuum Tubes

All electronic circuits—whether discrete-component or integrated—consist fundamentally of active devices (tubes or semiconductors) and passive devices (resistors, inductors, and capacitors). For many years the vacuum tube was considered as the heart of the circuit. In fact, it was not until the development of these tubes that the electronics industries began to grow in leaps and bounds. Then in the 1960s, with advancements in solid-state technology, transistors began to replace the vacuum tube. At first the change was limited to the switching or digital field, where the main requirement of the active device is to conduct or not conduct. The much smaller size and the lower power consumption of the transistor (plus its theoretically infinite life) made it ideal in computer applications requiring thousands of these devices. By the close of that decade, transistors had made deep inroads in the communications field. New radio receiver designs were all solid-state; some television receivers were "transistorized"; and even low-power transmitters were using transistors.

Because most new design uses solid-state active devices, some may consider the need to study vacuum tubes passé. (This could apply to those interested only in the computer field.) For such persons, this chapter could be skipped, and all references to vacuum-tube circuits in the succeeding chapters can be omitted without any loss in the understanding of solid-state circuitry. Yet, vacuum tubes are far from extinct. They are still being manufactured; they are still being used in some new designs; many products with vacuum tubes are still in use and need maintenance; and even more important—at high power levels (broadcasting, for example), the vacuum tube is still the only suitable active device. Obviously, the study of vacuum-tube characteristics—although not as important as in the early days of radio—should not be dropped, if time is available.

THERMIONIC EMISSION—CATHODE STRUCTURES

When a metal is heated, some of the heat energy is converted to kinetic energy, causing accelerated motion of the electrons. If the temperature rises sufficiently (and the increase in kinetic energy equals the *work function** of the material), electrons can overcome the attracting forces of the atomic structure and fly off into space.

In a vacuum tube, the element which is heated to emit electrons is called the *cathode*. If the cathode were heated to the required temperature in open air, it would burn up because of the presence of oxygen in the air. Thomas Edison learned this in his experiments with the incandescent lamp. For this reason the cathode is placed in a glass or metal envelope and the bulb is evacuated. Since the cathode is sealed in a vacuum, the simplest way to heat it is electrically. On this basis, cathodes may be divided into two types, depending on whether they are heated directly or indirectly:

1. *Filament type or directly heated cathode.* In this type the cathode consists of a metallic wire or ribbon. The heating current passes directly through this wire. The heating is a result of the I^2R loss due to the resistance of the filament wire. Typical filament structures are shown in Fig. 4-1a.

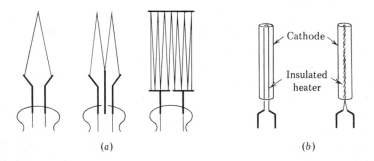

(a)   (b)

Figure 4-1 Types of cathode structures.

---

*The work function of a material is a measure of the work done to remove an electron against the opposing force due to the charge of the electron. Since work per unit charge is a measure of electric potential, work function (Φ) is given in *electron volts* (eV) or simply *volts*.

2. *Indirectly heated or separate heater-cathode.*   The indirectly heated cathode consists of a thin metal sleeve coated with the electron-emitting material. A filament or heater is enclosed within the sleeve and insulated from the sleeve. The cathode is heated by radiation and conduction from the heater. Useful emission does not take place from the heater itself. This type of construction is shown in Fig. 4-1*b*.

Separate heater construction has its advantages. Since the cathode is completely separated from the heating circuit, it can readily be connected to any desired potential as needed, independent of the filament or heater potential. Furthermore, in electronic equipment it is preferable to use ac in the heater circuit to simplify the power-supply requirements. With a separate heater circuit this presents no problem, but with filament tubes this may produce hum. On the other hand, filament heaters require less filament power than a similar material indirectly heated and can be more easily designed to handle very high current requirements. For these reasons filament tubes are often used in high-power applications (for example, as transmitting tubes).

Cathodes may also be classified by the type of material from which they are made. The high temperatures required to produce satisfactory thermionic emission limit the number of substances suitable as emitters to a very few. Of these *tungsten, thoriated tungsten,* and *oxide-coated* emitters are the only ones commonly used in vacuum tubes. Each material has its advantages.

Tungsten (work function 4.52 eV) is widely used for the filaments of large power-handling tubes where very high plate voltages are encountered. They are the most rugged of the three types, can withstand overload with minimum damage, and will give the longest life. On the other hand, they must operate at very high temperatures (2200°C, dazzling white) in order to emit sufficient electrons. A large amount of filament power is required. Since the filament efficiency is low, tungsten filaments are used only when other emitters cannot be employed. Tungsten (and thoriated tungsten) emitters are made only in the directly heated or filament type of cathode. Due to the lower heat transfer efficiency, too much filament power would be needed for indirect heating.

At the other extreme is the oxide-coated emitter. This type of cathode operates at comparatively low temperatures (750°C, dull red) and is the most efficient with regard to emission current per watt of heating power. It is therefore used wherever possible. Receiving tubes and transmitter tubes of lower power rating employ oxide-coated emitters, even up to plate potentials of several thousand volts. Oxide-coated emitters may be of the heater or the filament type.

SPACE CHARGE.   We have seen that as the temperature of the cathode is increased to the emission point of the material, electrons are thrown off into the space surrounding the cathode. Where do they go? They remain in a space surrounding the cathode in a sort of cloud. This effect is known as *space charge*. Since it is formed of electrons, the space charge is negative. Meanwhile the cathode, having lost electrons, acquires a positive potential. The combined action of the repelling force due to the negative space charge and the attractive force of the positive cathode tends to counteract the kinetic energy, which throws the electrons off the cathode. Some of the electrons from the space charge return to the cathode. But at the same time, more electrons are being emitted by the cathode. In other words, the electrons in the space charge are not static, but in continuous motion. An equilibrium condition is soon reached where the number of electrons emitted is equal to the number of electrons attracted back to the cathode.

Let us assume that the operating temperature of the cathode is increased, by increasing the filament or heater current. The kinetic energy of the electrons in the cathode will also be increased, destroying the equilibrium condition. The number of electrons forming the space charge will increase. The higher the temperature of the cathode, the more electrons it will emit; the greater the rate at which they are emitted, the greater the number of electrons in the space charge; but again a new equilibrium condition is reached where the number of electrons emitted is balanced by the number of electrons returning to the cathode. Obviously the number of electrons forming the space charge depends on the cathode temperature. However, if too high a voltage is applied to the filament or heater, the increase in current (and temperature) may cause the filament or heater to burn out. Tubes should be operated at a safe filament or heater voltage as determined by the manufacturer. They are sometimes operated at slightly below rating (if the maximum current rating is not needed) to increase the life of the tube. Too low a voltage will result in insufficient emission and prevent proper functioning of the tube.

DIODE CHARACTERISTICS

The simplest type of vacuum tube contains two elements—the cathode and an *anode* or *plate*. (The heater of the indirectly heated cathode is not considered as an element.) Such a tube is called a *diode*. Figure 4-2 shows a diode connected to a source of heater and plate supply. This circuit can be used to study the effects of cathode temperature and plate potential on the diode characteristics. The plate of the tube is made positive with respect to the cathode.

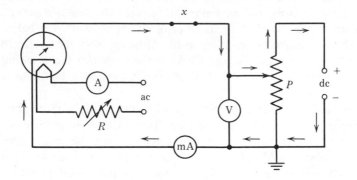

Figure 4-2 Circuit for determining diode characteristics.

CURRENT FLOW IN A DIODE.   If the cathode is at an operating temperature, electrons are being emitted to form a space charge. Remember that this space charge is in constant flux but that an equilibrium condition exists between the number of electrons being emitted and being returned to the cathode. Let us consider what will take place in this complete circuit of Fig. 4-2. We start with the plate potentiometer P set down to the very bottom. The plate potential is obviously zero. Until now we have considered the space charge to exist as a cloud around the vicinity of the cathode. This is essentially true. But some electrons are emitted with sufficient initial velocity to extend up to the plate. Now that we have a complete circuit, some of these electrons will return to the cathode via the external path (see arrows in Fig. 4-2). Current is flowing through the cathode-plate circuit even though no voltage is applied to the circuit! This phenomenon due to the initial velocity of the electrons is known as the *Edison effect* in honor of Thomas Edison, who first noticed this action. However, since this current is extremely low (microamperes) compared to normal action (milliamperes) it can be neglected.

Now let us raise the setting of the plate potentiometer, applying a positive potential to the plate. Since the plate is positive, an appreciable number of electrons will be pulled from the space charge to the plate. These electrons will flow through the circuit back to the cathode. This *plate current* will be measured by the milliammeter. (The meter could have been placed on the plate side of the circuit, but then it would have been at a high potential with respect to ground. This would be dangerous to the operator. Since the same plate current also flows in the cathode, it is common practice to locate the meter as shown.) Will this plate current destroy the equilibrium condition between space charge and cathode? Does the cathode emission change to supply these extra electrons to the plate? The answer to both these questions is No! Cathode emission depends only on cathode temperature—and we have not changed the heater current. As for equilibrium, we know that the number of electrons emitted is balanced by the number of electrons returning to the cathode. Now fewer electrons are returned from the space charge back to the cathode; the remainder are returned through the completed plate circuit as shown by the arrows in Fig. 4-2.

If we were to reverse the plate supply voltage, making the plate negative compared to cathode, current flow from cathode to plate would stop —the negative plate would repel the electrons. Even the small current due to Edison effect would cease as soon as the plate was made a few volts negative. Would current flow in the opposite direction? Definitely not! The plate is not hot enough to emit electrons, nor is it made of suitable emitter material. *Current can flow only in one direction, from cathode to plate, and only when the plate is positive compared to cathode.* This is true of all vacuum tubes. Does this mean that plate current will not flow in *any* circuit if the plate potential is for example -50 V? Think a minute before you answer. Notice also the italic phrase three sentences back: the plate must be positive *compared to cathode.* In other words, the plate potential can be negative if the cathode potential is even more negative. Remember that the operating voltage for any element of a tube is the voltage between that element and the cathode. So a plate potential of -50 V is permissible if the cathode potential is for example -150 V, since this makes the plate 100 V positive compared to cathode. Remember this point well, when in your later studies you analyze circuits. Many television receivers, radar units, and industrial electronic circuits use such seemingly odd potentials— yet the operating voltages are correct.

VOLTAGE SATURATION.   We have seen that a positive plate potential will cause a plate current to flow. Let us check the relation between plate voltage and plate current, as we raise the plate potential in steps. A more positive plate will attract electrons from the space charge at a greater rate. The plate current will increase. But this effect cannot continue indefinitely. For a given cathode temperature, the cathode emits electrons at a fixed rate. As the plate voltage is increased, the plate current keeps increasing as long as electrons are being supplied by the space charge. This continues until the electron supply forming the space charge is exhausted and electrons are being taken from the cathode as fast as they are given off. Beyond this point, further increase in plate potential cannot cause a corresponding increase in plate current. This condition is known as *voltage saturation.* This maximum current is called the *saturation current,* and because it is an indication of the total number of electrons emitted it is also known as the *emission current* or, simply, *emission.*

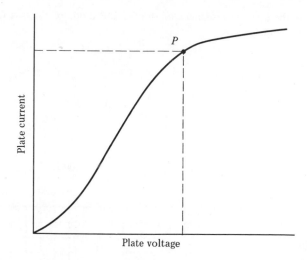

Figure 4-3   Diode static characteristic curve.

Figure 4-3 shows the variation of plate current with plate potential. Such a curve is known as a *diode static characteristic*. Notice the flattening of this curve at the higher plate potentials. Saturation is not sharply defined. It is considered to start at the knee of the curve, around point P.

Tubes are designed to operate well below the voltage saturation point. The normal cathode operating temperature will supply sufficient emission so that saturation will not occur under normal operating conditions. The plate current at saturation level would be excessive and would ruin the tube. (If it is desired to demonstrate saturation, the tube should be operated at considerably reduced heater voltage and current. In this way, voltage saturation can be made to occur within safe limitations of plate current.)

PLATE RESISTANCE.   Just as the semiconductor diode has internal resistance, so does the tube. Several factors contribute to the internal resistance of a tube. Physical structure—shape, size, and spacing—of the elements is important. When low plate resistance is desired, larger elements and closer spacing between elements are used. However, for a given physical structure the negative space charge is largely responsible for the internal resistance of the diode.

Calculations for plate resistance follow the procedures described for semiconductor diodes. The dc plate resistance R_p is found from

$$R_p = \frac{\text{total plate voltage}}{\text{total plate current}} = \frac{E_b}{I_b} \qquad \textbf{(4-1)}$$

EXAMPLE 4-1

Calculate the dc plate resistance of the 5U4GB diode of Fig. 4-4 for the operating points of P and Q.

Solution

1. For operation at point P,

$$R_p = \frac{E_b}{I_b} = \frac{10 \text{ V}}{18 \text{ mA}} \times 10^3 = 555 \ \Omega$$

2. At point Q,

$$R_p = \frac{E_b}{I_b} = \frac{40 \text{ V}}{200 \text{ mA}} \times 10^3 = 200 \ \Omega$$

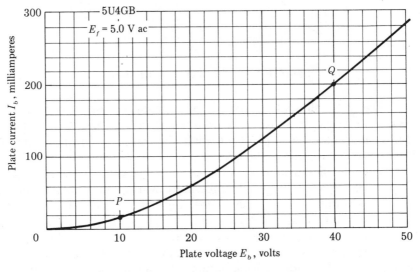

Figure 4-4 Average plate characteristic—5U4GB.

The ac plate resistance r_p is obtained from the ratio of a small change in plate potential to the resulting change in plate current:

$$r_p = \frac{\Delta E_b}{\Delta I_b} \tag{4-2}$$

Remember that the small change should be taken one-half to each side of the operating point.

EXAMPLE 4-2

Calculate the ac plate resistance of the 5U4GB of Fig. 4-4 at a plate potential of 20 V.

Solution

1. Using a swing of ± 2 V:

$$\text{At } E_b = 22 \text{ V}, \ I_b = 75 \text{ mA}$$
$$\text{At } E_b = 18 \text{ V}, \ I_b = 50 \text{ mA}$$

2. $r_p = \dfrac{\Delta E_b}{\Delta I_b} = \dfrac{4 \text{ V}}{25 \text{ mA}} \times 10^3 = \mathbf{160 \ \Omega}$

TRIODES

As the name implies, a triode has three active elements. Two of these we have already discussed—the plate and the cathode. The third element is called the *grid*. Figure 4-5 shows the schematic diagram of a triode and also a structural diagram. The grid structure consists of a mesh or helix of wire extending the full length of the cathode. The spacing between the turns of the mesh or helix is wide and the winding is made of fine wire, so that the passage of electrons from cathode to plate is practically un-obstructed by the physical size of the grid structure. The grid is located between the plate and cathode, but much closer to the cathode than to the plate.

The purpose of the grid is to control the amount of plate current flowing through the tube, independently of the plate potential. From this action it is often referred to as the *control grid*. This nomenclature is neces-sary to distinguish this grid from additional grid structures found in multi-grid tubes. When the grid is made positive, it counteracts some of the limiting action of the negative space charge and the plate current will in-crease. On the other hand, if the grid is made negative the electric field around the grid structure will assist the space charge in holding back elec-trons, thereby reducing the plate current.

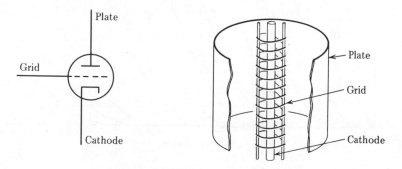

Figure 4-5   Triode—symbol and construction.

In normal operation of vacuum tubes, the cathode temperature is kept constant at the recommended value by applying rated voltage to the heater (or filament). The plate current flowing in a triode therefore depends on the respective grid and plate voltages (compared to cathode). The grid is usually operated at a lower potential than the cathode, making the grid negative compared to cathode. This grid-to-cathode voltage is referred to as the *grid bias*. The plate current will increase or decrease as the grid bias is made less negative or more negative.

TRIODE CHARACTERISTIC CURVES.   In order to study the operation of a triode and to pick suitable operating conditions, it is necessary to know the value of plate current that would flow for any value of plate and grid potentials. Such information is best presented by a group of plate characteristic curves known as a family of plate characteristics for the tube. Each curve shows how plate current varies with plate potential *for a given fixed value of grid bias*. These curves can be obtained by experiment or from a tube manufacturer's data sheets. Plate characteristic curves for a typical triode are shown in Fig. 4-6.

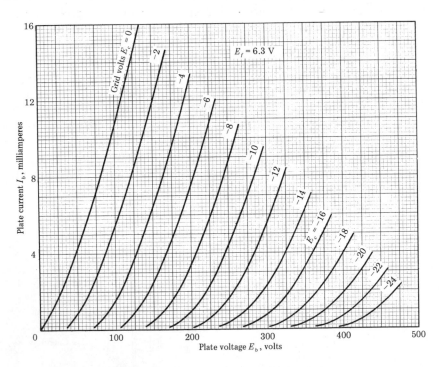

Figure 4-6   Plate characteristics of a triode.

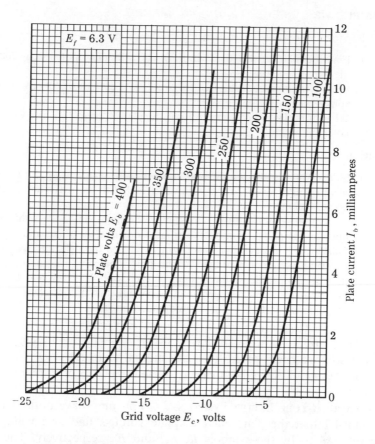

Figure 4-7 Mutual characteristics of the triode in Fig. 4-6. [*Note:* ⁵⁄₁₆ in. = 0.5 V (grid) or 0.2 mA (plate).]

   The same information can be presented in a sort of reverse procedure. That is, we can hold plate voltage constant while we vary the grid voltage. The resulting curve is called a *mutual characteristic*. The term *mutual* is used because it shows the effect of the grid (which is part of the input circuit) on the plate circuit (which is part of the output circuit). A family of mutual characteristic curves (for the typical triode of Fig. 4-6) is shown in Fig. 4-7. In general, the plate characteristics present the data in more usable form. Occasionally the mutual characteristics are preferable. However, it is a simple matter to construct the mutual characteristic curves from the plate family or vice versa. In either case, the curves are static characteristic curves. Now let us try some simple illustrations to show how these curves can be used.

EXAMPLE 4-3

For the triode whose characteristics are shown in Fig. 4-6:

1. Find the plate current that will flow when the plate potential is 200 V and the grid potential is -6 V.
2. What plate voltage is required to give a plate current of 4.0 mA with a grid voltage of -8 V?
3. What grid bias is necessary to limit the plate current to 8 mA with a plate potential of 250 V?

Solution

1. Starting on the abscissa (x axis) at 200 V, move straight up until you intersect the curve $E_c = -6$. Now move to the left to the y axis and read (interpolate) 7.5 mA.
2. Starting at 4.0 mA on the y axis, move horizontally to the right until you intersect the $E_c = -8$ curve. Then move down to the x axis and read 207 V.
3. Starting at 8 mA (y axis), move horizontally to the right until you intersect the 250-V plate potential line. At this point imagine a new E_c line (or draw it in lightly) parallel to the E_c lines on each side ($E_c = -8$ and $E_c = -10$). Notice that this line is approximately one-fifth the distance between the -8 and the -10 grid lines. The desired grid voltage can be estimated at -8.4 V.

These problems were solved by using the plate family curves. They could also have been solved from the mutual characteristics. In fact, the latter curves would have been easier to use for part 3 of Example 4-3.

VACUUM-TUBE LETTER SYMBOLS.   You may have noticed in Figs. 4-6 and 4-7 that the plate current, plate voltage, and grid voltage were also referred to by the symbols I_b, E_b, and E_c, respectively. The use of symbols I and E for current and voltage should already be familiar to you. The subscripts may need further explanation.

Originally, vacuum tubes were operated with batteries for the power source. Three sets of batteries were employed:

1. The A battery to supply the filament heating power. This was generally a lead-acid storage battery similar to an automobile battery.
2. The B battery, consisting of a bank of dry cells to supply the plate power.
3. The C battery, a much smaller bank of dry cells used for the grid-bias supply.

From these battery nomenclatures, the element currents and potentials got their subscripts—b for the plate circuit dc values and c for the grid circuit dc values. Quite often there is a resistance in series between the supply source and the element itself, causing a voltage drop. To distinguish the supply voltage from the element potential, a double subscript is assigned to the source symbol. For example, E_{cc} represents the grid-bias supply voltage and E_{bb} the plate supply voltage.

When a signal (ac) voltage is applied, new subscripts are used to distinguish the signal, or varying-component values, from the dc values. This time the first letter of the element involved is used as the subscript. Therefore, E_g represents the ac or signal voltage applied to the grid; I_p represents the ac component of plate current. Finally, in keeping with basic ac circuit principles, capital E and I are used when referring to fixed values such as dc, maximum, or rms values. Lowercase e and i are used to represent instantaneous values of the time-varying voltage and current. A graphic representation of these values is shown in Appendix C (Fig. C-2).

AMPLIFICATION FACTOR (μ, mu).   The addition of a third element, the grid, to a vacuum tube makes the tube usable as an amplifier. Since the grid is much closer to the cathode than the plate, a small change in the voltage applied to the grid has the same effect on the plate current as some large change in plate voltage. The *amplification factor* (μ) of a tube is a measure of the effectiveness of the grid voltage (relative to the plate voltage) in controlling the plate current. The value of the amplification factor of a tube depends on the relative size and location of its elements, but primarily on the grid structure itself. Larger grid wires, closer spacing of the grid wires, or greater distance between grid and plate all contribute to making the electric field of the grid more effective in controlling the plate current and therefore result in a higher amplification factor. Triodes are classified as low-, medium-, or high-mu tubes depending on their amplification factor, which varies from as low as 3 (low mu) to as high as 100 (high mu). The simplest method for computing the amplification factor of a triode is to find what *small* change in grid voltage is necessary to counteract a *small* change in plate voltage *so that the plate current remains constant*. Expressing this relation by formula,

$$\mu = \frac{\Delta E_b}{\Delta E_c} \quad \text{(for constant } I_b\text{)} \tag{4-3}$$

Are you wondering why the emphasis is on a small change? It is a good question, and covers an important basic principle applying also to transistors. So let us try to clarify the point.

From Figs. 4-6 and 4-7 we see that the characteristic curves of a typical triode are not straight lines. As a result, the amplification factor varies depending on where along the curves we evaluate this quantity. This means that we must first select the operating point. (As you will learn later, amplifiers are operated in the straight-line section of their characteristics for best results.) This operating point is fixed by selecting the dc grid bias and the dc plate potential applied to the tube. In commercial applications, a signal voltage (ac) is also applied to the grid circuit of the tube. The change in grid potential (above and below the dc bias) will cause a change in plate current. The plate current is therefore a complex wave consisting of a dc component due to the original operating point conditions and an ac component due to the ac grid signal. The amount of change in plate current— the magnitude of the ac component and consequently the degree of amplification—depends on the slope of the tube characteristic. A flat characteristic gives a small plate current change. On the other hand, a steeper characteristic results in a much larger ac component. But for a given tube the slope of the section of the characteristic curve used depends on the operating point chosen, and for small changes to either side of this operating point this section of the curve can be considered straight. So when evaluating the amplification factor, if we take small increments in grid bias—*above and below* the operating point—the effect of curvature of the characteristic curves is minimized and the resulting answer will be accurate. Let us apply this to a problem.

EXAMPLE 4-4

Find the amplification factor of the triode in Fig. 4-6 for an operating point of $E_c = -6$ V and $E_b = 200$ V.

Solution

We will use the plate characteristic curves (Fig. 4-6) and take a change in grid voltage from -4 to -8 (2-V swing on either side of the operating point). Draw a horizontal line passing through the point $E_c = -6$ V and $E_b = 200$ V. The plate current for this operating point is 7.5 mA. To hold the plate current constant at this value when E_c is -4 V requires a plate potential of 160 V. Similarly for a grid voltage of -8, the plate potential must be 240 V. Therefore $\Delta E_c = 4$ V and $\Delta E_b = 80$ V, and

$$\mu = \frac{\Delta E_b}{\Delta E_c} \ (I_b \text{ constant}) = \frac{240 \text{ to } 160}{-8 \text{ to } -4} = \frac{80}{4} = 20$$

For best accuracy the change in voltage should be as small as possible. However, owing to the limitations of our characteristic curve plots, the ± 2-V grid swing to either side of the operating point is most convenient.

PLATE RESISTANCE AND CONDUCTANCE. As in the case of the diode, a triode tube also has an "internal resistance" that limits the amount of current for a given plate voltage. But now the grid potential also affects the plate current. In the diode the ac plate resistance is given by

$$r_p = \frac{\Delta E_b}{\Delta I_b} \tag{4-4}$$

To eliminate the effect of the control grid all we need do is *hold the grid potential constant,* and the same equation applies. It should be obvious that the ac plate resistance will vary—again depending on the operating point. Plate resistances for triodes range from as low as 300 Ω for low-mu tubes to approximately 100,000 Ω for high-mu tubes.

EXAMPLE 4-5

Find the ac plate resistance for the tube of Example 4-4 (operating point $E_c = -6$ V and $E_b = 200$ V).

Solution

Let us take a plate current change of 1 mA above and below the operating value of 7.5 mA. Since the grid potential must be held constant, our data is readily taken from the plate characteristic curves along a constant E_c line. In this case, it will be along the $E_c = -6$ line.

1. For $E_c = -6$ V and $I_b = 6.5$ mA, $E_b = 192$ V
2. For $E_c = -6$ V and $I_b = 8.5$ mA, $E_b = 208$ V

3. $$r_p = \frac{\Delta E_b}{\Delta I_b} = \frac{192 \text{ to } 208 \text{ V}}{6.5 \text{ to } 8.5 \text{ mA}} = \frac{16}{2 \times 10^{-3}} = 8000 \; \Omega$$

Another value which is sometimes desirable is the converse of the ac plate resistance or the *plate conductance* (g_p). Since conductance is the reciprocal of resistance, it should be obvious that

$$g_p = \frac{\Delta I_b}{\Delta E_b} \qquad \text{(for constant } E_c\text{)}$$

In Example 4-5, the plate conductance would be

$$g_p = \frac{\Delta I_b}{\Delta E_b} = \frac{2 \times 10^{-3}}{16} = 125 \; \mu\text{mho}$$

TRANSCONDUCTANCE OR MUTUAL CONDUCTANCE.   We know what conductance means—the ability of a circuit to conduct current. We applied this to the *plate* circuit and saw that *plate* conductance is the ratio of change in *plate* current to change in *plate* voltage *for a constant grid potential.* Notice that both current and voltage changes are in the plate circuit. But we know that the grid voltage will also affect the plate current. There is therefore a *mutual* effect between plate and grid electrodes. *Transconductance, mutual conductance,* or *control grid-plate transconductance* (g_m) are all terms used to show the relation between plate current and grid voltage *for a constant plate voltage.* By formula:

$$g_m = \frac{\Delta I_b}{\Delta E_c} \qquad \text{(for } E_b \text{ constant)} \tag{4-5a}$$

If we remember the meaning of conductance and that of the term *mutual*, this formula should be obvious.

EXAMPLE 4-6

Calculate the mutual conductance for the tube in Example 4-4 (operating point $E_c = -6$ V and $E_b = 200$ V).

Solution

Let us use a grid swing of 2 V above and below the operating point. Since E_b must be kept constant, we can take our data along the vertical line corresponding (in this case) to $E_b = 200$.

1. For $E_b = 200$ V and $E_c = -4$ V, I_b 13.2 mA
2. For $E_b = 200$ V and $E_c = -8$ V, $I_b = 3.4$ mA

3. $$g_m = \frac{\Delta I_b}{\Delta E_c} = \frac{3.4 \text{ to } 13.2 \text{ mA}}{-8 \text{ to } -4 \text{ V}} = \frac{9.8 \times 10^{-3}}{4}$$

$$= 2.45 \times 10^{-3} \text{ mho} = 2450 \ \mu\text{mho}$$

The amplification factor of a tube is given by the mu of the tube. In practical circuits, you will learn that it is impossible to achieve the full amplification of the tube, due to the loss in its own internal resistance r_p. To realize maximum gain in a circuit, a tube should have a high amplification factor but also a low plate resistance. For best results, the ratio of amplification factor to plate resistance should be as high as possible. But this ratio is the mutual conductance of the tube! This can be readily shown as follows:

$$\mu = \frac{\Delta E_b}{\Delta E_c}$$

$$r_p = \frac{\Delta E_b}{\Delta I_b}$$

Dividing the upper equation by the lower, we get

$$\frac{\Delta E_b}{\Delta E_c} \times \frac{\Delta I_b}{\Delta E_b} = \frac{\Delta I_b}{\Delta E_c} = g_m$$

Therefore

$$g_m = \frac{\mu}{r_p} \qquad \text{(4-5b)}$$

Mutual conductance is one of the most important tube coefficients. It is widely used in the design of electronic equipment, both in the selection of suitable tubes and in the calculation of the ac component of the plate current.

It is obvious from the preceding relation for the three tube coefficients that if we know any two values we can find the third. For example, from the results of Examples 4-4 and 4-5 ($\mu = 20$ and $r_p = 8000$) we could solve for mutual conductance:

$$g_m = \frac{\mu}{r_p} = \frac{20}{8000} \times 10^6 = 2500 \ \mu\text{mho}$$

This checks fairly well with the value (2450) previously found from the curves. The reason for the slight discrepancy can be readily explained. In Examples 4-4 to 4-6 we used graphs and had to interpolate for values. A slight misjudgment in any interpolation would account for this small difference in the two values of the g_m coefficient.

INTERELECTRODE CAPACITANCE.   You know that capacitance exists between any two conducting materials separated by a dielectric.* From this it is obvious that capacitance effects must exist between grid and cathode, grid and plate, and plate and cathode. These capacitances are called *interelectrode capacitances*. They are represented by symbols as follows:

Grid-plate capacitance: C_{gp}
Grid-cathode capacitance: C_{gk}
Plate-cathode capacitance: C_{pk}

Interelectrode capacities are quite low, ranging from approximately 2 to 12 pF. At low frequencies their effects are negligible. However, these effects cannot be neglected at high frequencies.

At radio frequencies the grid-to-plate capacitance C_{gp} introduces serious complications. The reactance of even this small capacitance is low, and RF currents from the output circuit of the tube are fed back to the input circuit. This may cause the tube to act as an oscillator producing its own sine waves in addition to amplifying the input RF waves. These unwanted oscillations will prevent proper functioning of the circuit. Numerous "neutralizing" methods have been devised to compensate for the effect of this interelectrode capacitance. The real cure, however, is to reduce the grid-to-plate capacitance to negligible values. Research toward this solution led to the insertion of additional elements in the basic tube structure.

OTHER TUBES

TETRODES.   With the addition of a fourth element, the resulting tube is called a *tetrode*. The purpose of this new element is to shield the

---

*J. J. DeFrance, *Electrical Fundamentals*, Englewood Cliffs, N.J.: Prentice-Hall, Inc., 1969, part 1, chap. 20.

grid from the plate. Since electrons must flow through this new element to the plate, the element is made in the form of a grid. From its construction and purpose, it gets the name *screen grid*. A cross-sectional diagram of the tube structure and the schematic diagram for a screen grid tube are shown in Fig. 4-8.

Normally the screen is operated at a positive dc potential. However, its ac potential is held at ground level (zero) by connecting a suitable bypass capacitor from this element to ground. This grid serves as an electrostatic shield, effectively preventing any interaction between the electric fields of the plate and grid. This shielding action eliminates nearly all the capacitive feedback from plate circuit to grid circuit. The grid-plate capacitance of tetrodes is reduced to approximately 0.005 pF. In a tetrode the danger of unwanted oscillations is negligible except at ultrahigh frequencies.

Unfortunately, however, the screen grid introduces a new problem. Electrons moving toward the plate are often accelerated to such an extent that they hit the plate with sufficient velocity to knock *other* electrons off the plate itself. This effect is known as secondary emission. The secondary-emission electrons are directed toward the screen. At low signal levels, this does not affect operation. The secondary-emission electrons are reattracted to the plate. However, as you will learn when studying amplifier circuits, the instantaneous value of the plate (collector or drain) potential varies above and below its dc value by the magnitude of the ac output voltage.

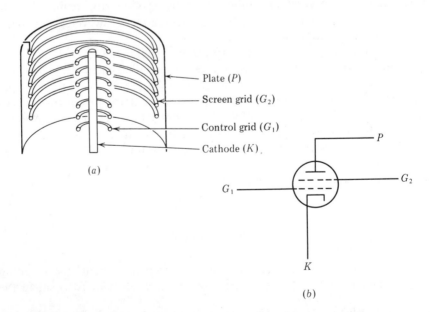

Figure 4-8   Cross-sectional and schematic diagrams of a screen-grid tube.

Therefore, for high output levels, the plate potential will swing below the screen value. Now the secondary-emission electrons are attracted to the screen, increasing the screen current. Meanwhile the plate current actually decreases. In fact, the secondary-emission effect may be so great that the plate current drops to zero. Obviously, the ac output is severely distorted or the power output from a tetrode is seriously limited. The range of operation could be greatly increased if the effects of secondary emission were eliminated.

PENTODES.   The addition of a third grid structure to a vacuum tube accomplishes the desired effect. This new grid, called a *suppressor* (G_3), is located between the screen grid and the plate. The tube now has five active elements and is therefore classified as a *pentode*. Figure 4-9 shows the arrangement of elements and the schematic diagram for a pentode. In normal operation the suppressor grid is connected to the cathode and operates at near-zero potential. The positive electric field of the screen grid is unable to reach through the barrier of the suppressor grid and affect the region between the suppressor grid and plate. The electrons from the cathode still produce secondary emission as they hit the plate at high velocity. But now these secondary electrons are no longer attracted to the screen grid. Temporarily they may gather around the suppressor grid to form a *virtual cathode*. Almost immediately these electrons are recaptured by the plate, since the plate potential is positive with respect to the suppressor grid. Consequently, the plate potential of a pentode can be allowed to swing far below the screen potential. Such operation results in greater maximum voltage and maximum power output ratings for a pentode as compared to an equivalent-sized tetrode.

A family of plate characteristics for a pentode is shown in Fig. 4-10. These curves apply only for a screen potential of +100 V. (If data at any other screen potential is desired, a new family of curves must be obtained.) Notice how flat these curves are. The screen and suppressor grids shield the

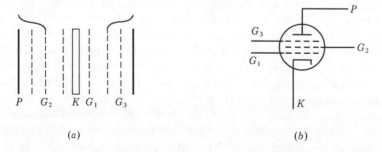

(a)   (b)

Figure 4-9   Electrode arrangement and schematic diagram of a pentode.

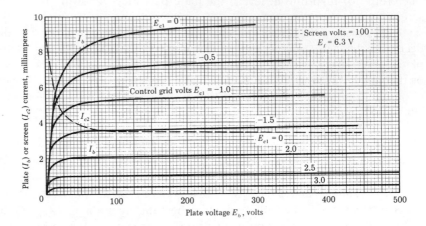

Figure 4-10   Pentode plate characteristics.

plate from the space-charge region. The electric field of the plate cannot pull electrons out of this area (as in a triode). Consequently, except at very low values, the plate current is almost entirely independent of the plate potential. As a result, pentodes have much higher amplification factors and plate resistances than triodes. Typical values for a pentode are:

1. Amplification factor: 1000 to 5000
2. Plate resistance: 0.5 to 2 MΩ
3. Transconductance: 1000 to 9000 μmho

POWER OUTPUT TUBES.   For large power outputs, tubes must be capable of passing high currents. Power tubes therefore have large emitting surfaces which may be of the direct or indirectly heated type. Figure 4-11 shows several high-power tubes. Note their large size and also that forced-air cooling and water-cooling techniques are used to increase their power-handling ability. In general, because they have only one "interfering" grid, triodes can pass higher currents and have higher power-output ratings than an equivalent-sized pentode. To maintain the shielding and sensitivity advantages of the pentode while still retaining the current-passing ability of the triode, a *beam-power tube* was developed. This tube should be classified as a tetrode since it has only four elements: cathode, control grid, screen grid, and plate. In addition this tube has a pair of *beam-forming plates* (internally connected to the cathode) that enclose the screen grid, giving the effect of a suppressor grid without impeding current flow. In performance the beam-power tube is superior to a power pentode.

Figure 4-11   Higher-power commercial tubes. **(a)** Water-cooled power tride, output 110 kW; $24\frac{1}{2} \times 9\frac{1}{2}$ in. **(b)** Forced-air-cooled triode, output 27 kW; $17\frac{3}{8} \times 14\frac{1}{2}$ in. **(c)** Transmitter diode, output 2700 W; $8^{17}/_{16} \times 4^{19}/_{32}$ in. **(d)** High-voltage mercury-vapor thyratron, peak rating 20,000 V, 6.4 A; $11\frac{1}{16} \times 3\frac{7}{8}$ in.

MULTIUNIT TUBES.   To save space, reduce initial cost, and also reduce operating cost, two or more tubes can be combined within a common envelope. Such combinations as two diodes (duo-diodes), twin triodes, triode-pentodes, and duo-diode-triodes are common. One manufacturer, under the trade name *Compactron,* encloses as many as four basic tubes in one shell.

UHF AND MICROWAVE TUBES.   Conventional tubes are not suited to operation above 100 MHz. This frequency limitation is due to *transit time* and interelectrode capacitance. The time required for an electron to travel from the cathode to the plate is known as the transit time. For low-frequency operation, this time interval is of little importance, but at the higher frequencies the transit time may be too long compared to the period, or time for one cycle of the signal frequency. At a television frequency such as 200 MHz, the time for one cycle is $\frac{1}{200}$ μs. A time delay of only 0.00125 μs would correspond to a 90° phase change in the signal voltage. So it is quite possible for an electron to leave the cathode with a given phase relationship and to find drastic changes in phase relations by the time it reaches the plate.

The other objection, interelectrode capacitance, should be obvious. At very high frequencies the shunting reactance of the input and output capacitances would be so low as to cause serious attenuation of the signal frequency. Both problems were solved by designing tubes specifically for operation at these higher frequencies. One result was miniature tubes such as the *Nuvistor,* which was usable to over 1000 MHz. With constructional changes, *disk-seal tubes* such as the *lighthouse tube* extended operation to about 3000 MHz. However, at microwave frequencies above 3000 MHz, transit time effects could not be eliminated. Instead new design techniques utilizing transit time were used, and they led to the development of *klystrons, magnetrons,* and *traveling wave tubes.* The study of such tubes is left to texts on ultrahigh frequencies.

REVIEW QUESTIONS

1. (a) Name the element in a vacuum tube that gives off electrons. **(b)** What is the name of the process by which electrons are given off?

2. Refer to Fig. 4-1: **(a)** Name the two types of cathode structures shown. **(b)** Describe each type briefly. **(c)** Which type is more efficient in heater power requirement? **(d)** Which is better suited for operation from an ac supply? Why?

3. (a) Name three materials used for cathode structures. **(b)** Which is most efficient in heater power requirement? Which is most rugged?

4. (a) What is a *space charge*? **(b)** How is it formed? **(c)** What happens to the space charge if the cathode temperature is increased?

5. Refer to Fig. 4-2: **(a)** What is the purpose of resistor R? **(b)** What is the purpose of potentiometer P? **(c)** Does the milliammeter (mA) read cathode or plate current? **(d)** Which is the better location for this meter—as shown, or at location x? Explain. **(e)** How is a characteristic curve obtained? **(f)** What do the arrows in this diagram represent? **(g)** Is this conventional current or electron flow? **(h)** Trace the direction of current (electron) flow if the dc supply voltage were reversed. Explain.

6. (a) In Fig. 4-3, what causes the curve to level off at the higher plate-voltage values? **(b)** What is this leveling off called? **(c)** In what region of this curve is a tube normally operated?

7. (a) What is the function of the grid in a triode? **(b)** In normal operation—and using the cathode as reference—what is the polarity of the plate? **(c)** What is the polarity of the grid? **(d)** What does the term *grid bias* signify? **(e)** What is the effect on plate current as the grid is made more negative?

8. (a) What does a *single* mutual characteristic curve show? **(b)** Which set of curves gives more information—the plate family or the mutual family? Explain.

9. Refer to the characteristic curves of Fig. 4-6:
(a) If $E_c = -2$ V and $E_b = 120$ V, find I_b.
(b) If $E_c = -5$ V and $E_b = 160$ V, find I_b.
(c) If $E_c = -7$ V and $E_b = 180$ V, find I_b.
(d) If $E_c = -12$ V and $I_b = 4$ mA, find E_b.
(e) If $E_b = 240$ V and $I_b = 6$ mA, find E_c.

10. Using the mutual curves of Fig. 4-7, find the missing quantity **(a)** for $E_b = 150$ V and $E_c = -5$ V. **(b)** for $E_b = 350$ V and $I_b = 3$ mA. **(c)** for $E_c = -10$ V and $E_b = 250$ V. **(d)** for $E_c = -7.5$ V and $E_b = 275$ V.

11. With reference to vacuum-tube letter symbols: **(a)** What do lowercase e and i signify? **(b)** Give three instances when the use of capital E or I is appropriate. **(c)** How do we distinguish dc grid values from ac grid values? **(d)** How do we distinguish dc plate values from ac plate values?

12. What do the following letter symbols represent: I_b; E_{bb}; E_b; E_g; i_p; e_g; I_{pm}; E_{pm}?

13. (a) Name and give the letter symbols for three important triode parameters. **(b)** Give the equation for evaluating these parameters from the static characteristic curves. **(c)** Why must small changes in E and I be used? **(d)** What are the units for each of the three parameters?

14. Give another name for mutual conductance.

15. (a) What do the following symbols represent: C_{gp}; C_{gk}; C_{pk}? **(b)** What causes these effects? **(c)** What is the typical range of values for these quantities? **(d)** Which of the three can cause oscillations at radio frequencies?

16. (a) What is the name of the "new" element in the tetrode? **(b)** Where is this element physically located with respect to the other elements? **(c)** What is the purpose of this added element? **(d)** At what dc potential is this element normally operated? **(e)** At what ac potential is it operated? **(f)** How are both these effects obtained?

17. (a) What is *secondary emission*? **(b)** Does it occur in a triode, a tetrode, or both? **(c)** What effect, if any, does secondary emission have on the power output available from a tetrode? **(d)** A tetrode is operated at screen voltage, $E_{c2} = +120$ V, and plate voltage, $E_b = +200$ V. What is the maximum value that the ac output voltage can have before distortion from secondary emission begins? Explain. **(e)** What would be the maximum value of the ac output voltage if there were no limitation due to secondary-emission effects?

18. (a) How many grids does a pentode have? **(b)** Name these grids. **(c)** What is the purpose of the third grid? **(d)** How does it accomplish this purpose? **(e)** For the same plate and screen voltage, which tube is capable of developing the higher output voltage, the tetrode or the pentode? Explain.

19. **(a)** Is a beam-power tube most similar in construction to a triode, a tetrode, or a pentode? Explain. **(b)** In operating characteristics, is it most similar to a triode, a tetrode, or a pentode? Explain. **(c)** In power output capability, does it compare to a tetrode or to a pentode?

20. **(a)** What is a multiunit tube? **(b)** What advantage is there in using such tubes in place of "regular" tubes? **(c)** Give two examples of multiunit tubes.

PROBLEMS AND DIAGRAMS

1. Draw the schematic diagram of a vacuum-tube diode and label the parts.

2. Calculate the dc plate resistance of the diode in Fig. 4-4 at a plate voltage of 20 V.

3. Figure 4-12 shows the characteristic curve of a diode used in high-voltage rectifier circuits of TV receivers. **(a)** Find the current flowing through the tube when the voltage drop across the tube is 25 V. **(b)** Find the dc plate resistance for this operating point. **(c)** Find the ac plate resistance at this same point.

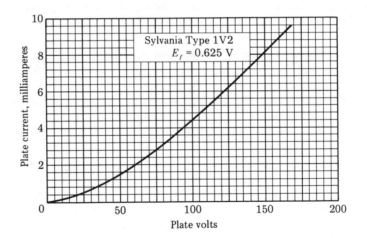

Figure 4-12

4. Calculate the ac plate resistance of the diode in Fig. 4-4 for operation at point Q.

5. From Fig. 4-12 and a plate voltage of 100 V, find **(a)** the plate current; **(b)** the dc plate resistance; **(c)** the ac plate resistance.

6. Draw the schematic diagram of a triode and label the parts.

7. For the triode of Fig. 4-6, corresponding to an operating point of 250 V plate and -10 V grid bias, find **(a)** the amplification factor; **(b)** the ac plate resistance; **(c)** the mutual conductance. **(d)** Repeat part **(c)** by calculation from the values obtained in (a) and (b), and compare your result with the answer to (c).

8. For the triode of Fig. 4-6, corresponding to an operating point of 150 V plate and −6 V grid bias, find **(a)** the amplification factor; **(b)** the ac plate resistance; **(c)** the mutual conductance. **(d)** Repeat part (c) by calculation from the values obtained in (a) and (b), and compare your result with the answer to (c).

9. Figure 4-13 shows two mutual characteristic curves for a high-mu triode. This tube is to be used as an amplifier with the operating point fixed by a plate potential of 250 V and a dc grid bias of −1.0 V. An ac signal of 0.141 V is applied in series with the bias voltage. Find: **(a)** the maximum value of the ac signal; **(b)** the maximum instantaneous grid potential; **(c)** the minimum instantaneous grid potential; **(d)** the maximum instantaneous plate current; **(e)** the minimum instantaneous plate current; **(f)** the change in plate current—or peak-to-peak value of the ac component of plate current.

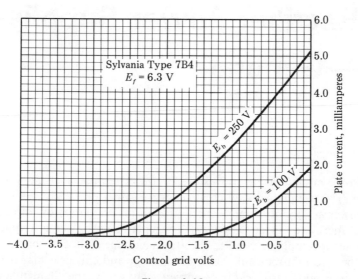

Figure 4-13

10. The tube of Fig. 4-13 is now operated at 250 V plate and −2.0 V grid. Again using an ac signal of 0.141 V, find (a) to (f) as in Problem 9.

11. The tube of Fig. 4-13 is now operated at $E_b = 250$ V and $E_c = -2.5$ V. Still using an ac signal of 0.141 V, find (a) to (f) as in Problem 9.

12. In Problems 9 to 11, the ac input signal is the same. Explain why the ac component of plate current is different. How does selection of an operating point affect gain?

13. Draw the schematic diagram for a tetrode and label all the elements.

14. Draw the schematic diagram for a pentode and label all the elements.

5

Rectifier Power Supplies
—General

Electronic circuits using transistors or tubes require a primary source of power. These circuits do not create energy; rather they convert energy from the power supply into signal energy (as in an *oscillator*) or into increased signal energy (as in an *amplifier*). In general, dc power is required. For example, in amplifier circuits the drain voltage and gate bias for FETs must be direct current. Similarly, the emitter and collector bias for bipolar transistors must also be direct current. However, the filament or heater power supplied to electron tubes can be obtained from either a dc or an ac source.

In the early days of radio, batteries were used for power supply. Three separate batteries were used. The battery that supplied the filament heating power was called the A battery. It was usually a lead-acid storage battery similar to the typical automobile battery. Plate voltage and screen voltage were supplied by B batteries. These consisted of stacks of many small dry cells (layer-built or minimax types) connected in series to form one commercial package. A common voltage for these batteries was 45 V. Two or more such batteries were then used (in series), depending on the voltage requirements of the particular circuit. For grid-bias supply, the same type of dry cell was used as for the B supply, but since bias voltages were generally much lower, fewer series cells were grouped into a package. A very common bias battery was the 4.5-V battery (with a tap at 3 V). Batteries used for bias voltages were called C batteries. This same nomenclature (A, B, and C supplies) is still used with tube circuits—and the term *B supply* is even used in some solid-state circuits.

However, battery supplies have several disadvantages. Unless a careful preventive maintenance program is followed, the batteries will invariably go dead when you need them most. How often has this happened to your portable radio or your flashlight? Peak performance of the equipment is obtained only when the batteries are fresh or fully charged; as they age or run down, the efficiency of operation decreases. Since the development of ac-operated power supplies, most electronic equipment has therefore been energized directly from the power lines. Portable equipment and devices such as hearing aids are, of course, still battery-operated.

REQUIREMENTS FOR POWER SUPPLIES.   Power requirements for electronic equipment vary widely with the size and complexity of the individual piece of equipment. Among the factors to be considered are these: (1) What voltages will be needed? (2) What is the maximum load demand? (3) How much ripple content can be tolerated? (4) Is constancy of the supply voltage important? Let us consider each of these requirements briefly. The details are discussed in succeeding chapters.

1. *Voltage requirements.*   Depending on the ratings of the active devices, the dc operating voltages needed may be higher than or lower than the nominal 120-V house supply. Low-power, small-signal transistors generally require a relatively low voltage (6 to 12 V). On the other hand, high-power transistors, as would be used in the deflection circuits of color TV receivers, are rated at several hundred volts. Finally, large power tubes for broadcast transmitters are operated at plate and screen voltages of many thousand volts. Speaking of tubes, it is common practice to use low-voltage ac for their heater supply. Two typical heater voltages are 6.3 V and 12.6 V. Consequently, a typical power supply may have to deliver alternating current at low voltages, direct current at low voltages, and direct current at high voltages.

2. *Load or current requirements.*   The term *load* as used in the electrical and electronics fields is synonymous with *current*. For example, a heavy load means high current and a light load signifies relatively low current. A power supply must be capable of meeting the full-load requirement of the equipment. If the drain (load) on the power supply is greater than the current rating for which it was designed, the components will overheat. If the overload is sustained, the power unit will be damaged.

3. *Ripple content.*   The dc voltage output from a power supply is not as smooth and steady as that which would be obtained from a

battery. It is basically a complex wave having a dc component and various ac components.* These ac components are referred to as the *ripple voltage*. This ripple content, expressed in percent, is a measure of the rms value of the ac components compared to the direct voltage. In some electronic applications a higher ripple content can be tolerated than in others. If ripple is objectionable, the ripple content of a power supply is an important factor.

4. *Regulation.*   All sources of power, whether they are generators, batteries, or rectifier power supplies, have internal resistance. Due to this internal resistance, there will be an *IR* drop within the power source and the terminal voltage will decrease as load is applied. The higher the load, the greater the internal *IR* drop and the lower the terminal voltage. The term *regulation* is used as an indication of the ability of a power source to maintain a constant terminal voltage. Regulation is considered good when there is little change in voltage as the load varies, whereas poor regulation implies that the voltage changes appreciably with load. Of course, if the load does not vary, the supply voltage does not vary and regulation is of no importance.

POWER SUPPLY BLOCK DIAGRAM.   Before we discuss actual circuits, let us examine a block diagram of a power supply (Fig. 5-1). From this we can get an overall view of the total circuit operation and the function of the individual components. The principles and circuits are discussed in succeeding chapters.

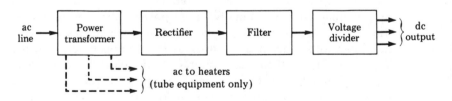

Figure 5-1   Block diagram of a typical power supply.

1. *Power transformer.*   The main purpose for using transformers in a power supply is to produce the proper operating voltages—high or low—required by the equipment. The voltage is still ac and must later be changed to dc. The transformer does not "transform" anything. It merely steps down (reduces) the applied voltage and/or steps up (increases) the voltage. (A third function of a transformer

*J. J. DeFrance, *Electrical Fundamentals*, Englewood Cliffs, N.J.: Prentice-Hall, Inc., 1969, part 2, chap. 4.

is *isolation*—to break any common connection between the primary and secondary circuits.)

2. *Rectifier.*   The first step in converting the ac output of the transformer to direct current is called *rectification,* and the device used for this function is called a *rectifier.* For this purpose, semiconductor diodes are most commonly used. The rectifier distorts the sine wave so that the areas under the positive and negative half-cycles are no longer equal, the average value of the wave is no longer zero, and a dc component has been produced.

3. *Filter.*   As indicated above, the process of rectification distorts the sine wave so as to produce a dc component. But the same process also produces harmonics of the original sine wave input and may include a weak component at the original (fundamental) frequency. These ac components were described earlier as the *ripple content.* The function of the filter is to reduce the ripple content to within tolerable limits.

4. *Voltage divider.*   When more than one value of direct voltage is needed, a series resistance chain is used across the total dc output. Appropriate taps in this voltage divider will supply the desired dc voltages. A typical voltage divider, supplying three positive voltages and one negative bias with respect to ground, is shown in Fig. 5-2. Four separate resistors, or one resistor with taps, could

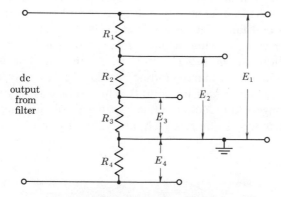

Figure 5-2   Typical voltage divider.

be used for this unit. Analysis of this type of circuit to determine resistance values and power ratings for the voltage-divider sections is an application of dc series-parallel circuit principles and is not discussed in this text.*

*J. J. DeFrance, *Electrical Fundamentals,* Englewood Cliffs, N.J.: Prentice-Hall, Inc., 1969, part 1, chap. 10.

Although all power supply units must contain a rectifier, it is possible to omit one or more of the other section of the block diagram (Fig. 5-1), depending on the specific circuit requirements. As an illustration, the power supply used with low-power units sometimes does not contain a transformer. A typical example of this is in ac/dc sets where use of a transformer is impossible. Transformerless supplies are discussed in Chapter 7.

POWER TRANSFORMERS

We are now ready for a detailed study of the power supply. We will start with the first block, the power transformer.

TRANSFORMER PRINCIPLE.   Basically a transformer consists of an iron core on which are wound a minimum of two windings. One winding, connected to the source of power, is called the *primary*. The other winding, the *secondary,* is connected to the load and delivers power to the load. A transformer may have more than one secondary winding, each secondary delivering power to its separate load. A basic transformer with only one secondary winding is shown in Fig. 5-3.

When the primary winding is connected to an ac source, current flows in the coil and a magnetic field is developed around the coil. The flux path is through the iron as shown in Fig. 5-3 by the dotted line. Since the primary is connected to an ac supply, the current and flux strength will rise to a maximum and fall to zero twice per cycle. In each cycle the current and flux will reverse—flowing in one direction for a half-cycle and in the opposite direction for the other half-cycle. As the flux lines expand and collapse, they cut the primary winding. By Lenz's law a voltage will be induced in this winding (self-induction). This voltage is in opposition to the applied voltage and reduces the "net" primary voltage, thereby keeping the primary current at some relatively low value. In an ideal transformer,

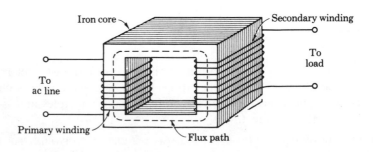

Figure 5-3   Basic transformer.

all the primary flux reaches and cuts the secondary winding. A voltage is induced in the secondary.

What will happen if the transformer primary is connected to a dc supply? Except for the initial surge, the primary current will be steady; no voltage will be induced in the secondary; and no counter EMF will be produced in the primary. The primary current will rise to an excessive value and the winding will be damaged.

VOLTAGE-TURNS RATIO.   For a given line frequency (speed of cutting), the *induced* voltage in each transformer winding depends on the number of turns. If the primary and secondary windings have the same number of turns, their induced voltages will be equal. The voltage ratio for any number of turns is given by

$$\frac{E_p}{E_s} = \frac{N_p}{N_s} \qquad (5\text{-}1)$$

In an *ideal* transformer, the resistance of the windings can be considered as zero. Also, there is no *leakage flux;* that is, all the flux produced by one winding reaches and cuts the other winding. Under these assumptions the primary induced voltage is equal to the applied voltage, and the secondary induced voltage is equal to the output voltage. Equation 5-1 can therefore be used to represent the ratio of input to output voltages. Even in an actual transformer, the error introduced by these assumptions is very small and Equation 5-1 is still valid as the ratio of input to output voltage.

When the primary and secondary have an equal number of turns (a 1:1 ratio), the secondary output voltage equals the applied line voltage. If a *step-down* transformer is desired so as to supply the low voltage for small-signal transistors, this can be achieved by using fewer turns for the secondary winding. Reversing the technique—using more turns for the secondary—produces a *step-up* transformer.

EXAMPLE 5-1

A combination plate and filament power transformer must deliver 400 V for plate supply and 5 V for heater supply from a 120-V line. The primary has 300 turns. Find the number of turns required for each secondary winding.

Solution

1. Plate winding (step-up):

$$N_s = \frac{E_s}{E_p} \times N_p = \frac{400}{120} \times 300 = \textbf{1000 turns}$$

2. Heater winding (step-down):

$$N_s = \frac{E_s}{E_p} \times N_p = \frac{5}{120} \times 300 = \textbf{12.5 turns}$$

CURRENT-TURNS RATIO.   To evaluate the current-turns relationship in a transformer, let us first consider the power relations. If a load is connected across the secondary winding, power is delivered to this load. In the general case this power will be $E_s I_s \cos \theta_s$. Assuming an ideal transformer (no losses or 100 percent efficiency*), the power taken from the line by the primary ($E_p I_p \cos \theta_p$) must equal the power delivered to the load. What is more, neglecting the very small magnetizing current, the primary power factor equals the secondary power factor and

$$E_p I_p = E_s I_s$$

Solving for current ratio,

$$\frac{I_p}{I_s} = \frac{E_s}{E_p}$$

and substituting turns ratio for voltage ratio,

$$\frac{I_p}{I_s} = \frac{N_s}{N_p} \tag{5-2}$$

Notice that *the winding with more turns has less current.*

EXAMPLE 5-2

A transformer delivers 650 V at 300 mA from a 120-V primary winding. Find the primary current.

Solution

1. Since this is a step-up transformer, the primary has fewer turns and will carry a higher current.

2. $I_p = \dfrac{E_s}{E_p} \times I_s = \dfrac{650}{120} \times 300 \times 10^{-3} = \mathbf{1.625\ A}$

TRANSFORMER CORE CONSTRUCTION.   Transformer cores are made of steel alloy, carefully selected and processed so as to reduce hysteresis losses. Silicon steel alloys are quite commonly used in power transformers. In addition, the core is laminated and each lamination is insulated from the next by an oxide and/or varnish coating. This in turn reduces eddy-current losses. To facilitate assembly of the transformer, the laminations are cut into sections and inserted through the coil openings of the prewound coils. The shape of these sections depends on the core construction. There are two fundamental designs used for the magnetic circuits—the *core* and the *shell* types:

---

*Transformers are very efficient devices. In well-designed units the efficiency at full load is in the high 90s.

1. *Core type.* This type is characterized by one magnetic path surrounding a "window." The iron of the basic transformer shown in Fig. 5-3 is of this type. Laminations for this type of core are generally cut into L-shaped sections. As the laminations are inserted through the coil opening, every other layer is reversed so that the air gap produced at the butt joints falls at opposite corners for successive layers (see Fig. 5-4). In cheaper designs every group of three layers is reversed.

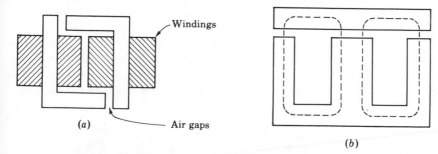

(b)

Figure 5-4   Cross sections of core-type and shell-type transformer construction.

2. *Shell type.* The shell-type construction is characterized by two magnetic paths surrounding two "windows." This construction, shown in Fig. 5-4b, uses E and I sections. Again, as for core-type units, successive layers (or groups of layers) are reversed so as to distribute the air-gap butt joints among the four corners. Notice that with shell-type cores only one coil form is used (containing both primary and secondary windings) and that it is mounted on the "tongue" or center leg of the core. Since they are more economical, shell-type cores are more commonly used in transformers for electronic equipment.

GRAIN-ORIENTED STEEL.   In their research to improve core materials, steel manufacturers have developed a steel with preferred grain orientation. The sheets are cut so that the magnetic flux flows in the direction of the structural grain of the material. One such silicon steel is known by the trade name of Hypersil. Grain-oriented steel has lower losses and higher permeability. To take full advantage of grain orientation, a new core construction—the type C core*—became necessary. This is shown in Fig. 5-5.

*R. Lee, *Electronic Transformers and Circuits,* New York: John Wiley & Sons, Inc., 1955, chap. 2.

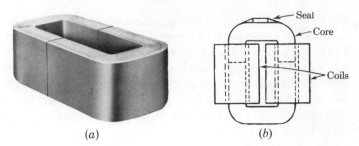

(a)   (b)

Figure 5-5   Type C grain-oriented core and its application to core-type transformers (*Westinghouse*).

TYPES OF CASES.   The physical appearance of transformers varies widely, depending on the type of covering (if any) used to protect the windings and on the mounting provisions. Figure 5-6 shows four external appearances. The transformer of Fig. 5-6a has no protection for the wind-

Figure 5-6   Typical transformer cases (*Merit Coil & Transformer Corp.; Freed Transformer Co.*).

ings except for the special (kraft) paper wrapping. In Fig. 5-6*b* and *c*, metal caps are placed over the coil form on either side. These caps not only provide physical protection but also act as magnetic shields to prevent the magnetic field of the transformer from reaching out into space. The transformer shown in Fig. 5-6*d* is a hermetically sealed unit designed to meet military specifications. This type of enclosure can be used with core-type or shell-type construction, with either stamped laminations or strip cores.

BASIC RECTIFIER CIRCUITS

The conversion from alternating to direct current begins in the rectifier section of the power supply. Here the sine-wave input voltage is deliberately distorted. The distorted output is a complex wave, and if the area under the positive half-cycle of output is greater than the area under the negative half-cycle (or vice versa), the average value of the wave is no longer zero. This average value is the dc component.

TYPES OF RECTIFIERS.   To produce a large dc component, the difference between the areas under each half-cycle should be as large as possible. This can be seen from Fig. 5-7. In Fig. 5-7*a*, a sine wave is shown, and since the two half-cycles are identical, the average value or dc component is zero. Figure 5-7*c* shows a complex wave with a very small negative area. Here the net area (the unshaded portion) is much greater and the average value is appreciably higher. Under what condition would we obtain the highest value of dc component? This would occur when one of the half-cycles (positive or negative area) is zero. We can obtain this effect by utilizing the current-voltage characteristics of the diode. The diode conducts when its anode is positive compared to cathode, and it offers almost infinite resistance to current flow when the anode is negative.

Since currents flows only in one direction in a diode, it is said to have the property of *unilateral* conductivity. (A semiconductor diode is classified as having unilateral conductivity because the inverse or leakage current flow, within normal operating conditions, is negligible compared to the conductivity in the forward direction.) Another term that is used to describe

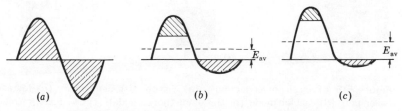

Figure 5-7   Effect of unbalanced half-cycle areas on the average value or dc component (unshaded portion represents the net positive area).

the effect of change in conductivity with amount or polarity of voltage is *nonlinear resistance* or, more generally, *nonlinear impedance*. A device having such a characteristic is called a *nonlinear circuit element*. A diode is definitely a nonlinear device since its resistance is very low with one polarity of applied voltage whereas it has infinite resistance (or nearly so) with reverse polarity.

To summarize this discussion, if a sine wave of voltage is applied to a nonlinear circuit element, the current waveform is distorted; this distortion results in a net average value for the cycle; this average value is the dc component. Unilateral conductivity is an extreme case of nonlinear impedance and produces the largest possible dc component; devices having this property of unilateral conductivity are diodes—of all types. Typical diode rectifiers are shown in Fig. 5-8.

Figure 5-8   Typical power rectifiers. **(a)** Silicon. **(b)** Germanium. **(c)** Metallic (selenium). **(d)** Vacuum diode. **(e)** Gas diode (hot cathode). **(f)** Gas diode (mercury pool cathode). (*Raytheon, General Electric, Radio Receptor Co., RCA, Sylvania*)

The balance of this chapter discusses basic rectifier circuits. Other circuits which require filter theory for analysis of their operation (such as voltage-doubler circuits) are covered in Chapter 7, after filter circuits have been considered.

HALF-WAVE RECTIFIER CIRCUIT.   The simplest rectifier circuit uses only one diode connected in series between the ac supply line and the load. Such a circuit is known as a *half-wave rectifier* and is shown in Fig. 5-9.*

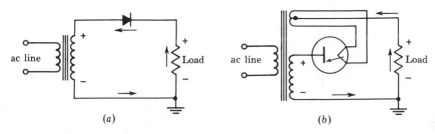

(a)   (b)

Figure 5-9   Half-wave rectifier circuits.

The solid-state circuit, shown in Fig. 5-9a, seems much simpler because the filament or heater wiring is unnecessary. The operation of the two circuits is otherwise identical, however, and can be explained as follows:

1. For one half-cycle of input, the upper terminal of the transformer secondary winding is positive. This instantaneous polarity is shown in Fig. 5-9.
2. Electrons will flow from the lower or negative terminal of the transformer, up through the load, through the rectifier, back to the positive terminal of the transformer secondary winding, and through the winding to complete the path. Since current flows up through the load, the voltage developed across the load resistor is positive on top and negative on bottom. By grounding the negative end (as shown in Fig. 5-9) the output of this rectifier circuit is positive compared to ground. With semiconductor rectifiers, the "bar" side of the rectifier symbol (right side in Fig. 5-9a) is usually identified with a plus sign, or red dot, to indicate that this end of the load will have positive polarity.

*Quite often in tube circuits, the rectifier filament winding is not center-tapped. (This reduces the transformer production cost.) In such cases the plate circuit is completed by returning the positive side of the load to either end of the filament winding. The imbalance, or ac component introduced by such a connection (usually approximately ± 2.5 V), is very low compared to the full secondary voltage and is readily removed by the filter circuits that follow the rectifier.

3. Since the applied voltage during the half-cycle varies from zero to maximum and back to zero, and since the load shown is pure resistance, the current will vary in like manner.

4. During the next half-cycle the polarity of the transformer secondary voltage is reversed. The upper terminal is negative. If this were a simple resistive circuit, current flow would normally be reversed. But current, neglecting leakage, cannot flow through a diode with reversed polarity. As a result of this unilateral conductivity of the rectifier, the current through the load is zero for this half-cycle.

5. The same current waveform is repeated cycle after cycle. This sequence is shown in Fig. 5-10.

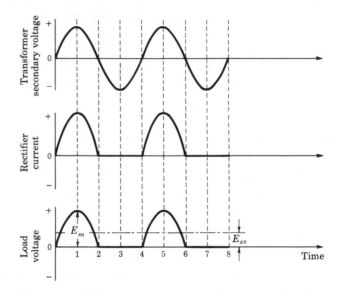

Figure 5-10 Waveshapes of current and voltage in a half-wave rectifier.

Since we are using a resistive load, the voltage across the load will have the same waveshape as the current waveform. In magnitude, the maximum value of this load voltage is *almost* equal to the maximum value of the transformer secondary voltage. Any difference in these two voltages is due to the voltage drop (IR) in the rectifier. However, the internal resistance of the rectifier when conducting is quite low and therefore the output voltage can be considered to be an exact replica of the positive half-cycles of the input voltage.

This output waveshape is a complex wave and consists of an average value (dc component) and many ac components of different frequencies.*

*J. J. DeFrance, *Electrical Fundamentals,* Englewood Cliffs, N.J.: Prentice-Hall, Inc., 1969, part 2, chap. 4.

Mathematical analysis or laboratory measurements with a wave analyzer show these components to be:

1. A dc component equal to the average value of the wave, $(1/\pi)E_m$, or $0.318E_m$
2. An ac component at the fundamental input frequency and an amplitude of $0.5E_m$
3. A second-harmonic component with an amplitude of $0.212E_m$
4. A fourth-harmonic component with an amplitude of $0.042E_m$
5. Additional *even* harmonics (although their amplitude is progressively lower and they are of negligible importance)

The total equation for this wave is given by

$$e = 0.318E_m + 0.5E_m \sin \omega t - 0.212E_m \cos 2\omega t$$

$$- 0.042E_m \cos 4\omega t \cdots \qquad \text{(5-3)}$$

Problem 12 at the end of this chapter demonstrates the validity of this mathematical analysis.

In the preceding discussion of rectifiers, the output voltage was positive with respect to ground. Sometimes a negative potential is needed; for example, to supply gate bias to an N-channel FET. There are two simple ways of reversing the output polarity. One technique is merely to ground the positive terminal of the power supply. Naturally, the remaining terminal is negative with respect to ground. This is shown in Fig. 5-11a. Compare these circuits with the previous corresponding circuits of Fig. 5-9.

The second technique for obtaining a negative output voltage is to reverse the rectifier connections. This is shown in Fig. 5-11b. Current will now flow only when the upper terminal of the transformer winding is negative or during the negative half-cycle of the input voltage. In this diagram, current flows *down* through the load, making the upper terminal negative. The output waveshape will consist of the negative half-cycles of the input voltage and the average value will therefore be negative.

The half-wave rectifier circuit has certain limitations or drawbacks. One is that the ratio of direct voltage to applied alternating voltage is low.

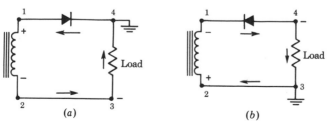

Figure 5-11   Rectifier circuits connected for negative output voltage. (For simplicity, the primary windings are not shown.)

(Remember that the dc component is equal to $0.318E_m$.) In other words, if 100 V is the desired dc output, the ac input must be:

1. $E_m = \dfrac{E_{dc}}{0.318} = \dfrac{100}{0.318} = 314 \text{ V}$

2. E (rms value) $= 314 \times 0.707 = 222 \text{ V}$

For a given rectifier (semiconductor or tube) the maximum direct voltage output obtainable is limited by the peak inverse voltage rating of the rectifier. When the unit is not conducting, the peak alternating voltage is applied across the unit in the inverse direction. As seen from the preceding calculation, for a dc output of 100 V the peak inverse voltage is 314 V. Therefore, with a half-wave circuit the rectifier unit must be capable of withstanding a peak inverse voltage of more than three times the required direct voltage. (It should be noted, however, that when filter circuits are added these relations will change. Filter circuits are discussed in the following chapter.)

A second drawback is that the use of half-wave circuits is limited to devices requiring relatively light loads. The reason for this can be seen from Fig. 5-10. Notice that the current waveshape is a complex wave (half sine wave) having an average value (dc component) of $0.318I_m$. The maximum steady current output from such a circuit is obviously less than one-third the peak current rating of the rectifier unit.

A third drawback of this circuit lies in its high ripple content. Analysis of the components of the complex output wave showed that *the first ac component is at the fundamental frequency and its magnitude* $(0.5E_m)$ *is actually greater than that of the dc component.* This places a greater burden on the filter circuit. Its design (for a given maximum percent ripple at a given load) becomes more complicated and more costly.

CENTER-TAP RECTIFIER CIRCUIT. Figure 5-12a shows a typical center-tap rectifier circuit using two solid-state diodes.* In contrast to the half-wave circuit, the center-tap circuit conducts on both halves of the input cycle. For this reason, it is also called a *full-wave rectifier.*

To analyze the operation of this circuit let us use the semiconductor diagram of Fig. 5-12a and the waveshapes of Fig. 5-13. During the first half-cycle of the input voltage (time interval 0 to 1) we will assume that the upper end of the transformer secondary (terminal 1) is positive compared to the center tap. Current (solid arrow in Fig. 5-12a) will flow from the center tap *up* through the load, through the diode rectifier D_1, and back to the positive terminal of the transformer. This makes the top of the load

---

*An equivalent tube diagram is shown in Fig. 5-12b. The tube is a twin diode (two diodes in one envelope).

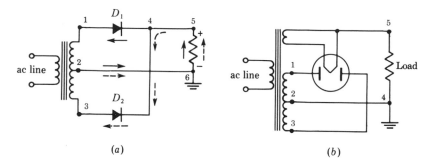

Figure 5-12   Single-phase center-tap rectifier circuits.

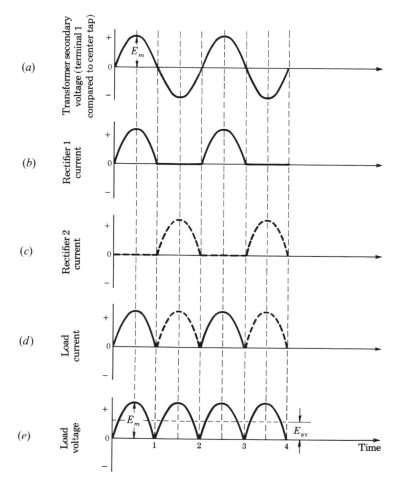

Figure 5-13   Wave shapes in a center-tap rectifier circuit.

resistor positive with respect to ground. During this time interval (0 to 1), will current also flow through the lower rectifier (D_2)? No. Terminal 3 of the transformer will be negative compared to the center tap. This biases the semiconductor in the inverse direction and, except for a negligible leakage current, the rectifier cannot conduct. (In a tube circuit, this polarity makes the plate negative with respect to cathode and the tube cannot conduct.)

Since the input voltage during this time interval (0 to 1) varies sinusoidally, the current flowing through rectifier D_1, the current flowing through the load, and the voltage across the load will also vary sinusoidally. These waveshapes are shown in Fig. 5-13. Neglecting the small voltage drop in the rectifier, the maximum value of the load voltage is equal to the peak value of the transformer secondary voltage *measured from center tap to either end.*

Now consider the circuit operation during the next half-cycle of input (time interval 1 to 2). The transformer secondary voltage has reversed. This makes terminal 1 negative with respect to center tap but terminal 3 is now positive. Rectifier D_2 conducts; rectifier D_1 cannot conduct. Current flow through the circuit is now shown by the dashed arrow. Notice again that current flows *up* through the load, maintaining the top of the load resistor positive. Waveshapes for currents and voltages during this time interval are also shown in Fig. 5-13. These waveshapes are repeated for each cycle of input voltage.

Notice from Fig. 5-13 that although each rectifier acts independently as a half-wave rectifier (conducts for only one half-cycle), current flows through the load for *both* half-cycles of input. That is why this circuit gives full-wave rectification.

Are there any advantages to full-wave rectification? To answer that question let us analyze the load voltage waveshape shown in Fig. 5-13. This again is a complex wave, but this time it consists of:*

1. A dc component (the average value) equal to $(2/\pi)E_m$, or $0.636E_m$
2. An ac component at twice the input frequency and a magnitude of $0.425E_m$
3. A fourth-harmonic component with an amplitude of $0.085E_m$
4. A sixth-harmonic component with an amplitude of $0.036E_m$

The equation for this wave is

$$e = 0.636E_m - 0.425E_m \cos 2\omega t + 0.085E_m \cos 4\omega t$$

$$- 0.036E_m \cos 6\omega t \cdots \qquad \text{(5-4)}$$

*J. J. DeFrance, *Electrical Fundamentals,* Englewood Cliffs, N.J.: Prentice-Hall, Inc., 1969, part 2, chap. 4.

When these values are compared with the half-wave output wave-shape, it is seen that:

1. The dc component in the full-wave circuit is double the half-wave value ($0.636E_m$ as compared to $0.318E_m$).
2. Each ac component, taken in order regardless of frequency, is *smaller* in the full-wave circuit; that is, 0.425 versus 0.5 and 0.085 versus 0.212, and so forth.
3. The first ac component in the full-wave circuit is at *twice* the input frequency, whereas the first ac component in the half-wave circuit is at the fundamental input frequency.

What is the significance of these comparisons? The first conclusion seems fairly obvious: the center-tap circuit has a higher dc component for a given peak *load* voltage. (However, to get this doubled dc component, the full transformer secondary voltage is twice that of the half-wave circuit.) From items 1 and 2 of the comparison, a second conclusion can be drawn: the center-tap circuit has much lower percent ripple (less than half). A third advantage of the center-tap circuit is easier filtering of the ripple voltages because of the higher frequency and lower amplitude of the ac components.

Unfortunately, the center-tap circuit also has drawbacks. The first, of course, is that the transformer secondary requires a center tap. This makes the transformer more expensive. The second drawback is that each rectifier must be capable of withstanding a peak inverse voltage equal to the peak value of the *full* transformer secondary voltage, or twice E_m. This can be seen from Fig. 5-14, which is really the secondary circuit of Fig. 5-12a redrawn. With the instantaneous polarities as shown, rectifier D_1 will conduct and develop a peak voltage of E_m across the load, making the right end of the load positive. Now let us check the voltage across rectifier D_2 at this same instant. Starting at the arrowhead of the rectifier (the anode) and tracing clockwise, we pick up E_m across the transformer winding (terminal 3 to 2) and another E_m across the load resistor before we get back

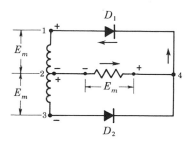

Figure 5-14   Peak inverse voltage in a center-tap circuit.

to the bar or cathode side of the rectifier. These two voltages are series-additive, so that the total inverse voltage is $2E_m$.

BRIDGE RECTIFIER CIRCUIT.   The bridge rectifier circuit achieves full-wave rectification without the need for a center tap in the transformer secondary winding. In fact, depending on the magnitude of the direct voltage required, the transformer may be omitted completely. The circuit for this type of rectifier is shown in Fig. 5-15. Let us analyze the operation of this circuit. During the half-cycle that terminal 1 of the transformer is negative, current will flow from terminal 1 to junction a and through D_1 (cathode to anode) to junction b. Will this current now flow through D_3? No, current cannot flow from anode to cathode. The current path must therefore continue to the bottom of the load and up through the load, making the top of the load resistor positive. Which path will the current follow from junction d? Through D_2 from cathode to anode? It cannot. The anode of D_2 is the starting point and is negative compared to its cathode. The correct current path is through D_4 to junction c and to the positive terminal 2 of the transformer. This current path is shown by the solid-arrow path in Fig. 5-15.

During the next half-cycle of input, the transformer polarity reverses. Terminal 2 is now the most negative point in the circuit. Let us trace the path of current flow by using the dotted arrow: from terminal 2 to junction c, through D_3 to junction b, to the bottom of the load, and up through the load to junction d. Notice that again the current is flowing up through the load, maintaining the top of the load resistor positive. Now from junction d the current flow is through D_2 to junction a and to the positive transformer terminal 1.

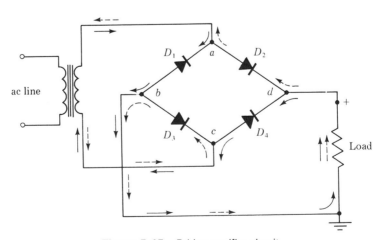

Figure 5-15   Bridge rectifier circuit.

From the above discussion, we see that the current flow through the load is always in the same direction and that there is current flow for each half-cycle of input. In other words, the bridge circuit gives full-wave rectification. The load voltage will therefore be the same as shown in Fig. 5-13 for the center-tap circuit and the dc component will be $0.636E_m$ with ripple components as given in the equation for full-wave rectification (Equation 5-2). But we already have a full-wave rectifier—the center-tap circuit. How do these two circuits compare?

The bridge circuit requires four diodes, while the center-tap circuit needs only two. On the other hand, the bridge circuit needs only a simple high-voltage secondary winding, whereas the center-tap circuit requires a center-tapped secondary with each half-winding equal to the bridge circuit secondary. Comparing these circuits on a voltage basis, for a given dc component (average value), the total secondary voltage in the center-tap circuit must be twice the secondary voltage needed by the bridge circuit. Transformer cost is higher in the center-tap circuit; rectifier cost is doubled in the bridge circuit.

Another basis for comparison is peak inverse voltage. In the center-tap circuit we saw that the rectifier must withstand twice the peak value of the rectified voltage waveshape in the inverse direction. The same is true for the bridge circuit, but since there are two diodes in series in either conduction path, the inverse voltage rating for each diode need only be E_m. This should suggest the specific application for each type. At lower voltages, within the inverse rating of the rectifier diode, the center-tap circuit is preferable. When the inverse voltage rating of the rectifier is exceeded, series units would be needed even with the center-tap circuit. In such cases the bridge circuit is preferable.

THREE-PHASE HALF-WAVE RECTIFIER.   Whenever it is necessary to satisfy heavy load demands (usually over 1 kW), it is preferable to supply these loads from a three-phase four-wire power line. This is true for motor loads as well as for rectified dc loads. A widely used three-phase four-wire system supplies 208 V between lines and 120 V between any line and neutral. Such a voltage supply is shown in Fig. 5-16 together with the phase relations of the three voltages. Notice that each voltage waveshape is a pure sine wave and that at the crossover points of these waves each wave is at half maximum value.

The circuit diagram for a three-phase half-wave rectifier is shown in Fig. 5-17. The transformer shown could be a single three-phase unit or three separate single-phase transformers. In either case, the voltages induced in the secondary windings will have the same phase relations as shown in Fig. 5-16b. Let us assume that these voltages, E_1, E_2, and E_3, are applied through the common load to D_1, D_2, and D_3, respectively, with the positive

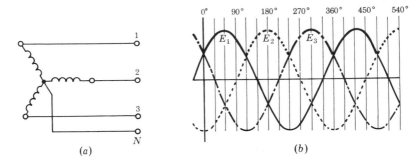

Figure 5-16 Three-phase power. **(a)** Supply. **(b)** Phase relations.

half-cycles making each anode positive. Which diode will conduct when? First let us select 200 V for E_m, so that at crossover the voltage is 100 V. Let us also assume for the moment that at just beyond 0° (Fig. 5-16b) only D_1 will conduct. Current will flow from the common transformer terminal, up through the load, through D_1 to the upper terminal of the E_1 transformer winding. As this voltage rises (0 to 60°, Fig. 5-16), the current through the load increases and the voltage across the load also increases. Neglecting voltage drops in the transformer and rectifier, the voltage across the load will be identical to the transformer secondary voltage E_1, rising from 100 to 200 V.

Now notice that the cathodes of each diode are connected to the positive load terminal. During this time interval, the cathode potentials of D_2 and D_3 rose from +100 to +200 V. Notice that their anodes are at lower potentials (see curves E_2 and E_3). Therefore the anodes are negative compared to cathodes and neither D_2 nor D_3 can conduct.

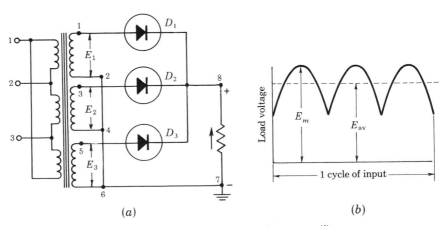

Figure 5-17 Three-phase half-wave rectifier.

During the time interval 60 to 120°, E_1 and the load voltage are decreasing. Just beyond 120°, E_2 rises above E_1. The anode of D_2 now becomes positive compared to the common cathodes. D_2 begins to conduct. Meanwhile voltage E_1 drops below cathode potential, making the anode of D_1 negative compared to its cathode. Conduction in D_1 is cut off. Diode D_2 conducts as long as its plate potential (E_2) remains positive with respect to the load voltage. When will D_3 conduct? From Fig. 5-16b you can see that just beyond 240°, E_3 rises above E_2. Diode D_3 will conduct from 240 to 360°. This completes one full cycle of input after which the process is repeated over and over.

In this three-phase half-wave rectifier, notice that each diode conducts for 120°, that there are three conduction peaks per cycle of input, and that the load voltage never drops below $\frac{1}{2}E_m$. Without going through a complete mathematical analysis, it can be seen that the average value (dc component) is higher than for single-phase circuits ($E_{av} = 0.826E_m$). Also, this circuit produces lower ripple content at higher frequency. The lowest frequency component is three times the frequency of the input voltage wave; the magnitude of this component is only $0.25E_m$.

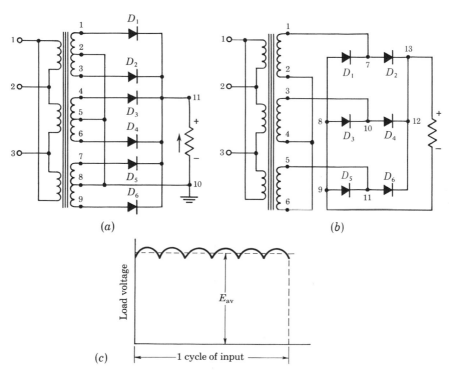

Figure 5-18   Three-phase rectifiers. **(a)** Center-tap. **(b)** Bridge.
(c) Output waveshape.

THREE-PHASE CENTER-TAP AND BRIDGE RECTIFIERS. In the discussion of single-phase rectifier circuits, it was pointed out that the center-tap and bridge circuits give higher dc components and lower ripple amplitudes at higher (double) frequencies. The same advantages can be achieved with three-phase supplies. Center-tap and bridge circuits for use with three-phase supply lines are shown in Fig. 5-18. With either of these circuits, since each rectifier conducts for only 60°, the ripple content is reduced. The lowest frequency component is now the sixth harmonic of the input frequency, at an amplitude of only $0.057E_m$. The dc component is increased to $0.955E_m$. (With the center-tap circuit, this E_m is with respect to the center tap of the transformer secondary windings.) The choice between bridge and center-tap circuit—as in single-phase systems—again depends on the desired voltage and peak inverse voltage rating of the rectifier. It should be noted that tube rectifiers can be used with any of these circuits.

Thus far we have discussed six fundamental rectifier circuits. Table 5-1 lists the salient features of each circuit for direct comparison.

REVIEW QUESTIONS

1. **(a)** What type of supply voltage is required for transistors? **(b)** What magnitude of voltage is needed?

2. **(a)** What type cf supply voltage do plates and screen grids of vacuum tubes require? **(b)** What range of voltage values may be needed to supply these tube elements?

3. **(a)** What type of supply voltage is generally used for vacuum-tube heaters? **(b)** What voltage value is commonly used for heaters?

4. **(a)** With reference to power supplies, to what does the term *load* refer? **(b)** What is meant by a heavy load? **(c)** What is meant by full load?

5. **(a)** How does the output from a power supply differ from that of a battery? **(b)** What term is used to describe these extra components?

6. **(a)** What is the effect of increasing load on the terminal voltage of a power supply? **(b)** In general, what causes this effect? **(c)** Explain how *regulation* ties in with this effect. **(d)** What is meant by poor regulation? **(e)** Under what operating condition is regulation of the power source of no importance?

7. Draw a simple block diagram for a power supply.

8. **(a)** State the main purpose for using a power transformer in a power supply. **(b)** What is a second purpose?

9. **(a)** What is the first step in converting alternating current to direct current? **(b)** What is the general name of the device used for this purpose? **(c)** Name the specific devices used.

10. **(a)** What is the name of the section of the power supply following the rectifier? **(b)** What is the purpose of this block?

11. When is a voltage divider used in a power supply?

12. Name the three basic components of a transformer.

Table 5-1
FUNDAMENTAL RECTIFIER CIRCUITS—RESISTIVE LOAD

	Single-phase			Three-phase		
	Half-wave	Center tap	Bridge	Half wave	Center tap	Bridge
Circuit						
Load voltage						
$\dfrac{E_{dc}}{E_m}$	0.318	0.636*	0.636	0.826	0.955	0.955
Lowest ripple frequency	F	$2F$	$2F$	$3F$	$6F$	$6F$
Approx. ripple content†	121%	48%	48%	18%	4%	4%

*E_m is measured to center tap of winding.

†Percent ripple = $\dfrac{\text{rms value of ac components} \times 100}{\text{Average value or dc component}}$

13. **(a)** Which winding of a transformer is identified as the *primary*? **(b)** What is a *secondary* winding? **(c)** Under what circumstances would a transformer have more than one secondary winding?

14. Explain briefly the principle of transformer action.

15. A transformer is connected to a dc supply at rated voltage. Explain what would happen.

16. What is meant by the term *1:1 ratio* as applied to a transformer?

17. **(a)** What is meant by a *step-up* transformer? **(b)** How is this feature obtained? **(c)** Give an example of where such a transformer could be used.

18. **(a)** What is meant by a *step-down* transformer? **(b)** How is this feature obtained? **(c)** Give an application of such a transformer.

19. **(a)** Can both step-up and step-down effects be achieved in a single transformer? **(b)** If so, how? **(c)** Give an application of such a transformer.

20. When a transformer secondary delivers power to a load, where does this power come from?

21. What is the general range of efficiencies of transformers?

22. In your own words, state the relationship between the primary and secondary currents of a transformer delivering current to a load. Do not give the equation.

23. If a transformer has several secondary windings, each supplying a load, how can we find the primary current?

24. **(a)** Which loss in a transformer is directly related to the quality of the core material? **(b)** Name a second type of core loss. **(c)** How is this second type of loss reduced?

25. **(a)** When assembling a core, why is each layer of laminations cut into sections? **(b)** Why are the laminations in each layer (or group of layers) reversed?

26. **(a)** Name the two fundamental designs used for transformer magnetic circuits. **(b)** Draw a cross-sectional view of one type of core construction. **(c)** Draw a cross-sectional view of the second type. **(d)** On which of these cores are the windings often placed on two legs? **(e)** Are the windings on either core generally placed on one leg only? Which core? Which leg?

27. **(a)** State two advantages of using grain-oriented steel in the core. **(b)** What type of core assembly is used with this steel?

28. State two advantages of the case construction shown in Fig. 5-6c over that of Fig. 5-6a.

29. Draw the schematic diagram of a transformer with two secondary windings.

30. What is the average value of a sine wave taken over one full cycle?

31. How can a dc component be produced from a sine wave?

32. **(a)** What is meant by nonlinear impedance? **(b)** What term is used to describe an extreme case of nonlinearity? **(c)** Name two devices that have this property. **(d)** Explain why this is so. **(e)** When used for power supply applications, what are these devices called?

33. Draw the circuit diagram for a half-wave rectifier using a semiconductor diode, supplying a resistive load, and with a positive output compared to ground.

34. **(a)** What symbol of a semiconductor diode represents the cathode? **(b)** In what direction does current (electrons) flow through a semiconductor diode, relative to its schematic representation? **(c)** Why is the bar side of the rectifier symbol often identified with a plus sign? **(d)** How else may it be identified?

35. Refer to the half-wave rectifier of Fig. 5-9: **(a)** When does the rectifier conduct? **(b)** How long is the conduction period compared to the input cycle? **(c)** If the input is a sine wave, describe the shape of the current wave. **(d)** Describe the waveshape of the voltage across the load resistor. **(e)** How does the maximum value of the load voltage compare to the transformer secondary voltage? Explain.

36. With reference to the output voltage from a half-wave rectifier as in Fig. 5-9, **(a)** what is the average value? **(b)** How does this compare with the dc component? **(c)** What is the magnitude of the fundamental frequency component? **(d)** What are the frequency and magnitude of the next component? **(e)** What other ac components are present in the output? **(f)** How do their amplitudes compare with the components mentioned above?

37. **(a)** State one method for reversing the polarity of the output voltage from the half-wave rectifier of Fig. 5-9a. **(b)** Draw this circuit diagram.

38. **(a)** State a second method for reversing the polarity of the output voltage from the half-wave rectifier of Fig. 5-9. **(b)** Draw this circuit diagram.

39. Compared to the dc output from a half-wave rectifier, what is the approximate peak inverse voltage that the rectifier unit must withstand?

40. Why are half-wave rectifier circuits usable only on relatively light loads?

41. State a third disadvantage of the half-wave circuit.

42. Draw the circuit diagram for a center-tap rectifier.

43. Trace the direction of current flow for the center-tap rectifier circuit of Fig. 5-12a **(a)** when the top of the secondary winding is positive and **(b)** when the bottom of this winding is positive.

44. With respect to a center-tap rectifier circuit, **(a)** for how long a period does each rectifier unit conduct? **(b)** What is the phase relation between their conduction periods? **(c)** Describe the waveshape of current that would flow through a resistive load. **(d)** What type of rectification is this called? Why? **(e)** In Fig. 5-13e, what portion of the time base of this wave corresponds to one cycle of input? **(f)** How does the maximum value of this output voltage compare to voltages from the transformer secondary?

45. With reference to the output voltage from a center-tap rectifier as in Fig. 5-12, **(a)** what is the relative value of the dc component? **(b)** What is the magnitude of the fundamental frequency component? **(c)** What are the frequency and magnitude of the first ac component? **(d)** What other ac components are present in the output? **(e)** How do their amplitudes compare with the component in part (c) above?

46. Compare the outputs produced by full-wave and half-wave rectification with respect to **(a)** magnitude of the dc component for a given peak voltage across the load; **(b)** frequency of the first ac component; **(c)** relative magnitude of their first ac components; **(d)** relative magnitude of their second ac components; **(e)** ripple content; **(f)** ease of filtering.

47. Compare the probable cost of power transformers for use with half-wave and center-tap circuits. Explain.

48. Compare the circuits of Question 47 with regard to the peak inverse voltage applied to the rectifier unit for a given peak load voltage.

49. (a) Name another type of rectifier circuit that produces full-wave rectification. (b) How many rectifier elements does this circuit use? (c) What is the major advantage of this circuit over the center-tap circuit?

50. Refer to Fig. 5-15: (a) Which diodes will conduct when terminal 1 of the transformer is positive? (b) Starting at terminal 1, trace the current flow point by point, returning to terminal 1. (c) Starting at terminal 1 when it is *negative,* repeat part (b).

51. Refer to Fig. 5-15: (a) Is the ac lead, terminal 1, connected to like or to unlike diode elements? (b) Repeat part (a) for ac terminal 2. (c) Is the dc output, terminal *b,* connected to like or to unlike diode elements? (d) Repeat part (c) for dc terminal *d.* (e) Make a generalization from these observations.

52. Refer to Fig. 5-15: (a) Considering diodes D_2 and D_4 interchanged, trace the current flow point by point when transformer terminal 1 is negative. (b) What would happen? (c) Now also consider diodes D_3 and D_1 interchanged, and again trace the current flow point by point. (d) What happens this time?

53. Compare the center-tap and bridge rectifiers for the same peak load voltage with respect to (a) transformer cost; (b) rectifier cost; (c) magnitude of the dc component; (d) inverse voltage rating of the diodes.

54. (a) Which of the circuits in Question 53 is preferable for high-voltage applications? (b) State two reasons.

55. When are three-phase rectifiers used in preference to single-phase units?

56. (a) What is the phase relation between the three voltages in a three-phase system? (b) What is the relative value of the voltage at the crossover points of these waves?

57. Redraw the circuit of Fig. 5-17a to show more clearly which of the transformer windings are connected in Y or delta.

58. If the three-phase supply of Fig. 5-16 is fed respectively to D_1, D_2, and D_3 of a three-phase half-wave rectifier, (a) which diode is conducting at the 150° instant? (b) At the 350° instant? (c) At what point does D_1 stop conducting?

59. In any three-phase half-wave rectifier, (a) what is the conduction period for each diode per cycle of input? (b) What is the frequency of the first ripple component? (c) What is the magnitude of this component? (d) What is the magnitude of the dc component? (e) What is the lowest instantaneous voltage across the load?

60. (a) Give the full name of the circuit shown in Fig. 5-18a. (b) Redraw this circuit to show clearly which of the transformer windings are in Y or delta.

61. (a) Give the full name of the circuit shown in Fig. 5-18b. (b) Redraw this circuit to show clearly which of the transformer windings are in Y or delta.

62. (a) What is the conduction period for each diode in the three-phase bridge circuit? (b) How does this compare with the three-phase center-tap circuit? (c) What is the lowest ripple frequency in the output from these circuits? (d) What is the magnitude of this ripple?

63. Compared to the peak value of the load voltage, what is the magnitude of the dc component in (a) a three-phase bridge rectifier circuit? (b) A three-phase center-tap circuit?

PROBLEMS

1. The secondary winding of a transformer has 150 turns. Its primary winding, for use on a 120-V line, has 400 turns. Find the secondary voltage.

2. If a secondary voltage of 600 V were required in the transformers of Problem 1, what change in the transformer design would be necessary? Give a specific numerical answer.

3. A filament transformer has a primary winding as in Problem 1, but only 8.3 turns in the secondary. Find the secondary voltage.

4. A 6.3-V output is required. The transformer primary has 520 turns and is energized from a 120-V line. How many turns should the secondary have?

5. How many primary turns are needed to get 275 V from a 1200-turn secondary if the primary voltage is 220 V?

6. A transformer has a power input of 42 W. The secondary load is 115 mA at 340 V. Find the efficiency of the transformer.

7. A transistor transformer supplies 30 V at 3 A. If its efficiency is 92 percent, find the primary current. (Line voltage is 120 V.)

8. A control transformer supplies the following loads: 1.5 A at 32 V, 3 A at 5 V, and 3 A at 12 V. The line voltage is 120 V. Assuming an efficiency of 90 percent, find the primary current.

9. A step-down transformer has a 300-turn primary and a 15-turn secondary. If the secondary load is 6 A, find the primary current.

10. A step-up transformer draws a primary current of 1.8 A from a 120-V line. It has a 400-turn primary and a 3330-turn secondary. Find **(a)** the secondary current and **(b)** the secondary voltage.

11. A power transformer has 360 turns in the primary winding. It has four secondary windings carrying the following loads: **(a)** 1000-turn secondary—240 mA; **(b)** 7.5-turn secondary—5 A; **(c)** 15-turn secondary—3 A; **(d)** 19-turn secondary—2.1 A. For a nominal line voltage of 120 V, find the output voltage from each secondary winding and the total primary current if all the loads are resistive loads. (Assume an efficiency of 90 percent.)

12. Using the equation for the output wave from a half-wave rectifier and 100 V for E_m, plot each of the first three components for $1\frac{1}{2}$ cycles of input. Solve graphically for the resultant waveshape. Explain any difference between this resultant waveshape and the load voltage waveshape of Fig. 5-10.

13. The secondary voltage applied to a half-wave rectifier with resistive load is 300 V at 60 Hz. Find: **(a)** the magnitude of the dc component; **(b)** the magnitude and frequency of the first ripple component; **(c)** the magnitude and frequency of the second ripple component.

14. A half-wave rectifier with resistive load is energized from a 120-V 60-Hz line through a 4:1 step-down transformer. Find (a), (b), and (c) as in Problem 13 above.

15. Find the peak inverse voltage that is applied across the rectifier of **(a)** Problem 13 and **(b)** Problem 14.

16. (a) What voltage (rms) must be applied to a half-wave rectifier circuit in order to develop a dc output of 100 V across a resistive load? (b) What is the peak inverse voltage?

17. Find the transformer ratio needed to obtain 250 V dc across a resistive load from a half-wave rectifier. The supply line is 120 V, 60 Hz.

18. Using the equation for the output wave from a center-tap rectifier and 100 V for E_m, plot each of the first three components for $1\frac{1}{2}$ cycles of input. Solve graphically for the resultant waveshape. Explain any discrepancy between this resultant wave-shape and the load voltage waveshape of Fig. 5-13.

19. A transformer secondary is rated at 300–0–300 V. If this transformer is used to supply a center-tap rectifier circuit with resistive load, calculate the dc voltage output.

20. The transformer and circuit of Problem 19 are used on a 60-Hz supply. Find: (a) the magnitude and frequency of the first ripple component; (b) the magnitude and frequency of the second ripple component; (c) the peak inverse voltage applied to the rectifier element.

21. (a) Find the transformer secondary voltage needed to develop a dc output of 180 V across a resistive load, using the center-tap circuit. (b) What is the peak inverse voltage across the diode?

22. Calculate the transformer-turns ratio needed to obtain 300 V dc across a resistive load from a center-tap rectifier. The supply line is 120 V, 60 Hz.

23. A bridge rectifier is energized from a 120-V 60-Hz line through a 4:1 step-up transformer. Find: (a) the magnitude of the dc component; (b) the magnitude and frequency of the first ripple component; (c) the peak inverse voltage applied to the diodes.

24. The 300–0–300-V transformer of Problem 19 is now used with its full secondary winding feeding a bridge rectifier circuit. Find: (a) the dc output voltage and (b) the peak inverse voltage per diode.

25. What transformer-turns ratio is needed to obtain 300 V dc across a resistive load from a bridge rectifier circuit on a 120-V line?

26. A three-phase half-wave rectifier supplying a resistive load has a peak output voltage of 1000 V. (a) What is the dc component of this load voltage? (b) What is the minimum value of load voltage? (c) What is the peak-to-peak value of the ripple voltage? (d) If this ripple voltage *by itself* is assumed to be similar in shape to the wave obtained from full-wave rectification (see Fig. 5-17), calculate the average value of the ripple component. (e) How does the sum of (b) plus (d) above compare to the dc component found in (a)?

27. If the voltage across any one secondary winding of a three-phase half-wave rectifier is 2000 V, find the dc output across a resistive load.

28. The full secondary voltage of each phase of a three-phase center-tap rectifier circuit is 1800 V. Find the dc output across a resistive load.

29. The transformer of Problem 28 is used in a three-phase bridge circuit. What is the dc output now?

30. What value of secondary voltage will be needed in each phase of a three-phase bridge rectifier circuit to produce 2000 V dc across a resistive load?

6

Power Supply—Filter and Regulator Circuits

In the previous chapter it was demonstrated that the output from a rectifier contains various ac components (ripple content) in addition to the desired dc output. In many applications these ac components cannot be tolerated. For example, a power supply with a ripple content of only 1 percent would produce a loud hum in a radio receiver or audio amplifier. The problem here is to reduce the ripple content to within tolerable limits. As a measure of the purity or smoothness of the dc output, the terms *ripple factor* or *percent ripple* are used. Both terms compare the amount of ripple to the magnitude of the direct current or voltage. By equation,

$$\text{Ripple factor} = \frac{\text{effective value of ac components}}{\text{average value or dc component}} \tag{6-1}$$

and

$$\text{Percent ripple} = \text{ripple factor} \times 100$$

The percent ripple for the basic rectifier circuits is given in Table 5-1.

Various types of filter circuits are used to reduce the ripple content of a rectified wave. Among these are the (1) simple capacitor filter, (2) R-C filter, (3) capacitor-input or π-type filter, (4) choke-input or L-type filter.

Each of these has specific advantages which make it preferable in certain applications.

SIMPLE CAPACITOR FILTER

A simple capacitor filter consists of a capacitor shunted across the load. Such a circuit is shown in Fig. 6-1 for a half-wave rectifier. R_L represents the equivalent resistance of the actual load. The effectiveness of this type of filter depends on the discharge time constant formed by the filter capacitor C and the equivalent load resistor R_L. (You will recall that the time constant is equal to the product of $C \times R$ and determines the time required for the capacitor to charge to 63 percent of its maximum value or to discharge by 63 percent from its maximum value.) Let us analyze the action of this filter for several conditions of load.

EFFECT OF LOAD ON OUTPUT VOLTAGE. The dotted curve in Fig. 6-1b shows the shape of the voltage wave that would appear across the load if there were no filter. This same voltage would therefore be applied across the filter capacitor. During time interval 0 to 1, the capacitor would charge. Assuming an ideal circuit (negligible resistance in the transformer windings and in the rectifier), the capacitor would charge to the peak value of the rectified wave E_m and the charge curve would follow the applied voltage curve. This is so because the charging path is through the transformer and rectifier and the charging time constant is zero. What will happen to the charge on the capacitor as the applied voltage drops to zero and remains at zero (time intervals 1 to 2 and 2 to 3)?

To answer this, first assume a condition of no load. Can the capacitor discharge back through the rectifier and transformer? No, because the rectifier is a unilateral conductor. The capacitor discharge path can only be through the equivalent load resistor. But at no load, the equivalent load resistance is infinite and the discharge time constant is also infinite. In other words, the capacitor cannot discharge and the output voltage remains constant at E_m. This is shown in Fig. 6-1b. Notice that there is no ripple in the output voltage.

Such a condition might exist in a negative bias supply if there were no drain (used with FETs or vacuum tubes operated class A), but what happens if current is drawn from the supply? Consider first a condition of light load so that R_L is large and the time constant is long. Such a case is shown in Fig. 6-1c. During the time interval 1 to 3, the capacitor is discharging slowly. Now at 3 the applied voltage starts to rise again; at 4 the rising applied voltage would tend to exceed the capacitor voltage. The capacitor recharges during time interval 4 to 5, following the applied voltage curve, until once again both are at peak value. This action is repeated

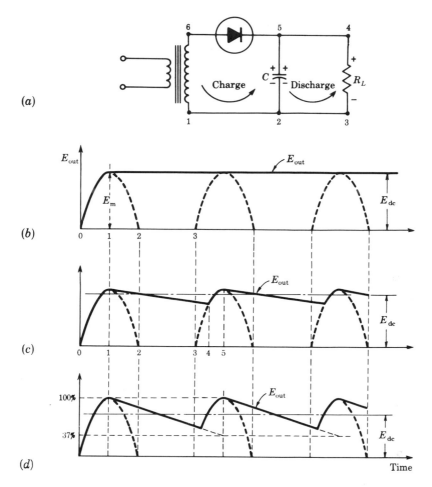

Figure 6-1   Effect of load on filter action (capacitor filter).

again and again. The output voltage now contains some ripple content, but much less than without the filter. This is shown by the heavy curve of Fig. 6-1c. Since the ripple is so much less, the average value rises from the 0.318 without filter to a much higher value as shown by the dot-dash line for E_{dc}.

As the load is increased, the value of the equivalent load resistance is decreased and the discharge time constant is also decreased. Obviously, the capacitor will discharge faster and to a lower voltage value. The ripple content will increase; the average value or dc component will be reduced. Fig. 6-1d shows these effects for a load condition where the time constant is approximately equal to the period (time for one cycle) of the ac input.

From this discussion, it can be seen that as the load is increased, the effectiveness of the filter decreases. Not only does the ripple content increase, but the magnitude of the dc component decreases. If the load is increased to the point where the time constant becomes less than the time for one quarter-cycle, the capacitor discharges practically along the applied voltage curve and no benefit whatever is obtained from this type of filter.

Is there any advantage in using a full-wave type of circuit for heavy loads? Yes, the capacitor is recharged sooner, the ripple content is reduced drastically, and the dc component is appreciably higher. This is demonstrated in Fig. 6-2, where the output voltage for a full-wave rectifier is shown for the *same load* as in Fig. 6-1*d*.

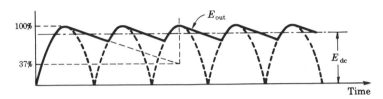

Figure 6-2 Improved filter action with full-wave input (capacitor filter).

EFFECT OF CAPACITOR ON OUTPUT VOLTAGE.   It was shown above that at heavy loads the effectiveness of the simple capacitor filter is impaired: the capacitor discharges too rapidly, because the equivalent load resistance is reduced. But the time constant can be increased by using larger values of capacitance. In other words, for any given load, the larger the filter capacitor, the less the ripple content and the higher the dc component. Carrying the idea to its ultimate conclusion, an infinite capacitor should give perfect filtering action regardless of load. This is theoretically true, but there are practical drawbacks. Larger capacitors are bulkier and more expensive; beyond a certain point it is wiser to use some other type of filter. Furthermore, the use of larger and larger capacitors would overload the rectifier unit.

Refer again to Fig. 6-1*a*. Notice that the capacitor is charged plus on top and minus on bottom. During the nonconducting half-cycle, the transformer voltage is minus on top and plus on bottom. At this instant the two voltages are series-additive, with the anode negative and the cathode positive. If the filtering action is good and the capacitor discharges only slightly, this peak inverse voltage will approach $2E_m$. Thus with a capacitor filter the peak inverse voltage in a half-wave circuit may be as high as for a center-tap circuit.

EFFECT OF CAPACITOR ON RECTIFIER CURRENT.   With a capacitor filter, when the applied (transformer secondary) voltage exceeds the capacitor voltage the capacitor will accept charge. The charging current comes through the rectifier. At the same time, the rectifier is also supplying current directly to the load. On the other hand, when the line voltage drops and the capacitor voltage exceeds the transformer secondary voltage, the capacitor no longer charges but instead discharges through the load. Current stops flowing from the rectifier through the load. The rectifier cannot conduct because its anode (at transformer potential) is negative compared to its cathode (at capacitor potential). Actually, a semiconductor rectifier would have a slight (negligible) conduction when it is biased in the inverse direction. These conditions are shown in Fig. 6-3a for applied voltage exceeding capacitor voltage and in Fig. 6-3b for applied voltage below capacitor voltage.

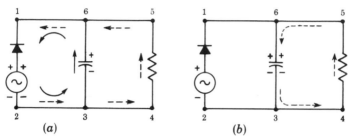

Figure 6-3   Current flow with capacitor filter. **(a)** Applied voltage exceeds capacitor voltage. **(b)** Capacitor voltage exceeds applied voltage.

In Fig. 6-3b the rectifier is cut off, but current is still flowing through the load. For what portion of the cycle is the rectifier cut off? That depends on the time constant of the filter circuit, and for any given value of load it depends on the size of the capacitor. Figure 6-4 shows the conduction period and the magnitude of the rectifier current for two values of the filter capacitor. In Fig. 6-4a the capacitor is small and discharges fairly rapidly.* The rectifier starts to conduct at time instant 1, since the applied voltage rises above the capacitor voltage. Due to the capacitive load, this first inrush of current is high, but as the peak voltage is reached and the capacitor approaches full charge, the charging current decreases. At time instant 2, the applied voltage falls below the capacitor voltage and conduction ceases. It is obvious that the conduction period must be less than one quarter-cycle of input (in this case approximately 60°). Yet, since the output voltage never drops to zero, current must flow through the load for the full cycle. With a resistive load, the load current waveshape must be a replica of the output voltage waveshape. This load current is shown as the dot-dash curve of Fig. 6-4. The average value of this load current can be read by a dc ammeter in the load circuit.

*The output voltage waveshape is similar to Fig. 6-1d. This time, however, the first half-cycle is not shown since it is a transient condition that occurs only when the circuit is energized.

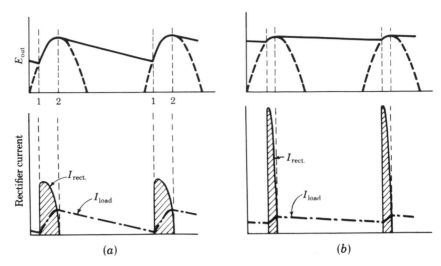

Figure 6-4   Effect of filter capacitor on rectifier current—constant load. **(a)** Small C. **(b)** Large C.

What is the peak value of the current through the rectifier? The load current must be supplied directly *or indirectly* (by recharging the capacitor) from the power line through the rectifier. Therefore, the average value of the current pulse through the rectifier (averaged over the full cycle) must equal the average value of the load current. Expressing this relation in another way, for each cycle of input the area under the pulse of rectifier current must equal the area under the load current wave. Mathematical evaluation of this relationship is quite involved (and beyond the level of this text). A graphic solution could be used, or the peak value of the rectifier current could be measured by laboratory experiment using a calibrated oscilloscope. However, to a close approximation the rectifier peak current can be evaluated as follows. Assume that the pulse of rectifier current is a perfect rectangle. The average value of the load current is another rectangle of duration 360°. If the duration of the rectifier pulse is only 60°, its height must be six times the height of the average value of current, or

$$I_{FM} = \frac{360}{\text{degrees of cond.}} \times \text{dc load current} \tag{6-2}$$

where I_{FM} is the maximum value of the rectifier forward current pulse. The peak value obtained by this method is always somewhat lower than the true peak current. This is so because the base of the current pulse is being used for the "rectangle," although the true pulse is narrower at the top.

Now examine Fig. 6-4*b*. These curves are obtained with the same value of load current as in Fig. 6-4*a*, but with a larger filter capacitor. The discharge time constant is longer in Fig. 6-4*b*. The capacitor discharges more slowly; the output waveshape has less ripple; the capacitor voltage and output voltage are maintained at a higher level. The conduction period through the rectifier in this illustration is reduced to 30°. With the approximate method, the peak value of the rectifier current pulse is now more than 12 times the average value of the load current. If a still larger capacitor were used to reduce the ripple content, the conduction period would be even shorter and the peak current through the rectifier even higher. At some large value of capacitance, the peak current through the rectifier could exceed the peak current rating and destroy the unit even with relatively light external loads. The seriousness of such a technique can be shown by an example.

EXAMPLE 6-1

A 1N646 silicon rectifier is used to supply a dc load of 200 mA. The forward current ratings for this diode are 400 mA for the dc (I_o) or average ($I_{F(av)}$) value and 1.25 A for the maximum peak repetitive current (I_{FM}).

1. Can this rectifier be used with a capacitor filter that would reduce the conduction period to 20°?
2. If not, what is the minimum allowable conduction period?

Solution

1. For 20° conduction and a dc load of 200 mA,

$$I_{FM} \cong \frac{360}{20} \times 200 = 3.6 \text{ A}$$

This is greater than the peak current rating of the diode.
2. For $I_{FM} = 1.25$ A, the ratio of peak current to average value of load current is $1.25/0.2 = 6.25$ and the minimum conduction angle is

$$\theta_{min} \cong \frac{360}{6.25} = 57.5°$$

It is interesting to note from Example 6-1 that although the rectifier is rated for a maximum dc forward current of 400 mA, the peak current rating of 1.25 A can be exceeded with a dc load of only half its rated value (that is, 200 mA) merely by using a filter capacitor of too large a value.

To summarize, the simple capacitor produces relatively smooth output at light loads and the direct voltage is high, approaching E_m. However, the effectiveness of this filter is inversely proportional to the load. As the load increases, the ripple content increases and the direct voltage level drops, until at heavy loads the capacitor serves no useful function. Application of this type of filter should be limited to light and fairly constant loads.

SURGE CURRENT ($I_{FM(surge)}$).   Another problem with capacitor-input filters is *surge current*. Each time the power supply is energized, the capacitor is probably fully discharged. Refer to Fig. 6-1a. Assume, as we have done before, an ideal transformer with zero-resistance windings and an ideal diode with zero resistance in the forward direction. Let us also assume an ideal capacitor with zero series resistance. When the secondary voltage is applied across the capacitor, there is no opposition to current flow and the inrush of current to charge the capacitor is infinite! Of course, in any practical circuit there is some resistance in each component. Still, this transient current may exceed the surge-current rating of the rectifier. To prevent such failures, a *surge resistor* can be inserted in series with the diode (between point 6 and the anode of the rectifier in Fig. 6-1a). Unfortunately, any series resistance will cause a power loss and will also cause the output voltage to drop, even more, with load. Therefore surge-resistance values should be kept low. The worst possible case of surge current will occur if the power supply is energized at the instant that the ac supply voltage is at its maximum. The minimum value of series resistance is therefore fixed by the peak value of the secondary voltage and the surge-current limitation of the diode. If the sum of the circuit component resistances is less than this value, an additional series surge resistor will be needed.

On the other hand, series resistance—whether due to surge resistor or component resistances—does serve another useful purpose: it increases the charging time of the capacitor, thereby decreasing the peak value of the charging current. This permits safe use of larger values of filter capacitor for better filtering action.

DESIGN FACTORS.   So far we have seen, qualitatively, that the output voltage is affected by the size of the capacitor and by the load, but this does not tell us what the dc output voltage or the percent ripple will be. An exact analysis is quite involved, but with a reasonable assumption these values can be determined with good accuracy. Notice from Figs. 6-1 and 6-2 that the ripple content is essentially a triangular wave. Note also that the charge time is rather short compared to the discharge time. Furthermore, recall that the better the filtering action (larger capacitor value and/or load resistance), the shorter this charge time. If we neglect this charging time, the discharge time becomes the time for one cycle of the lowest (major) ripple component.* This approximation converts the ripple content into a sawtooth waveform as shown in Fig. 6-5, with

E_m = peak value of the ac supply voltage
E_{dc} = dc component of the output voltage
$E_{r(p-p)}$ = peak-to-peak value of the ripple content

---

*$T = 1/f_r$, where f_r is the major ripple frequency. Remember that in a half-wave rectifier f_r is equal to the line frequency, whereas f_r is twice the line frequency in a full-wave circuit.

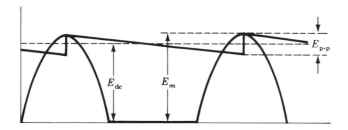

Figure 6-5   Approximation of ripple waveshape.

Obviously, the average value or dc component is

$$E_{dc} = E_m - \frac{E_{r(p\text{-}p)}}{2} \tag{6-3}$$

Also, from circuit analysis, the rms value of this sawtooth wave is

$$E_r = \frac{E_{r(p\text{-}p)}}{2\sqrt{3}} \tag{6-4}$$

Furthermore, since the ripple factor (the ratio of E_r to E_{dc}) increases with smaller capacitors and with heavier loads (R_L smaller), it can be shown that*

$$\text{Ripple factor} = \frac{1}{2\sqrt{3}f_r C R_L} \tag{6-5}$$

where f_r is the major ripple frequency component. Also, E_{dc} becomes

$$E_{dc} = \frac{E_m}{1 + \dfrac{1}{2f_r C R_L}} \tag{6-6a}$$

If the dc load current is known, this reduces to

$$E_{dc} = E_m - \frac{I_{dc}}{2f_r C} \tag{6-6b}$$

Finally, if we wish to determine the capacitor value needed for a given maximum ripple content, then

$$C \text{ (in } \mu F) \geq \frac{10^6}{2\sqrt{3}f_r R_L \text{ (ripple content)}} \tag{6-7}$$

*J. D. Ryder, *Electronic Fundamentals and Applications* (4th ed.), Englewood Cliffs, N.J.: Prentice-Hall, Inc., 1970, sec. 4-7.

EXAMPLE 6-2

In Fig. 6-1, the transformer secondary voltage is 80 V, the capacitor is 20 μF, and the load resistor is 2000 Ω. The circuit is energized from the 60-Hz power line. Find:

1. The percent ripple
2. The dc output voltage
3. The capacitor needed to reduce ripple to not over 5 percent

Solution

Since the circuit is half-wave, $f_r = f = 60$ Hz.

1. To find the percent ripple,

$$\text{Ripple factor} = \frac{1}{2\sqrt{3}f_r C R_L} = \frac{1}{2\sqrt{3} \times 60 \times 20 \times 10^{-6} \times 2000}$$

$$= 0.120$$

$$\text{Percent ripple} = 0.120 \times 100 = 12.0\%$$

2. To find the dc output voltage, first we must find the peak value of the ac input voltage:

$$E_m = \sqrt{2}E = \sqrt{2} \times 80 = 113 \text{ V}$$

Then

$$E_{dc} = \frac{E_m}{1 + [1/(2f_r C R_L)]} = \frac{113}{1 + [1/(2 \times 60 \times 20 + 10^{-6} \times 2000)]}$$

$$= \frac{113}{1.208} = 93.6 \text{ V}$$

3. To find the filter capacitor value for maximum ripple of 5 percent,

$$C = \frac{10^6}{2\sqrt{3}f_r R_L \text{ (ripple content)}} = \frac{10^6}{2\sqrt{3} \times 60 \times 2000 \times 0.05}$$

$$= \frac{10^6}{20{,}760} = 48.2 \ \mu F$$

R-C FILTER

The output waveshape with a simple capacitor filter has much less ripple content than the unfiltered rectifier output. However, even at light loads, unless the load is zero some ripple is still present. Much smoother output can be obtained by adding a resistor in series with the rectifier output and shunting a second capacitor across the load end of this resistor to form an

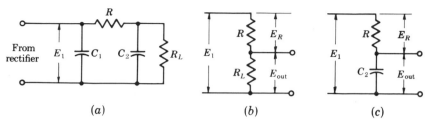

Figure 6-6   R-C filter circuit. **(a)** Actual circuit. **(b)** Equivalent dc circuit.
(c) Equivalent ac circuit.

R-C filter. Such a circuit is shown in Fig. 6-6. The first capacitor C_1 has
the same action as described above. This time, however, the value of the
series resistor R must be added to the equivalent load resistance when
figuring the discharge time constant. The voltage E_1 has the same waveshape as would be obtained with any simple capacitor filter.

The filtering action of the series resistor and second capacitor can be
readily explained from the equivalent diagrams. To the dc component, the
capacitor C_2 has infinite impedance and can be neglected. The equivalent
circuit is the voltage divider formed by R and R_L (Fig. 6-6b). The effect of
the series resistor is merely to lower the output voltage due to the IR drop
across itself. (This is not a serious drawback since compensation could be
made for this voltage drop by starting with a slightly higher voltage from
the rectifier.) With regard to the ac components, the equivalent circuit of
this filter is shown in Fig. 6-6c. C_2 is selected so as to have a low reactance
at the lowest frequency component of the ripple content. With respect to
this low reactance, the equivalent load resistance value R_L can be neglected.
This R-C circuit is a frequency-discriminating voltage divider. If R is large
compared to X_{C_2}, most of the ac voltage drop occurs across the series
resistor R and only a negligible amount appears across C_2 and across the
load. What is more, the higher the ripple frequency, the lower the ripple
content in the output waveshape. The effectiveness of the R-C filter can be
illustrated with an example.

EXAMPLE 6-3

The output from a half-wave rectifier with capacitor filter has a dc component
of 200 V and an rms ripple content of 30 V. The dc load is 10 mA. The power line
frequency is 60 Hz.

1. Find the percent ripple.
2. An R-C filter is formed by adding a series resistor of 1000 Ω and a shunt
   capacitor of 40 μF. Find the dc output voltage and the ripple content with
   the filter.

Solution

1. Percent ripple $= \dfrac{\text{rms value of ac component}}{\text{dc component}} \times 100$

$= \dfrac{30}{200} \times 100 = \mathbf{15\%}$

2. (a) First we must find the equivalent load resistance (dc):

$$R_L = \frac{E_{dc}}{I_{dc}} = \frac{200}{10 \times 10^{-3}} = 20{,}000 \ \Omega$$

(b) Total resistance $= 1000 + 20{,}000 = 21{,}000 \ \Omega$

(c) Output voltage E_{dc} with filter, by voltage-divider action:

$$E_{dc} = \frac{20{,}000}{21{,}000} \times 200 = \mathbf{190.5 \ V}$$

(d) Lowest ripple frequency is 60 Hz, and

$$X_C = \frac{1}{2\pi fC} = \frac{0.159 \times 10^6}{60 \times 40} = 66.2 \ \Omega$$

(e) Using the ac equivalent diagram,

$$Z = \sqrt{R^2 + X^2} = \sqrt{(1000)^2 + (66)^2} = 1000 \ \Omega$$

(This should have been obvious since R is greater than $10 \times X_C$.)

(f) Assuming that all the ripple content is at 60 Hz (worst possible case), the rms value of the ripple voltage in the output is

$$E_r = \frac{66.2}{1000} \times 30 = 1.99 \ V$$

(g) Ripple content $= \dfrac{E_r}{E_{dc}} \times 100$

$= \dfrac{1.99}{190.5} \times 100 = \mathbf{1.04\%}$

In Example 6-3 the addition of an R-C filter reduced the ripple from 15 to 1.04 percent. This is a very worthwhile improvement in filtering action. The 10-V drop due to the series resistor is not serious. But suppose that the load current had been 100 mA, instead of 10. Would the voltage drop in the series resistor still be unimportant? No, the drop this time would be $1000 \ \Omega \times 100 \ \text{mA} = 100 \ \text{V}$. Could we reduce the series resistor to below $100 \ \Omega$ to keep the dc voltage drop to within 10 V? Again, no. This would impair the ac filter action. In short, the R-C filter is suitable *only for very light loads*. In addition, since this circuit uses an input capacitor, the effectiveness of filtering will vary with load and the direct voltage level will also vary with load. This circuit finds its main application in power supplies for cathode-ray tubes, ac/dc receivers, or as an individual stage filter for small-signal (or voltage) amplifier stages.

CAPACITOR-INPUT OR π-TYPE FILTER

If we could maintain a high value of series "resistance" to the ac components while somehow reducing the resistance to the dc component, the R-C filter could be used with much higher load currents. An inductance has high opposition to alternating currents and if wound with the proper size of wire, its dc resistance is quite low. Therefore, by replacing the series resistor with an inductor we have a filter circuit suitable for moderately high values of load. Such a circuit (Fig. 6-7) is called a *capacitor-input filter*

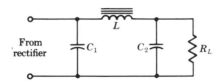

Figure 6-7   Capacitor input or π-type filter.

or a π-*type* filter, and the inductor used for filtering purposes is called a *choke coil*. (In construction and appearance a choke coil looks identical to the power transformers shown in Fig. 5-6, except that it has only one coil and only two terminals or leads.) The effectiveness of the π-type filter can be demonstrated by an example, using the same conditions as in Example 6-3.

EXAMPLE 6-4

The output from a half-wave rectifier with capacitor filter has a dc component of 200 V and an rms ripple content of 30 V. The dc load is 10 mA. The power line frequency is 60 Hz. A π-type filter is formed by adding a 30-H 90-Ω choke in series and a capacitor of 40 μF in parallel with the load. Find the dc output voltage and the ripple content with this filter.

Solution

1. The voltage drop (dc) in the choke is

$$E_{\text{drop}} = IR = 10 \times 10^{-3} \times 90 = 0.90 \text{ V}$$

2. The dc output voltage is

$$E_{\text{dc}} = E_T - E_{\text{drop}} = 200 - 0.90 = \textbf{199.1 V}$$

3. The ac reactance of the choke and the capacitor at the lowest ripple frequency is

$$X_L = 2\pi fL = 2\pi \times 60 \times 30 = 11,300 \ \Omega$$

$$X_C = \frac{1}{2\pi fC} = \frac{0.159 \times 10^6}{60 \times 40} = 66.2 \ \Omega$$

4. The total impedance of this L-C-R circuit by inspection is $X_L - X_{C_2}$ or 11,234 Ω. (The resistance of the choke is negligible in the *vector* addition.)
5. By voltage-divider action, the rms ripple voltage E_r across the output capacitor is

$$E_r = \frac{66.2}{11,234} \times 30 = 0.117 \ \text{V}$$

6. Ripple content $= \dfrac{E_r}{E_{dc}} \times 100$

$$= \frac{0.117}{199} \times 100 = \mathbf{0.089\%}$$

A comparison of Examples 6-3 and 6-4 reveals that for the same dc load, the ripple output is less than 10 times smaller with the 30-H choke and the dc voltage drop is less than 1 V, or again less than 10 times smaller than with the resistance filter. For the same dc voltage drop as in Example 6-3, the load current can be increased by a factor of 10. Due to the increased load the discharge time of the first capacitor will decrease and the ripple content will increase. However, the added filter action of the choke and second capacitor will still give very effective filtering. For example, before the ripple content with the choke increases to 1.99 V (as in Example 6-3), the ripple content *before* the choke would have to rise to

$$\frac{11,234}{66.2} \times 1.99 \quad \text{or} \quad 3.48 \ \text{V}$$

This is a ripple percentage of 174 percent, which is impossible. (The maximum ripple content for a half-wave circuit is 121 percent, as shown in Table 5-1. Such a condition corresponds to complete ineffectiveness of the first filter capacitor.)

Since the π-type filter is so much better than the R-C type, why is the latter used at all? Good design should consider not only which gives better filtering but also other factors such as cost, weight, and size. Obviously, the simple resistor is preferable with respect to these three factors. Therefore at light loads the R-C filter is selected, whereas at heavier loads the π-type filter may be necessary for adequate filtering.

The addition of the choke and second capacitor has made it possible to obtain a smooth output even at higher loads. Yet this circuit, too, has many of the drawbacks of the simple capacitor filter. Due to the charging current of the input capacitor, the peak rectifier current is much higher than the load current. Therefore if it is desired to use this circuit at higher loads, *full-wave rectification (center-tap or bridge) should be employed.* This will reduce the peak rectifier current per diode to approximately one-half, and this filter can be used with loads up to 200–400 mA. However, the effectiveness of filtering still varies with load, and the direct voltage output level also varies with load. For optimum results, the π-type filter should not be used with varying loads.

CHOKE-INPUT OR L-TYPE FILTER

All the filters discussed so far use a shunt capacitor as the first (or only) element, and the drawbacks of these filters are directly attributable to the charge-discharge action of this capacitor. Considering, in addition, that at heavy loads this capacitor may discharge so rapidly that it practically follows the fall in applied voltage, it may be wiser to eliminate the input capacitor, thereby also saving the expense, bulk, and weight of this unit. This results in the *choke-input* or *L-type filter* shown in Fig. 6-8. Since the advantages of this type of filter are primarily noticed with high-current loads, this circuit is used only with full-wave or polyphase rectifier circuits.

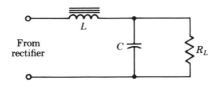

Figure 6-8   Choke input or L-type filter.

OUTPUT VOLTAGE (DC).   The smoothing effect of an *L-C* section due to voltage-divider action follows the filter analysis of Fig. 6-6. This time, however, it should be remembered that the voltage wave applied to the choke-input filter has the "raw" rectified waveshape with full ripple content as shown in Table 5-1.

To the ac components (the ripple content) of the applied rectified wave, the voltage divider consists of the X_L of the choke in series with X_C of the capacitor.* If the choke has *sufficient* inductance, its reactance to all

---

*Again, we can neglect the effect of the equivalent load resistance R_L in parallel with X_C because in any practical case the capacitor is large enough to make X_C very much lower than R_L.

ac components is very high and increases with frequency. If the capacitor has adequate capacitance, its reactance is low and decreases with frequency. As a result, very little of the ac component voltage appears across the capacitor; almost all this voltage is "filtered" as voltage drop across the choke coil; therefore the capacitor cannot charge to the peak value of the rectified voltage waveshape.

With regard to the dc component, the choke offers very little opposition, due only to its resistance. At light loads, the resistance of the coil is unimportant and the full dc component voltage can be considered as appearing across the capacitor and across the load. For use with heavy loads, chokes should be wound with large wire to keep the choke resistance and the resulting IR drop to low values. The dc output voltage with choke-input filter is approximately equal to the *average* value of the rectified waveshape and does not vary much with changes in load.

RIPPLE CONTENT.   The effectiveness with which the ripple components are removed from the output wave depends on the voltage-divider action between the coil and the capacitor, or more specifically on the ratio X_C/Z (where Z is the series impedance of the capacitor and choke). However, since the resistance of the coil is negligible compared to its reactance, and since X_L in practical circuits is high compared to X_C, $Z \cong X_L$ and the ratio can be simplified to X_C/X_L. Expressing this relation in words, the ripple content in the output of a choke-input filter decreases as (1) the inductance value is increased, (2) the capacitance value is increased, (3) the frequency of the ripple component is increased. Since frequency is involved in both X_L and X_C, ripple content varies inversely as the *square* of the frequency.

Notice that R_L does not appear in the discussion of ripple content. In other words, the effectiveness of this filter is independent of load and depends only on the filter values L and C. A simple relation between ripple factor and L-C values for a full-wave rectifier can be derived from basic ac circuit theory as follows:

1. The equation for the rectified wave is

$$e = 0.636E_m - 0.425E_m \cos 2\omega t + 0.085E_m \cos 4\omega t \cdots$$

2. Since the amplitude of the ripple component decreases rapidly as the frequency increases, and since the ripple content in the output of the filter decreases as the square of the ripple frequency, we need consider only the first ac component ($0.425E_m \cos 2\omega t$). Therefore, at the *input* to the filter

$$\text{Ripple voltage input (rms)} \cong 0.707 \times 0.425E_m$$

3. The ripple content (E_r) in the output of the filter depends on the ratio X_C/X_L, or

$$\text{Ripple voltage output (rms)} \cong 0.707 \times 0.425 E_m \times \frac{X_C}{X_L}$$

4. From the equation of the rectified wave, the dc component is

$$E_{dc} = 0.636 E_m$$

5. Solving for ripple factor gives

$$\text{Ripple factor} = \frac{E_r}{E_{dc}} \cong \frac{0.707 \times 0.425 E_m X_C}{0.636 E_m X_L}$$

$$\cong 0.472 \frac{X_C}{X_L} \tag{6-8}$$

6. For a 60-Hz full-wave power supply, the ripple frequency is 120 Hz, and converting to capacitance values in microfarads, the general equation above reduces to

$$\text{Ripple factor} \cong \frac{0.83}{LC} \tag{6-9}$$

where L is in henrys and C is in microfarads.

EXAMPLE 6-5

Find the percent ripple for a bridge-type rectifier using a choke-input filter and energized from a 60-Hz line, when:

1. $L = 5$ H and $C = 4$ μF
2. $L = 10$ H and $C = 8$ μF

Solution

1. Percent ripple $\cong \dfrac{0.83}{LC} \times 100 = \dfrac{0.83}{5 \times 4} \times 100 = 4.15\%$

2. Percent ripple $\cong \dfrac{0.83}{LC} \times 100 = \dfrac{0.83}{10 \times 8} \times 100 = 1.04\%$

TWO-STAGE FILTERS.   The equation for ripple factor shows that filtering action is improved by using larger values for filter components. This was also shown in Example 6-5 above, where the percent ripple was cut to one-fourth by doubling the values of L and C (using four times the previous LC product). Theoretically, it would follow that larger and larger values of L and C should be used if improved filter action is desired. However, for the same total values of L and C a smoother output voltage can be obtained if two *stages* of filtering are used, wherein the output of the first filter is fed to the second filter and then the output of the second filter is fed to the load. Such a two-stage filter is shown in Fig. 6-9. Of course, the

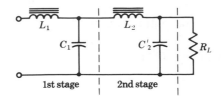

Figure 6-9   A two-stage L-type filter.

cost of two chokes and two capacitors is greater than that of single units of the same total value. Good designers compare cost versus larger L and C values when deciding whether to use single- or two-stage filters.

The improved effectiveness of this circuit can be shown as follows:

1. The ripple factor for the first stage is $0.472(X_C/X_L)$.
2. But we know, by definition, that the ripple factor is the ratio of E_r, the ripple voltage (rms value of the ac components), to E_{dc}, the average value or dc component.
3. Therefore, the ripple voltage output from the first filter stage is

$$E_{r1} = E_{dc} \times \text{ripple factor}$$

$$\cong \frac{0.472 X_C}{X_L} \times E_{dc}$$

4. This ripple voltage is fed to the second stage, and by ac voltage-divider action the ripple in the *output* of the second stage must be further reduced by the X_C/X_L ratio. Therefore the ripple output from the second stage is

$$E_{r2} \cong \frac{0.472 X_{C_1}}{X_{L_1}} \times \frac{X_{C_2}}{X_{L_2}} \times E_{dc}$$

5. If the capacitance and inductance values are the same for each stage, the impedance ratio becomes $(X_C/X_L)^2$. Then, solving for ripple factor by dividing E_r by E_{dc}, we get

$$\text{Ripple factor} \cong 0.472\left(\frac{X_C}{X_L}\right)^2 \qquad \text{(6-10)}$$

Comparing this to the general equation for a single-stage filter, we see that the only change is that the impedance ratio has been squared. (Similarly it can be shown that for a three-stage filter the impedance ratio would be cubed, and so forth.)

6. For 60-Hz full-wave supply, and with capacitance in microfarads, the ripple factor for the two-stage filter becomes

$$\text{Ripple factor (two-stage)} \cong \frac{1.47}{(LC)^2} \qquad \textbf{(6-11)}$$

EXAMPLE 6-6

Find the percent ripple for the bridge circuit of Example 6-5 if a two-stage filter is used, each section having $L = 5$ H and $C = 4\ \mu F$.

Solution

$$\text{Percent ripple} \cong \frac{1.47}{(LC)^2} \times 100 = \frac{1.47}{(5 \times 4)^2} \times 100 = 0.37\%$$

Comparing this result with part 2 of Example 6-5, we can see that for the same total values of L and C, this two-stage filter is almost three times more effective (1.04 to 0.37).

CRITICAL INDUCTANCE.   One drawback of the capacitor-input type of filter is due to the high peak currents flowing through the rectifier when the capacitor is charging. Example 6-1 dramatically showed how this can cause serious limitation in the amount of load (dc) that such a rectifier can supply. With choke-input filters, these peak currents are reduced drastically. As a result, rectifiers are given a higher dc rating for use with choke-input filters. The explanation for the elimination of current peaks is relatively simple. With capacitor-input circuits, rectifier current is supplied in short spurts per cycle, mainly to charge the capacitor. For the rest of the cycle, this capacitor supplies the load current. However, if sufficient inductance is used with choke-input filters, current flows through one rectifier or another for the full cycle. In a full-wave circuit each rectifier unit supplies current for 180°; in a three-phase half-wave circuit, each rectifier unit supplies current for 120°. Since there is no "storing of energy," there are no peak currents.

The minimum value of inductance necessary for such operation is known as the *critical inductance*. Let us analyze this limitation, using a full-wave rectifier. The rectified voltage waveshape has a dc component of $0.636E_m$ and a second-harmonic component of $0.425E_m$. Since the magnitude of the higher-order harmonic components is relatively small, they can be neglected. (This simplifies the discussion.) When this complex voltage is applied to the load, two component currents will flow: a dc component wherein

$$I_{dc} = \frac{E_{dc}}{R_L}$$

(where R_L is the equivalent load resistance) and an ac component equal to

$$I_{ac} = \frac{E_{ac}}{Z_L} \cong \frac{E_{ac}}{X_L}$$

These two components of the load current are shown in Fig. 6-10a. Notice that the ac component is shown smaller than the dc component. This is quite normal. The choke-input filter is generally used with high-current loads, while the ac component can be made as small as desired by using larger and larger inductance values. Now suppose a choke coil of lower inductance value is used. The ac component will increase. If this process is continued, a condition will be reached where the peak value of the ac component will equal the dc component value (see Fig. 6-10b). This value of inductance is the critical inductance *for that particular value of load.* If the inductance is reduced to an even lower value, the condition shown in Fig. 6-10c will occur, where the peak value of the ac component exceeds the dc component. For the brief instant t_1 to t_2, *cutout* has occurred. During this time interval, the rectifier does not supply current to the load, there is no voltage drop in the choke, and the filter capacitor charges to E_m, the peak value of the input waveshape. The output voltage rises sharply. Effectively the circuit has temporarily "switched over" to a capacitor-input type of circuit. This drastic change in output voltage is not desirable.

The same result could have been produced with a constant inductance value, if the load current had been reduced. In other words, the critical value of inductance needed to prevent cutout also depends on the amount

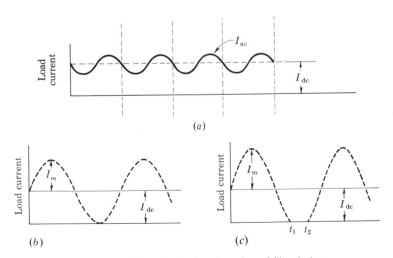

Figure 6-10   Effect of reducing the value of filter inductance.

of load to be supplied. This relation can be developed mathematically for a full-wave circuit as follows. For the condition of critical inductance,

$$I_{dc} = I_m$$

Replacing these currents by their Ohm's law values,

$$\frac{0.636E_m}{R_L} = \frac{0.425E_m}{X_L}$$

Solving for X_L,

$$X_L = 0.666R_L$$

For 60-Hz supply ($f = 120$ Hz),

$$L = \frac{0.666R_L}{2\pi f} = \frac{0.666R_L}{754} = \frac{R_L}{1130}$$

Since we have made some approximations, a slightly larger value of inductance is commonly used (as an added safety factor). The preceding equation is "rounded off" to

$$L = \frac{R_L}{1000} \tag{6-12}$$

where L is in henrys.

It follows from this equation that with heavy loads there is little danger of cutout even with low values of inductance. For example, in a power supply delivering 150 mA at 300 V ($R_L = 300/150 \times 10^{-3} = 2000\ \Omega$), a 2-H choke (2000/1000) would be sufficient. However, should the load drop to 10 mA, still at 300 V, what value of filter choke inductance would be needed?

$$R_L = \frac{E}{I} = \frac{300}{10 \times 10^{-3}} = 30,000\ \Omega$$

$$L \geq \frac{R_L}{1000} = \frac{30,000}{1000} = 30\ \text{H}$$

At no load, the value of inductance required is infinite. This, of course, is impossible. To prevent such a situation, the power-supply voltage divider is designed to provide a sufficiently high bleeder current so that the critical inductance is reduced to a reasonable value.

In the preceding illustration we saw that, at 150 mA, a 2-H choke was needed; whereas, at the lower current value, the inductance needed was 30 H. In practice a *swinging choke* is used to satisfy this requirement economically. The choke is designed with large enough wire to carry the maximum current requirement (in this case 150 mA). However, the number of turns, the type of iron, and the cross-sectional area of the core are designed to give the maximum inductance at the *low* current value. This results in a smaller, less expensive unit. Then, as the current increases, the iron begins to saturate and the inductance *decreases*. A typical swinging choke may have an inductance of 30 H at 10 mA but only 5 H at 150 mA.

REGULATED POWER SUPPLIES
(ELECTRONIC FILTERS)

In many electronic applications, it is important that the voltage supplied to the units remains constant. Variations in voltage may cause inaccurate and erratic operation or even complete malfunction. For example, in an oscillator or signal generator, the frequency will shift; in electronic instruments, calibration will change; in transmitters, distorted signal output will result; in computers, switching or trigger circuits may fail to operate. Power supplies for such applications must have good regulation. (Recall from Chapter 5 that the term *regulation* is used to indicate the ability of a power supply to maintain constant terminal voltage, and that regulation is considered good if there is little change in voltage as the load varies.) Coincidentally, any circuit that improves regulation will automatically also reduce the ripple content in the output waveshape. This should be obvious, since ripple causes the output voltage to vary (at the ripple frequency). If the regulating action is fast enough to react at the ripple frequency, the output variations are eliminated. When used specifically for such purpose, these circuits are called *electronic filters*. Also, when voltage regulators are to be used, there is no need for elaborate passive component filters and only a simple capacitor filter is used ahead of the regulator.

For quantitative evaluation or comparison, the term *percent regulation* is used, where

$$\text{Percent regulation} = \frac{\text{no-load value} - \text{full-load value}}{\text{full-load value}} \times 100 \qquad \text{(6-13)}$$

In electronic applications, the no-load and full-load conditions are not always obvious. To eliminate ambiguity, it is common practice to specify the minimum and maximum load current limits for which the regulation is desired.

EXAMPLE 6-7

The output voltage from a power supply drops from 325 V at 25 mA to 290 V at 120 mA. Find the percent regulation over this range.

Solution

$$\text{Percent regulation} = \frac{E_{NL} - E_{FL}}{E_{FL}} \times 100$$

$$= \frac{325 - 290}{290} \times 100 = 12\%$$

CAUSES FOR POOR REGULATION.   In the preceding discussion, it was stated that the direct-voltage output from a power supply changes as the load is varied. More specifically, Example 6-7 shows that it decreases as the load increases. This decrease can result from voltage drops in any portion of the circuit that has resistance. For example: (1) resistance in the primary and secondary winding of the power transformer, (2) resistance in the rectifier, (3) resistance in the series filter element (L or R).

These are not the only causes for poor regulation. Two other factors must be considered. If the filter is of the π type, any change in the discharge time constant of the input capacitor caused by changes in load will affect the output voltage drastically (see Fig. 6-1). The remaining factor is the line voltage. If the ac line voltage applied to the power transformer varies, the secondary voltage will vary and the dc output will vary. Line voltage variations may be caused by our own load changes, but more often they are caused by outside factors beyond our control.

Knowing the causes for poor regulation, we can take steps to improve regulation. Obviously, series resistance must be held to a minimum. Power transformers and chokes should be wound with larger gauge wire. This will reduce the winding resistances; unfortunately, it also increases the size, weight, and cost of the units. R-C filters or π-type filters should not be used—the first because of their high series resistance, the second because of the varying discharge time constant of the input capacitor. For good regulation, choke-input filters are superior. In many applications, these design techniques are sufficient to produce satisfactory regulation. On the other hand, it is often easier—especially with integrated circuitry—to use special voltage-stabilizing circuits.

ZENER-DIODE VOLTAGE REGULATOR.   One of the simplest stabilizing circuits uses a zener diode as the voltage regulator.* Examination of semiconductor diode characteristics will show that only a very small (negligible) current flows through such a diode when a small reverse voltage is applied. However, if the reverse voltage is progressively increased, a breakdown value will be reached. At this point, the electric field is strong enough to break the covalent pair bonds, producing additional—and copious cur-

---

*Older versions using cold-cathode gas diodes (glow tubes) may still be found.

rent carriers. The back resistance drops to a very low value, and the current increases rapidly while the voltage drop across the diode remains essentially constant. Since this is so, it is possible to use semiconductor diodes in voltage-stabilizing circuits, except that standard diodes are not used for this purpose. They are not designed for, and could not withstand, the high currents without permanent injury. Instead, specially designed silicon units called *zener diodes* are used. Typical reverse current characteristics for several zener diodes are shown in Fig. 6-11. When used for voltage regulator service, the reverse current through the diode must not be allowed to fall below the knee of its characteristic. The zener must always operate in the breakdown region. Otherwise, the diode voltage drops rapidly and regulation is lost. This minimum-current value differs for each diode type and rating.

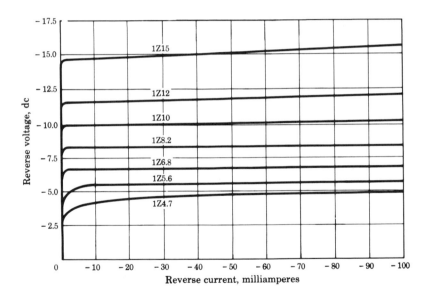

Figure 6-11 Reverse current curves of typical zener diodes (*adapted from International Rectifier Corp. graph*).

A typical regulator circuit is shown in Fig. 6-12. Notice that the circuit diagram shows current flowing through the zener diode *in the direction indicated by the arrow*. This is not an error, nor is it indicating "conventional" current flow. In a zener diode this is a *reverse* current flow due to breakdown action—electrons flow opposite to the normal diode direction. This circuit corrects not only for variations of input voltage but also for variations in the load current. For example, let us consider the regulating action when

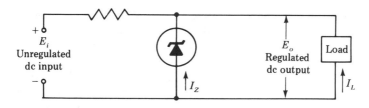

Figure 6-12   A zener-diode voltage regulator.

the load decreases. The decrease in current would cause the supply voltage to increase (poor regulation) and would also cause the voltage drop across resistor R to decrease. Both these effects would tend to increase the load voltage. But this also affects the voltage across the zener diode. Its current would start to increase sharply—so that the total current ($I_Z + I_L$) would tend to return to its previous value and the load voltage would remain constant. Similarly, if for any reason the supply voltage (the unregulated dc input) tends to decrease, this would tend to reduce the voltage across the diode. Its current would decrease, reducing the voltage drop across the series resistor R and maintaining the load voltage constant.

The design of a zener-diode voltage regulator has two aspects: (1) determining the voltage and power rating of the zener diode and (2) determining the resistance and power rating of the series resistor. To start the design it is necessary to know over what range the load (current) may vary and over what range the unregulated dc input voltage would swing. Finally, two limiting conditions must be satisfied:

1. The minimum value of the dc input voltage must not fall below the desired constant dc output voltage. (This should be obvious, since we must allow for the voltage drop across the resistor R.)
2. The minimum value of the diode current must not fall below the knee of its characteristic curve. (The diode must operate within the breakdown region.) Since this current value is not necessarily known, it is common design practice to select 10 percent of the maximum load current as a safe value for $I_{Z(min)}$.

The zener diode selected for a specific design should have a voltage rating as close as possible to the desired constant dc output voltage.* (Calculation of the power rating of the diode will have to wait for further circuit analysis.)

---

*The actual output voltage may differ from this nominal rating of the diode. Common-run zener diodes have tolerances of ±5, 10, or 20 percent. If more exact voltages are required, diodes with closer tolerance limits (±1 percent) are available—at higher prices.

Let us proceed with the selection of a suitable series resistor. Notice from Fig. 6-12 that the voltage drop E_R across this resistor is equal to $E_i - E_o$ and that the current I_R through the resistor is equal to $I_L + I_Z$. Then, from Ohm's law, $R = E_R/I_R$. But, simple as this seems, remember that three of these quantities vary, and so the question is—what values do we use for E_i, I_L, and I_Z? The resistance value is set by the minimum value of the dc input voltage. In turn, this occurs when the load current is a maximum, and under these conditions, the diode current is a minimum. Putting this all together, we get

$$R = \frac{E_R}{I_R} = \frac{E_{i(min)} - E_o}{I_{L(max)} + I_{Z(min)}}$$

$$= \frac{E_{i(min)} - E_o}{1.1 I_{L(max)}}$$

(6-14)

(Remember that we use 10 percent $I_{L(max)}$ as a safe value for $I_{Z(min)}$.) Of course, when ordering this resistor we must also specify its power rating. This is readily calculated from the E^2/R form of the power equation. But what value do we use for E_R? The worst-case condition occurs when the voltage drop across the resistor is a maximum, and this happens when the input voltage is a maximum, or

$$P_R = \frac{E_{R(max)}^2}{R} = \frac{(E_{i(max)} - E_o)^2}{R}$$

(6-15)

Now we are ready to calculate the power dissipated by the diode. The EI form of the power equation is most convenient. We know E: it is the breakdown voltage E_Z. However, we still have to evaluate the maximum value of current through the diode. We know that I_Z will be a maximum when the load current I_L is a minimum. We also know that $I_Z = I_R - I_L$ and that I_R is a maximum when E_R is a maximum. By Ohm's law we can find $I_{R(max)}$, and subtracting $I_{L(min)}$ we get $I_{Z(max)}$. An example will clarify this analysis.

EXAMPLE 6-8

An unregulated power supply has a voltage variation of from 30 V at 60 mA to 38 V at 10 mA. It is desired to maintain a dc output of 24 V over this range of currents.

Solution

1. For a 24-V output select a zener diode (such as the 1N4749) that has a nominal zener voltage rating of 24 V.
2. To calculate the series resistance value:

$$(a) \quad R = \frac{E_R}{I_R} = \frac{E_{i(min)} - E_o}{1.1 I_{L(max)}} = \frac{30 - 24}{1.1(0.06)} = 91 \ \Omega$$

(b) $P_R = \dfrac{E^2_{R(\text{max})}}{R} = \dfrac{(E_{i(\text{max})} - E_o)^2}{R} = \dfrac{(38 - 24)^2}{91} = 2.16 \text{ W}$

(A 5-W resistor should provide an ample safety factor.)

3. To calculate the power rating of the zener:

(a) $E_{R(\text{max})}$ from step 2 above is 14 V

(b) $I_{R(\text{max})} = \dfrac{E_{R(\text{max})}}{R} = \dfrac{14}{91} = 154 \text{ mA}$

(c) $I_{Z(\text{max})} = I_{R(\text{max})} - I_{L(\text{min})} = 154 - 10 = 144 \text{ mA}$

(d) $P_Z = E_Z I_{Z(\text{max})} = 24 \times 144 \times 10^3 = 3.46 \text{ W}$

The closest commercially available power rating is 5.0 W.

The breakdown voltage of a zener diode depends on the resistivity of the silicon material used, and it can be controlled to a great degree in the manufacturing process. It is therefore possible to manufacture these units for almost any desired voltage from just below 2 V to as high as 200 V. They are also available in a wide range of power ratings from 250 mW to as high as 50 W. A variety of zener diodes is shown in Fig. 6-13.

Because of their excellent stability, zener diodes are also used to develop reference voltages for measurements or control equipment. Such

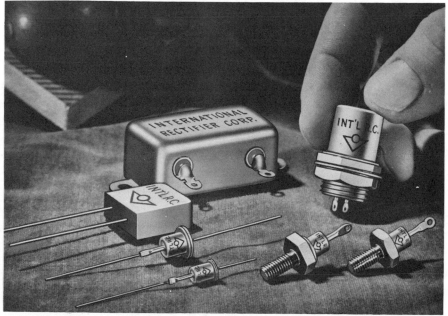

Figure 6-13   Silicon voltage regulators and reference zener diodes (*International Rectifier Corp.*).

applications are found in recorders, telemetering, guided missiles, and a variety of automatic machinery.

FEEDBACK REGULATORS. The simple zener-diode regulator has several drawbacks. First, the output voltage cannot be set at will. The diode must be changed, and a new matching resistor must be selected. Second, there is a practical limit to the number of steps available. Finally, these circuits are not suitable for high-current loads because of the power losses in the series resistor and in the diode. [Notice in Example 6-8 that the power lost in the resistor (2.16 W) and the power lost in the diode (3.6 W) are each greater than the power delivered to the load (1.44 W maximum). The series-resistance loss could be reduced by lowering the series-resistance value—but this would decrease the range of control.]

The losses would be minimized if the series resistance were variable and, automatically, varied inversely with the load. The product IR would then be constant, and the load voltage would be constant. This effect can be achieved by using a bipolar transistor as the variable-resistance element. Its conduction (and therefore its resistance) is controlled by changing the base bias. Such a circuit is shown in Fig. 6-14. It is called a *series-pass feedback regulator*. The series-pass element, Q_1, acts as the variable resistor; Q_2 is an amplifier; and D_1 is a zener diode supplying a reference voltage and holding the emitter potential constant at this value. Resistors R_3, R_4, and R_5 serve as a *sensing network* to detect any change in the dc output voltage E_o. Any change in output voltage is amplified by Q_1 and fed back to Q_2 to control its conduction, thereby restoring the output voltage to its original value—hence the term *feedback regulator circuit*.

To explain the regulating action of this circuit, let us assume that the dc output voltage tends to increase (either because of a rise in the ac line

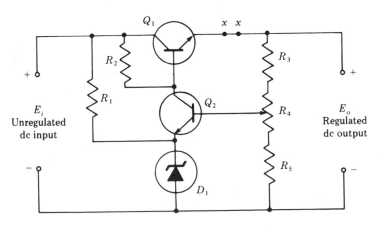

Figure 6-14   A feedback voltage regulator.

voltage or because of a decrease in the dc load current). The potential of point A becomes more positive, making the base of Q_2 more positive compared to its emitter. Since this is an *NPN* transistor, its conduction increases. The increase in collector current, flowing through R_3, drives the base of Q_1 in the negative direction. This reduces the conduction of Q_1, increasing its series resistance, offsetting the increase in voltage, and tending to restore the dc output voltage to its former value. Obviously, the correction can never be perfect, since then the "error" or changed sense voltage at A would be zero, and no corrective action would be applied to Q_1. A circuit such as that shown in Fig. 6-14 would hold regulation to approximately ± 2 percent. Regulation can be improved by using a high-gain amplifier in place of the single transistor Q_2. The load currents that can be controlled by this circuit are limited by the collector-current rating of transistor Q_1. For higher current loads, transistors in parallel or Darlington-connected pairs can be used in place of transistor Q_1.

Notice in Fig. 6-14 that R_4 is a potentiometer (instead of a fixed tap). This allows for setting the output voltage to any desired value, within limits (from a few volts below the unregulated dc input voltage down to slightly above the zener breakdown voltage of D_1). If we wish to decrease the dc output voltage (from its previously set value), we must move the potentiometer arm down. This drives the base negatively (compared to emitter) and, for an *NPN* transistor, reduces conduction. Less current flows through R_2, reducing the emitter-base forward bias for Q_1, decreasing its conduction, and increasing its resistance. The dc output voltage is lowered.

PROTECTIVE CIRCUITS.   Electrical power circuits—commercial or residential—are generally protected from overloads (and short circuits) either by fuses or by circuit breakers. Unfortunately, the reaction time of these devices is too slow for solid-state equipment: transistors and diodes will be permanently damaged before the fuse or breaker opens the circuit. Protection in solid-state circuits must thus come from electronic circuitry. A feedback-regulated power supply can readily incorporate overload protection. Take another look at Fig. 6-14. A resistor inserted in the emitter lead of Q_1 (at x–x) can sense an overload. Such a resistor is shown in the partial schematic of Fig. 6-15. (It has a low resistance value, and since it is included within the feedback loop of the total regulator circuit it will not impair the regulation of the power supply.) The load current, flowing through this resistor, develops a voltage drop providing base bias for the added transistor Q_3. The resistor value is chosen so that if the load exceeds some predetermined limit, the IR drop will exceed the threshold bias value of V_{BE} (approximately 0.6 V for a silicon unit). Transistor Q_3 is quickly driven into full conduction, shorting out the drive voltage for transistor Q_1. The series-pass resistance of Q_1 increases sharply, reducing the output voltage and load current to safe values.

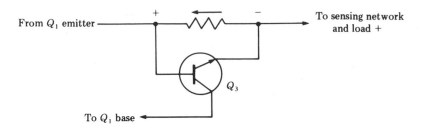

Figure 6-15 Overload protection for a feedback regulator.

Overvoltage protection can also be added to a series-regulator circuit. This may be especially necessary when the regulator is set for a low-voltage dc output. Remember that the unregulated dc input may be much higher. Should the regulator circuit malfunction, the full unregulated dc value could be applied to the low-voltage load and damage the equipment. Many circuit variations have been used to protect against accidental overvoltage. One commonly used circuit—called a *crowbar circuit*—uses a *silicon controlled rectifier* (SCR)* directly across the load. The SCR is normally nonconducting and so will not affect circuit operation. However, should the voltage rise above some predetermined value, the SCR is triggered into heavy conduction, shorting out the power supply (as if a crowbar were dropped across the output leads). The regulator circuit is assumed to have overload protection, and therefore the short circuit causes no harm.

MONOLITHIC VOLTAGE REGULATORS.   The regulator circuits discussed above use discrete components—individual separate units for each resistor, diode, and/or transistor. However, most if not all of these components can be deposited on a single integrated-circuit chip to form a monolithic voltage regulator. Some of the advantages claimed for these units are small size, low cost, and excellent performance characteristics. Their low cost and small size make possible point-of-use regulation—i.e., regulation of individual stages (or sections) of complex electronic equipment. Integrated-circuit products are available for operation at voltages from 0 to more than 1000 V and (with externally connected series-pass transistors) for currents to more than 50 A.

One of the simpler monolithic regulators is the Fairchild 78XX model. It looks like a conventional three-lead plastic power transistor and is used as shown in Fig. 6-16. It is available for output voltages of from 5 to 15 V and for currents up to 1 A. In spite of its apparent simplicity, in addition to good regulation it provides current-limiting protection and excess temperature-rise protection.

When using integrated-circuit (IC) techniques it is relatively easy to

*SCRs are discussed briefly in the next chapter.

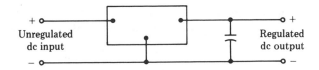

Figure 6-16   A simple monolithic voltage regulator.

add resistors, diodes, and transistors on one monolithic chip with only minor increase in cost. This makes possible complex regulator circuits with excellent performance characteristics. The schematic of one such unit is shown in Fig. 6-17a. All 23 components are housed in the popular TO-5 package (diameter is less than 3 in). It is connected as shown in Fig. 6-17b. The only external components required are the potentiometer and the 300-pF capacitor. A regulated output voltage ranging from 4 to 16 V can be set by the potentiometer control. At any setting, the regulation obtained with this unit is typically 0.2 percent. If currents greater than 150 mA are required, this IC can drive up to 5 A using one external transistor. When current limiting is desired, the output is taken from pin 7 through Q_7 and

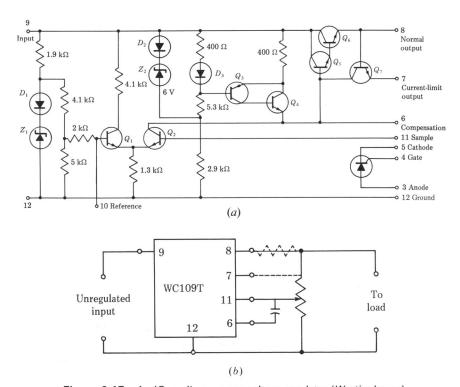

Figure 6-17   An IC medium-power voltage regulator (*Westinghouse*).

a sensing resistor is connected across the base-emitter junction of Q_7. (See the dotted connections in Fig. 6-17b.) The value of this resistor is determined by the load current limit ($R_s = 100/I_{L(\text{max})}$).

Notice that the SCR is not connected internally within the chip. It can be used (with external connections) for foldback current limiting or, in a crowbar circuit, for overvoltage protection.

Design of a power supply is greatly simplified when using monolithic regulators. The user need only select the most economical unit to meet the current and voltage requirements. The only chore is to calculate appropriate values for any voltage-setting resistors and any current-limiting resistors.

SWITCHING REGULATORS.   All the voltage-regulator circuits mentioned above have one fault in common—inherent low efficiency. Regulator action depends on the IR drop across a series-pass resistive element. This automatically means an I^2R power loss. For high-power loads—above several hundred watts—and especially if the desired dc output voltage is much lower than the unregulated input voltage, this power loss can be prohibitive.

Much higher efficiencies can be obtained by using switching techniques. The unregulated dc is fed to the load through a "switch" that is alternately on and off. (The SCR is generally used as the switch with high-power loads and low-frequency switching rates, while the power transistor is preferred for medium-power high-frequency switching applications.) The output from the switch is a series of pulses, and the dc level is the average value of these pulses. Regulation is achieved by a feedback circuit that senses any change in the desired output voltage and feeds back a signal to a corrective circuit that controls either the conducting time or the repetition rate of the switching action, or both. Obviously, the wider the pulse widths, or the more pulses of current in a given time, the higher the average value or dc component. Also, it should be obvious that this output now needs filtering to remove the ripple created by the switching action. This ripple is minimized by using switching frequencies in the 7 to 25 kHz range. Still, switching regulators have higher ripple output and poorer regulation than the series-pass feedback regulators described earlier. They are also more expensive. Their use is generally restricted to high-power applications where high efficiency is important or where their small size and low weight are a prime consideration.

THE FERRORESONANT TRANSFORMER.   Another technique for maintaining good regulation employs a specially designed power transformer known as a *ferroresonant* or *constant-voltage transformer*. Its ability to maintain a constant output voltage (within limits) is based on saturation effects, incremental inductance, and nonlinear ferroresonant circuit theory. The transformer uses a shell-type core, but with an air gap deliberately introduced into the center leg. With the primary and secondary windings on opposite legs of the core, this center leg acts as a magnetic shunt, so that only part of the primary flux reaches the secondary and vice versa. Also, a capacitor is connected across the secondary winding to form a broadly resonant circuit—the resultant circulating current and flux driving the secondary iron circuit into saturation. The magnetic shunt provides a flux path for the secondary so that the primary iron is not saturated. Now if the primary voltage increases (for example, by 20 percent), the primary current and the primary flux will also increase (by 20 percent), but since the secondary leg is saturated, most of this increase in primary flux will take the shunt path and only a small amount of this added flux (such as 4 percent) will pass through the secondary leg.

To improve regulation further, a compensating winding is added. This winding is placed on the *primary* leg but is connected in series opposition to the secondary. Since this winding is on the same leg as the primary, any change in primary voltage will cause a proportionate change in the compensating voltage.

This transformer also provides regulation for load change effects. The phase of the secondary current (and flux) will change. The compensating voltage and induced secondary voltage are no longer in direct phase opposition. Therefore, in spite of a decrease in the secondary-winding voltage with load, the combined phasor sum of induced voltage plus compensating voltage remains essentially constant. With the addition of a compensating winding, output regulation to better than 1 percent is obtained.

CURRENT-REGULATED SUPPLIES

The increasing use of semiconductors, magnetic components, and other current-sensitive devices has brought about the development and popularization of current-regulated power supply sources. Such a power source is also referred to as a *constant-current* supply. In this type of unit, the output voltage varies with the load resistance so as to maintain a constant current through the load. Ideally, a constant-current supply should be capable of providing an infinite output voltage so as to maintain the output *current* constant regardless of the load resistance. In a practical situation there are limitations to this ideal, and current regulation can be maintained only up to some maximum value of voltage.* This maximum voltage limitation is known as the *voltage compliance* of the power supply.

A common technique for obtaining current regulation is to connect a fixed "sampling" resistor in series with the external load. The voltage

---

*This is similar to a voltage-regulated power supply wherein the constant output voltage is maintained only up to some maximum current value.

drop across this resistor is then compared against a reference voltage. Any change in load current causes a variation in the voltage drop across the sampling resistor, and an error signal is produced. The error voltage is amplified and applied to a series regulating transistor which adjusts the output voltage and restores constant current.

Constant-current power supplies find their major application in tests and measurements with current-sensitive devices, particularly when high-precision current sources are necessary. Some of the many devices tested are semiconductor diodes and transistors, bolometers, thermistors, sole-noids, magnetic cores, relays, meters, fuses, and potentiometers.

REVIEW QUESTIONS

1. With reference to the rectified output of a power supply, **(a)** what is meant by the term *ripple content*? **(b)** Explain, in words, the meaning of the term *ripple factor*. **(c)** Give the equation for ripple factor. **(d)** Give the equation for percent ripple.

2. Name four basic types of filter circuits.

3. Describe the capacitor filter.

4. Draw the circuit diagram for a transformer power supply using a capacitor filter and supplying a resistive load.

5. Refer to Fig. 6-1*a*: **(a)** Trace the charging path. **(b)** Is this a long or short time constant circuit? Explain. **(c)** Explain why the capacitor cannot discharge through this same path (in reverse). **(d)** Trace the discharge path. **(e)** At very light loads, is the discharge time constant long or short? Explain.

6. Refer to Fig. 6-1*b*: **(a)** During what time interval does the capacitor charge? **(b)** During what interval does it discharge? **(c)** For what value of load is this diagram drawn?

7. Refer to Fig. 6-1*c*, neglecting the starting transient condition: **(a)** During what time interval does the capacitor charge? **(b)** During what interval does it discharge? **(c)** To what value of the complex output waveshape does the horizontal dash-dot line correspond? **(d)** With respect to this dash-dot line, what component of the complex wave output does the heavy solid line represent? **(e)** How would you evaluate the ripple factor of this rectifier output?

8. In a power supply using a capacitor filter, explain the effect of increased load on **(a)** the output voltage (dc) and **(b)** the ripple content.

9. If a given capacitor filter and load are supplied from a full-wave rectifier instead of a half-wave source, what happens (approximately) to **(a)** the capacitor discharge time interval? **(b)** The magnitude of the ripple content? **(c)** The magnitude of the dc component?

10. **(a)** Sketch, with dotted lines, the output wave from a full-wave rectifier for two cycles of input. **(b)** For some moderate load show the output that would be obtained from a capacitor filter. **(c)** For the same load and filter show the output wave that would be obtained if the supply were from a half-wave rectifier.

11. For a given load, using a capacitor filter, what is the effect of increasing the capacitance value on **(a)** the capacitor discharge rate? **(b)** The magnitude of the ripple content? **(c)** The magnitude of the dc component?

12. **(a)** Compare the peak inverse voltage applied to the rectifier element in a half-wave rectifier without a filter and with a capacitor filter. **(b)** Explain the effect of the filter on this value.

13. In a half-wave rectifier with capacitor filter, **(a)** what determines when the rectifier starts to conduct? **(b)** What determines when the rectifier stops conducting? **(c)** State two functions of the rectifier current during the conduction period. **(d)** When the rectifier stops conducting, does the load current stop flowing? Explain.

14. In a half-wave rectifier with capacitor filter, if the load is increased what happens to **(a)** the capacitor discharge rate? **(b)** The rectifier conducting period?

15. Refer to Fig. 6-4*a* and *b*: **(a)** What circuit difference (if any) is there between the two cases? **(b)** Why does Fig. 6-4*b* have the shorter rectifier conduction period? **(c)** Why does Fig. 6-4*b* have the higher peak rectifier current value? **(d)** What would be the effect of using a larger filter capacitor?

16. In a rectifier with capacitor filter, what is the approximate relation between the peak rectifier current and the dc load current?

17. Why do manufacturers specify both a dc current rating and a peak current rating for rectifier diodes?

18. For what type of load is the simple capacitor filter best suited?

19. **(a)** What is meant by the *surge current* in a power supply? **(b)** What limits the value of this surge current? **(c)** Under what condition is this surge current the worst? **(d)** When is this limiting action insufficient? **(e)** How is it then held to a safe value?

20. **(a)** What components are used in an R-C filter? **(b)** Describe how they are interconnected.

21. **(a)** In an R-C filter, how does the action of the first capacitor compare with the action of the capacitor filter? **(b)** How does the rest of this circuit further reduce the ripple content of the output? **(c)** What disadvantage does this filter have? **(d)** For what type of loads is this circuit best suited?

22. What is a choke coil?

23. **(a)** Why is use of a choke coil preferable in place of the resistor in the R-C filter? **(b)** What is the filter circuit now called? **(c)** Give another name for this circuit.

24. Draw the schematic diagram for a center-tap rectifier using a capacitor input filter.

25. Compare the R-C filter and the capacitor-input filter with regard to **(a)** effectiveness of filtering; **(b)** maximum load it is suitable for; **(c)** service with varying loads.

26. **(a)** Draw the circuit diagram of a choke-input filter. **(b)** For what type of load is this circuit primarily used? **(c)** Is this filter used with half-wave rectification? Explain.

27. For a given rectified waveshape, what is the approximate dc output voltage obtainable from **(a)** a capacitor filter? **(b)** A capacitor-input filter? **(c)** A choke-input filter?

28. As the load is increased, what is the effect on the magnitude of the dc output voltage from **(a)** a capacitor filter? **(b)** A capacitor-input filter? **(c)** A choke-input filter?

29. What is the main advantage of the L-type filter over the previously studied filters? Why is this so?

30. In an L-type filter what is the effect on the ripple content **(a)** if the choke value is increased? **(b)** If the capacitor value is increased? **(c)** If both the choke and the capacitor values are *doubled*? (Give a numerical answer.) **(d)** If the load is increased?

31. High-voltage power supplies in television receivers are usually energized from a 15,750-Hz source (instead of 60-Hz line). State one advantage of using such a supply source. Explain.

32. **(a)** A full-wave rectifier is to be operated from a 60-Hz supply line and will use a choke-input filter. Very low ripple content is desired. In general, would doubling the filter component values or using a two-stage filter give better filtering action? Explain. **(b)** Would this apply for *any* values of L and C? Explain.

33. Using the ripple content equations for the single-stage and the two-stage choke-input filter, for what LC product (L in henrys and C in microfarads) in the basic choke-input circuit will the effectiveness of filtering of a two-stage filter be the same as doubling the original L and C values? (Compare with Question 21.)

34. **(a)** What is the conduction period for each rectifier element in a bridge circuit when used with a choke-input filter? **(b)** How is the dc rating of a rectifier diode affected when used with capacitor-input or choke-input filter? **(c)** Explain.

35. With reference to a full-wave rectifier, **(a)** name the two major components of the unfiltered rectified wave. **(b)** What are their respective magnitudes? **(c)** For a given load and filter how can we find the magnitude of the dc component of the load current? **(d)** What determines the magnitude of the ac component of the load current?

36. **(a)** What is meant by *cutout* in a rectifier with L-type filter? **(b)** Under what condition can it occur? **(c)** What is the objection to this effect? **(d)** Without changing the filter circuit, how can this effect be prevented? Explain.

37. **(a)** What is meant by *critical inductance*? **(b)** What determines this value? **(c)** Does this term apply to a π-type filter? Explain. **(d)** If the condition of cutout is occurring in a rectifier circuit, how can it be corrected without changing the load? **(e)** For a given load give the equation that fixes the minimum value of inductance for use in a choke-input filter.

38. **(a)** What is a *swinging choke*? **(b)** When is this type of choke used? **(c)** What is the advantage of using this type of choke?

39. Why is it sometimes necessary to maintain constant voltage in electronic equipment?

40. Why may a regulator circuit also be called an electronic filter?

41. **(a)** What term is used to indicate the ability of a power supply to maintain constant voltage output? **(b)** Give the equation for evaluating this quantity.

42. **(a)** What is the underlying cause for variation of output voltage with load in any power supply? **(b)** What components in the power supply can contribute to this effect?

43. **(a)** Name two types of filters that have inherently poor regulation. **(b)** Explain why this is so for each type of filter.

44. **(a)** Name a semiconductor device that is specifically intended for use as a voltage regulator. **(b)** Is this unit connected with forward or with reverse bias? **(c)** Why is a standard unit not so used?

45. In the circuit of Fig. 6-12, current flow is shown in the direction of the semi-conductor arrowhead. Is this correct? Explain.

46. In Fig. 6-12, assume that the load resistance is decreased. **(a)** What happens to the load current I_L? **(b)** What generally happens to the unregulated dc input voltage? **(c)** What tends to happen to the voltage drop across resistor R? **(d)** What tends to happen to the voltage across the zener diode? **(e)** What happens to the current I_Z? Why? **(f)** What happens to the current I_R? **(g)** What happens to the regulated output voltage E_o?

47. In designing a circuit such as the one shown in Fig. 6-12, **(a)** what determines the voltage rating of the diode? **(b)** If a dc output voltage of 24 V is desired, should a zener diode of at least 36-V rating be used as a safety factor? Explain. **(c)** If the load current may vary from 5 to 25 mA, what may we use as a safe minimum value for I_Z? **(d)** What voltage and current values fix the value for resistor R? **(e)** Give the equation used to calculate the power rating for this resistor.

48. Is the regulator circuit of Fig. 6-12 suitable for high-current loads? Explain.

49. Refer to Fig. 6-14: **(a)** What is the function of D_1? **(b)** What is the function of Q_1? **(c)** Of Q_2? **(d)** Of resistors R_3, R_4, and R_5? **(e)** Why is a potentiometer used for R_4?

50. In Fig. 6-14, assume that the output voltage tends to decrease. **(a)** What happens to the potential at A? **(b)** Does this increase or decrease the forward bias V_{BE} of Q_2? Explain. **(c)** What happens to the current through R_2? Why? **(d)** What happens to the base bias of Q_1? **(e)** What happens to the conduction and series resistance of Q_1? **(f)** What happens to the output voltage? **(g)** For better regulation, should we use a low-gain or a high-gain amplifier for Q_1? Explain.

51. In Fig. 6-12, the potentiometer arm is moved up. **(a)** What happens to the collector current of Q_2? Why? **(b)** What happens to the conduction of Q_1? Why? **(c)** What happens to the output voltage?

52. Are fuses or circuit breakers suitable for protection of solid-state circuits? Explain.

53. Refer to Fig. 6-15: **(a)** Is this a complete circuit? **(b)** Within what other circuit is it incorporated? **(c)** At low values of load, does transistor Q_3 conduct? Explain. **(d)** What determines when it does conduct? **(e)** What happens when it does conduct?

54. **(a)** What is the purpose of a *crowbar* circuit? **(b)** What component is used as a crowbar? **(c)** Briefly, how does it function?

55. **(a)** What is a *monolithic* voltage regulator? **(b)** Give three advantages over its discrete-component counterpart.

56. Refer to Fig. 6-16: **(a)** Which components are not part of the monolithic chip? **(b)** Explain why the potentiometer is not incorporated within the chip. **(c)** When would the dotted connections be used?

57. **(a)** What is the function of a *switching regulator*? **(b)** What is its main advantage over series-pass resistive regulators? **(c)** What devices are used as "switches" in switching regulators? **(d)** When a power transistor is used as the switching element, is it subject to the same power loss as the series-pass element of the previously discussed regulators? Explain.

58. (a) What is a *ferroresonant* transformer? **(b)** Why does an increase in applied primary voltage cause only a small increase in secondary induced voltage? **(c)** What is the purpose of the compensating winding? **(d)** Where is this winding physically located? **(e)** How is it connected electrically?

59. (a) What is a constant-current supply? **(b)** What does the term *voltage compliance* signify in this type of supply? **(c)** What is the function of the *sampling resistor*? **(d)** Where is this resistor connected electrically? **(e)** Explain how regulation is obtained.

PROBLEMS AND DIAGRAMS

1. The output from a power supply has a dc component of 250 V and an ac component (rms value) of 32 V. Find the percent ripple.

2. An electronic device requires 320 V dc for proper operation, but the ripple must not exceed 3 percent. **(a)** What is the maximum tolerable value of ripple voltage? **(b)** Considering the ripple as a sawtooth, calculate the peak-to-peak value of the ripple voltage.

3. Figure 6-2 represents the output waveshape from a power supply as shown on an oscilloscope. The dc output, measured by a dc voltmeter, is 135 V. By graphic measurement, find **(a)** the peak value of this wave in volts; **(b)** the peak-to-peak value of the ripple voltage in volts; **(c)** the effective value of the ripple voltage (assuming it is a sawtooth wave); **(d)** the percent ripple.

4. A half-wave rectifier with capacitor filter is energized from a 120-V 60-Hz line through a 3.5:1 step-up transformer. Find the dc output voltage at no load.

5. The power supply in Problem 4 uses a filter capacitor of 2 μF. A resistor of 2000 Ω is connected across the output. **(a)** What is the discharge time constant of the filter? **(b)** How does this compare with the period of the supply source? **(c)** What is the approximate dc output voltage with this load?

6. The secondary voltage supplying a full-wave rectifier with capacitor filter is 300–0–300 V. What is the approximate dc output voltage at very light loads using a capacitor filter?

7. A half-wave rectifier with capacitor filter is supplied with 320 V from the power transformer. Find the peak inverse voltage applied to the rectifier.

8. The rectifier of Problem 7 is supplying a dc load of 35 mA. If the conduction period is 25°, find the peak current through the rectifier.

9. A silicon diode rectifier has a peak forward current rating of 750 mA and an average (dc) current rating of 125 mA. If it is supplying a load current of 80 mA, what is the minimum conduction period that can be used without overloading the diode?

10. A bridge rectifier with capacitor filter is supplying a dc load of 150 mA. Each rectifier element conducts for 70°. What is the peak rectifier current?

11. A dc load of 120 mA is supplied from a center-tap rectifier with capacitor filter. Find the peak value of current through the diodes if each unit conducts for 60°.

12. A half-wave rectifier circuit is supplied from a 40-V, 60-Hz supply. The filter is a 40-μF capacitor and the load has an equivalent resistance of 1500 Ω. Find: **(a)** the percent ripple and **(b)** the dc output voltage. **(c)** What value capacitor will reduce the ripple to 5 percent or less?

13. Repeat Problem 12(a) and (b) for a full-wave rectifier.

14. A full-wave, bridge-circuit power supply is to deliver 300 V at 80 mA with not more than 2 percent ripple to the load. **(a)** What value of capacitance should be used in a simple capacitor filter? **(b)** What should the supply voltage to the rectifiers be?

15. Repeat Problem 14 for a half-wave rectifier with a load of 30 V at 30 mA and a permissible ripple of 4 percent.

16. A half-wave rectifier with capacitor filter is energized from a 120-V, 60-Hz supply. Its output, 300 V dc with a 10 percent ripple, is applied across a 15,000-Ω load. To reduce the ripple content, a series resistor of 800 Ω and a second shunt capacitor of 20 μF are added. Find the approximate value of **(a)** the new dc output across the load (assume that all the ripple content is at the first ac component frequency) and **(b)** the percent ripple with this R-C filter.

17. Repeat Problem 16 for a dc load of 60 mA in place of the 15,000-Ω load.

18. Repeat Problem 17 with the dc load of 60 mA but with the 800-Ω filter resistor replaced by a 12-H choke having a dc resistance of 80 Ω.

19. A center-tap rectifier with choke-input filter is supplied from a transformer rated at 280–0–280 V at 90 mA. What is the approximate dc output voltage at this load?

20. Give the approximate transformer secondary voltage rating needed to supply 350 V dc from the type of rectifier given above.

21. A bridge rectifier is used with choke-input filter. **(a)** Give the approximate transformer secondary voltage rating needed for a load voltage of 800 V. **(b)** If the rectifier is supplied from a 120-V, 60-Hz line, calculate the required transformer ratio.

22. The filter in Problem 21 uses $L = 5$ H and $C = 8\ \mu$F. Calculate the percent ripple in the output.

23. In the choke-input filter of Problems 21 and 22, if the maximum allowable ripple is 0.02 percent, what value of capacitor should be used with the 5-H choke?

24. A capacitor of 4 μF is used with a 1.75-H choke, in a choke-input filter, with full-wave rectification and 60-Hz supply. **(a)** Find the percent ripple in the output. **(b)** If each filter component is doubled in value, find the percent ripple. **(c)** Find the percent ripple using a two-stage filter with the original component values. **(d)** Compare the LC product of the basic filter in this problem with the "break-even" product of Review Question 33.

25. Repeat Problem 24(a) to (c), using 6 H and 10 μF.

26. A bridge rectifier with choke-input filter operating from a 60-Hz supply has an output voltage of 580 V. The load current varies from 30 to 850 mA. What is the minimum value of inductance needed to prevent cutout?

27. A choke-input filter is used with a center-tap rectifier energized from a 120-V, 60-Hz line. The choke has an inductance of 10 H. The load varies from 0 to 200 mA. What value of voltage-divider bleeder current must be used to prevent cutout if the dc output is 360 V?

28. A power supply has an output voltage of 450 V at 20 mA and 400 V at 150 mA. Find the percent regulation over this range.

29. Repeat Problem 28 for a full-load voltage of 580 V and a no-load voltage of 680 V.

30. A power supply has an output voltage of 600 V at its no-load condition. If its regulation is 12 percent, find the output voltage at full load.

31. Draw the circuit diagram for a voltage-regulator circuit using a zener diode and supplying a positive output voltage. Show input and output polarities.

32. Repeat Problem 31 for a negative output polarity.

33. It is planned to use a 1N4739 zener diode to obtain a regulated output of 9 V from an unregulated power supply. The load will vary between 5 and 40 mA. The unregulated voltage swings from 19 to 14 V over this current range. The diode is rated at 9.1 V and 1 W. Find: **(a)** the series resistance required for a circuit as in Fig. 6-12; **(b)** the power rating for this resistor (use a safety factor of 2). **(c)** Is the power rating of this diode adequate?

34. Repeat Problem 33 for a load of 20 to 100 mA, a load voltage of 180 V, and an unregulated dc input that drops from 300 to 220 V over this current range. The zener selected is a 1N5386 rated at 180 V and 5 W.

35. The unregulated power supply of Problem 33 (19 to 14 V) is to be used with zener diode 1N4742 (12 V, 1 W). The load, as before, will vary between 5 and 40 mA. Find (a) through (c) as in Problem 33.

36. Draw the schematic diagram for a series-pass feedback voltage regulator, using *NPN* transistors, for a positive polarity output.

37. Repeat Problem 36, using *PNP* transistors, for a negative polarity output.

7

Additional Power Supply Circuits

Previous chapters have covered the basic power supply circuits. However, there are other circuits in common use that have not been mentioned. Among them are several types of transformerless units and supplies used in mobile equipment. Each of these types is discussed in this chapter.

AC/DC SUPPLIES

The small radio receivers often used in kitchens and bedrooms are usually of the ac/dc variety, meaning that they operate from an ac or a dc supply line. To this same category can be added the three-way portables, which operate on internal battery supplies as well as from ac or dc lines. Can power transformers be used with this type of equipment? When used on a dc power line, there is no changing flux in the transformer winding, no reactance or induced counter EMF to limit the primary current, and the transformer will burn out. Obviously, then, power supplies for ac/dc receivers must be of the transformerless type. Since ac/dc equipment usually draws relatively light (approximately 50 mA maximum) and steady loads, half-wave rectifier circuits with R-C filters are generally used. The rectifier is supplied directly from the power line. Such a circuit is shown in Fig. 7-1.

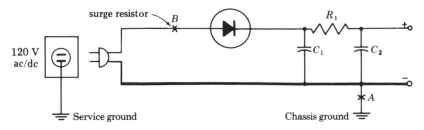

Figure 7-1 Ac/dc power supply.

When used on an ac power line, this circuit will deliver output regardless of which way the plug is inserted into the receptacle. The rectifier conducts on that half-cycle when its anode is positive with respect to cathode. However, there is a safety hazard inherent in all transformerless power supplies. In most communities, one side of the power line is grounded (connected to a water main or earth ground) at the service entrance. If the receiver is plugged in as shown in Fig. 7-1, the chassis ground and service ground are tied together and all is well. When the plug is reversed, however, the receiver still functions but now the chassis is connected to the "hot" side of the power line. Should the chassis accidentally touch a water pipe, radiator, or other grounded structure, a direct short circuit results. Possibly even more dangerous is that if a person touches the chassis with one hand while leaning on a radiator with the other, the circuit is completed through his body. In a number of cases the resulting electric shock has proved fatal.

This safety hazard can be reduced materially in these receivers by connecting all return lines to a common conductor or bus bar insulated from the chassis. Such a bus bar is shown in Fig. 7-1 as the heavy line. Then a capacitor is inserted between this bus and the chassis (at A in Fig. 7-1). The capacitor should have low reactance at the 60-Hz power line frequency (0.1 μF is a commonly used value).

Since a hazard does exist, you might justly wonder why this type of supply is used at all. This technique is used as a competitive measure—it saves the cost of a power transformer and it greatly reduces the size and weight of the complete unit. The power transformer would probably be the most expensive, heaviest, and bulkiest single item in such equipment.

On a dc power line, the power supply will function only if the line cord is plugged in with correct polarity, so that the anode of the rectifier is connected to the positive line lead. If the plug is reversed, the rectifier anode will be negative compared to cathode and the tube will not conduct. The grounding hazard, however, still exists because although some receptacles may have the negative side grounded, others may be connected to the opposite side of a three-wire system and will have the positive side grounded. Here again the capacitor between a common bus bar and the chassis will materially reduce the hazard. Another word of caution must be added when using semiconductor rectifiers in transformerless receivers. The semiconductor has low resistance. This can produce high instantaneous surge currents that can damage the rectifier and the capacitor. To limit these surge currents to safe values, it is necessary to add a series resistance between the line and the rectifier (at *B* of Fig. 7-1).* Commonly used values range from approximately 40 Ω for low-current supplies (20 mA) to less than 5 Ω for moderate-current supplies (200 mA). When power transformers are used, surge-limiting resistors are not necessary because the transformer winding has enough impedance to keep surge currents within safe values.

VOLTAGE-MULTIPLYING RECTIFIER CIRCUITS

Another group of power supply circuits that generally do not use a power transformer are known as *voltage-multiplying* circuits. Most common of these are the voltage doublers, triplers, and quadruplers. Unlike the ac/dc supplies, these circuits are usable only on an ac line. The basic principle in these circuits is to charge one or more capacitors during each half-cycle and then use the capacitor voltage either in series with the line voltage to boost the rectifier input voltage or in series with each other for increased output voltage.

FULL-WAVE VOLTAGE DOUBLER.   Figure 7-2 shows one type of voltage-doubler circuit: a full-wave voltage doubler.* To study the opera-

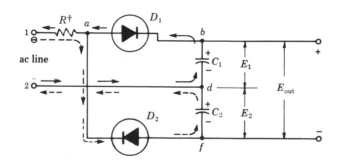

Figure 7-2 A full-wave voltage-doubler circuit.

*Special surge-limiting resistors are often used. They have a high resistance at room temperature that drops to a low value as they heat up. Typical trade names are *Surgistors* and *Glo-bar*.

†Resistor *R* (Fig. 7-2) is a surge-limiting resistor. Its use is desirable in any transformerless power supply.

tion of this circuit, assume that for the first half-cycle of input, line 1 is positive with respect to line 2. Current will flow (solid arrow) from line 2 to the junction between the two capacitors, up through capacitor C_1 to the cathode of D_1. Since D_1 has its anode positive with respect to its cathode, the diode conducts and current flows from cathode to anode and back to the positive line terminal 1. Since the resistance of this path is low, the charging time constant is low and capacitor C_1 *charges to the peak value* of the input voltage with polarity as shown.

During the next half-cycle of input, the line polarity is reversed. Terminal 1 is negative; terminal 2 is positive. Current flows (dotted arrow) from line 1 through D_2, up through capacitor C_2 and back to line terminal 2. In this time interval, C_2 charges to the peak value of the input voltage. Notice particularly the polarity of the charge on each of the two capacitors. Since in each case current flow is *up* through the capacitors, the voltages are additive and the output voltage E_o is equal to the sum of the voltages across each capacitor.

In the circuit of Fig. 7-2, no load is shown across the output of this rectifier. Capacitors C_1 and C_2 cannot discharge. What will the output voltage be? Since each capacitor is charged to the *peak* value of the input voltage, the output voltage will be equal to twice this peak value, or *almost three times the input voltage (rms)*. This condition is shown in Fig. 7-3. As

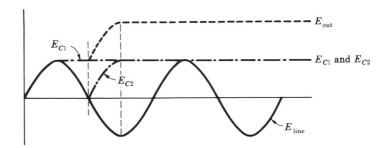

Figure 7-3 Waveshapes in a full-wave voltage doubler at no load.

load is applied, the capacitors will discharge, introducing a 120-Hz ripple (full wave) in the output voltage and also lowering the average value or direct-voltage output. The amount of such ripple and the value of the output voltage will vary with load. The higher the load, the lower the discharge time constant and, consequently, the greater the ripple and the lower the dc output voltage. Therefore, depending on load, the voltage doubler may have an actual output voltage of from almost triple to appreciably less than double the input voltage. To maintain a high output voltage, capacitors C_1 and C_2 are usually 40 μF or even larger.* Filtering is further improved

---

*The size of capacitor is limited by the peak current rating of the rectifiers.

by adding other filter elements as discussed in Chapter 6. However, since C_1 and C_2 act similarly to the input capacitor of the π-type filter, use of this type of circuit (and all voltage-multiplier rectifiers) should be limited to applications having moderate and steady load conditions.

CASCADE (OR HALF-WAVE) VOLTAGE DOUBLER.   Another voltage-doubler circuit is shown in Fig. 7-4. Since the voltage multiplying is done in steps—first one capacitor is charged and then its charge is used in turn to charge another capacitor—the circuit is known as a *cascade* voltage doubler. To analyze the action in detail, start with line 1 negative compared with line 2. Current (solid arrows) will flow from line 1 through C_1, through rectifier D_1, back to the positive line 2. Capacitor C_1 will charge to the peak value of the input line voltage, with polarity as shown. Meanwhile, rectifier D_2 is biased with reverse polarity and cannot conduct. Capacitor C_2 does not charge.

During the next half-cycle line 2 is negative. Diode D_1 cannot conduct. Current flow (dotted arrow) is from line 2 through capacitor C_2, rectifier D_2, capacitor C_1, and back to line 1 (which is now positive). Two voltages are effective in this path. We can check this with reference to Fig. 7-4. The line polarity for this half-cycle is marked. Starting at the midpoint of the rectifiers and tracing to the left, we pick up + to − across C_1 and again + to − for the line voltage. Therefore capacitor C_2 must charge to twice the peak value of the input voltage. Since capacitor C_2 is charged only on alternate half-cycles, this circuit is also known as a *half-wave voltage doubler*. As for the previous circuit, when load is applied the output voltage decreases and the ripple content increases.

On a comparison of the two doubler circuits, the full-wave circuit should give better regulation and should be easier to filter (higher frequency components). Yet this discharge circuit contains two capacitors in series, so that for equal values of C and load the full-wave circuit has a lower discharge time constant. This in turn means a faster discharge rate and offsets the advantages of full-wave operation. Consequently, the two circuits have approximately equal characteristics. One disadvantage of the cascade circuit is that capacitor C_2 must have a voltage rating of at least twice the

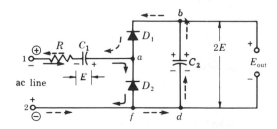

Figure 7-4   Cascade voltage-doubler circuit.

peak input voltage, whereas in Fig. 7-2 the rating of either capacitor need only equal the peak line voltage. On the other hand, the cascade circuit of Fig. 7-4 has a common input and output terminal. In some cases, this feature is a decided advantage.

VOLTAGE TRIPLERS AND QUADRUPLERS.   The principle of voltage doubling can be extended theoretically to any number of capacitors and rectifiers to get higher and higher output voltages. However, practical considerations regarding capacitor ratings and regulation usually limit the application to *triplers* and *quadruplers*. Figure 7-5 shows one form of

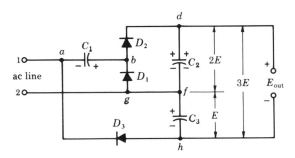

Figure 7-5   Voltage-tripler circuit.

voltage-tripler circuit. The portion of this circuit containing capacitors C_1 and C_2 and rectifiers D_1 and D_2 is the cascade doubler discussed above. This time, in addition, during the half-cycle when C_1 charges, C_3 also charges through rectifier D_3. Since the load is connected across the combination of C_2 and C_3, the output voltage (at light loads) approaches three times the peak input voltage.

By using two cascade voltage doublers with their outputs in series, the total output voltage can be made to approach four times the peak value of the ac input. Such a voltage quadrupler is shown in Fig. 7-6, where C_1

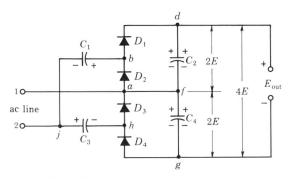

Figure 7-6   Voltage-quadrupler circuit.

and C_2 with their associated rectifiers D_1 and D_2 form one cascade doubler while C_3 and C_4 in conjunction with D_3 and D_4 form the second doubler circuit.

Notice that neither of these last two circuits (Figs. 7-5 and 7-6) has a common input and output line. When such a feature is desired, it can be obtained with a slight circuit modification as shown in Fig. 7-7. This circuit is an extension of the general cascade principle started in Fig. 7-4 for the doubler. There is a penalty for this feature: with each added stage of multiplication, the voltage rating for that capacitor increases. At the tripler stage, for example, the rating for C_3 must be equal to at least three times the peak input voltage, and for the quadrupler stage C_4 must be capable of withstanding four times the peak value of the input voltage. Also, since the charge on the final (or output) capacitor is dependent in turn on the charge and discharge of each and every previous capacitor, the regulation is not as good as for the circuits shown in Figs. 7-4 and 7-5, respectively.

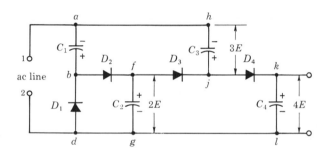

Figure 7-7   Voltage multiplier with common input and output line.

COLOR TV HIGH-VOLTAGE MULTIPLIER.   The concern over X-ray radiation led to TV receiver designs that eliminated the two culprits—the high-voltage rectifier and the shunt regulator tubes. Instead, voltage-multiplying circuits are now commonly used. A typical circuit is shown in Fig. 7-8. Notice that it consists of three sections, each of which is a cascade (half-wave) doubler circuit. The input signal, a 15,750-kHz pulse wave from the horizontal output transformer, is coupled through the upper capacitors to each section. On the negative portion of the wave, the odd-numbered diodes conduct charging capacitors C_1, C_3, and C_5 with polarities as shown. On the positive portions, the even-numbered diodes conduct charging capacitors C_2, C_4, and C_6 to the peak-to-peak value of the input signal. For an 8.3-kV peak-to-peak input, the dc output voltage is 25 kV. By taking an output from the first section, this circuit also supplies dc for the focus electrode of the cathode-ray tube.

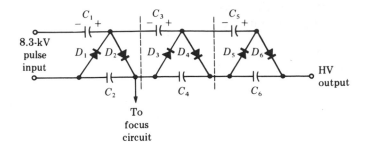

Figure 7-8 A color TV high-voltage circuit.

MOBILE UNITS—POWER SUPPLIES

All the power supplies discussed up to this point obtain their primary energy from the ac power line. Such a source is not always available. You cannot connect the police radio in a roving patrol car to the local power company lines. The same applies to pleasure cars, trucks, small boats, and aircraft. However, these vehicles are equipped with low-voltage battery power—from 6 and 12 V in cars to 28 V in many aircraft. For most transistorized equipment, this may be completely adequate. For high-power transistors and for vacuum-tube equipment, these voltages would not be suitable. The problem in mobile equipment is therefore one of supplying an adequate source for high-voltage dc power. Three techniques have been used: the dynamotor, the vibrator power supply, and the transistorized power supply. Each of these is discussed below.

Dynamotor.   The dynamotor is an electric machine similar to a motor or generator. In fact it is really a combination of the two within a common frame. This machine is provided with two armature windings on a common core, rotating in a single field structure. Each winding has its own commutator and a separate set of brushes. The commutators are located one at each end of the armature shaft. One armature winding has relatively few turns for operation from a low-voltage source and is intended as the motor armature. The second armature winding (the generator winding) has many more turns (depending on the output voltage desired).

When the field winding and the brushes of the low-voltage armature winding are connected to the battery supply (6, 12, or 28 V), this armature will operate as a motor. At the same time, the second armature also rotates within the magnetic field, and a voltage is generated in this winding. Since both armatures rotate at the same speed (S) and must cut the same flux (ϕ), the induced voltage in each winding is directly proportional to its respective number of turns. Also, neglecting the small $I_a R_a$ drop in the windings, the ratio of output voltage to battery voltage is

$$\frac{\text{dc output}}{\text{dc input}} = \frac{N_2}{N_1}$$

To obtain a desired supply voltage, the turns ratio of the dynamotor windings must be equal to the voltage ratio. For any given unit, this ratio is fixed by the original design. Dynamotors are available for operation from any of the common battery sources and for a wide variety of output voltage and current ratings.

VIBRATOR SUPPLIES.   The basic principle in any vibrator supply is to apply the low-voltage direct current to the primary of the power transformer. However, the current is interrupted frequently so as to cause the magnetic field (flux) to rise and fall rapidly—much the same as would happen if we had used an ac supply. Then, as with true ac operation, the changing flux induces a voltage in the transformer secondary winding. By using a sufficient number of turns in the secondary, any desired secondary voltage can be developed. This voltage is in turn rectified and filtered as in the previous circuits. Unfortunately, vibrators have an inherent weakness: the interruption of current as the contacts open and close causes arcing and pitting of the contacts. Malfunctions—due to contacts sticking either closed or open—are quite common. Consequently, with the development of solid-state "choppers," vibrator power supplies have fallen into disuse.

TRANSISTORIZED POWER SUPPLIES.   Both types of mobile power supplies discussed above have their drawbacks. In the vibrator types, the vibrator itself is the weakest link. The unit has a relatively short life, while the sparking "hash" it produces necessitates thorough shielding of the power supply as well as the electronic equipment. The dynamotor, although more rugged, requires appreciable servicing of brushes and bearings; it also causes spark interference due to its commutator and is in addition noisy, bulky, and quite expensive. The development of power transistors gave rise to a new technique in mobile power units, one which eliminates these drawbacks. Transistorized power supplies use no tubes, have no moving parts, and are compact and lightweight. Operating advantages claimed for transistorized power supplies are improved reliability, ruggedness, high efficiency (up to 90 percent), and easier filtering. In these power units, the transistors (two) are used as switches to alternately reverse the direction of current flow through the transformer primary winding. In other words, the transistors perform the same function as the vibrator. Since there are no mechanical moving parts, much higher switching rates are possible.

"Frequencies" of as high as 3500 Hz have been used. These higher "line" frequencies reduce the bulk and cost for the power transformer and filter components.

Figure 7-9a shows a typical circuit for a transistorized power supply using *PNP* transistors. The dc aspects of this circuit as they apply to each transistor can be seen more readily from the simplified diagram of Fig. 7-9b. Notice particularly that the emitter is connected to the positive battery terminal, the collector to the negative terminal; the base, through the voltage-divider network (R_1, R_2), is slightly less positive than the emitter. Since the transistors are of the *PNP* variety, the emitter-base junctions are forward-biased whereas the collectors are reverse-biased. These are the normal polarities for transistor operation; therefore, as far as the dc aspects are concerned, both transistors would conduct. Resistors R_1 and R_2 are chosen so as to produce the proper bias for heavy conduction. (They are not equal.)

Now consider the action of the full circuit to see the effect of the transformer windings. When Q_1 conducts, current (electrons) will flow *down* through the upper half of the transformer primary winding (L_1). Conduction of the other transistor Q_2 causes current to flow *up* through the lower half of the primary. If the transistor emitter currents were *exactly* equal (and the transformer windings perfectly balanced), the two fluxes produced by each of these currents would be equal and opposite. The net flux would be zero. However, this is never the case. No two transistors are ever so perfectly matched. (This is also true of vacuum tubes.) For this discussion let us assume that Q_1 tends to conduct just the least bit more.

As soon as the circuit is energized, because of the higher conduction of Q_1 an unbalanced current flows in the transformer primary and a net

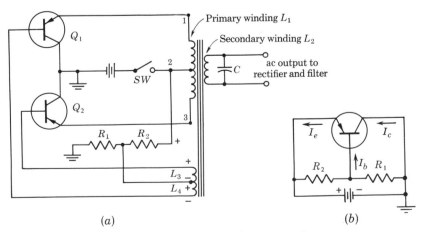

(a) (b)

Figure 7-9 Transistorized power supply.

flux is produced. Since this current is increasing—each emitter current is rising from zero toward full conduction—the net flux is also increasing. The changing flux induces a voltage in the "feedback" winding, $L_3 + L_4$. The feedback coil is so phased with respect to this flux as to make the polarity of the induced voltage positive on top and negative on bottom. Notice also that the L_3 portion of this winding is in series *opposition* with the base bias for Q_2. No matter how small this induced voltage, it will reduce the base-bias current and reduce the emitter and collector currents. Simultaneously, the L_4 portion of the feedback winding is phased series-*aiding* with the base bias for Q_1. This transistor is driven into heavier conduction. The combined effect increases the unbalanced current in the transformer primary. Furthermore, the net flux increases; the induced voltage in the feedback winding increases; Q_2 is driven further toward cutoff and Q_1 into heavier conduction. The process is cumulative and almost instantaneous.

As the transformer approaches saturation, the induced voltage in the feedback winding levels off. Also, as the transistor approaches saturation, the emitter current ceases to rise, making the unbalanced primary current steady. What happens to the feedback voltage? If the current is steady, the flux is steady and the induced voltage drops to zero.

This drop in feedback voltage brings Q_2 out of cutoff and reduces the conduction of Q_1. The primary unbalanced current decreases; the net flux decreases; and the *induced voltage* reverses. This time the induced voltage forward-biases transistor Q_2. Again the action is cumulative. Q_2 is driven to heavy conduction and Q_1 into cutoff. This sequence of events is repeated again and again. The waveshape of the induced voltage approaches a square wave.

This action also produces a square-wave alternating voltage across L_2, the secondary winding. The required high voltage is obtained by using the proper step-up ratio. The ac output is then rectified and filtered. A semiconductor bridge-type rectifier is most commonly used. The capacitor C across the secondary suppresses the excessively high-voltage transient spikes that might result from the rapid reversals of the square-wave currents.

SILICON-CONTROLLED RECTIFIERS

A diode conducts whenever its anode is positive compared to its cathode. Current flow is in one direction only—from cathode to anode—and so if the input is ac it acts as a rectifier. In this respect, the silicon-controlled rectifier (SCR) also acts as a diode. However, the SCR has a third element, a gate, that controls whether or not the SCR will conduct even when the anode is positive. Figure 7-10 shows the constructional features and the

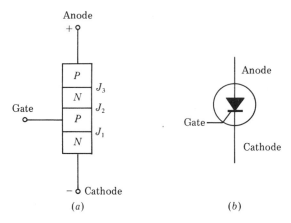

Figure 7-10 Silicon-controlled rectifier.

schematic diagram of an SCR. Notice that this is a four-layer, three-junction device. When connected with the polarity as shown in Fig. 7-10a, junctions 1 and 3 are forward-biased but junction 2 is reverse-biased. Current cannot flow. The SCR is in the off state even though the anode is positive.

Now let us feed a small positive trigger pulse to the gate, just enough to overcome the threshold voltage for the cathode-to-gate P-N junction (approximately 0.6 V for a silicon junction). A small current (I_G) will flow from cathode to gate. But this is similar to a forward-biased transistor (with the gate as the base), and an amplified current αI_G will flow through. A feedback effect takes place from the action of the upper three layers (PNP), and the SCR is driven deep into conduction. Once conduction has started, the gate no longer has control over conduction. Conduction can be stopped only if the anode voltage is reduced almost to zero (so that the current falls below some minimum holding value I_H) or is reversed.*

The great advantage of an SCR over a plain diode is its ability to control large amounts of power delivered to a load, with only a minimum of gate-control power. This is achieved by varying the point (in each cycle) when the SCR conducts. For example, the voltage and power delivered to the load are decreased by delaying the application of the trigger pulse to the gate until late in the input voltage cycle. (This can be done with a phase-shift trigger circuit.) This reduces the average voltage and power applied to the load. Obviously, the SCR is a half-wave rectifier. Full-wave action can be obtained by using two SCRs or by using a bidirectional thyristor† such

---

*Anyone familiar with tube technology will recognize the similarity between the SCR and the thyratron gas triode.

†Thyristor is the generic name for all solid-state controlled rectifiers. (The SCR is one type of thyristor.)

as the *triac*. Since thyristors are essentially used in industrial electric controls (lights, motors, etc.), they are not discussed further in this text.

HIGH-FREQUENCY POWER SUPPLIES

In the late 1960s and early 1970s a new trend in power-supply design began. Although this design is more complex, the end product is smaller, lighter, and cheaper. It can be used either with 60-Hz power lines or with battery sources. Starting with a 60-Hz line, the ac power is first rectified. The dc is then chopped by using power transistors or SCR switches—but the switching, or square waves so produced, is at a rate of 7 to 25 kHz. This higher-frequency ac is applied to the power transformer for step-up or step-down to the desired voltage level. This is where the first cost, size, and weight saving is effected. At this high frequency, a transformer with ceramic, ferrite, or powdered-iron core is used. The output from the secondary is then rectified and filtered in the usual manner. The filter component values, at this frequency, are also very much smaller. For example, at 24,000 Hz a capacitor (or choke) "400 times smaller" will give the same filtering action as its 60-Hz brother (that is, a 0.1 μF in place of a 40 μF). Very efficient regulation is readily obtained with this type of power supply by controlling the pulse width of the square waves. Overvoltage protection (SCR crowbar) is also readily applicable. A TV manufacturer, using this technique in a 1972 model, reduced the weight of his power supply from 13 to $1\frac{3}{4}$ lb.

REVIEW QUESTIONS

1. **(a)** What is meant by an ac/dc receiver? **(b)** Can a power transformer be used in equipment intended for ac/dc operation? Explain.

2. **(a)** In the circuit shown in Fig. 7-1, does a safety hazard exist? Explain. **(b)** If the line plug is reversed, what effect will it have on safety? Explain. **(c)** How can the safety hazard be materially reduced, regardless of plug polarity?

3. Does the safety hazard of Question 2 exist when the equipment is used **(a)** on an ac line? Explain. **(b)** On a dc line? Explain.

4. In Fig. 7-1, what is the effect of plug polarity on the output voltage if the circuit is used **(a)** on an ac line? Explain. **(b)** On a dc line? Explain.

5. **(a)** When semiconductor diodes are used in transformerless power supplies, state a reason for high surge currents. **(b)** How is this effect minimized? **(c)** Does this effect occur when power transformers are used? Explain.

6. **(a)** What is a *voltage-multiplying* circuit? **(b)** Are these circuits used on ac, on dc, or on either type of power source? **(c)** Name three commonly used multiplying circuits.

7. Refer to Fig. 7-2: **(a)** Trace the current flow during the half-cycle that line 1 is positive. **(b)** Will current flow through D_1? Explain. **(c)** Trace the current flow for the other half-cycle. **(d)** To what value does each capacitor charge? **(e)** For the load shown, compare the output (dc) voltage with the rms value of the input voltage.

8. (a) Why do voltage doublers have poor regulation? **(b)** Can such circuits be used with choke-input filters to improve regulation? Explain.

9. Refer to Fig. 7-4: **(a)** Why is it called a *cascade* doubler? **(b)** During which half-cycle of input does capacitor C_1 charge? **(c)** Through which rectifier does it charge? **(d)** Explain why the polarity is as shown. **(e)** To what value does capacitor C_1 charge? **(f)** Trace the current flow when line 1 is positive. **(g)** What voltages are effective in producing conduction during this half-cycle? **(h)** To what value does capacitor C_2 charge? Why?

10. (a) Give one distinct advantage of the cascade doubler over a full-wave doubler. **(b)** Give a distinct disadvantage.

11. Refer to Fig. 7-5: **(a)** During which cycle of input does capacitor C_1 charge? **(b)** Trace the current flow. **(c)** During which cycle of input does capacitor C_2 charge? **(d)** Trace the current flow. **(e)** During which cycle of input does capacitor C_3 charge? **(f)** Trace the current flow.

12. Refer to Fig. 7-6: **(a)** What basic type of multiplier circuit is being used? **(b)** How many such circuits are used? **(c)** Identify the components pertaining to each.

13. Refer to Fig. 7-6: **(a)** Trace the current flow for the half-cycle of input when capacitor C_1 charges. **(b)** Repeat for capacitor C_2. **(c)** Repeat for capacitor C_3. **(d)** Repeat for capacitor C_4.

14. (a) What advantage does the quadrupler circuit of Fig. 7-7 have over the circuit of Fig. 7-6? **(b)** What disadvantage does the circuit have with respect to the components needed? **(c)** Give another disadvantage.

15. Refer to the circuit of Fig. 7-7: **(a)** Trace the current flow, showing how capacitor C_1 charges. **(b)** Repeat for capacitor C_2. **(c)** Repeat for capacitor C_3. **(d)** Repeat for capacitor C_4.

16. Refer to Fig. 7-8: **(a)** What is the basic circuitry used here? **(b)** Is this a doubler, a tripler, or a quadrupler circuit? **(c)** To what value do capacitors C_2, C_4, and C_6 charge? **(d)** What voltage is fed to the focus coil? **(e)** What is the output voltage at the HV terminal?

17. (a) Describe a dynamotor. **(b)** What is it used for? **(c)** Can the output voltage from its generator armature be varied? Explain.

18. Power supplies in mobile radios may use power transformers. Yet these units are energized from a dc source. Explain why the transformer does not burn out.

19. (a) In a transistor power supply, how many transistors are used? **(b)** What is their specific function? **(c)** What switching rates are used? **(d)** State two advantages of this switching rate. **(e)** State two other advantages of this type of power supply. **(f)** What additional semiconductors are generally used in these power supplies? **(g)** For what function? **(h)** How many?

20. Refer to Fig. 7-9: **(a)** What types of transistors are used? **(b)** What is the polarity of the emitters compared to base? **(c)** How is this polarity obtained? **(d)** What is the polarity of the collectors compared to base? **(e)** Will these transistors normally conduct? **(f)** What is the function of the windings L_3 and L_4? **(g)** How are the voltages across these windings obtained? **(h)** With the polarities shown, what is the effect of winding L_3? **(i)** What is the effect of winding L_4? **(j)** Explain how switching action is obtained. **(k)** What voltage waveshape is developed across winding L_2? **(l)** What is the function of capacitor C?

21. **(a)** What does the abbreviation SCR stand for? **(b)** How many elements does it have? **(c)** Name the elements. **(d)** Does an SCR function essentially as a diode or as a transistor?

22. Refer to Fig. 7-10: **(a)** How many P-N junctions does the SCR have? **(b)** Will the SCR conduct if the anode-cathode polarity is reversed from that shown? **(c)** Will it conduct with the applied voltage shown? Explain. **(d)** What else must be done to turn the SCR on? **(e)** A positive polarity signal is applied to the gate to make the SCR conduct. The gate circuit is now opened. What happens to the SCR conduction? Explain. **(f)** How can conduction be stopped?

23. With reference to an SCR, what is meant by the term *holding current*?

24. **(a)** Using a single SCR, can we construct a full-wave rectifier circuit? Explain. **(b)** How many SCRs would be required for such duty? **(c)** Name a solid-state device that can give control and full-wave operation using only one unit.

25. **(a)** What is a *thyristor*? **(b)** Name two specific thyristors.

PROBLEMS AND DIAGRAMS

1. During the servicing of an electronic unit using a voltage doubler and operating from a 120-V line, the positive supply lead was disconnected and the power supply unit was checked for output. The voltmeter indicated 336 V dc. Account for this high voltage.

2. Draw the circuit diagram for a full-wave voltage doubler. Mark the polarity of charge on each capacitor.

3. Draw the circuit diagram of a cascade voltage-doubler power supply with R-C filter.

4. Draw the circuit diagram for a voltage tripler.

5. Draw the circuit diagram for a voltage quadrupler.

6. Draw the circuit diagram for a transistor power supply operating from a 6-V battery and using *NPN* transistors.

7. Draw the schematic symbol for an SCR and label the elements.

8

The Decibel

Back in the early days of telephony, communication engineers discovered that the human ear responds in a logarithmic manner to changes in sound levels. That is, to double the apparent sound level (intensity) from a sound source, the acoustic power level must be squared. Similarly, to increase a sound intensity to three times its previous sensation, the sound power level must be cubed. In other words, the increase in acoustic power necessary to create the impression of a change in sound intensity depends not only on the actual *amount of change* but also on the *original sound level*. For example, increasing the power level of a sound source from 2 W to 32 W makes the sound seem not sixteen times louder but only five times louder ($2^5 = 32$). However, using the same *amount* of change (30 W), but from an original power level of 6 W, the sound will now be only twice as loud ($6^2 = 36$).

ORIGIN OF THE DECIBEL. A desirable unit for sound-level measurements should reflect this exponential or logarithmic relationship. The unit adopted by telephone engineers was made equal to the logarithm of the ratio of the power levels involved. In honor of Alexander Graham Bell, inventor of the telephone, this unit was named the *bel*. Expressed mathematically,

$$\text{bel} = \log \frac{P_2}{P_1} \qquad \text{(8-1)}$$

In practical applications, it was found that the bel was too large a unit (similar to using the ton for measuring household grocery weights).

186

It was therefore decided to create a smaller unit, the *decibel* (dB), so that 10 dB would be needed to equal 1 bel. Since the bel $= \log P_2/P_1$, it follows that

$$dB = 10 \log \frac{P_2}{P_1} \qquad \text{(8-2)}$$

where dB is the power gain (or loss) in decibels. Incidentally, to get a physical notion of the decibel, a difference of 1 dB between two sound intensities is just about the minimum change noticeable by ear.

USE OF DECIBELS.   From Equation 8-2, it should be obvious that the decibel is a *relative* unit of measurement used to compare two sound power levels. However, the use of decibels has expanded to many fields. The unit is used to describe the performance of amplifiers (not just audio amplifiers, but all types), microphones, antennas, filter networks, transformers, and other types of equipment or components. Since use of this term is so widespread, it is important to learn to use it properly—otherwise serious confusion can arise. One cause of such trouble is failing to recognize the dual use of the decibel for:

1. Comparison of two power levels (gain or loss)
2. Power rating of equipment or components (using some arbitrary level as the reference)

To add to the possible complications, *voltage* gains are often also expressed in decibel units. All these complications are discussed in the pages to follow, but since decibels involve logarithms, a brief review of the pertinent mathematics is desirable first.

LOGARITHMS—BASIC PRINCIPLES.   Logarithms and exponential values are closely related. Earlier in this chapter the expression $2^5 = 32$ was used. This is an exponential term meaning that if 2 is raised to the fifth power, the resulting number is 32. Let us call 2 the *base,* 5 the *exponent* to which this base 2 will be raised, and 32 the *number.* Expressing this same relation in logarithmic terms,

$$\log_2 32 = 5$$

This is read as: "The logarithm of 32 to the base 2 is 5." As before, this still means that 2 raised to the fifth power equals 32.

A little later in the introduction to decibels, $6^2 = 36$ was referred to. Expressed as a logarithm, 6 is the base, 2 is the exponent, and 36 is the number, or

$$\log_6 36 = 2$$

This is read as: "The logarithm of 36 to the base 6 is 2."

EXAMPLE 8-1

Convert each of the following exponential terms into logarithmic expressions.

1. $2^6 = 64$
2. $4^3 = 64$
3. $8^2 = 64$

Solution

1. $\log_2 64 = 6$
2. $\log_4 64 = 3$
3. $\log_8 64 = 2$

Example 8-1 brings out another point. A number can be expressed as any logarithm *merely by changing the base*. This would give rise to an infinite number of systems and would require an infinite number of log tables for solutions. In practice only two systems are in use: (1) the Napierian or natural system invented by Lord John Napier, a Scottish astronomer, which uses 2.71828 as its base,* and (2) the common or Briggs system, which uses 10 as its base.

Since decibels are used with the common system of logarithms, the remainder of this discussion deals only with this system. With specific reference to the common system, to find the logarithm of a number means "to find the exponent to which the base 10 must be raised in order to equal the given number." Obviously, the logarithms of some simple numbers are directly related to powers of 10.

EXAMPLE 8-2

Find the logarithms of each of the following numbers, using base 10:

1. 1,000,000   4. 100
2.   100,000   5. 10
3.   1000   6.   1

Solution

The first step is to express the given numbers in powers of 10. Once this is done you will immediately realize that you have the base 10; the power of 10 is the exponent and therefore the log.

1. $1,000,000 = 10^6$; $\log 1,000,000 = 6$
2. $100,000 = 10^5$; $\log 100,000 = 5$
3. $1000 = 10^3$; $\log 1000 = 3$
4. $100 = 10^2$; $\log 100 = 2$
5. $10 = 10^1$; $\log 10 = 1$
6. $1 = 10^0$; $\log 1 = 0$

*This value is commonly known as base *e*.

It should be obvious from Example 8-2 that the logarithms of numbers between 1 and 10 must be decimal values between 0 and 1. Similarly the logs of numbers between 10 and 100 must be mixed numbers between 1 and 2. Between what limits would you find the logarithm of 37,162? Expressing this number in powers of 10, we get 3.7162×10^4. Its logarithm would therefore be some number between 4 and 5.

Now all we need to know are the logarithms for numbers between 1 and 10 to whatever accuracy is desired. This data has been calculated and is available in tables of logarithms. (The technique for establishing such tables is not pertinent to our discussion.) Logarithms for numbers between 1 and 10 can also be obtained from the *D-to-L* scale of any slide rule. Remember that the logarithms for numbers between 1 and 10 are always less than unity. This portion of a logarithm is called the *mantissa*. The portion of the logarithm found directly from the power of 10 is called the *characteristic*.

EXAMPLE 8-3

Find the logarithm of each of the following numbers:

1. 27.3
2. 27,300
3. 8910

Solution

In each case apply the two steps. Rewrite in powers of 10 to get the characteristic; then find the mantissa by using slide rule or tables.

1. $27.3 = 2.73 \times 10^1$
   The characteristic is 1.
   The mantissa for 2.73 is 0.4362 (table).
   log 27.3 = 1.4362
2. $27,300 = 2.73 \times 10^4$
   The characteristic is 4.
   The mantissa is again 0.4362.
   log 27,300 = 4.4362
3. $8910 = 8.910 \times 10^3$
   The characteristic is 3.
   The mantissa for 8.910 is 0.95 (slide rule).
   log 8910 = 3.95

If we know the logarithm of a number, how can we find the number itself? This is called finding the *antilog* and is merely the reverse of the previous procedure. The antilog for any logarithm between 0 and 1 must be a number between 1 and 10. Such numbers can be found directly from a table or slide rule. If the logarithm is greater than unity, the characteristic tells you the power of 10 that must be used as a multiplier.

EXAMPLE 8-4

Find the antilog of
1. 0.8463
2. 2.290

Solution

1. Corresponding to the mantissa 0.8463 the number is 7.02. Since the characteristic is 0, the multiplier is 10^0 or 1. Therefore the antilog of 0.8463 is 7.02.
2. Antilog 2.290: characteristic $2 = 10^2$
   mantissa $0.290 = 1.95$
   antilog $= 1.95 \times 10^2 = 195$

POWER GAIN OR LOSS IN DECIBELS.   The basic application for the decibel unit is in the evaluation of power gain or loss. For example, if a certain type of filter is inserted in an electronic unit, will the power output increase or decrease? An additional stage is added to an amplifier. By how much will the power increase? A new antenna is used with a radar transmitter. How much more power is directed into a specific direction? Evaluating these power changes as a ratio of old power level to new power level (or vice versa), finding the logarithm of this ratio, and multiplying this log by 10 will give us the power gain or loss in decibels. Obviously, if the power level increased, there was a gain; if the power level decreased, there was a loss. To represent the loss, the decibel figure is preceded by a negative sign; for example, -12 dB.

Earlier in the chapter the equation for decibels was given as

$$dB = 10 \log \frac{P_2}{P_1}$$

P_1 and P_2 were not identified. This time we will be specific by adding "where P_2 is always *the larger power level*."* Gain or loss can be handled by common sense, and the proper sign is then added.

EXAMPLE 8-5

Tests at a UHF relay station showed that the power directed toward the main station increased from 2 W with a basic dipole antenna to 124 W with a parabolic antenna. What is the power gain or loss (in decibels) of the parabolic antenna?

Solution

1. Since the power increased, it is a gain.

2. $dB = 10 \log \dfrac{P_2}{P_1}$

*This technique avoids the mathematical complication of finding the logarithm of numbers smaller than unity.

$$= 10 \log \frac{124}{2}$$

$$= 10 \, (1.792)$$

$$= 17.9 \, \text{dB}$$

EXAMPLE 8-6

The insertion of a vestigial sideband filter in a television transmitter reduced the power output from 40 kW to 32 kW. Find the insertion loss in decibels.

Solution

1. Since the output decreased, there is a loss and the final answer should be marked negative.

2. $dB = 10 \log \dfrac{P_2}{P_1}$

$$= 10 \log \frac{40}{32}$$

$$= 10 \, (0.097)$$

$$= -0.97 \, \text{dB}$$

POWER GAIN FROM VOLTAGE OR CURRENT.   Since power, voltage, current, and resistance are interrelated, the power gain (or loss) can be calculated from voltage or current values if the resistance of the associated circuit is known. The procedure can be understood most readily if we use two steps:

1. Calculate P_1 and P_2, using E^2/R if voltages are known or I^2R if current values are known.
2. Once the two power levels are computed, use the technique already established for finding decibel gain or loss.

A simplified technique can be used if the components for which the powers P_2 and P_1 are calculated are identical units or have equal resistances. Since power is proportional to the square of the voltage,

$$\frac{P_2}{P_1} = \left(\frac{E_2}{E_1} \right)^2$$

and

$$dB = 10 \log \frac{P_2}{P_1}$$

$$= 10 \log \left(\frac{E_2}{E_1} \right)^2$$

But the logarithm of any value squared is equal to twice the logarithm of the value itself, or

$$10 \log \left(\frac{E_2}{E_1} \right)^2 = 20 \log \frac{E_2}{E_1}$$

and

$$\textit{Power gain} \text{ in dB} = 20 \log \frac{E_2}{E_1} \qquad \text{(8-3)}$$

Notice the italics in the last equation. *Voltage ratios* are used to calculate *power gain* in decibels. This equation applies only when *the resistance values are equal.*

EXAMPLE 8-7

A power amplifier (with a 16-Ω output impedance) has an output voltage of 24 V at 1000 Hz and only 16 V at 40 Hz. Using the 1000 Hz as reference, calculate the decibel (power) gain or loss at 40 Hz.

Solution

1. Since the voltage at 40 Hz decreased, there is a loss.

2. (*a*) $P_2 = \dfrac{E_2^2}{R} = \dfrac{24 \times 24}{16} = 36 \text{ W}$

   $P_1 = \dfrac{E_1^2}{R} = \dfrac{16 \times 16}{16} = 16 \text{ W}$

   $\text{dB} = 10 \log \dfrac{P_2}{P_1} = 10 \log \dfrac{36}{16} = -3.52 \text{ dB}$

(*b*) Since the output voltage in each case is measured across the same impedance, the simplified formula can be used:

$$\text{dB} = 20 \log \frac{E_2}{E_1} = 20 \log \frac{24}{16} = -3.52 \text{ dB}$$

Example 8-7 shows that in either case the same answer is obtained for the power loss in decibels. Yet notice how much simpler the voltage ratio technique is in this case. Remember, however, that this equation—power gain or loss in decibels—can be used only if the resistance values (across which E_2 and E_1 are measured) are equal.

By similar development it can be shown that power gain can also be calculated from

$$\text{dB} = 20 \log \frac{I_2}{I_1} \qquad \text{(8-4)}$$

As before, this equation applies only if the resistances through which the currents I_2 and I_1 flow are equal.

EQUIPMENT SPECIFICATIONS OR RATINGS IN DECIBELS. It is quite common for manufacturers of electronic equipment or components to rate the power output or power-handling ability of their products in decibel units. For example, a power amplifier may have an undistorted power output rating of $+37$ dB, a microphone may have an output rating of -74 dB, or an audio transformer may have a power-handling ability of $+32$ dB. What do these figures signify? Since the term *decibel* is normally used to compare *two* values, the other value with which these figures are being compared must be understood. Actually, it is the *reference* or *zero-level* value. Here is another complication. This reference value could be 1 mW; it could also be 6 mW, or even 12.5 or 50 mW. In fact, with microphones it could also be 1 V. At one time 6 mW was the most common reference level. In recent years, the shift has been toward the 1-mW reference level. However, *any* value *may* be used. Good practice would be to state the zero level used. If doubt exists, the manufacturer should be consulted. In this text, unless otherwise stated, the 1-mW reference level is used for 0 dB. One technique which is being adopted to reduce misunderstandings is to use "dBm" to signify this reference level and "dB 6m" for the older 6-mW reference level. Similarly, "dBV" (for microphones) is often used to signify decibels referred to 1 V. (More is presented on microphones in Chapter 17.)

A few examples should clarify this type of application.

EXAMPLE 8-8

A power amplifier is rated at $+44$ dB. Find its rated power output.

Solution

1. The reference level is understood as 1 mW.
2. Since the rating is positive, the output is greater than the reference level. Therefore, in the decibel equation, output is the numerator and reference power is the denominator.

3. $10 \log \dfrac{P_o}{P_r} = 44$

$\log \dfrac{P_o}{P_r} = 4.4$

$P_o = P_r(\text{antilog } 4.4) = 0.001(2.51 \times 10^4)$

$ = 25.1 \text{ W}$

EXAMPLE 8-9

A high-impedance (crystal) microphone has a rating of -54 dBV. Find its open-circuit output voltage.

Solution

1. The reference level is 1 V (for a sound pressure of 1 dyne per square centimeter).
2. Since the rating is a negative decibel value, the output level is *less* than the reference level. It is therefore the denominator in the decibel equation.
3. Since voltage is used for reference,

$$20 \log \frac{E_r}{E_o} = 54$$

$$\log \frac{E_r}{E_o} = 2.7$$

$$\frac{E_r}{E_o} = \text{antilog } 2.7$$

Inverting and transposing,

$$E_o = \frac{E_r}{\text{antilog } 2.7} = \frac{1}{5.01 \times 10^2}$$

$$= 0.002 \text{ V}$$

Another application area that uses voltage as a reference is the cable TV (CATV) and master antenna (MATV) fields. These industries agree that 1 mV is a pretty good signal to feed to a TV receiver, and they accordingly design their master antenna and cable systems to deliver at least a 1-mV signal level (per channel) to each TV receiver in the system. To simplify their gain and loss calculations, they use decibels, with 1000 μV (that is, 1 mV) as the reference level. To distinguish this unit from other reference levels, they use the abbreviation dBmV.

Figure 8-1 shows a fairly simple master antenna system feeding 19 TV receivers. The signals picked up by the antenna proper are too weak to feed directly to the receivers. They are therefore amplified by the *broadband amplifier*. The amplified signal is fed to a *two-way splitter*, and the two outputs from the splitter are in turn fed to a series of *tap-offs* (*A* through *S*). Each tap-off feeds the signal to its TV receiver and also passes the signal to the next tap-off.

Now let us examine the gains and losses in this system:

1. The amplifier has a gain. A typical value is 30 dB.
2. The interconnecting cable has a loss (attenuation). The better the quality of the cable, the lower the attenuation. This loss is given

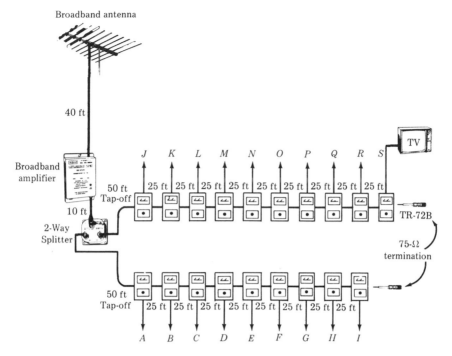

Figure 8-1 A typical master antenna system (*Jerrold Electronics*).

in decibels per 100 ft (dB/100 ft). This loss is also affected by the operating frequency, increasing with frequency. For example, a high quality cable may have only a 1.6-dB loss at 60 MHz (channel 2) but a 3.2-dB loss at 216 MHz (channel 13) and a 6.7-dB loss at 890 MHz (channel 83).

3. Splitters divide the incoming signal between two (or more) feeder lines. In Fig. 8-1, each feeder line gets only one-half the incoming signal *power*. This is a 3-dB "loss" for each feeder. To this we must add the actual loss in the splitter itself (approximately 0.5 dB at VHF), so that a two-way splitter introduces a total loss of 3.5 dB. (A four-way splitter has a 7.0-dB loss.)

4. Tap-offs produce two losses: an *insertion loss* and an *isolation loss*. In a well-designed tap-off, the insertion loss is relatively low—typically 0.5 to 0.75 dB at VHF and up to 1.5 dB at UHF. However, notice from Fig. 8-1 that since the tap-offs are in series the insertion loss affecting any one TV receiver is the sum of the insertion losses of all the tap-offs ahead of its location. For example, the insertion loss at location *G* would be six times (*A* to *F*) the loss of one unit. Isolation is necessary to reduce interaction between

receivers but, in general, the better the isolation the higher the loss. This loss ranges from about 12 to 24 dB, but there is only one such loss for any location.

EXAMPLE 8-10

Using the MATV system of Fig. 8-1, calculate:

1. The worst-case total distribution system loss, using a cable attenuation of 6 dB/100 ft, an insertion loss (per tap-off) of 1.0 dB, and an isolation loss of 15 dB.
2. If the signal level developed at the antenna is 7600 μV, what amplifier gain will be needed to obtain a 1-mV signal level at the receiver?

Solution

1. The worst-case loss would apply to the TV receiver furthest away—location S.
   (*a*) The total cable length from antenna to S is

   $$40 + 10 + 50 + (9 \times 25) = 325 \text{ ft}$$

   This loss is $3.25 \times 6 = 19.5$ dB.
   (*b*) The splitter loss is 3.5 dB.
   (*c*) The tap-off insertion loss is

   $$(0.9 \text{ dB per unit}) \times (9 \text{ units}) = 8.1 \text{ dB}$$

   (*d*) The insertion loss is 15.0 dB.
   (*e*) The total distribution system loss is

   $$19.5 + 3.5 + 8.1 + 15.0 = 46.1 \text{ dB}$$

2. To find the amplifier gain needed:
   (*a*) The antenna signal level is 7600 μV = 7.6 mV.
   (*b*) Converting to dB = 20 log 7.6/1 = 17.6 dB.
   (*c*) For 0 dBmV (that is, 1 mV):

   $$17.6 - 46.1 + \text{amp gain} = 0$$

   $$\text{Amp gain} = 46.1 - 17.6 = 28.5 \text{ dB}$$

VOLTAGE GAIN IN DECIBELS.   It was demonstrated earlier that the power gain or loss in decibels can be calculated from voltage measurements by taking the resistance(s) of the circuit(s) into account. It was also explained how the decibel gain or loss can be obtained directly from voltages if the resistance values are equal. The equation used in this latter case was 20 log (E_2/E_1).

Now we are going to use the same equation, but we are going to forget about the resistance values. Yet if the resistance values are not equal, this is not a true power comparison! This accounts for the heading to this discussion, Voltage Gain in Decibels. In other words, since it is not a power comparison, it has been given a new name. A typical application is in specifying the gain of an amplifier—from input to output. The input and output impedance are seldom (if ever) equal. Therefore, if the decibel gain (power) is desired, it is necessary to calculate power input, power output, and then decibels, using the power form of the equation (see Example 8-7, solution 2*a*). More often the voltage form of the equation is used. However, since the resistance (input and output) are not equal, this should be labeled *voltage gain in decibels,* not *power gain* or just *gain in decibels.*

Unfortunately, this distinction is not clearly made in specifications. This is another cause for confusion in decibel terminology. Let us clearly differentiate these cases:

1. The term *power gain in decibels* is quite clear. It involves power ratios and can also be obtained from voltage values either by first calculating true powers or by using 20 log (voltage ratio) if the resistances are equal.
2. The term *decibel gain* should be used only when referring to power gain and can be calculated as in case 1 above.
3. The term *voltage gain in decibels* is used when the resistances are *not* equal, when no correction is made for the unequal resistance values, and when 20 log (voltage ratio) is involved.

An example at this point may make the distinction clearer.

EXAMPLE 8-11

An input of 0.3 V is sufficient to drive an amplifier to its full rated output of 36 W. The input and output resistances are 250,000 Ω and 16 Ω, respectively. Find:

1. Power gain in decibels.
2. Voltage gain in decibels.

Solution

1. To find the power gain, since the resistances are unequal we must find the power input:

$$P_i = \frac{E_i^2}{R_i} = \frac{0.3 \times 0.3}{250,000} = 3.6 \times 10^{-7} \text{ W}$$

$$\text{Decibel gain} = 10 \log \frac{P_o}{P_i} = 10 \log \frac{36}{3.6 \times 10^{-7}}$$

$$\text{Power gain} = 80 \text{ dB}$$

2. To find voltage gain we need the voltage output:

$$E_o = \sqrt{P_o R_o} = \sqrt{36 \times 16} = 24 \text{ V}$$

$$\text{Voltage gain (in dB)} = 20 \log \frac{E_o}{E_i} = 20 \log \frac{24}{0.3}$$

$$= 38 \text{ dB}$$

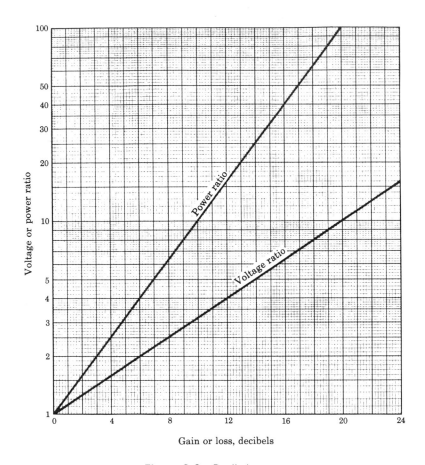

Figure 8-2 Decibel curve.

DECIBEL CURVES.   A shortcut often used in practice is to obtain decibel values from curves rather than by calculation. Figure 8-2 shows such a curve, plotted on semilogarithmic paper. Notice that the decibel values are plotted on an arithmetic scale, while the ratios are on a logarithmic scale. This is similar to the equation $dB = k \log (\text{ratio})$. Use of this semilogarithmic paper converts the curve to a straight line.

Within the limits of the graph, use of the curve should be obvious. In addition, this curve can also be used for voltage ratios greater than 10, power ratios greater than 100, or, conversely, decibel values greater than 20. The logarithm of the product of two (or more) factors is equal to the sum of their individual logarithms. This principle can be applied directly or in reverse. Let us see how this is done in a practical case by repeating Example 8-8.

EXAMPLE 8-12

A power amplifier is rated at +44 dB. Find its rated power output by using the decibel curve of Fig. 8-2.

Solution

1. 44 dB is beyond the range of the graph. However, it can be broken up into 20 + 20 + 4.
2. The corresponding power ratios are $100 \times 100 \times 2.5 = 2.5 \times 10^4$.
3. Since the reference level is 1 mW (understood) and there is a power gain, the rated power output is

$$P_o = 1 \times 10^{-3} \times 2.5 \times 10^4 = 25 \text{ W}$$

(The answer in Example 8-8, by equation, was 25.1 W.)

EXAMPLE 8-13

Solve for the voltage gain of the amplifier of Example 8-11 by using the decibel curve (input 0.3 V and output 24 V).

Solution

1. The voltage ratio is $24/0.3 = 80$.
2. This is beyond the range of the graph but can be broken up into two factors: 10×8.
3. The corresponding decibel values are 20 + 18 = 38 dB.
   (This is the same answer obtained in Example 8-11.)

REVIEW QUESTIONS

1. If the acoustic power level of a sound source is doubled, will it sound twice as loud? Explain.

2. (a) What was the original unit used for the comparison of sound-intensity levels? (b) Give the mathematical expression for calculations using this basic unit. (c) Why was a logarithmic relationship used?

3. (a) Why is the decibel unit used in preference to the original basic unit? (b) What is the relationship between the decibel and this basic unit? (c) Give the mathematical expression for calculations involving the decibel. (d) What physical significance can be attached to a sound-level change of 1 dB?

4. Is the decibel unit used only with relation to sound levels? Explain.

5. (a) Give one *general* area in which decibel evaluations are used. (b) Give a second *general* application.

6. (a) What base is used for the logarithmic equations in decibel calculations? (b) What is the first step in finding the logarithm of a number?

7. (a) Under what condition will the logarithm of a number be a whole number? **(b)** What is the name given to the whole-number portion of a logarithm (to the left of the decimal point)? **(c)** To what other value does this quantity correspond?

8. (a) Within what range of values will the logarithms of numbers between 1 and 10 be found? **(b)** Repeat for numbers between 10 and 100. **(c)** Repeat for numbers between 100 and 1000.

9. (a) What is the name given to the decimal portion of a logarithm? **(b)** State two ways by which this quantity can be obtained.

10. What is meant by the antilog of a number?

11. (a) Give the general equation for evaluating power gain or loss in decibels. **(b)** Which power value is used as the numerator? **(c)** How is a power loss in decibels indicated?

12. (a) Give the equation for evaluating power gain or loss from voltage values. **(b)** Under what condition does this equation apply? **(c)** If this condition is not met, how can the power gain or loss be evaluated?

13. (a) Give the equation for obtaining power gain or loss from current values. **(b)** Under what condition does this equation apply?

14. What values are being compared when equipment is rated in decibels?

15. (a) What is meant by zero level in decibel ratings? **(b)** What power level is most commonly used? **(c)** Give another power level that may be encountered as a reference. **(d)** In specifying equipment ratings, show how to distinguish which of the above levels has been used as reference. **(e)** What reference level is often used with microphones? **(f)** How is use of this reference indicated in a specification? **(g)** What reference level is used in the MATV field? **(h)** How is the use of this reference indicated?

16. (a) What equation is used for evaluating voltage gain in decibels? **(b)** In this evaluation, what correction is made for unequal resistances? **(c)** Give an example of this type of application.

17. Is correction for unequal resistances necessary when evaluating **(a)** power gain in decibels? **(b)** Decibel gain? **(c)** Voltage gain in decibels? **(d)** Power gain from voltage ratios in decibels?

18. (a) What are the coordinates of a decibel curve? **(b)** On what kind of graph paper is this curve drawn? Why? **(c)** Which coordinate value is plotted on which scale?

19. Refer to Fig. 8-2: **(a)** Can this curve be used for decibel values over 20? Explain. **(b)** Can this curve be used for power ratios over 100? Explain.

PROBLEMS

1. A sound source has an acoustic power level of 3 W. **(a)** What power level is needed to triple the sound intensity? **(b)** How much change in power is required?

2. If the original acoustic power level in Problem 1 were 5 W, **(a)** find the power level needed to triple the sound intensity. **(b)** How much change in power is required? **(c)** Explain why a larger change is now required.

3. Find the characteristics of the following numbers: **(a)** 483,127; **(b)** 4.83127; **(c)** 48.3127; **(d)** 321.0; **(e)** 321,000; **(f)** 32.10.

4. Find the mantissas for the numbers in Problem 3.

5. Find the logarithms of the following numbers: **(a)** 42.6; **(b)** 3770; **(c)** 1.25; **(d)** 726; **(e)** 412,000; **(f)** 37.2.

6. Find the antilog of the following logarithms: **(a)** 4.3222; **(b)** 0.8848; **(c)** 0.8851; **(d)** 2.48; **(e)** 1.752; **(f)** 3.6924.

7. A power amplifier delivers a 36-W output from an input of 3 W. Find the decibel gain or loss.

8. A preamplifier delivers 3 W from a 5-mW input. Find the decibel gain or loss.

9. A filter circuit has an insertion loss of 15 dB. If the power input is 5 W, find the power output.

10. An amplifier stage has an output of 60 V when the input is 0.7 V. If the input and output impedances are alike, find the power gain in decibels.

11. An amplifier with equal input and output impedances has a gain of 24 dB. Find the input voltage required to produce 30 V output.

12. The output voltage from an amplifier (for a constant input level) varies from 15 V at 80 Hz to 75 V at 400 Hz to 25 V at 8000 Hz. Using the 400-Hz output as reference, calculate the decibel gain or loss at the other two frequencies.

13. If zero level is taken as 1 mW, find **(a)** power output corresponding to +40 dB; **(b)** power output corresponding to −52 dB; **(c)** rating of an amplifier having an output of 12 W; **(d)** rating of a microphone having an output of 2 μW.

14. A transformer is rated at +36 dB. It was bought for use with a 30-W amplifier. Was this a proper selection? Explain.

15. A microphone has a rating of −70 dB. It is to be used with an amplifier to deliver 30 W output. What amplifier gain (in decibels) is needed to develop the desired output?

16. A signal survey shows that a TV antenna develops the following signal levels: **(a)** for channel 2 (60 MHz)—63,000 μV; **(b)** for channel 4 (72 MHz)—5000 μV; **(c)** for channel 13 (216 MHz)—360 μV. Express each of these signals in dBmV units.

17. A TV antenna develops a signal level of 6000 μV. This signal is fed through a 200-ft lead-in with a loss of 6 dB/100 ft. Find the signal level at the receiver in decibels and in millivolts.

18. The MATV system of Fig. 8-3 uses 16 tap-offs, a four-way splitter, and a broadband amplifier. The cable attenuation is 6 dB/100 ft; the splitter loss is 7.0 dB; each tap-off insertion loss is 0.7 dB; and the isolation loss is 20 dB. The weakest signal level at the antenna is 8000 μV. Find the amplifier gain (in decibels) needed for the worst-case location.

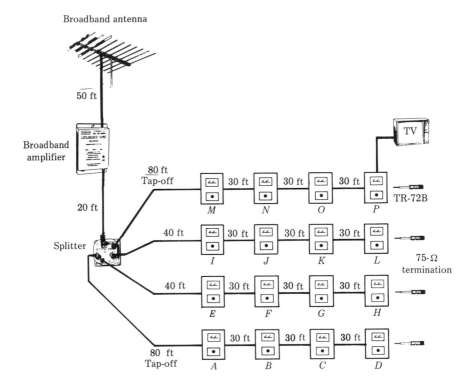

Figure 8-3 MATV system for Problem 18.

19. An amplifier is rated at +40 dB. Its gain is 120 dB. The input and output impedances are 500 Ω and 8 Ω, respectively. Find: **(a)** rated power output; **(b)** voltage output at rated power; **(c)** voltage input required to deliver rated output; **(d)** voltage gain in decibels.

20. An amplifier has an input impedance of 600 Ω and an output impedance of 50 Ω. The power output is 24 W when the input voltage is 1.5 V. Find: **(a)** decibel gain; **(b)** voltage gain in decibels; **(c)** rating in decibels if 24 W is the rated output.

9

Amplifiers—General

One of the most important functions of electronic circuitry is amplification In fact, were it not for this property, many other specific circuit functions would not be possible. For example, oscillators to produce sine waves, square waves, pulses, or any other desired waveshapes would not be possible except for the amplification properties of the circuit components. It is therefore fitting that any complete study of electronic circuits should cover amplifiers. A good starting point is to discuss briefly the various classifications in order to provide an overall view of the amplifier field.

The heart of an amplifier is the active device (transistor or tube). We have already studied the static characteristics of these devices and have seen the basic circuits used for obtaining these characteristics. Now let us see how such a circuit must be modified to act as an amplifier. For this illustration we will use an N-channel JFET.

NEED FOR A LOAD IMPEDANCE.   Figure 9-1a shows a basic circuit suitable for determining the static characteristics of the FET. To obtain a transfer (mutual) characteristic curve, a fixed value of drain supply voltage (V_{DD}) is used; the gate bias supply (V_{GG}) is varied in steps from zero to some negative value, and the drain current (I_D) is read for each step. During this process will the drain potential (V_D) change? It cannot, because the drain is connected directly to the supply voltage and the supply voltage was not changed.

Now assume we have a weak sine-wave signal voltage (V_g) that we wish to amplify. We can connect it in series with the bias supply (Fig. 9-1b). The

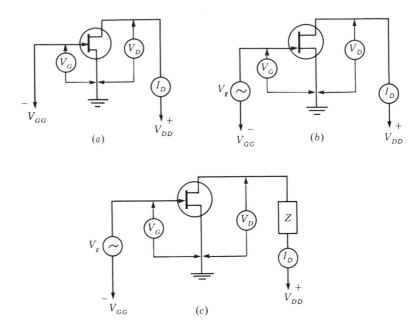

Figure 9-1 Development of a simple amplifier circuit. **(a)** Basic circuit for static characteristics. **(b)** Addition of an input signal source (V_g). **(c)** Addition of a load (Z).

total gate voltage at any instant (v_G) will obviously be the sum of the constant supply voltage (V_{GG}) and the varying signal voltage ($v_g = V_{gm} \sin \omega t$):

$$v_G = V_{GG} + v_g \tag{9-1}$$

The instantaneous gate voltage must vary from the original bias supply value (V_{GG}) to some maximum negative value ($V_{G(max)}$) when the signal voltage is at negative maximum, and to some minimum negative ($V_{G(min)}$) value when the signal voltage is at positive maximum. This is shown in Fig. 9-2. (We are using a signal voltage whose maximum value is lower than the bias voltage.)

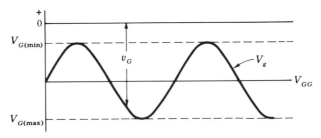

Figure 9-2 Effect of gate signal voltage on instantaneous gate bias.

The addition of this ac signal voltage causes the drain current to change. The drain current will decrease as the instantaneous gate voltage becomes more negative and will increase as this voltage becomes less negative. Since the drain is connected directly to the power supply and V_{DD} was not changed, V_D remains constant. Have we produced an amplified signal voltage? No, only the drain current has changed. However, if we make the drain current flow through some impedance in the drain circuit, a voltage IZ will be produced, and since the current changes in accordance with the signal frequency, this voltage will be a replica of the signal frequency. By proper design the voltage IZ will be greater than the input signal voltage (V_g). The circuit is now acting as an amplifier. Such a circuit is shown in Fig. 9-1c. It is not an actual, commercially used circuit, but it shows quite simply why the load impedance is necessary. This load impedance could be one of several things—a resistor, an inductor, or a transformer. In fact, one means of classifying amplifier circuits is by the type of load used. This will be discussed later.

Figure 9-1 shows an N-channel FET. It should be obvious that had we used a P-channel device, the polarity of the gate and drain voltage supplies would have been reversed. The action would have remained the same. Furthermore, bipolar transistors (or tubes) could have been used in the basic amplifier circuit—and again a load would have been necessary. Therefore, to complete the picture, Fig. 9-3 shows a basic circuit for each of these devices.

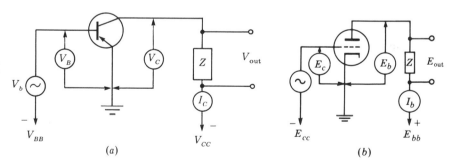

Figure 9-3   Basic amplifier circuit for bipolar transistors and vacuum tubes.

VOLTAGE VERSUS POWER AMPLIFIERS.   Amplifiers can be divided into two broad groups: voltage amplifiers and power amplifiers. The ultimate aim of an amplifier is to operate some device. A few possible examples are (1) to drive the loudspeaker in a public address system; (2) to sweep the picture-tube beam side to side and up and down in a television receiver; (3) to drive the turret motors in an automatic fire control system; (4) to energize the relay controlling the motor drive to the door in an electronic garage door opener.

There are many more applications, but each of these has one thing in common—power must be delivered to the load device, whatever it is. This is the function of a *power amplifier*. To develop high power levels, large changes must be produced in the output current of the power amplifier. To produce these large current changes, large variations in the instantaneous input voltage (v_i) are needed. This in turn requires large input signal voltages (V_i). But in many cases the signal voltage starts out as an extremely low voltage. For example, the voltage picked up by a television antenna is only a few microvolts. Similarly, the output from a microphone feeding a public address system or the "error" voltage in an automatic control system are also very low voltages. *Voltage amplifiers* are used to build up these weak signals until they are strong enough to drive the power amplifier. In most applications several stages of voltage amplification are needed ahead of the power amplifier stage.*

In a voltage amplifier, the aim is to get the maximum possible voltage output for a given input signal voltage. (We will see later that there are limitations to this ambition.) This results in maximum voltage gain. The amount of power developed in the output of voltage amplifiers is relatively unimportant. The reverse is true in a power amplifier. Here voltage gain is comparatively unimportant, and we strive to get the maximum possible power output from a given input signal voltage.

CONDITION FOR MAXIMUM VOLTAGE OR POWER OUTPUT.   Figure 9-1 shows that a load impedance is needed in the output circuit of a transistor (or tube) in order to get amplification. The value of this load impedance will affect the amount of voltage or power output we can get from the device. Zero impedance (Fig. 9-1a) gives no output. Does that mean that infinite impedance will give maximum output? Before we try to answer that question, let us analyze what happens to the output voltage and power when several values of load resistance are connected across a dc supply source *having internal resistance*. This can be seen best from an example.

EXAMPLE 9-1

A dc power source has an EMF of 600 V and an internal resistance of 1000 Ω. Find the voltage across the load and the power dissipated by the load for various values of load resistance from 0 to 20,000 Ω. Plot curves of voltage output and power output versus load resistance.

Solution

1. For each value of load resistance (R_L), the total resistance (R_T) is equal to $R_L + 1000$ (the internal resistance).

---

*The early stages of such an amplifier—because the signal level is low—are also referred to as *small-signal amplifiers*. This is especially true in circuits using bipolar transistors.

2. The current (I) in the circuit is E/R_T (where $E = 600$ V).
3. The voltage across the load or output voltage (E_o) is IR_L.
4. The power output (W_o) is $I^2 R_L$ (or $E_o I$; or E_o^2/R_L).
5. Using these relations, let us tabulate answers for convenient values of R_L (Table 9-1).
6. The curves of voltage and power output versus load resistance are shown in Fig. 9-4.

Table 9-1

R_L (OHMS)	R_T (OHMS)	I (MA)	E_o (VOLTS)	W_o (WATTS)
0	1000	600	0	0
200	1200	500	100	50
500	1500	400	200	80
1000	2000	300	300	90
2000	3000	200	400	80
5000	6000	100	500	50
11,000	12,000	50	550	27.5
19,000	20,000	30	570	17.1

From Table 9-1 and Fig. 9-4 several important facts should be noted:

1. The voltage across the load E_o never reaches the full supply voltage. Even for higher values of R_L, it will never quite make it. At $R_L = 99,000\ \Omega$, for example, $R_T = 100,000$, $I = 6.0$ mA, and

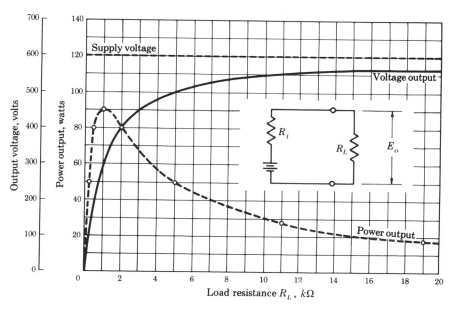

Figure 9-4   Effect of load resistance on voltage and power output.

$E_o = 594$ V. In general, the higher the load resistance value, the higher the voltage across the load.

2. Notice, however, that the increase in output voltage becomes slower and slower as higher values of R_L are used (such a curve is called *asymptotic*). At $R_L = 19,000$ we are only 30 V or 5 percent from the maximum possible value. By increasing R_L to 99,000 Ω, only an additional 24 V is picked up to reach within 1 percent of possible maximum. A tremendous increase in R_L (500 percent) produces only a slight increase (4 percent) in output voltage.

3. Now examine the power output curve. A maximum value (90 W) is reached at $R_L = 1000$ Ω. Below or beyond this load the power output decreases. But 1000 Ω is also the value of the internal resistance of the supply. This is not a coincidence. Mathematically it can be proved (using calculus) that *maximum power is delivered to the load when the load resistance equals the internal resistance of the supply source.*

4. One more point of interest is that the power output curve rises rapidly to the maximum power point and falls much more gradually as the maximum is passed. Using Fig. 9-4, for example, two-thirds of maximum power output (60 W) can be obtained either at $R_L \cong 300$ Ω or at $R_L \cong 3700$ Ω. Notice that whereas the first value is only 700 Ω less than the optimum R_L value, the latter is 2700 Ω more. This relation will be explored further when we discuss power amplifiers.

Now let us consider how this discussion applies to amplifiers. The amplifier delivers voltage and power to its load. Also, the amplifier has internal resistance (for example, r_d in a FET circuit). The situation is therefore analogous to the dc power supply with internal resistance. From Example 9-1 and to the best of our *present* knowledge, we can therefore conclude that:

1. For maximum voltage output (voltage amplifiers), the load impedance should be as high as possible.

2. For maximum power transfer from the source to the load (power amplifiers), the load impedance should equal the internal impedance of the source. This is called *impedance matching*.

OTHER CLASSIFICATIONS OF AMPLIFIERS

Thus far we have seen that amplifier circuits can fall into two major categories—voltage or power amplifiers. Each of these in turn is further classified depending on details of use or circuitry. Some of the more common classifications are discussed briefly in this chapter. The details are left to actual applications in the chapters to follow.

FREQUENCY HANDLED. Depending on the frequencies—or better yet, frequency range—for which amplifiers are designed, they are generally classified as:

1. *Audio amplifiers* (AF). These are used in radio receivers, public address systems, and other applications involving audible frequencies. The frequency range of such units can be as narrow as 400 to 4000 Hz for amplifiers intended for speech only; or 50 to 5000 Hz for average-quality music systems; or as wide as 20 to 20,000 Hz for full high-fidelity systems. These values should not be treated as gospel. They are given as rounded-off approximations. For example, 20,000 Hz is used as the high end of the high-fidelity spectrum, yet few people can hear beyond 16,000 Hz.

2. *Video amplifiers.* These are used in television receivers, radar receivers, and other devices producing visible pictures. The frequency range of these units is much greater than for audio amplifiers. For this reason they are often also referred to as *wideband* amplifiers. For use in television, the typical frequency band needed is from approximately 30 Hz to 4.5 MHz. For use in some oscilloscopes or in radar and other pulse-type circuits, even higher frequency ranges are often needed.

3. *Pulse amplifiers.* These are wideband amplifiers quite similar to the video amplifiers; in fact, the names are often used interchangeably. The term *video* is usually reserved for applications such as television; the term *pulse* is more general and is used for other industrial electronics applications.

4. *Radio-frequency amplifiers* (RF). These amplifiers, in general, are used to amplify a relatively narrow band of frequencies to each side of some mean value or "carrier frequency." For example, in a broadcast-band radio receiver an RF amplifier may be set to operate at a frequency of 1200 ± 5 kHz. In an FM receiver, the frequency band may be ±100 kHz around a carrier frequency of 100 MHz. These amplifiers use some form of *L-C* circuit to tune or resonate at the desired mean frequency.

5. *Intermediate-frequency amplifiers* (IF). These amplifiers also use resonant circuits and are used to amplify a relatively narrow band of frequencies to each side of a mean or carrier frequency. In fact they are RF amplifiers. The term *intermediate* frequency is used because somewhere in the complete unit there are one or more other amplifiers operating at some higher radio frequency. IF amplifiers are found in all *superheterodyne*-type receivers. Broadcast-

band radio receivers usually employ an intermediate frequency of approximately 456 kHz. Common IF values for FM and television receivers are 10.7 and 44 MHz, respectively. UHF receivers have used intermediate-frequency values around 60 MHz.

6. *DC amplifiers.*   Most amplifier circuits will amplify only varying or ac signals. The reason for this will become apparent when typical circuits are studied. If it is necessary to amplify a steady-state phenomenon, a direct voltage, or a complex waveform having a dc component, special amplifiers must be used. These amplifiers (usually) will also amplify ac signals. However, since they are more expensive and more critical in design, they are used only when dc levels need be amplified.

Two of these subdivisions (RF and IF) use resonant circuits to select the mean carrier value and the narrow band of "frequency swing" to each side of this value. Such amplifiers are grouped into a classification of *tuned* or *resonant-circuit* amplifiers. The remainder of the amplifiers (audio, video, pulse, and dc) really amplify a continuous-frequency spectrum from as low as possible (zero for the dc amplifier) to as high as possible, or as needed. They do not use tuned circuits and are therefore known as *nonselective, nonresonant,* or *untuned* amplifiers.

TYPE OF AMPLIFYING DEVICE.   In this connection there are two broad classifications—solid-state and vacuum-tube amplifiers.* Each of these can be further subdivided into specific types of transistors (or tubes) —for example, bipolar, FET, *NPN*, or *P*-channel amplifiers.

TYPE OF COUPLING.   The output of an amplifier must be fed either to another stage (cascade) or to the final device to be operated. In either case, the output of one stage must be coupled properly to its load. Amplifiers are often classified by the type of coupling circuits used. Among the more common are (1) R-C coupled (resistance-capacitance), (2) impedance coupled, (3) transformer coupled, and (4) direct coupled. Each of these circuits is discussed in the chapters to follow. However, it can be noted here that R-C coupling and direct coupling are used in untuned amplifiers while impedance and transformer coupling can be used in either tuned or untuned amplifiers. Also, whereas the last three can be used with either voltage or power amplifiers, R-C coupling should never be used for power amplifiers.

DURATION OF CONDUCTION PERIOD.   Another means of classifying amplifiers is by the duration of the conduction period in relation to one

---

*Masers and magnetic amplifiers are not considered in this text.

cycle of the input signal. In this respect, circuits are divided into four categories:

1. *Class A:* Output current (collector, drain, or plate) flows for the full cycle of input.
2. *Class AB:* Current flows for less than one cycle but for more than 180° (half-cycle).
3. *Class B:* Current flows for approximately one half-cycle of the input signal.
4. *Class C:** Current flows for (appreciably) less than one half-cycle of input. This class of operation is used only with RF power amplifiers and therefore is not discussed further in this text.†

DISTORTION

A characteristic desired in amplifier circuits is that the waveshape of the output signal be an exact replica (amplified, of course) of the input signal waveshape. If a perfect sine wave is fed into the amplifier, a perfect sine-wave output is desired; if a square wave is fed in, a perfect square-wave output is desired. Any deviation whatever between the input and output waveshape is charged against the amplifier as *distortion.* Amplifier distortion may occur in one of three forms or combinations thereof. Each of these types of distortion is discussed in general below. The details are left to the actual circuit discussions in the chapters that follow.

FREQUENCY DISTORTION.   This type of distortion occurs when the gain of the amplifier varies depending on the frequency of the input signal. For example, an audio amplifier would suffer severely from frequency distortion if it had a gain of 30 at around 1000 Hz and the gain dropped to 3 at 15,000 Hz. (This would correspond to a 20-dB loss.) With this amplifier, you probably could not hear cymbals or other high-frequency sounds. The term *frequency response* of an amplifier is generally used to describe the frequency limits within which distortion is not serious. Unless otherwise specifically mentioned, these limits are taken at the frequencies where the loss equals 3 dB (voltage drops to 0.707 of its constant value). This type of distortion is caused by variation in the impedance of the load and is due to inductive or capacitive components in the load.

*In vacuum-tube amplifiers, this classification may be further identified with subscripts 1 or 2. Subscript 1 signifies that grid current does not flow; subscript 2 indicates that grid current flows for some portion of the input cycle.

†See J. J. DeFrance, *Communications Electronics Circuits* (2nd ed.), New York: Holt, Rinehart and Winston, 1972, chap. 4.

DELAY OR PHASE DISTORTION.   In many amplifier circuits, the output and input waveshapes are 180° apart. In other words, a phase shift of 180° has occurred in the amplifier. This in itself is no drawback. *Phase distortion* occurs only when the amount of phase shift varies with the frequency of the incoming signal and the amount of shift is not proportional to the frequency. For example, you will learn that in R-C coupled amplifiers the phase shift of the output voltage may vary from the ideal value of 180° at a mid-frequency, to some lagging value (such as 130°) at a low frequency, and to a *leading* angle (such as 120°) at a high frequency. Since a shift in phase is also a change in time, this type of distortion is also known as *delay* distortion. In video amplifiers (television), delay distortion causes smearing of large objects (low frequencies) and blurring of fine details (high frequencies).

Figure 9-5 gives a visual representation of the effect of phase distortion on the waveshape of a signal. The input signal consists of a fundamental and a third harmonic in phase. The composite input signal and its components are shown in Fig. 9-5a. For the output waveshapes (Fig. 9-5b) it is assumed that the fundamental component has the same phasing as in the input signal but that the third harmonic is shifted 180°. (This can occur

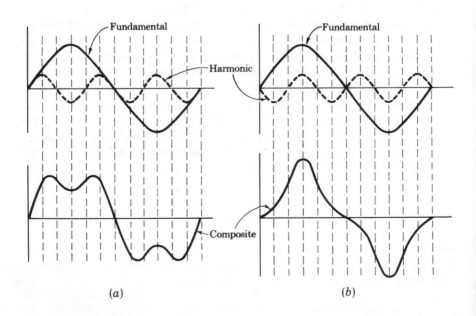

Figure 9-5   Effect of phase distortion on signal waveshape. **(a)** Input signal. **(b)** Output signal.

in a multistage amplifier.) Notice the drastic difference in the composite waveforms.

AMPLITUDE DISTORTION.   If the gain of an amplifier varies depending on the amplitude of the input signal, that amplifier has (produces or suffers from) *amplitude distortion*. As a numerical illustration, examine Table 9-2. Notice that although the input signal level is increasing steadily, the output for the last three values is definitely leveling off and the gain is decreasing. Beyond the 3-V input level, this amplifier has amplitude distortion.

Table 9-2

INPUT LEVEL (VOLTS)	OUTPUT LEVEL (VOLTS)	GAIN
1	10	10
2	20	10
3	30	10
4	34	8.5
5	35	7.0
6	36	6.0

Figure 9-6 shows a dynamic transfer characteristic* with exaggerated curvature. Notice that any input signal, such as V_{g1} and V_{g2}, that falls within the straight-line section of this dynamic characteristic (*a* to *b*) will be amplified without distortion. On the other hand, signals that fall outside the straight-line section, such as V_{g3}, will have their peaks flattened. Since amplitude distortion is caused by operation in the nonlinear areas, it is also called *nonlinear distortion*.

When amplitude (or nonlinear) distortion occurs, the output waveshape is no longer a sine wave. The resulting complex wave consists of the original frequency and harmonics of the original frequency. In general, if the distortion is slight and only one side of the output wave is flattened, only even harmonics are produced. In the more severe cases, or where both sides of the wave are flattened, odd harmonics are also produced. The exact harmonic content (amplitude and frequency) could be evaluated mathematical analysis, such as Fourier series. However, a graphic analysis will be used here to show how addition of harmonic content can flatten or

*Dynamic characteristics are discussed in detail in Chapter 10. For this preliminary discussion, all we need to know is that it shows—for a specific circuit—how the drain, collector, or plate current will vary for any input signal value, above and below the original dc bias. This original bias point is shown as point *Q* and is known as the operating point or quiescent point.

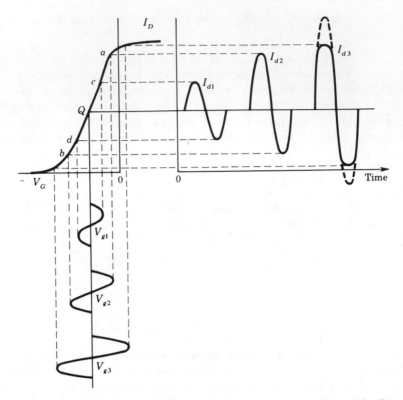

Figure 9-6   Distortion caused by operation of a FET amplifier beyond the linear portion of the dynamic characteristic.

peak a sine wave. This is shown in Fig. 9-7. From the point-by-point analysis you can readily see the distortion effect of harmonics. Since amplitude distortion—caused by nonlinear operation—results in the production of harmonics, this form of distortion is also known as *harmonic distortion.*

In high-fidelity audio systems, another type of distortion, *intermodulation distortion,* is often considered. This is not a fourth type: it is only another version of nonlinear distortion and can occur only because more than one input frequency is fed to the amplifier at the same time. To explain this, let us first digress a little and briefly discuss musical sounds.

All musical instruments, even when they play a single note, produce a fundamental frequency and harmonics thereof. These harmonics are not unpleasant sounds. On the contrary, it is the order and magnitude of these harmonics that give the sound its quality and permit you to distinguish one instrument from another. If a chord is played, or several instruments

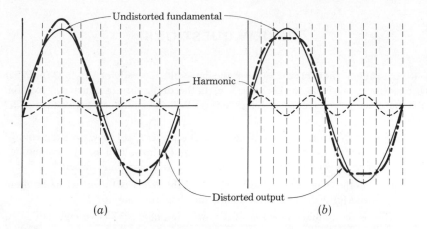

Figure 9-7   Harmonic content in amplitude-distorted waves.

play at the same time, several fundamental frequencies and the respective harmonics of each of these are produced at the same time. Again, if properly arranged, these combinations of frequencies are still pleasant (in fact, more so). However, if some odd frequency is selected, the discord or "sour note" is easily detected.

The electronic correlation can now be shown. A basic principle of all amplifiers is this: If two or more frequencies are fed to a nonlinear amplifier, the output will contain (1) the original frequencies, (2) harmonics of all original frequencies, (3) dc component, and (4) sum and difference frequencies. The fourth item is the key factor. In many instances nonlinear operation is deliberately used to take advantage of this factor,* but in audio amplifiers this produces unwanted, discordant frequencies. For example, assume that a 400-Hz note (with its harmonics—800, 1200, 1600) and an 1800-Hz note (also with its harmonics—3600, 5400, 7200) enter as input to an amplifier. Let us further assume that such a chord or combination is harmonious. If the amplifier is nonlinear, each of the frequencies and harmonics of all the above frequencies will appear in the output. This may still be quite harmonious, even though the new harmonics will change the quality of the sound. However, sum and difference frequencies (such as 1400 and 2200) will also be produced. These intermodulation products (the 400-Hz note modulating the 1800-Hz note) are not harmonically related to any of the original frequencies and may very well be discordant.

*Modulation in amplitude-modulated (AM) transmitters, demodulation or detection in AM receivers, and frequency conversion or mixing in superheterodyne receivers all depend on this basic principle.

REVIEW QUESTIONS

1. Refer to Fig. 9-1b: **(a)** Is the value of the gate voltage steady or varying? **(b)** What two quantities contribute to this value? **(c)** Is the drain current constant? Explain. **(d)** Is the drain potential constant? Explain. **(e)** Will the circuit have a voltage gain? Explain.

2. Refer to Fig. 9-1c: **(a)** What does the block Z represent? **(b)** How does the voltage across this component compare with the input voltage in waveshape? In magnitude? **(c)** Why is some form of load impedance necessary in an amplifier circuit?

3. **(a)** Give another name for voltage amplifiers. **(b)** When is this other name more commonly used? **(c)** Why is this other name more appropriate now?

4. **(a)** What is the primary function of a power amplifier? **(b)** Is voltage gain, in a power amplifier, of equal importance as power output? **(c)** Give several types of loads that may be used with power amplifiers.

5. **(a)** Why are voltage amplifiers needed? **(b)** How does the aim for voltage amplifiers differ from that for power amplifiers? **(c)** Is power output of equal importance in these amplifiers?

6. A load is supplied from a source having appreciable internal resistance. **(a)** Can the full supply voltage be applied across the load? Explain. **(b)** Under what condition will maximum power be delivered to the load?

7. From theoretical aspects, explain briefly what value of load impedance is best for **(a)** voltage amplifiers and **(b)** power amplifiers.

8. **(a)** What is meant by impedance matching in relation to amplifier loads? **(b)** In which type of amplifier is this of advantage? Why?

9. In a power amplifier, if the optimum value of load impedance cannot be obtained, which would be better, a value 50 percent less, or 50 percent more, than optimum? Explain.

10. Give a brief description and (where possible) the frequency range for each of the following classifications of amplifiers: **(a)** audio; **(b)** video; **(c)** wideband; **(d)** pulse; **(e)** radio frequency; **(f)** intermediate frequency; **(g)** dc.

11. Which of the above types can also be classified as **(a)** tuned or resonant circuit and **(b)** nonresonant or untuned?

12. How does an IF amplifier differ from an RF amplifier?

13. Name four types of amplifiers based on the method of coupling used.

14. Which of the four types of coupling can be used with **(a)** untuned amplifiers? **(b)** Tuned amplifiers?

15. Which of the four types of coupling can be used with **(a)** voltage amplifiers? **(b)** Power amplifiers?

16. **(a)** Name four classes of amplifiers based on the duration of their conduction period. **(b)** Give the distinguishing feature for each class.

17. **(a)** In vacuum-tube power amplifiers, what notation is used to further describe the class of operation? **(b)** What does each specific notation mean?

18. **(a)** What is meant by frequency distortion? **(b)** Give an example.

19. Unless more specifically stated, what is meant by the *frequency response* of an amplifier?

20. **(a)** In general, what is the phase relation between the input and output voltage of an amplifier? **(b)** Is this phase distortion? Explain. **(c)** If the phase angle of the output voltage shifts by 0.1° at 100 Hz to 1.0° at 1000 Hz and 10° at 10 kHz, is this a case of serious phase distortion? Explain. **(d)** Define phase distortion.

21. Refer to Fig. 9-5: **(a)** What frequency components are present in the input signal? **(b)** What is the phase relation (starting) between these components? **(c)** What frequency components are present in the output signal? **(d)** Why do the resultant waveshapes differ?

22. **(a)** What is meant by amplitude distortion? **(b)** Give an example.

23. Refer to Fig. 9-6: **(a)** Does distortion result from signal V_{g1}? **(b)** Does distortion result from signal V_{g3}? **(c)** What causes this distortion? **(d)** What is another name for this type of distortion? **(e)** On the dynamic characteristic shown, over what limits may the input signal swing and still not produce distortion? **(f)** If the signal zero level were along line d, over what limits could the signal swing and still not produce distortion?

24. Why is amplitude distortion also referred to as harmonic distortion?

25. What harmonics are generated if the distortion **(a)** affects only one alternation of the signal wave? **(b)** Both half-cycles of the wave?

26. **(a)** What is meant by intermodulation distortion? **(b)** To which of the three basic classes does it belong? Why? **(c)** Give a numerical illustration.

10

R-C Coupled FET (and Vacuum-Tube) Amplifiers

One of the more common voltage amplifier circuits is the R-C (resistance-capacitance) coupled circuit. It is used in audio amplifiers, in video amplifiers, as a wideband or pulse amplifier in radar units or oscilloscopes, and in most places where a nonresonant voltage amplifier is needed. What makes it so popular? It is cheap, it is simple, and with proper design it has good stability and wide frequency response. The circuit for an R-C coupled amplifier using an N-channel JFET is shown in Fig. 10-1. The section between dotted lines is considered as one stage. The other components, C_2 and R_{g2}, belong to the following stage.

FUNCTION OF CIRCUIT COMPONENTS

Before we study the overall electrical characteristics of this circuit, let us analyze the circuit, part by part, and discuss the function of each component (or combinations).

SELF-BIAS CIRCUIT.   In any amplifier, freedom from amplitude distortion is highly desired. This requires operating the FET within the straight-line portion of its dynamic characteristic. For an N-channel FET this means operating with the gate at some negative value *compared to*

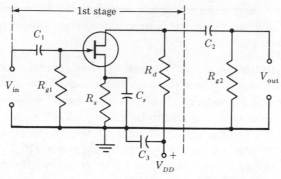

Figure 10-1   Typical FET R-C coupled amplifier. *Note:* The other terminal (negative) of the power supply is grounded. Unless otherwise shown, this connection is always implied.

source. In the previous chapter (see Fig. 9-1), the proper negative bias was obtained from a negative supply source. Such a method is expensive and is not generally used. A more common technique is shown in Fig. 10-1. Components R_s and C_s provide the gate bias for the FET in what is known as a *self-bias* or, more specifically, a *source-bias* circuit.

Actually, the gate is at ground potential (since it is connected through R_g to ground and no gate current is flowing). Drain current flows from ground (negative terminal of the power supply), through R_s, through the FET (source to drain), through R_d, and back to the positive terminal of the power supply. Because of the voltage drop in resistor R_s, the source is positive with respect to ground. Using the positive source as reference, the gate is in turn negative compared to source. The proper bias voltage is obtained by selecting a suitable value for R_s (Ohm's law).

Why is the resistor shunted by the capacitor C_s? When ac signals are applied at the input of this amplifier, the instantaneous gate voltage will change and the drain current will vary accordingly. But if the drain current changed, the bias (drop across R_s) would also change, and this effect is generally not desired. Capacitor C_s serves as a bypass capacitor to prevent these bias variations. For proper action, the reactance X_C of this capacitor should be very low (less than one-tenth the resistance value R_s) at the lowest input frequency. Then, since IX_C for the ac components of drain current is very low, there is negligible change in the bias voltage.

COUPLING CAPACITOR (C_1).   Capacitor C_1 in the input coupling network serves two functions: it allows the "feeding" or application of an ac signal to the gate of the FET, but it blocks any dc component in the input signal from the gate. Application of direct voltages would upset the bias of the FET. If high positive dc components are present (which is often the case), breakdown of this capacitor would also ruin the FET. C_2 functions similarly to couple the ac component of the FET's output voltage to the next stage. Typical values for coupling capacitors range from approximately 0.003 to 0.1 μF.

GATE RESISTOR (R_g).   This resistor has several functions. With regard to the dc bias, it keeps the gate at ground potential and therefore negative compared to source. If this resistor should open, the gate has no specific dc potential and is said to be *floating*.* With respect to the input signal, R_g and C_1 form a voltage divider. If the resistance value is large compared to the reactance of the coupling capacitor (10 X_C or greater), practically all the ac signal voltage will appear across R_g and the loss— voltage drop across C—will be negligible.

This might seem to indicate that this resistor should be made as large as possible. But there is a limitation. Leakage current in a JFET, although normally considered negligible, could upset the gate-source bias if R_g is too large or if for any reason (such as excess temperature) the leakage current should increase. With MOSFETs, on the other hand, too large a resistor may cause breakdown of the gate insulation. Resistor R_{g2} serves in like fashion for the next stage. Typical values for gate resistors range from approximately 0.1 to 2.0 MΩ; the larger values are needed if a low-value coupling capacitor is used.

DRAIN RESISTOR (R_d).   As was shown in the previous chapter, to achieve amplification a load impedance must be connected between the drain of the FET and the power supply. In R-C coupled amplifiers the *drain resistor* (R_d) acts as the load, *as far as the dc path is concerned*. Drain current flowing through this resistor will produce a voltage drop and the drain potential (V_D) will be less than the supply voltage (V_{DD}). If the drain current varies (with signal input), the drain potential will also vary.

In Fig. 10-1 the lower end of the drain resistor (R_d) is connected through capacitor C_3 to ground. This capacitor is generally large and serves primarily to return the ac component of drain current flowing through R_d directly to ground and back to the source without going through the power supply. Since the capacitor is large, its reactance at signal frequencies is low, and the power-supply end of the resistor is effectively at ground potential to all signal frequencies. In many diagrams this capacitor is not shown at all. Its function is now performed by the regular power supply filter capacitors. On the other hand, at very high frequencies, since electrolytic capacitors may have appreciable inductance, capacitor C_3 is often shunted by a small mica capacitor to maintain this point at ground potential at these frequencies.

---

*In MOSFETs, this could cause buildup of a high charge and breakdown of the insulation between gate and channel.

LOAD IMPEDANCE.   Elsewhere you may have seen resistor R_d labeled R_L and called the "load resistor." Yet we have carefully avoided calling this resistor the load resistor, *except as far as the dc path or static operation is concerned*. The obvious implication is that the true value of load to ac or under dynamic operating conditions is not just R_d but must be affected by other factors. That is true. Reexamine Fig. 10-1. The dc component of drain current can flow only through R_d and back to the power supply. However, the ac component of drain current has two paths —through R_d and C_3 to ground and back to source; or through C_2 and R_{g2} to ground and again back to source. Since these are parallel paths, the ac load impedance is less than R_d and is found by solving for the equivalent impedance of $(C_2 + R_{g2})$ in parallel with R_d.*

ANALYSIS AND PHASE RELATIONS

To get a clear picture of the action and phase relations in an R-C coupled amplifier, let us use a numerical example. The values used are purely nominal; they may not apply to any specific FET or operating conditions. They were selected merely for simplicity—the evaluations can be done mentally, by observation. How the true values would be obtained in a normal circuit will be shown later.

We start with some given circuit values and conditions:

1. Power supply $V_{DD} = 40$ V.
2. Drain resistor $R_d = 7000\ \Omega$.
3. Source-bias components (R_s and C_s) are chosen so that the dc bias is 3 V.
4. Quiescent (zero-signal) drain current = 3 mA.
5. Signal voltage (V_g) is a sine wave and has a maximum value of 2 V.
6. The FET operates on the linear portion of its dynamic characteristic such that a 1-V change on the gate produces a 1-mA change in drain current.

For ease of comparison, the changes that occur (with time) because of change in gate signal are presented in Table 10-1 and the pertinent waveshapes are shown in Fig. 10-2.

The first column of Table 10-1 is obvious—the gate signal (v_g) varies from zero to a positive maximum of +2 V and to a negative maximum of −2 V. Next, the instantaneous gate potential (v_G) must be the phasor

---

*The input impedance of the next stage is also a paralleling impedance. With FETs, however, this value is so high that it can be neglected. This is not true if the next stage is a bipolar transistor.

Table 10-1

INSTANTANEOUS VALUES IN AN R-C COUPLED AMPLIFIER

$V_{DD} = 40$ V; $V_{GG} = -3$ V; $R_d = 7.0$ kΩ

SIGNAL VOLTAGE v_g (volts)	GATE POTENTIAL v_G (volts)	DRAIN CURRENT R_d (mA)	VOLTAGE DROP ACROSS i_d (volts)	DRAIN POTENTIAL v_D (volts)	AC COMP. OF DRAIN POTENTIAL v_d (volts)	OUTPUT VOLTAGE v_o (volts)
0	−3	3	21	+19	0	0
0	−3	3	21	+19	0	0
+1	−2	4	28	+12	−7	−7
+2	−1	5	35	+5	−14	−14
+1	−2	4	28	+12	−7	−7
0	−3	3	21	+19	0	0
−1	−4	2	14	+26	+7	+7
−2	−5	1	7	+33	+14	+14
−1	−4	2	14	+26	+7	+7
0	−3	3	21	+19	0	0

(algebraic) sum of the bias and signal voltages and must swing from a low of −1 (when the signal is +2 V) to a maximum negative of −5 (when the signal is −2 V). This accounts for column 2. Column 3 is easily understood since a 1-mA drain current change per volt of gate swing was specified. The only other point to remember is that drain current increases as gate potential becomes less negative (and vice versa). The voltage drops across the drain resistor R_d (column 4) are readily calculated from Ohm's law. Obviously, as the current increases (or decreases) this drop must increase (or decrease) in step. Now examine Fig. 10-1 again; notice that the FET and the drain resistor are connected in series across the power supply (V_{DD} and ground). At any instant, as in any series circuit, the sum of the voltage drop v_d across the FET (from drain to ground) and the voltage drop across the drain resistor R_d must equal the supply voltage V_{DD}, or in terms of drain potential,

$$v_d = V_{DD} - i_d R_d \tag{10-1}$$

Obviously, as the drain current increases and the voltage drop across R_d increases, the drain potential must decrease. These relationships are shown graphically in Fig. 10-2. Study the notation used. Capital letters are used for fixed values—rms, average, maximum, or minimum. Lowercase letters signify instantaneous or changing values. Subscripts D and G are used for total values measured from the zero-axis drain and gate circuit, respectively; d and g subscripts are used for ac or signal component values measured from the average value as reference axis—again drain and gate circuit, respectively.

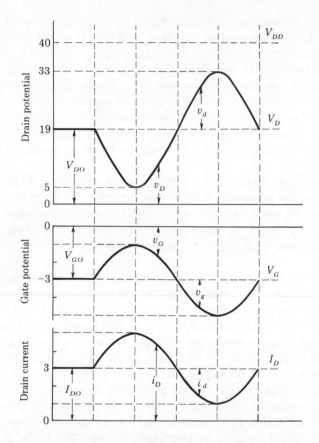

Figure 10-2   Phase relations in an R-C coupled amplifier.

Several aspects of the waveshapes in Fig. 10-2 should be stressed:

1. Each quantity (v_G, i_D, v_D) is a complex wave consisting of a dc component (V_G, I_D, V_D), which is the average value, and an ac component (v_g, i_d, v_d), which is the signal value.
2. Since the tabulated figures were predicated on a linear dynamic characteristic, there is no amplitude distortion and the no-signal values (I_{DO}, V_{DO}) are identical to the average values.
3. With respect to the ac or signal components, the drain current is in phase with the gate voltage and the drain voltage is 180° out of phase with the input gate signal voltage.

ACTION OF COUPLING NETWORK.   This network consists of R_d, C_2, and R_{g2}. The independent function of each of these components has already been covered; yet it is desirable that we review the combined action and add more details. This network has a twofold purpose: (1) to produce a varying drain potential (v_D) containing an amplified replica of the input signal (v_g); (2) to remove or block the dc component of the drain potential (V_D) from the output, thereby producing an output voltage (V_o) containing only the ac component. We have already seen how the varying drain potential is produced. The question now is: "How is the ac output voltage obtained?" One solution is to say that the capacitor C_2 blocks the dc component and allows only the ac component to flow through. However, this simple answer may lead to serious circuit analysis trouble later, particularly with pulse circuits, so it would be better to analyze this action in detail.

Figure 10-3 shows the coupling network section of Fig. 10-1. It is repeated here for convenience in further analysis. We have also added to the diagram the power supply value (40 V) and the waveshape values of the drain potential v_D, as determined in our analysis of phase relations. First let us study in detail what happens during time interval 1 to 2 of Fig. 10-3.

At the instant the circuit is energized, the FET conducts and the drain potential drops to $+19$ V because of the voltage drop in the drain resistor R_d. As long as the signal voltage is *not* applied, the drain-to-ground voltage remains constant at this value. But the coupling capacitor and gate resistor are connected in series across this voltage. The capacitor must charge as in any R-C dc circuit. There will be a high inrush of current at the start, and the current will drop off exponentially as the capacitor approaches full charge. This charging current flowing through R_g will cause a corresponding voltage drop across this resistor. When the capacitor is fully charged, the voltage across it is 19 V, with the polarity as shown in Fig. 10-3. Now what is the output voltage or, more specifically, what is the potential of

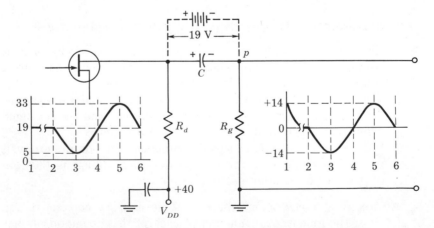

Figure 10-3 Action of coupling capacitor in an R-C amplifier.

point p? Is it $+19$? -19? Neither. It is zero. Except for the *transient* current flow that charged the capacitor, there is no current flow through R_g, no voltage drop across this resistor, and point p must be at ground potential. Therefore the output voltage is now zero.

How long will it take to establish this zero level? That depends on the time constant RC of the coupling capacitor and gate resistor; to be more exact, it will be approximately $5RC$ (assuming commonly used values of $C = 0.02 \mu$F and $R_g = 0.47$ MΩ, then $5RC \cong 0.05$ s). For the present discussion, assume that the time interval 1 to 2 is longer than $5RC$ and therefore the potential of point p is zero before the end of this interval.

When the signal voltage is applied to the FET, the drain potential varies above and below the $+19$ V value at the signal frequency. The capacitor would *tend* to discharge when the drain potential falls, and charge to a higher value when the drain potential rises above 19 V. However, the discharge time constant is long compared to the period of the ac signal. The amount of change in charge is insignificant and the voltage across the capacitor remains at 19 V. The capacitor therefore acts as if it were a 19-V battery, in opposition to the drain potential. This is indicated in Fig. 10-3 by the dotted battery above the capacitor. The rest is easy.

1. When the drain potential is $+19$ V (time instants 2, 4, 6), the net voltage (drain potential minus "battery" voltage) is zero, no current flows through R_g, and the output voltage is zero.
2. The current through R_g (and the output voltage) is a maximum whenever this net voltage is a maximum. This in turn occurs when the drain potential is a minimum (time instant 3) or a maximum (time instant 5).
3. When the drain potential falls *below* the "battery" voltage (time interval 2 to 4), current flows *down* through R_g (as the capacitor starts to discharge) and the output voltage is negative.
4. When the drain potential rises *above* the "battery" voltage (time interval 4 to 6), current flows *up* through R_g (as the capacitor tries to charge to a higher value) and the output voltage is positive.

What happens if the signal voltage is applied immediately and there is little or no time interval 1 to 2? The average value of the sine-wave ac signal is zero. Therefore the capacitor neither gains nor loses any charge over the cycle due to the ac variation. However, it does gain a little charge *due to the dc component* during each ac cycle until finally it reaches full charge. Therefore, although the ac output voltage starts out with the full dc component, after a few *transient* cycles, as the capacitor charges, the dc level is removed. This is shown in Fig. 10-4.

What happens if the ac signal is not a sine wave but has a dc component of its own? The capacitor does not distinguish one source of dc from

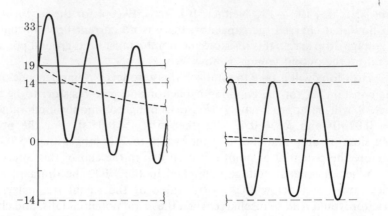

Figure 10-4   Transient output waveshape as the coupling capacitor removes the dc component.

another. Therefore, just as it removed the dc component V_{DO} due to the no-signal drain potential of the FET, it will also remove any dc component of the signal itself. The zero level of the output voltage is always at the average value of the ac signal. This is shown in Fig. 10-5.

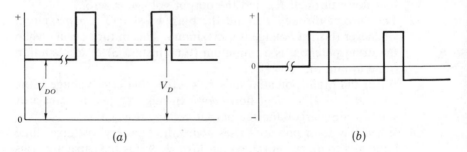

Figure 10-5   Effect of coupling capacitor on a complex wave with dc component. **(a)** Input. **(b)** Output.

GAIN—EQUIVALENT CIRCUIT.   An important parameter of a FET is the transconductance g_{fs}.* It shows how the drain current is affected by changes in the gate voltage (with the drain-source voltage held constant):

$$g_{fs} = \frac{\Delta I_D}{\Delta V_{GS}} \quad (V_{DS} \text{ constant}) \qquad \textbf{(10-2)}$$

---

*Recall that this is the real portion of the transadmittance, y_{fs}, and that the susceptance component is negligible except at very high frequencies.

In a practical circuit, the changes in current and voltage are the ac components. Rewriting Equation 10-2 in terms of these ac components, we get

$$g_{fs} = \frac{i_d}{v_{gs}} \qquad \text{(still with } V_{DS} \text{ constant)} \tag{10-3}$$

However, we know that the drain potential will vary because of the dc load resistor R_d. Adding the effect of change in V_{DS} and rewriting Equation 10-3 in terms of drain current,

$$i_d = g_{fs}v_{gs} + \frac{1}{r_d}v_{ds} \tag{10-4}$$

where r_d is the drain resistance or the reciprocal of the output conductance g_{os}.

Equation 10-4 gives rise to the equivalent circuit shown in Fig. 10-6. This is a simplified diagram that omits interelectrode capacitance effects.*

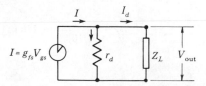

Figure 10-6 Constant-current equivalent circuit.

Notice that no direct voltages (drain potential or gate bias) are shown. This is because an equivalent diagram represents the action of a circuit with respect to the *varying or ac components only.* Since the input signal voltage affects the output current, it is replaced by a current source, $g_{fs}v_{gs}$. This circuit is a Norton-type or *constant-current* equivalent circuit. Notice that the output voltage is a result of this current flowing through an impedance consisting of a load impedance Z_L and r_d in parallel:

$$V_o = (g_{fs}V_{gs})\left(\frac{\dot{Z}_L \dot{r}_d}{\dot{Z}_L + \dot{r}_d}\right) \tag{10-5}$$

To calculate voltage gain from this equivalent circuit,

$$A_v = \frac{V_o}{V_i} = \frac{g_{fs}V_{gs}}{V_{gs}}\left(\frac{\dot{Z}_L \dot{r}_d}{\dot{Z}_L + \dot{r}_d}\right) = g_{fs}Z_e \tag{10-6}$$

where Z_e is equal to the parallel impedance of $\dot{Z}_L$ and $\dot{r}_d$.

*FET capacitances are quite low, and this equivalent circuit is valid up to very high frequencies. (The complete circuit, with capacitances, is discussed later under frequency distortion.)

This is a general equation applicable to any frequency. However, as before, *under ideal conditions* (mid-frequency) when circuit reactances can be neglected, the load impedance Z_L can be replaced by a load resistor R_L. Obviously, the impedance Z_e becomes a pure resistive quantity R_e and the gain equation for this ideal condition becomes

$$A = g_{fs} R_e \qquad (10\text{-}7)$$

where R_e is the parallel resistance of R_d, R_g, and r_d.

EXAMPLE 10-1

A 2N4868 JFET has an output conductance g_{os} of 4 μmho and a transconductance g_{fs} of 3000 μmho. It is used in an R-C coupled amplifier circuit with $R_d = 33.3$ kΩ and $R_g = 1$ MΩ. Find the gain of this stage by using the constant-current equivalent-circuit technique.

Solution

1. To find r_d:

$$r_d = \frac{1}{g_{os}} = \frac{1}{4 \times 10^{-6}} = 250 \text{ k}\Omega$$

2. To find R_e:

$$\frac{1}{R_e} = \frac{1}{r_d} + \frac{1}{R_d} + \frac{1}{R_g}$$

$$= \frac{1}{250,000} + \frac{1}{33,300} + \frac{1}{1,000,000} \qquad (1{,}000{,}000 R_e)$$

$$1{,}000{,}000 = 4R_e + 30R_e + R_e$$

$$R_e = 28{,}000 \ \Omega$$

3. $A_v = g_{fs} R_e = 3000 \times 10^{-6} \times 28{,}600 = 85.8.$

This technique for calculating the gain of an amplifier circuit seems quite simple and direct. Yet it has several limitations and weaknesses. The equivalent-circuit technique is limited to analyses for small-signal amplifiers because parameter values change over wide excursions. Even with small signals, the parameter values should apply to the specific operating point selected. What is the operating point in Example 10-1? It is not stated, and no data (such as V_{DD}, V_{GS}, or I_D) is available for fixing the quiescent condition. The parameters used in Example 10-1 were taken from the data sheet for $V_{DS} = 15$ V and $V_{GS} = 0$. What correction, if any, should have been made? Finally, we always have the problem that no two semiconductors— even of the same type—are alike, and wide parameter variations are possible from "minimum" to "maximum" characteristics.

GRAPHIC ANALYSIS

Two of the above limitations—errors due to signal swing, and not using parameter values for the specific operating point—can be overcome by using a graphic analysis technique. For this method we need the output characteristic curves, and we must know the component values and the dc operating voltages. The characteristic curves show the relation between drain current, drain-source voltage, and gate-source voltage, and are therefore a measure of g_{fs} and g_{os}. But we know that the value of these parameters depends on the operating point. What is the operating point? How do we find it?

The first question is readily answered. The operating point corresponds to the quiescent (no-signal) values of drain current I_D, drain-source voltage V_{DS}, and gate-source bias V_{GS}. But finding this operating point is a little more involved.

DC LOAD LINE.   Refer to Fig. 10-1 and consider only the dc conditions pertaining to the circuit between V_{DD} and ground. This is a simple series circuit containing R_s, the FET (source to drain), and R_d. By series-circuit rules (Kirchhoff's voltage law), it is obvious that

$$V_{DS} = V_{DD} - I_D R_s - I_D R_d$$
$$= V_{DD} - I_D(R_d + R_s) \tag{10-8a}$$

or, replacing $I_D R_s$ by the gate-source bias, it produces

$$V_{DS} = V_{DD} - V_{GS} - I_D R_d \tag{10-8b}$$

If R_s is small compared to R_d, or if V_{GS} is small compared to V_{DD}, these equations can be simplified to*

$$V_{DS} = V_{DD} - I_D R_d \tag{10-8c}$$

These equations (since V_{DD}, V_{GS}, and R_d are fixed values) are first-degree equations and can be represented as a straight line on a plot of I_D versus V_{DS}. This line is often referred to as the *load line*. More exactly, it should be called the *dc load line*. To draw a dc load line, only two points are necessary. These two points are readily established as follows:

*This simplification may be necessary when neither R_s nor V_{GS} is known. Should the error be significant when these values *are* found, the solution can be repeated by using Equation 10-8a or 10-8b. Conversely, a probable value for V_{GS} can be assumed and substituted into Equation 10-8b.

1. When the drain current is zero, the drain-source voltage is equal to the drain supply voltage. This point will lie on the drain voltage axis corresponding to $V_{DS} = V_{DD}$.
2. The drain-source voltage will be zero at a current that makes $I_D(R_s + R_d) = V_{DD}$; or $I_D R_d = V_{DD} - V_{GS}$ (or, using the simplified equation, $I_D R_d = V_{DD}$). This second point is located on the drain current axis at the calculated value.
3. By joining these two points we get the dc load line.

EXAMPLE 10-2

A transistor is to be operated from a 38-V supply using a drain resistor of $34,000 \, \Omega$ and a source resistor of $4000 \, \Omega$. Draw the dc load line.

Solution

The two points for the load line are

1. $V_{DS} = V_{DD} = 38$ V at $I_D = 0$.

2. $V_{DS} = 0$ at $I_D = \dfrac{V_{DD}}{R_s + R_d} = \dfrac{38}{38,000} = 1$ mA.

The dc load line is shown in Fig. 10-7 as the solid line.

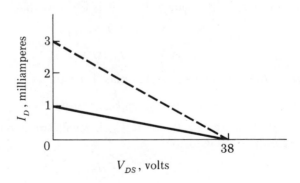

Figure 10-7   Load lines for Examples 10-2 and 10-3.

EXAMPLE 10-3

The transistor of Example 10-2 is again operated from the 38-V supply, but with $R_d = 12,000 \, \Omega$ and a V_{GS} of -2 V obtained by source bias. Draw the dc load line.

Solution

The two points for this load line are

1. $V_{DS} = V_{DD}$ at $I_D = 0$.

2. $V_{DS} = 0$ at $I_D = \dfrac{V_{DD} - V_{GS}}{R_d} = \dfrac{38 - 2}{12,000} = 3 \text{ mA}.$

This load line is shown in Fig. 10-7 as the broken line.

Notice from these two examples that the higher the drain resistance, the flatter (more nearly horizontal) the dc load line.

OPERATING POINT.   At the start of this discussion of graphic analysis, we said that we need the output characteristics of the FET. The axes for these curves (I_D versus V_{DS}) are the same as those for the dc load line. Therefore, once we select a specific transistor we can plot the dc load line on the output characteristics sheet for that transistor. The operating point must then be a point on that dc load line and is fixed by the intersection of this line with the bias condition. An example will clarify this.

EXAMPLE 10-4

Figure 10-8 shows the output characteristics and dc load line for the FET circuit of Example 10-3 ($R_d = 12,000 \ \Omega$; $V_{GS} = -2.0$ V; $V_{DD} = 38$ V).

1. Locate the operating point.
2. Specify the quiescent values.
3. Find the value of source resistor needed to produce the -2.0-V bias.

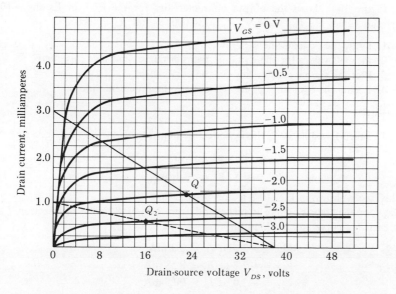

Figure 10-8   Construction of dc load line and evaluation of operating point.

Solution

1. Since the bias is given as $V_{GS} = -2.0$ V, the operating point must be at the intersection of the $V_{GS} = -2.0$ curve and the load line. This point is shown on the plot as point Q.
2. The quiescent values are $I_{DQ} = 1.7$ mA and $V_{DSQ} = 23$ V.
3. The required value of source resistor is

$$R_s = \frac{V_{GS}}{I_D} = \frac{2.0}{1.7 \times 10^{-3}} = 1710 \ \Omega$$

Locating the dc operating point is a little more involved when the value of R_s is given instead of the bias value V_{GS}. Although we know that the bias is the same as the voltage drop across the source resistor ($V_s = V_{GS} = I_D R_s$), we do not know the drain current—and we cannot find it, because we do not know the bias. The solution to this merry-go-round is to use a second graphic solution for the equation $V_{GS} = I_D R_s$. One point for this line is $V_{GS} = 0$ at $I_D = 0$. For the second point, select any value for I_D and calculate V_{GS}. But notice that we are plotting I_D versus V_{GS}. These are the coordinates for a *transfer* characteristic. Therefore, we must first plot the dynamic transfer curve from the output characteristic and the load line. An example will illustrate the procedure.

EXAMPLE 10-5

Figure 10-8 shows the output characteristics for the FET of Example 10-2. The circuit values are $V_{DD} = 38$ V; $R_d = 34,000 \ \Omega$; and $R_s = 4000 \ \Omega$.

1. Draw the dc load line.
2. Pick off values for the dynamic transfer curve, and plot this curve.
3. Plot the source-resistor line on the curve sheet.
4. Locate the operating point and specify the quiescent values.

Solution

1. From Example 10-2, the dc load line points are

$$V_{DS} = 38 \text{ V at } I_D = 0$$
$$V_{DS} = 0 \text{ V at } I_D = 1 \text{ mA}$$

This line is shown in Fig. 10-8 as the broken line.
2. Data for the transfer curve is picked off from Fig. 10-8 at the intersection of each V_{GS} curve and the load line:

$$V_{GS} = -2.0 \text{ at } I_D = 0.9 \text{ mA}$$
$$V_{GS} = -2.5 \text{ at } I_D = 0.58 \text{ mA}$$
$$V_{GS} = -3.0 \text{ at } I_D = 0.25 \text{ mA}$$

This portion of the transfer curve is shown in Fig. 10-9.
3. For the source-resistor line, with $R_s = 2000 \ \Omega$,

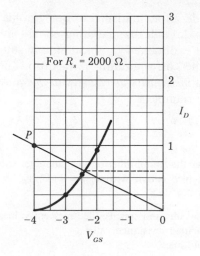

Figure 10-9   Transfer curve for Example 10-5.

$$V_{GS} = 0 \text{ V at } I_D = 0$$
$$V_{GS} = 4 \text{ V at } I_D = 2 \text{ mA}$$

These points are plotted in Fig. 10-9, and the slant line is the $R_s = 2000\ \Omega$ line.

4. At the intersection (Fig. 10-9) note that the drain current is 0.62 mA. This is the quiescent value. Locate this point as Q_2 in Fig. 10-8, where 0.62 mA of drain current intersects the dc load line. Notice that V_{DSQ} is 16 V.

AC LOAD LINE.   The sole purpose of establishing a dc load line is to fix the operating point (or to select a suitable one). However, when an ac input signal is applied, circuit operation takes place along an *ac load line*. Refer once again to Fig. 10-1. As mentioned earlier, the ac component of drain current can flow not only through R_d (the dc load) but also through C_2 and R_{g2} and whatever is connected across the output terminals. Therefore, the ac load impedance Z_L consists of R_d in parallel with $C_2 + R_{g2}$, in parallel with the next stage input impedance, and in parallel with any shunting interelectrode and/or stray capacitances. Fortunately, with FET circuits (except at very high frequencies) the shunting capacitive reactances are very high and the input impedance of a next FET stage is also very high. These effects can be neglected, and impedance Z_L simplifies to a load resistance R_L equal to R_d and R_{g2} in parallel.*

---

*Notice that we have neglected the source resistor R_s in this discussion of the ac load impedance. This is so because R_s is bypassed by C_s and its impedance to signal frequencies should approach zero.

Now we must add an ac load line to our graphic construction. This ac load line must pass through the operating point—giving us one plotting point. A second point is obtained by selecting any suitable value of drain current, calculating the corresponding $I_D R_L$ drop, and then adding this current and voltage *as increments or changes to the quiescent values.* Let us illustrate this with an example.

EXAMPLE 10-6

Figure 10-10 shows the (minimum) output characteristics of a 2N3688 JFET. It is used in an R-C coupled circuit with $V_{DD} = 24$ V, $V_{GS} = -1.5$ V, $R_d = 27$ kΩ, and R_g of the next stage $= 91$ kΩ.

1. Draw the dc load line and specify the quiescent values.
2. Calculate the ac load resistance value.
3. Draw the ac load line.

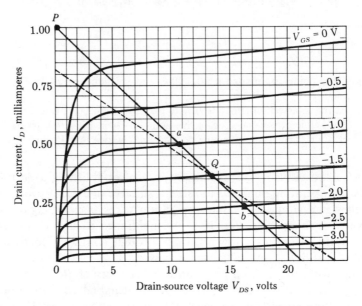

Figure 10-10   Graphic construction for Example 10-6.

Solution

1. To draw the dc load line, the points are

$$V_{DS} = 24 \text{ V at } I_D = 0$$

$$V_{DS} = 0 \text{ V} \quad \text{at } I_D = \frac{V_{DD} - V_{DS}}{R_d} = \frac{24 - 1.5}{27,000} = 0.833 \text{ mA}$$

Locate these two points and join them for the dc load line. (See the broken line in Fig. 10-10.) The operating point is at the intersection of this dc load line and the static characteristic curve for $V_{GS} = -1.5$. (See point Q.) The quiescent values are picked off as $I_{DQ} = 0.36$ mA and $V_{DSQ} = 13.5$ V.

2. The ac load (R_L) consists of R_d and R_{g2} in parallel, or

$$R_L = \frac{R_d \times R_{g2}}{R_d + R_{g2}} = \frac{27 \times 91}{118} \, k = 20.8 \, k\Omega$$

3. One point for the ac load line is the operating point Q. For a second point, let us select an I_D increment of $+0.64$ mA (so that the total I_D value is 1.00 mA). The added $I_D R_L$ drop is

$$\Delta I_D R_L = (0.64 \times 10^{-3})(20.8 \times 10^3) = 13.3 \, V$$

Subtracting this from the quiescent value, $V_{DS} = 13.5 - 13.3 = 0.2$ V. The coordinates for the second point are

$$V_{DS} = 0.2 \, V \text{ at } I_D = 1.0 \, mA$$

Locate this point (P) on the graph and draw the ac load line.

OUTPUT VOLTAGE AND GAIN.   The ac load line can now be used to calculate the output voltage and the gain for a given input signal voltage.

EXAMPLE 10-7

The FET of Example 10-6 (Fig. 10-10) has an input signal of ± 0.5 V peak. Find the output voltage and gain for this circuit.

Solution

1. The circuit operation follows along the ac load line—starting at the operating point ($V_{GS} = -1.5$ V and $V_{DS} = 13.5$ V).
2. When the input signal is $+0.5$ V, the instantaneous total gate voltage is $(-1.5 + 0.5)$ or -1.0 V. The corresponding V_{DS} is 10.8 V (ΔV_{DS} is -2.7 V).
3. When the input signal is -0.5 V, V_{GS} is -2.0 V and V_{DS} is 16.2 V (ΔV_{DS} is $+2.7$ V).
4. The gain is

$$\frac{\Delta V_o}{\Delta V_i} = \frac{5.4 \, V \, (p\text{-}p)}{1 \, V \, (p\text{-}p)} = 5.4$$

EFFECT OF R_L VALUE.   From Table 9-1 in the previous chapter, we saw that the higher the load resistance, the higher the voltage across the load (or the higher the output). From Equation 10-6 we saw that the higher the load value, the greater the gain. Yet, in a practical situation, this is true—only up to a point. Figure 10-11 shows the output characteristic curves of a FET and five ac load lines (A to E). This FET has a BV_{DGO} of 30 V. To prevent breakdown, the maximum swing on the ac load line—in each case—is fixed at 25 V. The ac load resistances are 5000, 12,500, 25,000, 50,000, and 125,000 Ω. In each case, the operating point has been centered (approximately) around the maximum gate-voltage swing. Using load line A, an input voltage swing of ± 0.5 V will produce an output voltage swing from its quiescent value of 21.1, to 19.8, and then to 22.4, or a total swing of 2.6 V. Note also that the output swing to each side is balanced (-1.3 V and $+1.3$ V).

In similar fashion let us examine the effect of a ± 0.5-V input with each of the load lines. The results are shown in Table 10-2.

Table 10-2 shows that the gain does increase as the load resistance is increased. However, it also shows that the imbalance (or distortion) also increases. A compromise must be made between higher gain and acceptable distortion level. Of course, even with load line E, if the input signal is quite small (millivolts), the imbalance to either side becomes negligible and the higher gain condition (with high R_d value) can be used without distortion. Also, it should be obvious that with a higher supply voltage (and a higher FET BV rating), we could still get high-enough drain currents to stay out of the cutoff region even with large R_d values. Such FETs would give us higher gain, higher output voltages, and lower distortion.

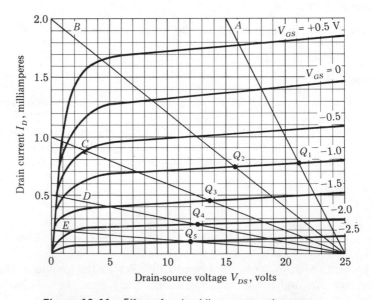

Figure 10-11   Effect of ac load line on operating range.

Table 10-2
EFFECT OF AC LOAD ON GAIN FOR ±0.5 V PEAK INPUT

R_L (kΩ)	QUIESCENT V_{DS} (volts)	$V_{DS(min)}$ (volts)	$V_{DS(max)}$ (volts)	TOTAL SWING AND GAIN (volts)	IMBALANCE (volts)	
A: 5	21.1	19.8	22.4	2.6	$(+1.3)$ & (-1.3) = 0	
B: 12.5	15.8	12.5	19.0	6.5	$(+3.2)$ & (-3.3) = 0.1	
C: 25	13.6	7.4	18.2	10.8	$(+4.6)$ & (-6.2) = 1.6	
D: 50	12.6	5.0	18.5	13.5	$(+5.9)$ & (-7.6) = 1.7	
E: 125	12.0	1.7	?	*	?	?

*The swing on the positive half-cycle of the input signal cannot be determined from these curves because the $V_{GS} = -3.0$ curve is not shown. Obviously, the curves are very crowded in this region as we approach cutoff. The imbalance (distortion) would be very great.

DESIGN CONSIDERATIONS.   When planning circuits, the economics of the design must always be considered. For example, FETs with break-down voltages of 20 V are cheaper than units with BV ratings of 50 V or 150 V. Also, the power supply for a low-voltage, small-signal FET would be less expensive. On the other hand, it is obvious from our graphic anal-yses that we cannot get an output voltage of 30 V peak to peak from a FET with a 20-V breakdown rating. So when selecting a FET one item to consider is how much output voltage we need.

Once we select a suitable FET (and power supply), another item to consider is how large (or small) the input signal voltage is. This is important in the selection of the load resistance and the operating point. Remember that the FET should operate beyond the pinch-off region. Also remember that type-A FETs—even of the junction variety—can operate with low positive gate potentials (+0.5 V).*

For this discussion, let us assume that a FET with a BV rating of 30 V is satisfactory. Therefore we can use the FET curves of Fig. 10-11. Reexamine this diagram, noting this time the range of input signal swing. With load lines A or B, the input signal could be a full 3.0 V peak to peak (from +0.5 to −2.5 V). Actually there is no advantage in using load line A,† since load line B gives us the same input signal-handling ability with appreciably more gain. At the other extreme, load line E could be used only for signals of at most 0.5 V peak to peak (from just over −2.0 V to just over −2.5 V).

*With a JFET, any higher voltage would forward-bias the junction above the threshold value and excessive gate current would flow.
†Yet load line A has zero imbalance. An optimum load line for minimum imbalance and maximum signal-handling ability would be a load line of about 11,500 Ω, so that it crosses the $V_{GS} = +0.5$ V curve at $V_{DS} = 6.0$ to 7.0 V.

Let us consider the merits of load lines *B*, *C*, and *D* in more detail. All three lines are redrawn in Fig. 10-12 as *dynamic transfer characteristics*. For minimum imbalance, operation should be restricted to the straight-line portion of these curves. Examine the highest curve, corresponding to load line *B*. This curve is reasonably linear for V_{GS} values between +0.5 and −2.0 V. Operating point *Q* should be selected in the center of this

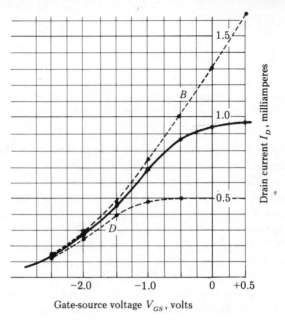

Figure 10-12   Dynamic transfer curves for Fig. 10-11.

range, or (approximately) at −0.75 V. The quiescent drain current is now picked off as (approximately) 0.9 mA. Since this is the current through the source-bias resistor, we can calculate:

$$R_s = \frac{V_{GS}}{I_{DQ}} = \frac{0.75}{0.9 \times 10^{-3}} = 830 \; \Omega$$

Going back to the output characteristics (Fig. 10-11), at $I_{DQ} = 0.9$ mA on load line *B* we find the quiescent value of drain voltage, $V_{DSQ} = 13.7$ V.

At this point we should select a power supply voltage—somewhat higher than the 25 V used for the ac load line (for example, 28 V). Now, using series circuit principles, we can solve for the value of the dc load resistance:

$$R_d + R_s = \frac{V_{DD} - V_{DSQ}}{I_{DQ}} = \frac{28 - 13.7}{0.9 \times 10^{-3}} = 15,900 \ \Omega$$

$$R_d = 15,900 - R_s = 15,900 - 830 \cong 15,000 \ \Omega$$

And since the ac load resistance (R_L) is 12,500 Ω,

$$R_{g2} = \frac{R_d \times R_L}{R_d - R_L} = \frac{15,000 \times 12,500}{2500} = 75,000 \ \Omega$$

If the input-signal swing is less than 3.0 V peak to peak, we can consider using a higher ac load resistance than the 12,500 Ω of load line B and get higher gain. For example, load line C could be used over a fairly linear range of -0.5 to -2.0 V, with a quiescent point at -1.25 V. This will handle an input signal of 1.5 V peak to peak. For lower input-signal levels, consider load line D. The swing on this load line should be restricted to the range of about -1.25 to -2.5 V (that is, 1.25 V peak to peak), with an operating point at about -1.9 V. For still lower input levels, ac load lines between D and E can be investigated. In each case, once the load line and operating point are established, the resistance values for R_s, R_d, and R_{g2} are then found as explained above for load line B.

OTHER BIAS CIRCUITS

The FET circuit of Fig. 10-1 uses self-bias or source bias to obtain the desired dc gate-source voltage. This is one of the most commonly used techniques with JFETs and type-A MOSFETs. However, it can result in appreciable drain current differences if the FET is replaced. Since the transadmittance (and g_{fs}) varies as the square of the drain current (see Equation 3-9), this could cause excessive gain variation. Figure 10-13 can be used to study this effect. Both diagrams show the minimum and maximum transfer characteristics for a 2N4341 N-channel JFET.* Let us assume that we plan to use a 500-Ω source resistor for bias. This bias line is shown in Fig. 10-13a going through point P. Notice that the drain current could range from a low of 1.6 mA to as high as 3.6 mA, depending on the individual FET used. This variation can be reduced by using a higher resistance value for R_s. For example, the broken line in Fig. 10-13a is for $R_s = 2000 \ \Omega$. The current variation with this bias line is reduced to 0.7–1.5 mA. However,

---

*Since the drain current of a FET is practically constant (in the operating region beyond V_p), it is permissible to use the manufacturers' static curves in place of the *dynamic* transfer curves.

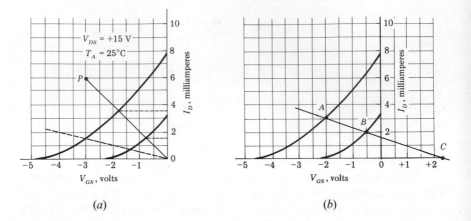

Figure 10-13   Comparison of source bias and combination bias.

we are now operating at rather low current values (for both cases), where the gain is lower and the distortion level is higher.

A solution to this situation is to use a combination of fixed (voltage-divider) bias in addition to source bias. Let us assume that we can tolerate a drain current variation of from 3.0 mA (maximum) to 2.0 mA (minimum). Figure 10-13b shows these two points located on their respective transfer characteristics as points A and B. By drawing a line through these two points we intersect the horizontal axis at $V_{GS} = +2.6$ V. This is the value of fixed bias needed to stabilize the drain current changes to within the specified limits. The value of the source resistor can now be obtained by Ohm's law, using the current and voltage changes between any two points on this line.

EXAMPLE 10-8

Find the source resistance needed for the bias operation shown in Fig. 10-13b.

Solution

Better accuracy is obtained by picking points far apart. We will use points A and C.

1. The current change from C to A is from 0 to 3 mA.
2. The voltage change is from $+2.6$ to -2.0 or 4.6 V.

$$R_s = \frac{\Delta V}{\Delta I} = \frac{4.6}{3 \times 10^{-3}} = 1530 \ \Omega$$

Two combination-bias circuits are shown in the partial schematics of Fig. 10-14. The ratio for the voltage-divider resistors, R_1 and R_2, is fixed

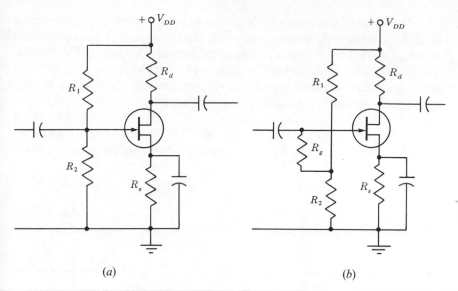

(a) (b)

Figure 10-14 Combination-bias circuits.

by the ratio of the desired fixed-bias value to the supply voltage value V_{DD}. In Fig. 10-13b, if the supply voltage is $+30$ V, then for a fixed bias of $V_{GS} = +2.5$ V, the ratio R_1/R_2 should equal 2.5/30 or 1:12. This means that R_2 should be 12 times larger than R_1. But notice that R_1 also takes the place of the former resistor R_g, which is already a large value. It may be difficult to obtain the proper ratio with such high resistance values. This situation is avoided in Fig. 10-13b. R_1 and R_2 can be any values (large or small) that will provide the desired ratio. The high impedance from gate to ground is then maintained by connecting the usual gate resistor, R_g, between the gate and the voltage-divider tap point.

ENHANCEMENT FETs. The enhancement FET requires a gate bias of the same polarity as its drain. An N-channel type-C FET would have a positive drain and a positive gate potential. (In a P-channel unit, both drain and gate would have negative polarities.) Can we use source bias with these devices? Consider an N-channel MOSFET. The drain is positive. Drain current flowing up through the source resistor will make the source positive compared to ground—and the gate will be negative *compared to source*. This is the wrong polarity. Similarly, with a P-channel unit the drain is negative; drain current flows down through R_s, making the source negative with respect to ground; the gate is positive compared to source; and again the wrong polarity. Obviously, we cannot use source bias with enhancement FETs.

The simplest way to obtain a bias of proper polarity is to connect the usual gate resistor, R_g, between gate and drain (instead of gate to ground as in Fig. 10-1). This change is shown in the partial schematic of Fig. 10-15a. Resistor R_g should have a high resistance (5 to 10 MΩ). Since there is no gate current flow, the gate bias V_{GS} is the same as the drain-source voltage V_{DS}. This is generally a satisfactory operating condition. However, if for any reason some lower value of V_{GS} is desired, it can be obtained by using the voltage-divider connection in Fig. 10-15b. To maintain a high input impedance, both resistors, R_1 and R_2, should have high-resistance values.

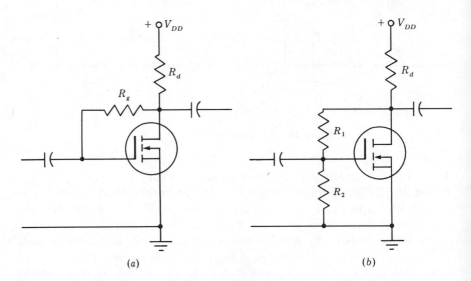

(a)   (b)

Figure 10-15   Bias circuits for enhancement MOSFETs.

There is one other type of MOSFET—the type B. You will recall that these units operate in both the depletion and the enhancement modes. The zero-bias condition ($V_{GS} = 0$) is just about in the center of their transfer characteristics. Therefore, these units are often operated with the gate resistor grounded and without a source resistor—in other words, at zero bias. If some other operating point is desired, we can use any of the preceding bias circuits to fix the quiescent point to either side of the zero-bias condition.

VACUUM-TUBE AMPLIFIERS

Vacuum tubes—as small-signal amplifiers—are being replaced by solid-state circuitry. However, even as of 1975, new equipment models (for example, monochrome television receivers) were still being marketed with vacuum tubes. In addition, there is much vacuum-tube equipment, both consumer and industrial, in operation and requiring maintenance. Obviously, the study of vacuum-tube circuits cannot be ignored. Since tubes are quite similar in action to FETs, it is appropriate to discuss them both in this chapter.

Figure 10-16 shows a typical R-C coupled triode amplifier circuit. Notice that—except for the tube replacing the FET and the corresponding change in component subscripts—this circuit is identical to Fig. 10-1 for

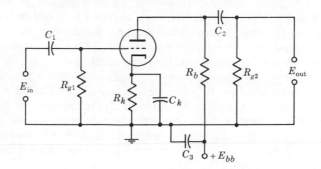

Figure 10-16   Typical triode R-C coupled amplifier. (*Note:* The negative terminal of the power supply is grounded. Unless otherwise shown, this connection is always implied.)

the FET. The function of these components follows the FET description, and no further discussion is necessary. One major difference is in the value of the plate supply voltage E_{bb}. This value may range from a low of about 95 V to as high as 300 V. Other common circuit values are plate resistor R_b, from 2 to 10 times the internal plate resistance r_p of the tube; coupling capacitor, from 0.005 to 0.1 μF; and grid-leak resistor R_g, from 0.1 to 1.0 MΩ.

By using pentodes in place of triodes in R-C coupled amplifiers, it is possible to obtain higher gain per stage and an extended high-frequency response. A typical pentode circuit is shown in Fig. 10-17.

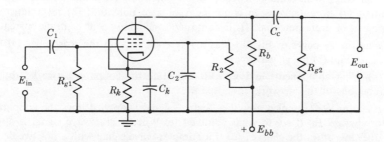

Figure 10-17   Typical pentode R-C coupled amplifier.

REVIEW QUESTIONS

1. (a) Name a circuit that is commonly used for voltage amplification. (b) Give four applications for this circuit. (c) Give four reasons for its wide usage.

2. Refer to Fig. 10-1: (a) Is this one complete stage? Explain. (b) What is the function of resistor R_s? (c) Explain how it performs this function. (d) What two quantities determine the value of this resistor? (e) What is the function of capacitor C_s? (f) What determines the minimum value of this capacitor? (g) What name is given to the combination of R_s and C_s? (h) What name is given to capacitor C_2? (i) State two functions of this capacitor. (j) What range of values is typical for this component? (k) What other circuit component is of importance when selecting this capacitor value? (l) If this capacitor were to become short-circuited, what effect would it have on the circuit? Explain. (m) What name is generally given to resistor R_{g2}? (n) State two basic functions of this resistor. (o) Explain why large resistance values are desirable for R_g. (p) What limits the maximum usable value for R_g? (q) What are typical values for this resistor? (r) Why is resistor R_c necessary? (s) What is the function of capacitor C_3? (t) Is this capacitor always used? Explain. (u) In some circuits, in place of C_3 two capacitors— a 10-μF and a 0.001-μF capacitor—are connected in parallel. Of what significance is the extra 0.001-μF capacitor?

3. Refer to Fig. 10-1: What constitutes (a) the dc load resistance? (b) The ac load resistance?

4. Refer to Table 10-1, when the signal voltage (column 1) is +1 V. (a) Explain why the gate potential is -2 V. (b) Explain why the voltage drop across R_d is 28 V. (c) Explain why the drain potential is $+12$ V. (d) Explain why the ac component is -7 V. (e) Explain why the output voltage is -7 V.

5. In an R-C coupled amplifier, what is the phase relation between (a) the input signal and the drain current; (b) the drain current and the drain potential; (c) the input signal and the output voltage?

6. Give the equation for the instantaneous drain potential.

7. What is meant by *quiescent* values?

8. Give the symbol used to designate (a) drain power supply voltage; (b) gate-bias supply voltage; (c) average drain potential; (d) average gate potential; (e) quiescent value of drain current; (f) quiescent value of drain potential; (g) instantaneous *total* drain current; (h) instantaneous total drain potential; (i) instantaneous total gate potential; (j) instantaneous ac component of gate potential; (k) instantaneous ac component of drain potential; (l) instantaneous ac component of drain current.

9. Using the v_G curve of Fig. 10-2 as a reference, explain the reason for (a) the i_d waveshape and (b) the v_D waveshape.

10. (a) What components in Fig. 10-1 constitute the coupling network? (b) State two functions of this network.

11. (a) In an R-C coupled amplifier, before signal is applied does the coupling capacitor charge? (b) If so, to what voltage? (c) When a sine-wave signal is applied, how does this affect the charge? (d) If a complex-wave signal with a positive dc component is applied, how does this affect the charge? **(e)** If the complex-wave signal had a negative dc component, how would it affect the charge on the capacitor?

12. Refer to Fig. 10-2: If the drain potential waveshape is applied to a coupling capacitor and gate resistor, what is the output voltage value across R_g corresponding to **(a)** V_{D0}; **(b)** $V_{D(min)}$; **(c)** $V_{D(max)}$?

13. In Question 12, if the gate potential waveshape is fed to the coupling network, find the output voltage corresponding to **(a)** V_{G0}; **(b)** $V_{G(min)}$; **(c)** $V_{G(max)}$.

14. Refer to Fig. 10-3: **(a)** What is the significance of the battery shown across the capacitor? **(b)** Why is this voltage 19 V? **(c)** Account for the -14 V of the output waveshape at time instant 3.

15. Refer to Fig. 10-4: **(a)** What causes the waveshape to "fall downhill"? **(b)** Why does it level off at the end of the time period shown?

16. Refer to Fig. 10-5: **(a)** What is the significance of the level marked V_D? **(b)** What does the zero level in diagram (b) correspond to?

17. **(a)** What is the letter symbol for transconductance? **(b)** What does this parameter show? **(c)** Except at very high frequencies, what other parameter can be used in its place? Why?

18. **(a)** Why are dc potentials omitted from an equivalent circuit? **(b)** How does the load for an ideal case differ from the general case? **(c)** Why is this so? **(d)** What circuit components determine the value of the load under ideal conditions?

19. Refer to the constant-current equivalent circuit: **(a)** Give the equation for the current source. **(b)** Give the general equation for gain, applicable to any frequency. **(c)** Give the gain equation for ideal (mid-frequency) conditions. **(d)** What determines the "equivalent load" value for this ideal condition?

20. State two precautions that must be observed when using analysis by the equivalent-circuit method.

21. What is the effect on the gain of a stage **(a)** if R_d is reduced? **(b)** If R_g is reduced? **(c)** In a practical FET circuit, which effect is greater? Why?

22. **(a)** What is the equation for the dc load line? **(b)** What component value determines the slope of this line? **(c)** If this component value is decreased, how does the slope change? **(d)** What two points are most commonly used to fix the dc load line? **(e)** Why is the dc load line necessary in starting a graphic analysis?

23. What two circuit values must be known before a dc load line can be constructed?

24. Refer to Fig. 10-7: **(a)** Specifically, what does the number 38 represent? **(b)** What do the two slant lines represent? **(c)** How is the 3 mA value obtained for the broken line?

25. In Example 10-3 (step 2), why is V_{GS} subtracted from V_{DD} when solving for I_D?

26. Refer to Fig. 10-8 and the solid load line: **(a)** For what value of drain supply voltage is this dc load line drawn? **(b)** For what value of drain resistor does this apply? **(c)** Give the quiescent value of drain current and drain-source voltage for a gate bias of -1.0 V. **(d)** Repeat part (c) for a gate bias of 2.5 V. **(d)** Can you see any objection to fixing the operating point at a bias of -2.5 V? Explain.

27. Refer to Fig. 10-9: **(a)** What does the curved line represent? **(b)** How are the points for this curve obtained? **(c)** What does the straight line represent? **(d)** How is the value for point *P* obtained?

28. How is the operating point fixed on the dc load line if the source resistor value is known instead of the bias voltage?

29. **(a)** What is the distinction between the dc load line and the ac load line? **(b)** What additional factor is considered in obtaining the value used for the ac load line? **(c)** Are any points in common between these two load lines? Explain. **(d)** Which of these two lines generally has the steeper slope? Why? **(e)** Under what condition could we consider the ac load line the same as the dc load line in an R-C coupled amplifier?

30. What is another name for the ac load line?

31. Refer to Fig. 10-10: **(a)** Which of these load lines is the ac load line? **(b)** How can you tell? **(c)** For what drain supply voltage are these load lines drawn? **(d)** How can you tell?

32. **(a)** Can the voltage gain of a stage be evaluated from an ac load line? Explain. **(b)** Will a change in the FET parameters—with signal swing—affect the accuracy of such evaluations? Explain.

33. Refer to Fig. 10-10: **(a)** What could cause the operating point to shift instantaneously from *Q* to *a* and to *b*? **(b)** What would be the peak-to-peak value of such an input signal? **(c)** What would be the corresponding drain voltage swing? **(d)** How could we calculate the voltage gain of the circuit?

34. Refer to Fig. 10-11: **(a)** What do the lines marked *A* to *E* represent? **(b)** Which is the highest resistance load line? **(c)** Which of these lines would produce the highest circuit gain? Explain. **(d)** Which would produce the lowest gain? **(e)** Which of these lines produces the least distortion? **(f)** Which produces the highest distortion?

35. **(a)** How can we use higher load resistance values for higher gain and still stay out of the low-current, high-distortion region? **(b)** What limits this type of solution?

36. Refer to Fig. 10-11: **(a)** Rate these five ac load lines with regard to their ability to handle high input-signal levels. **(b)** How do you arrive at these relative ratings? **(c)** For maximum signal-handling ability—on any load line—where should the quiescent point be located? **(d)** What happens to the signal-handling ability on load line *D* if its V_{GSQ} bias point is made the same as for load line *B* (that is, -1.0 V)? Explain.

37. Refer to Fig. 10-12: **(a)** How are the points for these curves obtained? **(b)** Over what range of gate-voltage swing will curve *B* give distortionless operation? **(c)** Repeat part (b) for curve *C*. **(d)** Repeat part (b) for curve *D*. **(e)** For maximum signal-handling ability, how should the operating point be selected on each of these curves? **(f)** For operating-point selection, which type of graphic representation is better: Fig. 10-11 or Fig. 10-12? Explain.

38. Refer to Fig. 10-13*a*: **(a)** What do the two curved lines represent? **(b)** What does the solid straight line represent? **(c)** What is the disadvantage of using so low a value of source resistance? **(d)** What resistance value does the broken line represent? **(e)** What is a disadvantage of so high a resistance value?

39. Refer to Fig. 10-13*b*: **(a)** How are points *A* and *B* obtained? **(b)** Compare the resistance value of the line through *A-B* with each of the source resistance values in diagram *(a)*. **(c)** What does the $+2.5$-V intersection at point *C* mean?

40. Refer to Fig. 10-14: **(a)** What is the purpose of resistors R_1 and R_2? **(b)** What is the advantage, if any, of diagram (b) over diagram (a)?

41. Can source-bias circuits be used with enhancement-mode FETs? Explain.

42. Refer to Fig. 10-15: **(a)** In diagram (a), how does the gate potential compare with the drain potential? **(b)** Why is this so? **(c)** How do these two potentials compare in diagram (b)? **(d)** Are these N-channel or P-channel units? **(e)** What change, if any, is necessary to adapt these circuits to P-channel MOSFETs?

43. What is the simplest bias method suitable for a type-B FET? Explain.

44. What difference, if any, is there between the tube diagram of Fig. 10-16 and the FET diagram of Fig. 10-1?

45. Refer to Fig. 10-17: **(a)** Specify two components not found in the triode circuit. **(b)** What is the purpose of R_2? **(c)** What is the purpose of C_2? **(d)** What advantage, if any, does a pentode circuit have over a triode?

PROBLEMS AND DIAGRAMS

1. Draw the circuit diagram for an R-C coupled P-channel JFET amplifier.

2. Repeat Problem 1 for an N-channel type-A MOSFET.

3. The drain current in an R-C coupled FET amplifier is 0.98 mA. What value of source resistor is needed to produce a bias of 1.3 V?

4. The source resistor in an R-C amplifier is 1800 Ω. The drain current is 1.05 mA. Find the bias voltage.

5. In Problem 4, if the circuit is to be used at frequencies from 40 Hz to 2 MHz, find the minimum desirable bypass capacitor value.

6. What value of source bypass capacitor should be used with an 800-Ω source resistor if the circuit operating frequency ranges between 200 and 4000 Hz?

7. An R-C coupled amplifier has the following coupling network values: $R_d = 47,000\ \Omega$, $R_g = 270,000\ \Omega$, and $C_C = 0.05\ \mu\text{F}$. **(a)** What is the value of the dc load impedance? **(b)** Find the value of the ac load impedance at 1000 Hz. **(c)** Repeat part (b) for a frequency of 20 Hz.

8. Repeat Problem 7, using $R_d = 100,000\ \Omega$, $R_g = 0.33\ \text{M}\Omega$, and $C_C = 0.02\ \mu\text{F}$.

9. An R-C coupled amplifier has a drain supply voltage of 28 V. The coupling network has $R_d = 0.01\ \text{M}\Omega$, $R_g = 0.47\ \text{M}\Omega$, and $C_C = 0.03\ \mu\text{F}$. Find the drain potential for drain currents of 0.6, 1.3, and 2.1 mA.

10. An R-C coupled circuit is used as a video amplifier in a television receiver. The waveshape (v_{ds}) developed at the output of the FET is shown in Fig. 10-18. Assuming that the average value is 12 V, draw the output voltage wave as it will appear across the gate resistor. Show all key voltage values.

11. Draw the constant-current equivalent circuit for an R-C coupled FET amplifier.

12. A constant-current equivalent circuit for an R-C coupled FET amplifier has $y_{fs} = 1200\ \mu\text{mho}$ and $g_{os} = 15\ \mu\text{mho}$. In the circuit, $R_d = 68,000\ \Omega$ and C_C, R_g (to the next stage) are 0.05 μF and 0.33 MΩ, respectively. Under ideal conditions, find **(a)** the

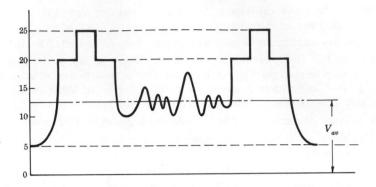

Figure 10-18   Typical TV composite video signal.

equivalent load resistance R_e; **(b)** the ac component of drain current for an input voltage of 0.25 V; **(c)** the output voltage; **(d)** the voltage gain.

13. A U184 JFET has the following typical parameter values: $g_{fs} = 4500 \ \mu\text{mho}$ and $y_{os} = 75 \ \mu\text{mho}$. It is used as an R-C coupled amplifier with $R_d = 25 \ \text{k}\Omega$ and R_g (to next stage) of 250 kΩ. Calculate the stage gain under ideal conditions.

14. The manufacturer's data sheet for a 2N4869 FET lists $g_{fs} = 4000 \ \mu\text{mho}$ and $g_{os} = 10 \ \mu\text{mho}$ at $V_{DS} = 20 \ \text{V}$ and $V_{GS} = 0$. The pinch-off voltage is given as $-5 \ \text{V}$. The FET is used as an R-C coupled amplifier with $R_d = 100 \ \text{k}\Omega$ and next stage $R_g = 1 \ \text{M}\Omega$. It is biased at $V_{GS} = -1.25 \ \text{V}$. Calculate the gain of this stage **(a)** using the data sheet value of g_{fs}; **(b)** correcting g_{fs} for the actual operating bias condition.

15. Using the output characteristic curves of Fig. 10-19, draw the dc load line for a drain supply voltage of 45 V and a drain resistor of 27,000 Ω. On this load line, find the drain current value corresponding to **(a)** $V_{GS} = -1.0 \ \text{V}$; **(b)** $V_{GS} = -3.0 \ \text{V}$; **(c)** $V_{GS} = -4.5 \ \text{V}$.

16. Again using Fig. 10-19, draw the dc load line for a drain resistor of 18,000 Ω and a supply voltage of 34 V. Find the drain current for operating points at V_{GS} values of **(a)** zero; **(b)** $-2.0 \ \text{V}$; **(c)** $-6.0 \ \text{V}$.

17. The FET of Problem 16 is to be used with a source resistor of 2700 Ω. Neglecting any change in the dc load line, find **(a)** the quiescent value of drain current and **(b)** the bias voltage. (*Hint:* First obtain the dynamic transfer characteristics; then draw the bias resistor line.)

18. The FET of Problem 15 is used with a 3900-Ω source resistor. Correct the load line for this resistance value, and find **(a)** the quiescent value of drain current and **(b)** the bias voltage.

19. Figure 10-20 shows the output characteristics of a 2N2608 *P*-channel JFET. It is used in an R-C coupled amplifier with $R_d = 27,000 \ \Omega$, $R_g = 0.1 \ \text{M}\Omega$, and a 27.5 V supply. The bias is set at 0.8 V (with a source resistor). **(a)** Draw the dc load line and find the quiescent values of drain current and voltage. **(b)** Calculate the value of source resistor needed. **(c)** Draw the ac load line. **(d)** For an input gate-voltage swing of $\pm 0.2 \ \text{V}$ (peak) find the peak-to-peak drain-voltage swing. **(e)** Calculate voltage gain.

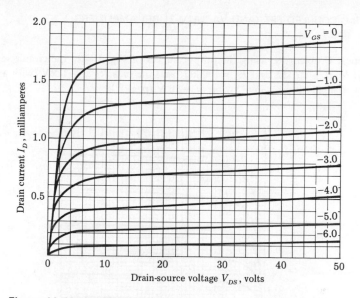

Figure 10-19   FET output characteristics (for Problems 15 to 18).

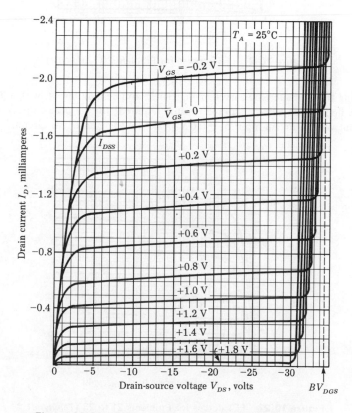

Figure 10-20   FET curves for Problem 19 (*Siliconix*).

20. Repeat Problem 19, using the output curves of Fig. 10-21; the circuit values of $V_{DD} = 45$ V, $R_d = 22$ kΩ, and $R_g = 100$ kΩ; and the bias set at -1.5 V (with a source resistor). Use an output signal swing of ± 50 V (peak).

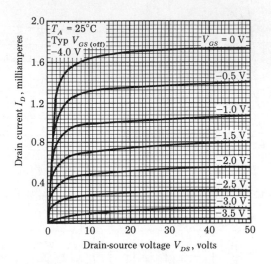

Figure 10-21   FET curves for Problem 20 (*Teledyne*).

21. Using the characteristic curves of Fig. 10-22 and ac load line A, with a gate bias of 0.6 V, find **(a)** the value of source resistor required; **(b)** the peak-to-peak value of the output voltage for an input of ± 0.6 V peak; **(c)** the gain of the circuit.

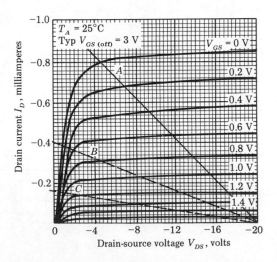

Figure 10-22   FET curves for Problems 21 to 23 (*Teledyne*).

22. Repeat Problem 21, using ac load line *B* in Fig. 10-22, a V_{GSQ} of 1.2 V, and an input voltage of ± 0.4 V peak.

23. Repeat Problem 21, using ac load line *C*, $V_{GSQ} = 1.4$ V, and an input of ± 0.2 V peak.

24. Figure 10-23 shows the minimum and maximum transfer curves for a 2N4341 FET. A FET of this type is used in a self-biased R-C coupled amplifier with $R_s = 500\ \Omega$. Draw the source resistance line and find the range of drain currents that may be obtained at 25°C.

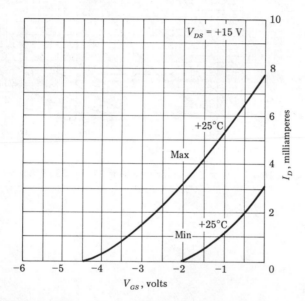

Figure 10-23   FET curves for Problems 24 to 26 (*Siliconix*).

25. Using the FET of Fig. 10-23, it is desired to limit the range of possible drain currents to between 2 and 3 mA (at 25°C). **(a)** What fixed value of bias must be used? **(b)** If V_{DD} is 30 V, what ratio should be used for the voltage-divider resistors? **(c)** What value of source resistor is needed?

26. Repeat Problem 25, limiting the drain current range to between 1.5 and 2.0 mA.

27. Figure 10-24 shows the plate characteristics for a 12AT7. This tube is to be used in an R-C coupled amplifier with $R_b = 27,000\ \Omega$, $R_g = 270,000\ \Omega$, plate supply = 400 V, and grid bias = -1.0 V. **(a)** Draw the dc load line. **(b)** Specify the quiescent values. **(c)** Draw the ac load line. **(d)** Why is there little difference between load lines?

28. A sine-wave signal having a maximum value of 1 V is applied to the input of the 12AT7 of Problem 27. Find: **(a)** the maximum and minimum values of the plate current swing; **(b)** the maximum and minimum values of the plate potential swing; **(c)** the gain of this stage.

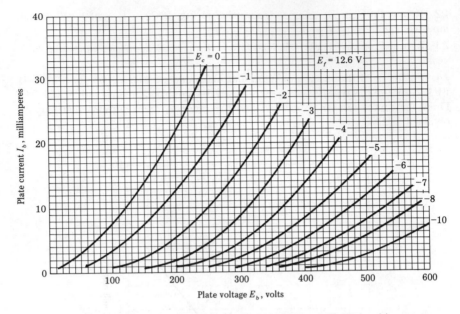

Figure 10-24   Tube curves for Problems 27 and 28 (*Sylvania*).

11

R-C Coupled BJT Amplifiers

There are so many interrelated aspects to bipolar junction transistor (BJT) amplifiers that it is very difficult to decide where to start. However, since the most generally used transistor circuit employs the common-emitter configuration with R-C coupling, that circuit will be discussed first. R-C coupling is used for the same reasons that made it popular with vacuum-tube circuits—light weight, compactness, low cost, a wider frequency range compared to transformer coupling, and less critical in design compared to direct coupling. The reason for *common-emitter* circuitry will be explained presently.

BASIC CIRCUIT AND COMPONENT FUNCTIONS.   For transistor action, the base-to-emitter junction must be biased in the forward direction. For an *NPN* transistor, the base must be positive compared to the emitter. For reverse bias of the collector junction, the collector in an *NPN* transistor must be positive compared to base or even more positive compared to emitter. These polarities are most readily obtained by use of two separate power sources, V_{BB} and V_{CC}. This technique is used in the circuit diagram of the R-C coupled amplifier of Fig. 11-1. However, it should be noted that one advantage of the common-emitter connection is that proper bias can also be obtained using only a single power supply. In practice, single-bias-source circuits are more commonly used. Since this introduces other complications, such circuit variations are considered later. For this introductory discussion we will focus on the basic dual-bias supply.

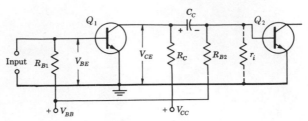

Figure 11-1   Common-emitter R-C coupled transistor amplifier using two supplies.

Referring to Fig. 11-1, the power sources V_{BB} and V_{CC} could be rectifier power supplies (for operation from an ac power line) or they could be batteries. It should be realized that in this diagram the negative terminals of these power sources, although not specifically shown, are connected to ground. (This interconnection is implied. The same technique was used with FET and vacuum-tube circuitry.) The actual voltage values that can be used depend on the specific transistor ratings and the resistance values R_b and R_c. In general, the collector supply voltage is much higher than the base supply voltage. Because of the forward-bias connection between elements, the actual base-to-emitter voltage V_{BE} is generally very low— around 0.1 V. On the other hand, depending on the *inverse* voltage rating, the collector-to-emitter voltage V_{CE} can be as high as 100 V.

Now let us turn our attention to the actual circuit and study briefly the function of each component. Start from the input side with resistor R_{b1}: one of its functions should be obvious. In conjunction with the base bias supply, the value of R_{b1} determines the bias current I_B. (In fact V_{BB} is usually large enough, compared to V_{BE}, that the bias current can be considered equal to V_{BB}/R_{b1}). The other function of R_{b1} will be seen in a moment.

On the output side, we find the coupling network R_c, C_C, R_{b2}. Their functions are identical to the equivalent components in a FET circuit. R_c is the dc load resistor. Variation of the collector current through this resistor creates an output voltage. Capacitor C_C blocks the dc component due to the supply source V_{CC}, but allows the ac component or signal "to pass through." R_{b2}, similar to R_{b1} in the input side, has two functions, one of which we have seen—to set the proper bias current. The second function, in combination with C_C, is to develop the output voltage that is applied to the next stage. This much is identical to the operation of the FET circuit.

Notice the additional resistor r_i, shown with dotted connections. It represents the input resistance of the next transistor Q_2. Its value for a common-emitter circuit is between 300 and 2000 Ω. Such a resistor was not shown in FET or vacuum-tube circuits because the input resistance for these devices (except at UHF) is practically infinite. The FET draws no gate current. The effects of the BJT input resistance can be analyzed from three aspects. One effect is due to the possible signal current division between R_{b2} and r_i. Remember that the output from transistor Q_2 (or any BJT) depends on the change in its base *current*. Therefore most of the signal current (output from Q_1) must be made to flow through r_i, with only a small value diverted through R_{b2}. Consequently, R_{b2} should be large compared to r_i. Depending on the dc bias considerations, this may not always be possible and we may be faced with the necessity for a compromise.

The second effect of this low-value input resistance is that in shunting resistor R_{b2}, their combined resistance will be less than 2000 Ω. This, in turn, reduces the time constant of $C_C R_b$. For good low-frequency response, this time constant should be long (R_g large compared to X_C). Because of the low value of r_i, C_C must be made much larger. A commonly used value is of the order of 10 μF. Electrolytic capacitors are often used to meet the need for such high capacitance values. Capacitor leakage current is not critical in transistor circuits, since the circuit resistance is already quite low. However, proper polarity of the capacitor terminals must be observed. Note the polarity shown in Fig. 11-1.

The third effect of this low input resistance is probably the most serious. It causes a drastic reduction in voltage gain. In FET operation, you will recall that maximum *voltage* gain was obtained by making the load resistance value as high as possible, while maximum *power* output required matching the load resistance to the internal resistance of the FET. This maximum power consideration sacrificed voltage gain. With transistors, since each next stage (as well as the final load) draws current, each stage must supply power. They must therefore be treated as power amplifiers, and optimum operation is obtained by matching the load to the internal output impedance of the transistor. For the common-emitter connection, the output impedance may range from 20,000 to 100,000 Ω. It is impossible to achieve a satisfactory match even if we use a resistance value of that same order of R_c. Resistor R_c represents only the dc load resistance. For the ac load resistance (neglecting X_C of the coupling capacitor), we must also consider the shunting effect of R_{b2} and r_i. The ac load resistance R_L is less than the input resistance of the next transistor and is therefore much less than the output resistance value. Operation will be far below optimum. This is the penalty that must be paid if R-C coupling is desired. Quite often an extra stage is needed to compensate for the low gain.

GRAPHIC ANALYSIS—R-C COUPLING

To study the action of a BJT circuit when signal voltage is applied, a graphic analysis can be used. Such an analysis is identical to the technique studied earlier with FET amplifiers. Let us apply this technique to a problem.

EXAMPLE 11-1

A 2N35 *NPN* (germanium) transistor, whose characteristic curves are shown in Fig. 11-2, is used in an R-C coupled amplifier with $V_{BB} = 3$ V (from a separate bias supply), $V_{CC} = 24$ V, $R_b = 75,000 \ \Omega$ each, $R_c = 6000 \ \Omega$, and $r_i = 2000 \ \Omega$.

1. Draw the dc load line on Fig. 11-2.
2. Find the quiescent values of collector current and voltage.
3. Draw the ac load line.

Solution

1. To draw the dc load line for $R_c = 6000 \ \Omega$, we need two points:
   (a) One point is at $I_C = 0$ when $V_{CE} = V_{CC}$, or

   $$I_C = 0 \quad \text{and} \quad V_{CE} = 24 \text{ V}$$

   (b) A second point is at $V_{CE} = 0$, at $I_C = \dfrac{V_{CC}}{R_c} = \dfrac{24}{6000}$, or

   $$V_{CE} = 0 \quad \text{and} \quad I_C = 4 \text{ mA}$$

   (c) Draw the dc load line in Fig. 11-2 through these two points.
2. To find the quiescent values:
   (a) The operating point must lie on the above dc load line at some specific value of base-bias current I_B. This bias current can be found from V_{BB}

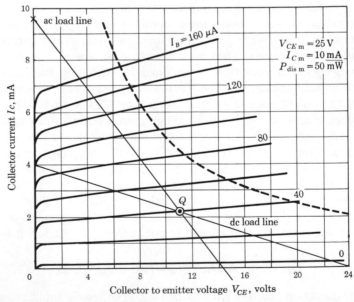

Figure 11-2 Construction of load lines for an R-C coupled transistor amplifier (*Sylvania*).

and R_b if we neglect the resistance of the base-to-emitter junction.* Compared to the 75,000-Ω value of R_b, this simplification will cause negligible error. By Ohm's law,

$$I_B = \frac{V_{BB}}{R_b} = \frac{3}{75,000} = 40 \ \mu\text{A}$$

(b) Now locate the intersection of the load line and the collector characteristic curve for $I_B = 40 \ \mu\text{A}$. This is the operating point. The quiescent values are

$$V_{CEQ} = 11 \text{ V} \qquad I_{CQ} = 2.2 \text{ mA}$$

3. To obtain the ac load line (R_L):
   (a) The ac load resistance is formed by R_c, R_b, and r_i in parallel. Obviously, this value must be less than the smallest component value, or less than 2000 Ω. By calculation, R_L is found to be 1470 Ω.
   (b) One point for the ac load line is the operating point Q. A second point can be obtained by finding at what value of collector current the collector voltage would drop to zero. Since the quiescent voltage is 11 V, the *increase* in collector current can be found from

$$\Delta I_C = \frac{\Delta V_{CE}}{R_L} = \frac{11}{1470} = 7.48 \text{ mA}$$

The total collector current would therefore be

$$\Delta I_C + I_{CQ} = 7.48 + 2.20 = 9.68 \text{ mA}$$

   (c) Locate this point ($I_C = 9.68$ and $V_{CE} = 0$) on the characteristic curves of Fig. 11-2. Draw the ac load line from this point through the operating point to the x axis.† This ac load line can then be used in evaluating output and gain.

PHASE RELATIONS—*NPN* TRANSISTOR.   Referring to the circuit diagram of Fig. 11-1, notice that the base is made positive compared to emitter. This established the forward bias, and with the values given in Example 11-1, the base current was set at 40 μA. Now suppose we apply a

*If a more accurate equation is desired, we can use

$$I_B = \frac{V_{BB}}{R_b + r_i} \tag{11-1}$$

or

$$I_B = \frac{V_{BB} - V_{BE}}{R_b} \tag{11-2}$$

where V_{BE} is the threshold voltage V_γ ($\cong 0.2$ V for germanium or 0.6 V for silicon units).

†The dotted curve on this data sheet shows the maximum power dissipation limit for this unit. The load line should not extend to the right of this curve. Power dissipation aspects are discussed further in later chapters.

sine-wave signal to the input of this circuit. During the positive half-cycle of this signal, the base is made more positive and the base current will increase. On the negative half-cycle, the signal opposes the bias and the base current will decrease. Let us assume that the signal level is of such magnitude as to cause a 40-μA swing to either side of the operating point. The base current will vary from a maximum of 80 μA to a minimum of zero. These conditions are shown in Fig. 11-3. The signal voltage and the ac component of base current are in phase. Applying this base current swing to the characteristic curves of Fig. 11-2, we can see that as the base current rises to 80 μA, the collector current also rises (to 4.15 mA). Similarly, when the base current drops to zero, the collector current drops to 0.25 mA. This variation is plotted in Fig. 11-3. Notice that the collector current varies in phase with the base current and the signal voltage.

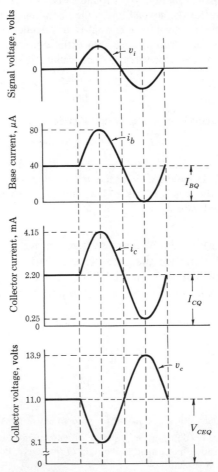

Figure 11-3   Phase relations in a common-emitter R-C coupled amplifier.

Now let us turn our attention to the collector voltage. Because of the load resistance, variations in collector current cause variations in collector potential. As the collector current increases, the drop across the load increases and the collector potential must *decrease*. This effect can be seen from the load line and characteristic curves of Fig. 11-2. Corresponding to the maximum instantaneous value of collector current of 4.15 mA, the collector voltage has dropped to 8.1 V. When the collector current decreases to 0.25 mA, the collector voltage rises to a maximum of 13.9 V. The change in collector voltage is opposite to the change in collector current. In other words, the ac component of the collector voltage is 180° out of phase with the ac component of the collector current. Since the change in collector voltage is also the output voltage, and comparing this to the input signal voltage, we see that in a common-emitter circuit there is a phase reversal (180° phase shift) between input and output voltages. These phase relations are similar to the conditions existing in a FET amplifier (see Fig. 10-2).

CURRENT GAIN.   The current gain of this transistor can be found by taking the ratio of the ac components of the output (collector) current to the input (base) current, or

$$A_i = \frac{I_c}{I_b} \tag{11-3}$$

This *operational* current gain (A_i) should not be confused with the current gain specified earlier as β. The β factor was obtained on the basis of V_{CE} being maintained constant. This would require zero load resistance and can be considered as the gain figure for the transistor *alone*. In evaluating the operational or circuit gain, we can use peak-to-peak, maximum, or rms values for our currents. Both currents must be expressed as similar quantities.

EXAMPLE 11-2

The BJT of Example 11-1 and Figs. 11-1 and 11-2 is used with an ac base current swing of $\pm 40\,\mu A$. Find:

1. The peak-to-peak value of the ac component of collector current
2. The current gain of the circuit

Solution

1. Corresponding to a base current swing of 40 μA above and below the quiescent value, Fig. 11-2 shows a collector current swing of 4.15 mA to 0.25 mA. This is a peak -to-peak value of 3.90 mA.
2. The current gain is

$$A_i = \frac{I_c}{I_b} = \frac{3.90 \times 10^{-3}}{80 \times 10^{-6}} = 48.8$$

VOLTAGE GAIN.   The BJT is calssified as a current-operated device, and as we saw in Example 11-2, if we feed an input signal *current* into the base, we can readily obtain the output signal *current*. We can also see from Fig. 11-2 that for any given input signal current swing, we can also get a corresponding collector-emitter voltage swing—or output voltage. In Example 11-2, for instance, for a base current swing of ± 40 μA above and below the operating point Q, the collector voltage swings from 8.1 V to 13.9 V. The peak-to-peak value of this output voltage is 5.8 V.* (See also Fig. 11-3.) If we are interested in voltage gain, we can obtain it from the ratio of output voltage to input voltage by using the basic equation:

$$A_v = \frac{V_o}{V_i} \qquad (11\text{-}4)$$

The voltages used can be rms, peak, peak to peak, or instantaneous—as long as both are in similar form. However, notice that in Example 11-2, although we have an output voltage, at no time did we say what input voltage produced the input current swing of ± 40 μA. Obviously, an input voltage is necessary. But how much? For a first approach, let us consider the simplest case,† wherein:

1. The input resistance r_i of the circuit is constant.
2. The signal source feeding the transistor has good regulation (negligible internal resistance).

Now examine the input circuit of Fig. 11-1, using the circuit values of Example 11-1. Resistor R_{b1} is in series with the bias voltage V_{BB}, and since it is large (75,000 Ω) compared to the transistor input resistance ($r_i = 2000$ Ω), it is the determining factor in arriving at the dc condition, or operating bias of 40 μA. On the other hand, with respect to the ac input, or signal voltage, resistor R_{b1} is effectively *in parallel* with the transistor input resistance. Therefore it does not enter into the calculation for input voltage. The input voltage required to produce a base current swing of 80 μA through the 2000-Ω input resistance must be $80 \times 10^{-6} \times 2000$ or 0.16 V peak to peak. Now we have both the input and the output voltages (in peak-to-peak value) and the voltage gain is

$$A_v = \frac{V_o}{V_i} = \frac{5.8}{0.16} = 36.2$$

If you reexamine the above discussion, you will notice in each case that the voltages are the products of the respective currents and resistances.

---

*This value of output voltage could also have been obtained from the product of output current (3.9 mA) and ac load resistance (1470 Ω).

†A more extensive discussion is given later in this chapter.

In obtaining the voltage ratio we are actually getting the product of the current ratio and resistance ratio. Since the current ratio is the current gain, we could have simplified the evaluation by using

$$A_v = A_i \frac{R_L}{r_i} \qquad \text{(11-5)}$$

or

$$A_v = 48.8 \times \frac{1470}{2000} = 35.9$$

COMMON-EMITTER BIAS METHODS

In the discussion so far, separate bias sources have been used for the input and output sides. This, at best, is an inconvenience. One advantage of using the common-emitter connection is that a single battery or power source can be used to supply both the forward bias for the input side and the reverse bias for the output side.

SIMPLE SINGLE-BIAS CIRCUIT.   Let us review for a moment several aspects of the common-emitter circuit used in Fig. 11-1. For proper bias, the base was made positive compared to the emitter (*NPN*) and the collector was also made positive with respect to the emitter. However, the collector was more positive than the base. This was achieved mainly by using a higher voltage source for V_{CC} (24 V for V_{CC} and 3 V for V_{BB}). The base-bias current was fixed by the bias supply value and the resistance value of R_b. A 75,000-Ω resistor was used to obtain a base current of 40 μA from the 3-V supply source. If we double the voltage of this supply source, we can still maintain the 40-μA base current merely by doubling the resistance value of R_b. Obviously, then, we can use only one supply, V_{CC}, and increase R_b so as to maintain the desired base-bias current. Such a circuit is shown in Fig. 11-4 and is known as a *fixed bias circuit*. It is so called

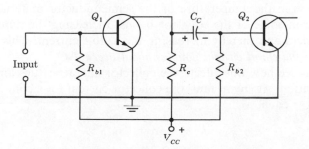

Figure 11-4   Fixed bias, common-emitter circuit.

because the bias current depends solely on the supply voltage and resistor R_b, which are normally fixed values. Change in transistor characteristics will have negligible effect on the bias current. To maintain a 40-μA base-bias current with a supply voltage of 24 V, the value of R_b must be increased to $24/(40 \times 10^{-6})$ or 0.6 MΩ.

With vacuum-tube circuits, constancy of bias is often a very desirable feature. This is not true with transistors. Unfortunately the fixed *bias current* —particularly when used with BJTs in the common-emitter configuration —leads to instability of the operating point, resulting in malfunction and even destruction of the transistor itself. One cause of instability is variation in the transistor cutoff current. Since the cutoff current in the common-emitter circuit (I_{CEO}) is much higher than in the common-base circuit (I_{CBO}), the condition is aggravated when using the CE circuitry. Variation of the cutoff current can be caused by replacement of the transistor. The characteristics of various transistors—even of the same type and from the same manufacturer—differ somewhat from unit to unit. In critical circuits, or if the transistor is operated at near-maximum ratings, and particularly if lower-priced units with wider tolerance limits are employed, circuit components may have to be changed to restore proper operation.

Another cause of instability is variation of the current gain, h_{FE} or β, of the transistor. This can happen when the transistor is replaced. For example, a 2N4036 has an h_{FE} range (min–max) of from 20 to 200! With a fixed bias current of 40 μA, the collector current ($h_{FE}I_B$) could be anywhere from 800 to 8000 μA, depending on the h_{FE} of the specific unit selected. This is a very drastic shift in operating point. Most BJTs do not have such wide h_{FE} variation limits. (A 3-to-1 range is more common.)

TEMPERATURE EFFECTS.   Instability of the operating point is also caused by temperature effects. You will recall that the cutoff current I_{CBO} or I_{CEO} is the current that flows in the collector circuit when the input circuit (base or emitter) is open. Since the collector-base junction is reverse-biased, this current is a result of minority carriers. At low temperatures, there are relatively few minority carriers and this cutoff current is low. Any increase in the temperature of the semiconductor materials causes a breakdown in some of the covalent bonds, increasing the number of minority carriers and thereby increasing the cutoff current. This increase is very rapid: *the cutoff current doubles about every 9°C.*

Let us see how this affects the collector characteristics and the transistor operation. At any instant, the collector current for a common-emitter connection is equal to

$$I_C = I_{CEO} + h_{FE}I_B$$

For this discussion, let us neglect the change in h_{FE} with temperature and consider that temperature changes do not affect the second term of this equation ($h_{FE}I_B$). Also, $I_C = I_{CEO}$ when the base current I_B is zero. There-fore, the location of the lowest $V_C - I_C$ characteristic curve (I_C values for $I_B = 0$) is dependent on temperature, but all other curves maintain their relative spacing *with respect to this lowest curve*. A change in temperature displaces the entire family of curves up or down, depending on whether the temperature increases or decreases. This displacement of the characteristics is shown in Fig. 11-5. Figure 11-5a shows the characteristics at the "recom-mended" operating temperature; Fig. 11-5b shows the effect of operating at a higher temperature. Notice the rise in the cutoff current curve (I_C for $I_B = 0$), and notice that the other curves (for $I_B = 20\ \mu A$ to $I_B = 80\ \mu A$) maintain their relative spacing with respect to the $I_B = 0$ curve.

In Fig. 11-5a the slant line represents the load line for a particular condition of operation, and the operating point is set at the intersection of this load line and a base-bias current of 40 μA. This operating point cor-responds to a collector current of approximately 4.5 mA. The load line is determined by the load resistance (R_c, R_{b2}, and r_i) and the collector supply voltage. The base current is fixed by this same voltage and the value of resistor R_b. As long as these values are not changed, the load line and base current remain the same. A rise in the transistor junction temperature does not affect any of these values. Therefore in Fig. 11-5b the same load line and the same base current must apply. Also, the operating point is still at the intersection of this load line and the collector current characteristic curve corresponding to $I_B = 40\ \mu A$. But now, because of the upward dis-placement of the collector characteristic curves, notice the drastic change in the operating point Q. For a 20-μA base current signal swing, an undis-torted output wave is obtained at normal operating temperature—whereas at the higher temperature condition of Fig. 11-5b, the transistor is driven into the saturation region and the output waveshape is severely distorted. In a practical situation, this effect is aggravated because h_{FE} also increases with temperature, thereby causing a still greater upward displacement of the collector characteristic curves.

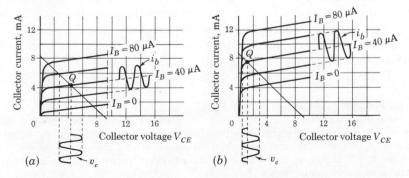

Figure 11-5   Effect of temperature on operating point and output waveshape.

Distortion is not the only drawback. Notice also that the quiescent value of collector current rises from approximately 4.5 mA to approximately 7.5 mA. This increase in collector current flowing through the base-collector junction causes a further rise in the junction temperature. This in turn raises I_{CEO} and shifts the entire family of curves still higher. You can now see what may happen: I_C increases again, resulting in more heating, and on and on. This effect is known as *thermal regeneration*. Carried to an extreme condition, *thermal runaway* results and the transistor is destroyed.

BIAS STABILIZATION—CURRENT FEEDBACK.   By examining Fig. 11-5*b* we can see that thermal runaway can be prevented if we counteract the rise in collector current, caused by increase in temperature (and h_{FE}), by simultaneously reducing the base current. If this occurs automatically we can stabilize the circuit operation. Actually, then, we are introducing *degeneration* (or *inverse feedback*)* to counteract thermal regeneration. One simple means for doing this is to insert a resistor in the emitter leg, as shown in Fig. 11-6.† Because of the emitter feedback resistor R_E, the quiescent (or average value) of collector current flowing through this resistor makes the emitter positive with respect to ground. Yet a proper forward bias can still be maintained by selecting a suitable value for R_b. Notice that the base is still maintained positive with respect to the emitter—as it should be for an *NPN* transistor. Now should the average value of collector current try to increase (because of replacement unit or temperature effects), the current through the emitter resistor will also increase and the emitter potential will become slightly more positive. This action opposes the forward bias; the base current decreases and the rise in collector current is inhibited. The larger the value of resistor R_E, the better the stability.

Notice the capacitor C_E shunting this resistor. If it were removed, *any* change in collector current would be opposed. This means that changes in collector current due to the input signal will also be minimized and the gain will be reduced. The inverse feedback now also affects the ac operation. Use of the shunting capacitor C_E bypasses the ac components of current around the resistor, preventing ac degeneration. If ac degeneration is also desired (in order to reduce distortion or to improve frequency response), this capacitor can be omitted entirely or only a portion of the resistor R_E can be bypassed.

In the circuit of Fig. 11-6, if the normal base current is 40 μA, resistor

---

*This topic is discussed in detail in Chapter 16.

†In this and in previous diagrams, resistors connected to each leg of a transistor were identified by using the appropriate *lowercase* subscript letter (e.g., R_d for drain, R_b for base). However, R_e was previously used to represent the equivalent load resistance at the output of the transistor. Therefore, to distinguish from this equivalent load value, a capital subscript letter is used for the emitter resistor; hence, R_E.

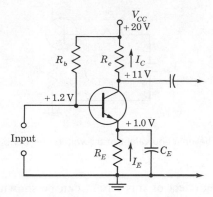

Figure 11-6   Bias stabilization by current-feedback resistor R_E.

R_b must be very large (approaching 0.5 MΩ) in order to drop the base potential to 1.2 V. Now suppose that the collector and emitter currents should double so as to raise the emitter potential to 2.0 V. Because of the large value of R_b, only a slight reduction in base current would be sufficient to make the base 0.2 V positive again with respect to emitter. Since the reduction in base current is small, only a slight improvement in stability is realized. In order to create larger changes in base current—and better stability—the value of resistor R_b must be reduced without upsetting the original dc bias conditions. This can be done by using the voltage-divider circuit of Fig. 11-7. By making the bleeder current through R_1 and R_2 large compared to the base current, the base potential is relatively unaffected by changes in the base current. (A good design rule is to make the voltage-divider current at least ten times greater than the desired base-bias current.) Now any change in the emitter degenerative voltage will cause a much larger change in the corrective base current.

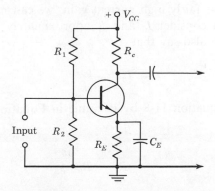

Figure 11-7   Improved bias stabilization with voltage-divider base-bias circuit.

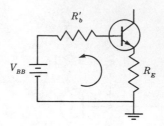

Figure 11-8 Thévenin equivalent of the voltage-divider base-bias circuit of Fig. 11-7.

The stabilizing effect of this circuit can be shown mathematically by applying Thévenin's theorem* to the input (base) circuit. If we disconnect the lead between the base and the voltage divider, we can see that the open circuit voltage developed across R_2 is

$$V_{oc} = V_{CC}\frac{R_2}{R_1 + R_2}$$

Then, to find the equivalent series resistance, we short-circuit (ground) the supply voltage. This places resistor R_1 in parallel with R_2. This equivalent circuit is shown in Fig. 11-8, where:

$$V_{BB} = V_{CC}\frac{R_2}{R_1 + R_2} \qquad \textbf{(11-6)}$$

and

$$R_b' = R_1 \parallel R_2 \qquad \textbf{(11-7)}$$

Applying Kirchhoff's voltage law to this equivalent circuit, we get

$$V_{BB} = I_E R_E + V_{BE} + I_B R_b' \qquad \textbf{(11-8)}$$

If the transistor has fairly high current gain, we can neglect I_B compared to I_C, and we can consider I_C and I_E approximately equal. Then, since $I_C = h_{FE}I_B$, we can also say that

$$I_B \cong \frac{I_E}{h_{FE}} \qquad \textbf{(11-9)}$$

Replacing I_B in Equation 11-8 by its value in Equation 11-9 and solving for I_E, we get

$$I_E = \frac{V_{BB} - V_{BE}}{R_E - R_b'/h_{FE}} \qquad \textbf{(11-10a)}$$

*See J. J. DeFrance, *Electrical Fundamentals,* Englewood Cliffs, N.J.: Prentice-Hall, Inc., 1969, chap. 24.

Earlier in this discussion, we said that the voltage-divider current should be large compared to I_B. This means that the lower the values of R_1 and R_2,* or the lower the value of R'_b, the better the stability. Now, from Equation 11-10a, we can also add that the higher the dc current gain, the better the stability. Notice that as R'_b is made lower, and as transistors with higher h_{FE} are used, the second term in the denominator, R'_b/h_{FE}, becomes smaller and smaller, and negligible compared to R_E. When a circuit is so designed, then

$$I_E \cong \frac{V_{BB} - V_{BE}}{R_E} \qquad (11\text{-}10b)$$

Notice in Equation 11-10b, that the emitter current—and therefore the collector current—is independent of the transistor parameters, depending only on the external circuitry, R_1, R_2, R_E, and V_{CC}. This circuit will therefore correct not only for temperature effects; but more important, it holds the operating point stable in spite of variations of the transistor h_{FE}. Two numerical examples of the stabilizing effect of this circuit are given at the end of the chapter. (See Problems 15 and 16.)

EMITTER-SUPPLY BIAS CIRCUIT.   When two power supplies are available, the circuit of Fig. 11-9 can be used to obtain even better stability than is possible with the voltage-divider action of Fig. 11-7. The two supply voltages V_{CC} and V_{EE} are of opposite polarities. Since the transistor in Fig. 11-9 is an *NPN*, the collector supply voltage is connected with its positive side to the collector resistor, and its negative terminal (not shown) is grounded. (This is the schematic convention we have been using.) The second supply source V_{EE} has its negative output fed to the emitter resistor, and its positive terminal is grounded. It should be carefully noted that the $+ V_{CC}$ and the $- V_{EE}$ are *not* the plus and minus terminals of one power supply, but instead represent two supplies with their opposite polarity terminals grounded.† Obviously, with a *PNP* transistor the supply polarities would be reversed.

If we apply Kirchhoff's voltage law to the input circuit of Fig. 11-9, we get

$$V_{EE} = I_E R_E + V_{BE} + I_B R_b \qquad (11\text{-}11)$$

---

*If these values are made too low, the gain of the amplifier is seriously reduced, because of low input impedance.

†Quite often a dual power supply is used to supply a + and a − output from a common ground. This is readily obtained using a bridge rectifier circuit, fed from the ends of a center-tapped power transformer. The bridge output delivers the + and − polarities, and the center tap of the transformer provides the common ground. (The drawing of such a power supply circuit is left as a student exercise.)

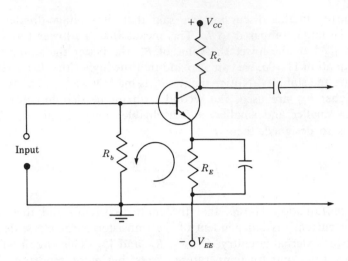

Figure 11-9   Emitter-supply bias.

But this is exactly the same type of equation as Equation 11-8 for the voltage-divider circuit. Therefore, as before, we conclude with

$$I_E = \frac{V_{EE} - V_{BE}}{R_E - R_b/h_{FE}} \qquad \text{(11-12a)}$$

And if R_E is large compared to R_b/h_{FE},

$$I_E \cong \frac{V_{EE} - V_{BE}}{R_E} \qquad \text{(11-12b)}$$

Then, since V_{EE} is often much larger than V_{BE} (which is the cut-in or threshold voltage), we can neglect V_{BE} compared to V_{EE}, and the equation simplifies further to

$$I_E \cong \frac{V_{EE}}{R_E} \qquad \text{(11-12c)}$$

The advantage of using emitter-supply bias is that since V_{EE} can have a much higher value than V_{BB} (of Fig. 11-8), the R_E value can be made much higher than in Figs. 11-7 and 11-8. Then, even with moderate values of h_{FE}, the term R_b/h_{FE} can be negligible compared to R_E. Consequently, the approximate forms of the I_E equation become more valid, making I_E— and I_C—truly independent of the transistor parameters. The disadvantage of the circuit is, of course, the need for a more complex power supply source.

BIAS STABILIZATION—VOLTAGE FEEDBACK.   Voltage feedback can be obtained by feeding part of the output voltage into the input circuit. This is most readily done by tying the base-bias resistor R_b directly to the collector side of the load instead of to the supply source. Figure 11-10 shows

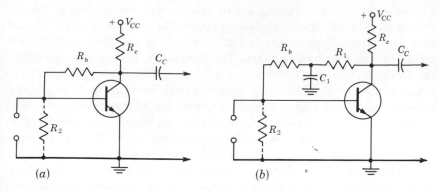

Figure 11-10   Voltage-feedback bias stabilization circuits.

this connection. In these circuits, should the average collector current tend to rise, the collector potential will fall, in turn reducing the base-bias current.

As with current feedback, a large value of series resistance (R_b or $R_1 + R_b$) tends to minimize the desired change in I_B. Therefore the voltage-divider form of these circuits (shown by adding the dotted resistor R_2) is often used. The circuit of Fig. 11-10a not only produces dc feedback for bias stabilization but also results in ac feedback and reduction of gain. Of course, the advantages of such inverse feedback are also realized. However, if ac feedback is not desired, the circuit of Fig. 11-10b can be used. The low-pass filter R_1–C_1 will bypass any ac signal to ground, yet still allow dc stabilization.

BIAS STABILIZATION—TEMPERATURE-SENSITIVE ELEMENTS.   The voltage-divider base-bias circuit, as used in Fig. 11-7, can be made to give stabilization directly if the resistor R_2 is replaced by a temperature-sensitive element. Ideally suited for this purpose are thermistors and junction diodes. Each of these elements has a negative temperature coefficient; that is, its resistance decreases with increase in temperature. Circuits using these elements are shown in Fig. 11-11. Resistor R_2 in Fig. 11-11a is a thermistor.

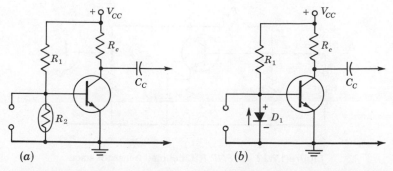

Figure 11-11   Bias stabilization—using temperature-sensitive elements.

The ratio R_1 to R_2 is initially set to produce the proper bias at normal temperature. Then, should the temperature rise and the collector current tend to increase, simultaneously the resistance of the thermistor will decrease, lowering the base potential. This reduces the base current and offsets the tendency of the collector current to rise. The diode of Fig. 11-11*b* functions in a similar fashion to achieve stabilization because its junction resistance decreases with rise in temperature. Use of semiconductor diodes has one advantage over themistors in that by "matching" the diode junction materials to the transistor base-collector junction, equal but opposite temperature effects can be achieved, resulting in better stabilization.

This covers the basic techniques of bias stabilization. Combinations and variations of these methods are also used. These special cases are not covered in this text.

PNP TRANSISTOR—R-C COUPLING

Thus far, the circuit discussions have concerned the *NPN* transistor using the common-emitter configuration. The *PNP* transistor can also be used in similar circuitry. There is only one basic principle to remember: regardless of which type of transistor is used, the emitter-base junction must have a forward bias while the collector-base junction must have a reverse bias. With a *PNP* unit, this means that the base must be negative with respect to the emitter and the collector must be even more negative with respect to base. These polarities are used in the *PNP* transistor R-C coupled amplifier of Fig. 11-12. Notice that these polarities are opposite to those used earlier with *NPN* units. Actually, except for the use of a negative supply source (V_{CC}) (or reversal of the battery), there is no difference between this circuit and previous circuits for the *NPN* transistor. In this one circuit, however, we have combined dc feedback, voltage feedback, and voltage-divider tech-

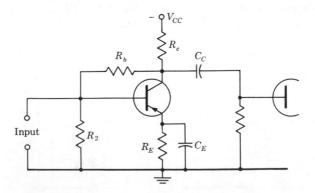

Figure 11-12 A *PNP* R-C coupled transistor stage.

niques for improved stabilization. Graphic analysis for *PNP* circuits would be identical to *NPN* circuits and need not be discussed further.

COMMON-BASE R-C COUPLING

When studying the R-C coupled amplifier using the common-emitter configuration, we saw that the loading effect of the next stage made it impossible to obtain proper matching for optimum gain. The output impedance for a common-emitter connection ranges upward from 20,000 Ω while the effective load resistance cannot exceed 2000 Ω because of the shunting effect of the input impedance of the next stage. Still, a voltage gain (A_e) is obtained, although it is somewhat less than the current gain (A_i). The actual voltage gain of the stage was shown to be

$$A_v = A_i \frac{R_L}{r_i}$$

With a common-base circuit, conditions would be even worse. The output resistance of such a stage is five to ten times higher (see Table 2-1). The input resistance of the next stage is very much lower (30 to 200 Ω). Proper matching is impossible; the ac load resistance would be approximately equal to the input resistance of the next stage. When we also consider that the current gain for the common-base circuit is less than 1, it is apparent that such operation would result in reduced voltage output instead of a voltage gain. Obviously, then, R-C coupling cannot be used with the common-base configuration.

MATHEMATICAL ANALYSIS

Accurate mathematical analysis of BJT circuits can be made by using equivalent circuits and appropriate transistor parameter values—in much the same manner as was shown for FET circuits. As before, such analyses are suitable only for small-signal operation and for a specific operating point. Unfortunately, the equivalent circuit for a BJT is appreciably more complex, and the solution of the BJT equivalent circuits is appreciably more laborious, because:

1. The input circuit draws current, and the current-voltage relation is nonlinear.
2. There is interaction between input and output circuits, and the input circuit parameters are affected by the output loading and circuit conditions.

The parameter values from the manufacturers' data sheets give either typical values or minimum and maximum values, at one specified operating point. However, any one transistor may not be a typical, or a minimum, or a maximum condition unit. Furthermore, what parameter changes do you make if the unit is used at a quiescent point other than the one specified in the data sheets?

Of course, you can always determine experimentally the actual parameter values for the given transistor at the desired operating point. But once you go to the trouble of a laboratory setup to measure parameters, why not instead measure the desired operational data of the circuit itself? In fact, the procedure for evaluating parameter values is more difficult than for measuring output and gain of the actual circuit. Thus, although accurate equivalent-circuit analyses have a place (and are discussed in Chapter 12), it is felt that practical approximation technique is more meaningful at this point.

When the circuit component values are given, dc analysis is fairly straightforward, but the steps vary somewhat, depending on the base-biasing method used. Let us analyze three of these basic circuits for the dc conditions.

OBTAINING QUIESCENT VALUES. Figure 11-13a shows a simple fixed bias-current circuit. The dc base current is

$$I_{BQ} = \frac{V_{CC} - V_{BE}}{R_b} \cong \frac{V_{CC}}{R_b}$$

Then

$$I_C = h_{FE}I_B \quad \text{and} \quad V_{CE} = V_{CC} - I_C R_c$$

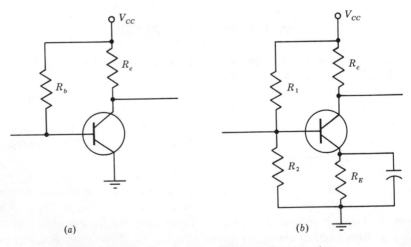

(a)

(b)

Figure 11-13 Components affecting quiescent values.

EXAMPLE 11-3

Find the quiescent values in the circuit of Fig. 11-13a if $R_b = 470\text{ k}\Omega$, $R_c = 5600\ \Omega$, $V_{CC} = 20$ V, and the transistor is a silicon unit with an h_{FE} of 50.

Solution

1. $I_B = \dfrac{V_{CC} - V_{BE}}{R_b} = \dfrac{20 - 0.6}{470\text{ k}\Omega} = 41.3\ \mu\text{A}$

2. $I_C = h_{FE}I_B = 50 \times 41.3 \times 10^{-6} = 2.06\text{ mA}$

3. $V_{CE} = V_{CC} - I_C R_c = 20 - (2.06 \times 10^{-3} \times 5600) = 8.5$ V

Figure 11-13b shows a bias circuit using voltage-divider base bias and emitter stabilization. Neglecting the small value of base current (compared to the much larger value of voltage-divider current),

$$V_B = V_{CC}\left(\frac{R_2}{R_1 + R_2}\right)$$

Also,

$$V_B = V_{BE} + V_E \quad \text{or} \quad V_E = V_B - V_{BE}$$

From current relations,

$$I_C = h_{FE}I_B$$

and again neglecting the small value of I_B (compared to I_C),

and

$$I_E \cong I_C \cong h_{FE}I_B$$
$$V_E = I_E R_E \cong I_C R_E$$

Solving for I_C,

$$I_C \cong \frac{V_E}{R_E} \cong \frac{V_B - V_{BE}}{R_E}$$

Finally,

$$V_{CE} \cong V_{CC} - V_E - I_C R_c$$

Let us put this all together in an example.

EXAMPLE 11-4

A silicon transistor with $h_{FE} = 50$ is used in the circuit of Fig. 11-13b. The circuit values are $R_1 = 12{,}000\ \Omega$, $R_2 = 6800\ \Omega$, $R_c = 3900\ \Omega$, $R_E = 3300\ \Omega$, $V_{CC} = 20$ Find the quiescent values.

Solution

1. $V_B = V_{CC} \dfrac{R_2}{R_1 + R_2} = 20 \dfrac{6800}{12,000 + 6800} = 7.24 \text{ V}$

2. $V_E = V_B - V_{BE} = 7.24 - 0.6 \cong 6.6 \text{ V}$

3. $I_C \cong I_E \cong \dfrac{V_E}{R_E} \cong \dfrac{6.6}{3300} \cong 2.0 \text{ mA}$

4. $I_B = \dfrac{I_C}{h_{FE}} \cong \dfrac{2.0}{50} \cong 40 \ \mu\text{A}$

5. $V_{CE} = V_{CC} - V_E - I_C R_c = 20 - 6.6 - (2.0 \times 10^{-3} \times 3900) = 5.6 \text{ V}$

A third commonly used base-bias circuit is the voltage-feedback method shown in Fig. 11-8 and repeated here as Fig. 11-14. The current

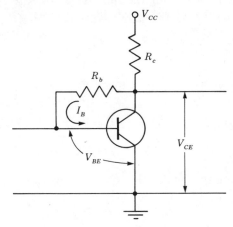

Figure 11-14   DC analysis—voltage-feedback bias circuit.

flowing through R_c is the combination of I_B and I_C. The analysis can be simplified if we neglect I_B, compared to the much larger I_C. Then:

Step 1: $$V_{CE} = V_{CC} - I_C R_c$$

But $V_{BB} = V_{CE}$, and so we can find I_B from

Step 2: $$I_B = \frac{V_{CE} - V_{BE}}{R_b}$$

Substituting for V_{CE} from step 1,

Step 3: $$I_B = \frac{(V_{CC} - I_C R_c) - V_{BE}}{R_b} = \frac{V_{CC} - V_{BE}}{R_b} - \frac{I_C R_c}{R_b}$$

And replacing I_C by its equivalent I_B,

Step 4:
$$I_B = \frac{V_{CC} - V_{BE}}{R_b} - \frac{h_{FE}I_B R_c}{R_b}$$

Now, solving for I_B,

$$I_B\left(1 + \frac{h_{FE}R_c}{R_b}\right) = \frac{V_{CC} - V_{BE}}{R_b}$$

$$I_B* = \frac{V_{CC} - V_{BE}}{R_b + h_{FE}R_c} \tag{11-13}$$

Current I_C can then be found from

Step 5: $I_C = h_{FE}I_B$

EXAMPLE 11-5

Find the quiescent values in the circuit of Fig. 11-14. R_b is 270 kΩ, R_c is 4700 Ω, V_{CC} is 20 V, and the transistor is a silicon unit with $h_{FE} = 100$.

Solution

1. $I_B = \dfrac{V_{CC} - V_{BE}}{R_b + h_{FE}R_c} = \dfrac{20 - 0.6}{270,000 + 100(4700)} = 26.2\ \mu A$

2. $I_C = h_{FE}I_B = 26.2 \times 10^{-6} \times 100 = 2.62$ mA

3. $V_{CE} = V_{CC} - I_C R_c = 20 - 2.62 \times 10^{-3} \times 4700 = 7.7$ V

SIGNAL (AC) ANALYSIS—VOLTAGE GAIN.   Now that the dc (quiescent) conditions have been fixed, we are ready to analyze the circuits with respect to their ac or signal behavior. Our interest at this time is in the voltage output and gain. For this we use a simplified equivalent circuit. As in any equivalent circuit, dc power sources are omitted and capacitors are assumed to have negligible effects. This analysis is similar to the FET discussion in the previous chapter. However, since the BJT draws current in the input side, the equivalent circuit must now include an input side. For simplicity, we are neglecting the interaction (feedback) between input and output circuits. Figure 11-15 shows such an equivalent circuit.

Since this represents a small-signal amplifier, it should be realized that its output is fed to a next stage. Therefore, part of the load on this stage is the *total input resistance* $r_{T(in)}$ *of the next stage.* (This next stage is most likely another BJT, so that the input resistance is fairly low and can-

---

*Notice that in solving for I_B, the effect of R_c is increased by a factor of h_{FE}—compared to a simple series connection value of $R_b + R_c$ if the BJT were not there. If an emitter resistor is also used, its effect on current I_B is also increased by the factor h_{FE}, so that

$$I_B = \frac{V_{CC} - V_{BE}}{R_b + h_{FE}(R_c + R_e)} \tag{11-14}$$

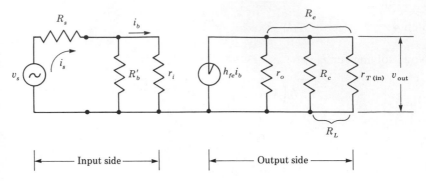

Figure 11-15 Simplified equivalent circuit—BJT amplifier.

not be neglected.) R_c is the collector circuit resistor, while r_o is the *output* (collector) resistance. This value is listed in Table 2-1 as ranging from 20 to 100 kΩ. These are general values. Depending on the unit used and on the operating point, this value can be as low as 3000 Ω or higher than 200 kΩ. How do we get the value? We may find it in the manufacturers' data sheets as r_o, as r_{oe} (the second subscript indicating common-emitter connection), or as h_{oe}. This last designation is a *hybrid* parameter and gives the *output conductance* (*o*) for the common-emitter connection (*e*). Unfortunately, quite often it is not listed at all. Luckily, both R_c and $r_{T(\text{in})}$ are generally much lower than r_o, and we may neglect r_o without serious error when calculating the equivalent resistance. (We are then actually using R_L instead of R_e.)

Now let us turn our attention to the input side. The source is represented by the generator symbol and v_s. It has an internal resistance R_s (which may or may not be negligible).* The resistor R_b' represents the effect of all the bias resistors (as is shown later), and r_i is the input resistance of the transistor itself. Since we are neglecting the interaction between input and output, this input resistance is the resistance of the forward-biased junction between base and emitter. As in any diode (see Chapter 2) this resistance is approximately

$$r_i = \frac{26}{I_B} \quad \text{(for a CE connection)} \tag{11-15}$$

where I_B is the quiescent (dc) value of the base current *in milliamperes*. (Remember that this value can vary from unit to unit from a low of about $20/I_B$ to a high of about $50/I_B$.)

From Fig. 11-15, it is obvious that the output voltage is $h_{fe}i_b \times R_e$.

---

*For example, in audio systems the source could be a microphone. A dynamic microphone may have an impedance as low as 5 Ω, whereas the output impedance of a ceramic microphone is around 1 MΩ.

The equivalent resistance value is the parallel resistance of the three resistors shown. Usually $r_{T(in)}$ and R_c are much lower than r_o. We can use our judgment as to whether r_o should or should not be neglected. Since h_{fe} is given in the data sheets, the problem now is to find i_b. This is done from the input side. For r_i we use Equation 11-15. It was mentioned above that R'_b represents the effect of all the bias resistors. To clarify this we must examine the possible bias circuits. For a simple fixed-bias circuit (Figs. 11-4 and 11-13a), there is only one bias resistor and R'_b is simply R_b. With a voltage-divider base bias (Fig. 11-11b), as far as ac operation is concerned the dc supply is at ac ground potential and the two bias resistors are in parallel from base to ground. Therefore, $R'_b = R_1 \| R_2$.

Once we have R'_b and r_i (and the source voltage and its internal resistance are given), we can calculate i_b by using series-parallel circuit theory (R'_b is in parallel with r_i, and their combination is in series with R_s.) We can simplify this calculation, depending on whether R_s and/or R'_b can be neglected. For example, if R_s is small compared to $R'_b \| r_i$, then we can neglect R_s, and i_b is simply v_s/r_i. On the other hand, if R'_b is large compared to r_i it can be omitted, and we have a simple series circuit with $i_b = v_s/(R_s + r_i)$. Let us try one example of each case.

EXAMPLE 11-6

A 2N3242A silicon *NPN* transistor has the following characteristics: $h_{fe} = 175$; $h_{oe} = 30$ μmho. It is used in the R-C coupled amplifier circuit shown in Fig. 11-16, with $V_{CC} = 16$ V, $R_c = 4.7$ kΩ, $R_1 = 68$ kΩ, $R_2 = 33$ kΩ, $R_E = 3.3$ kΩ. The input is from a 0.05-V, 50-Ω source. The next stage is identical. Find the output voltage and the voltage gain.

Solution

1. $V_B = V_{CC} \dfrac{R_2}{R_1 + R_2} = 16 \dfrac{33 \text{ k}\Omega}{(33 + 68) \text{ k}\Omega} = 5.23$ V

2. $I_C \cong I_E \cong \dfrac{V_B - V_{BE}}{R_E} = \dfrac{5.23 - 0.6}{3300} = 1.71$ mA

3. $I_B = \dfrac{I_C}{h_{fe}} = \dfrac{1.71}{175} = 9.85$ μA

4. $r_i = \dfrac{26}{I_B \text{ (mA)}} = \dfrac{26}{9.85 \times 10^{-3}} = 2640$ Ω

5. $r_{T(in)}$ of $Q_2 = r_i \| R_1 \| R_2 = 2640 \| 33,000 \| 68,000 = 2360$ Ω

6. $r_o = \dfrac{1}{h_{oe}} = \dfrac{1}{30 \times 10^{-6}} = 33.3$ kΩ

7. $R_e = r_o \| R_c \| r_{T(in)} = (33.3 \| 4.7 \| 2.36) \text{ k}\Omega = 1500$ Ω

   (*Note:* We could have neglected r_o in this case, with only a slight effect on R_e; that is, $R_e = 1570$ Ω.)

8. To find I_b (the signal current):

 (a) We know that $r_{T(in)} = r_i \| R_1 \| R_2 = 2360 \ \Omega$.

 (b) In comparison, the source resistance R_s can be neglected, so that

$$I_b = \frac{v_s}{r_i} = \frac{0.05}{2640} = 18.9 \ \mu A$$

9. The output voltage is

$$V_o = h_{fe}I_bR_e = 175 \times 18.9 \times 10^{-6} \times 1500 = 4.96 \ V$$

10. The voltage gain is

$$A_v = \frac{V_o}{V_s} = \frac{4.96}{0.05} = 99.2$$

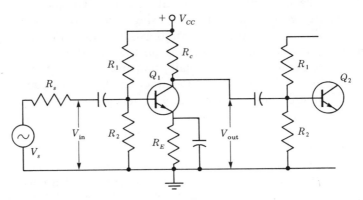

Figure 11-16   Circuit for Example 11-5.

In this example note that since the source resistance R_s is considered negligible (compared to $r_{T(in)}$), the signal voltage applied to the transistor (V_i) is the same as the source voltage. But V_i is in any case equal to $i_b r_i$ and V_o is $h_{fe}i_bR_e$. Therefore, the gain of the transistor (neglecting source resistance) is

$$A_{v(trans)} = \frac{V_o}{V_i} = \frac{h_{fe}i_bR_e}{i_br_i} = h_{fe}\frac{R_e}{r_i} \qquad (11\text{-}16)$$

Had we used this equation in Example 11-6, we could have ommitted steps 8 and 9 of the solution. Now let us consider a case where the source resistance is not negligible.

EXAMPLE 11-7

The source feeding the circuit in Fig. 11-16 has been changed. The signal voltage is still 0.05 V, but the source resistance is now 1000 Ω. All other values are unchanged. Find the output voltage and the voltage gain.

Solution

Steps 1 to 7 are the same as in Example 11-6.

8. To find I_b (the signal current):

    (a) $r_{T(in)} = r_i \| R_1 \| R_2 = 2360 \ \Omega$ (as before).

    (b) R_s cannot be neglected because there will be an appreciable voltage drop across this resistor. From voltage-divider action,

    $$V_i = V_s \frac{r_{T(in)}}{R_s + r_{T(in)}} = 0.05 \frac{2360}{1000 + 2360} = 0.0351 \text{ V}$$

    (c) $I_b = \frac{V_i}{r_i} = \frac{0.0351}{2640} = 13.3 \ \mu\text{A}$

    (The value for r_i was obtained in step 4 of Example 11-6.)

9. The output voltage is

    $$V_o = h_{fe} I_b R_e = 175 \times 13.3 \times 10^{-6} \times 1500 = 3.49 \text{ V}$$

    (The value for R_e was obtained in step 7 of Example 11-6.)

10. The voltage gain is

    $$A_v = \frac{V_o}{V_s} = \frac{3.49}{0.05} = 69.8$$

If we check the gain of the transistor alone in Example 11-7, we get

$$A_{v(trans)} = \frac{V_o}{V_i} = \frac{3.49}{0.0351} = 99.4$$

This is the same as in Example 11-6 and it *should* be, because the transistor circuit itself has not been changed—only the source changed. So we could have used Equation 11-16 to find the transistor gain, the new V_i to find the output voltage, and the V_o/V_s ratio to find the overall gain.

DESIGN CONSIDERATIONS.   So far we have considered given circuits —the component values were known. However, what if we have to design the circuit? To a great extent, the component values we use will depend on our design goals. Several guiding factors can serve as a starting point:

1. The maximum possible output voltage is obtained when the transistor operation swings from almost saturation to almost cutoff. At saturation, the collector-emitter voltage approaches zero; at cutoff, $V_{CE} = V_{CC}$. Therefore the maximum peak-to-peak output voltage approaches—but can never equal—the supply voltage V_{CC}.

2. To allow for maximum signal swing with minimum distortion, the quiescent value of collector current (I_{CQ}) should be approximately halfway between the saturation and cutoff values, or

$$I_{CQ} = \frac{V_{CC}}{2(R_c + R_E)}$$

3. To prevent thermal runaway—when I_C increases with temperature — V_{CE} should not exceed $\frac{1}{2}V_{CC}$.

4. The voltage gain of a transistor (CE connection and not considering source resistance) is

$$A_{v(\text{trans})} \cong h_{fe}\frac{R_e}{r_i}$$

5. And considering the source resistance, the overall gain is

$$A_v = A_{v(\text{trans})}\left(\frac{R_s}{R_s + R_b' \| r_i}\right)$$

Let us see how these guidelines can be applied to a practical case.

EXAMPLE 11-8

It is desired to obtain a 5.0-V output from a 0.1-V source with a 500-Ω internal resistance. Design a suitable CE R-C coupled amplifier.

Solution

1. The overall circuit gain needed is $5/0.1$ or 50. Since this source resistance is probably higher than the r_i of the transistor, the transistor h_{fe} should exceed 100 to allow for the loss across R_s.

2. For a 5-V output (rms), the peak-to-peak value is $2(5 \times 1.414)$ or 14.2 V. The supply voltage should exceed 14.2 V, and the transistor should have a BV rating of over 20 V.

3. Let us select a 2N4074 *NPN* silicon transistor. The manufacturer's data sheet lists $V_{(BR)CEO} = 40$ V, $h_{fe} = 175$ (min), $h_{ie} = 600\ \Omega$, and $h_{oe} = 75\ \mu\text{mho}$—all at $V_{CE} = 12$ V and $I_C = 10$ mA.

4. We will use the circuit of Fig. 11-16. To use the listed parameters, the operating point is fixed at $V_{CE} = 12$ V and $I_C = 10$ mA.

5. Since V_{CE} should not exceed $\frac{1}{2}V_{CC}$, let us use a V_{CC} of 30 V.

6. But from Fig. 11-16, $V_{CE} = V_{CC} - I_C(R_c + R_E)$; and for $I_C = 10$ mA,

$$R_c + R_E = \frac{V_{CC} - V_{CE}}{I_C} = \frac{30 - 12}{10 \times 10^{-3}} = 1800\ \Omega$$

For good stability, select a 2:1 ratio for the two resistors:

$$R_c = 1200\ \Omega \quad \text{and} \quad R_E = 600\ \Omega$$

7. We know that

$$I_E = \frac{V_B - V_{BE}}{R_E} \cong I_C$$

so that

$$V_B = I_C R_E + V_{BE} = (10 \times 10^{-3} \times 600) + 0.7 = 6.7\ \text{V}$$

8. The ratio of R_1 and R_2 should reduce the 30-V V_{CC} to 6.7 V for V_B, or

$$\frac{R_1}{R_2} = \frac{23.3}{6.7} \quad \text{and} \quad R_1 = 3.48 R_2$$

To avoid loading down the input, let us select a 16-kΩ resistor for R_2. Then $R_1 = 3.48 \times 16 = 55.7$ kΩ. (A 56-kΩ resistor is commercially available.)

9. Now, as a check, let us calculate the gain and output voltage by using the above circuit components.

(a) For $I_C = 10$ mA:

$$I_B = \frac{I_C}{h_{fe}} = \frac{10 \text{ mA}}{175} = 57.2 \ \mu\text{A}$$

$$r_i = \frac{26}{I_B \text{ (mA)}} = \frac{26}{57.2 \times 10^{-3}} = 455 \ \Omega$$

$$r_o = \frac{1}{h_{oe}} = \frac{1}{75 \times 10^{-6}} = 13,300 \ \Omega$$

$$R_c = 1200 \ \Omega$$

(b) $R_e = R_c \| r_o \| r_i = 1200 \| 13,300 \| 455 = 322 \ \Omega$

(c) $A_{v\text{(trans)}} \cong h_{fe} \dfrac{R_e}{r_i} = 175 \dfrac{322}{455} = 124$

(d) $A_{v\text{(overall)}} = A_{v\text{(trans)}} \left(\dfrac{R_b' \| r_i}{R_s + R_b' \| r_i} \right)$

But since $R_b' = 16$ kΩ $\| 56$ kΩ, this value can be neglected, compared to r_i of 455 Ω, and the equation for overall gain simplifies to:

$$A_{v\text{(overall)}} = A_{v\text{(trans)}} \left(\frac{r_i}{R_s + r_i} \right) = 124 \left(\frac{455}{955} \right) = 59.2$$

This is close enough to the desired gain of 50 and can be adjusted by slight circuit modifications, if necessary.

REVIEW QUESTIONS

1. Refer to Fig. 11-1: (a) What type of transistor is being used? (b) State two functions of resistor R_{b2}. (c) State two functions of resistor R_c. (d) What is the significance of the dotted resistor r_i? (e) For the circuit shown, what value might this be? Why? (f) State two functions of capacitor C_c. (g) What range of values is generally used for this capacitor? Why? (h) What type of capacitor is generally used? (i) How does capacitor leakage affect the operation of this circuit? Explain. (j) Which is the higher supply voltage, V_{BB} or V_{CC}? Explain. (k) What is a typical value for V_{BE}? Why? (l) What determines the maximum value for V_{CE}?

2. In the R-C coupled circuit of Fig. 11-1, how does the input resistance of the next transistor affect (a) the value used for R_{b2}? Explain. (b) The frequency response of Q_1? Explain. (c) The gain of Q_1? Explain.

3. In an R-C coupled, common-emitter amplifier, as in Fig. 11-1, what value or values determine **(a)** the dc load line? **(b)** The ac load line?

4. Can optimum operation be obtained from an R-C coupled, common-emitter circuit? Explain.

5. Refer to Fig. 11-2: **(a)** What two points are used to fix the dc load line? **(b)** How is the location of point Q determined? **(c)** Why is the ac load line steeper than the dc load line? **(d)** What component and/or parameter values contribute to the ac load resistance? **(e)** What point (other than Q) is used for drawing the ac load line? **(f)** How is the current value for this second point obtained? **(g)** What operational data can be obtained from the ac load line? **(h)** What is the significance of the broken-line curve?

6. In a common-emitter circuit using an *NPN* transistor, what is the phase relation between **(a)** signal voltage and base current? Explain. **(b)** Base current and collector current? **(c)** Collector current and collector voltage? **(d)** Signal voltage and output voltage?

7. Repeat Question 6 for a *PNP* transistor.

8. (a) What is the difference *in significance* between the factors A_i and β? **(b)** What accounts for this difference? **(c)** How is A_i evaluated?

9. Give the relation between voltage gain A_e and the above current gain A_i.

10. Refer to Fig. 11-4: **(a)** What circuit value or values determine the bias current? **(b)** Why is this called fixed bias? **(c)** How does change in transistor characteristics affect the bias current? Explain.

11. What is the disadvantage of the fixed bias circuit?

12. Refer to Fig. 11-5: **(a)** What is the bias current in diagram (a)? **(b)** What is the bias current in diagram (b)? **(c)** How do the load impedances (a) and (b) compare? Explain. **(d)** How do the signal inputs i_b in each case compare? **(e)** What is the major difference between these two diagrams? **(f)** What causes this difference? **(g)** Explain why this is so. **(h)** What effect does this have on the quiescent value of the collector current? Explain. **(i)** What effect does this have on the output waveshape? Explain.

13. (a) Explain thermal regeneration. **(b)** What is its effect on the operation of a transistor circuit? **(c)** In its extreme condition, what is this effect called?

14. (a) What is the purpose of bias stabilization? **(b)** What basic principle is employed in such circuits? **(c)** Explain the *general* action common to all these circuits.

15. Refer to Fig. 11-6: **(a)** What is the function of R_b? **(b)** What is the function of R_E? **(c)** Assuming that the collector current should tend to increase, explain the stabilizing action of this circuit. **(d)** What is the function of capacitor C_E? **(e)** What order of value is used for resistor R_b? **(f)** What disadvantage does this cause?

16. Refer to Fig. 11-7: **(a)** Does this circuit use current-feedback stabilization? **(b)** What component(s) is (are) responsible for this? **(c)** How does this circuit differ from Fig. 11-6? **(d)** What advantage does this change produce?

17. Refer to Fig. 11-8: **(a)** What does V_{BB} represent? **(b)** How is its value obtained? **(c)** What does R_b' represent? **(d)** How is its value obtained?

18. Under what condition will Equation 11-9 be valid?

19. (a) Under what condition will Equation 11-10b be valid? **(b)** Why is the circuit then immune to temperature effects and to variations in the transistor current gain?

20. Refer to Fig. 11-9: (a) How many power supplies does this circuit show? (b) Which side of each power source is grounded?

21. Under what conditions is Equation 11-12c valid?

22. (a) Which circuit gives better stabilization, Fig. 11-7 or Fig. 11-9? (b) Give two reasons for your choice. (c) What is a disadvantage of the circuit in Fig. 11-9?

23. Refer to Fig. 11-10: (a) What component is mainly responsible for stabilization? (b) Explain how stabilization is effected. (c) What is the purpose of resistor R_2? (d) Explain its effect.

24. In a comparison between Fig. 11-10a and b, which of these circuits has the lower gain? Why? (b) What components in the other circuit overcome this effect? (c) Explain how this is done.

25. Refer to Fig. 11-11a: (a) What component is responsible for bias stabilization? (b) Describe this component. (c) What special property does it have that makes it suitable for stabilization? (d) Explain the circuit action.

26. Repeat Question 25 for Fig. 11-11b.

27. Refer to Fig. 11-12: (a) What type of transistor is used? (b) What circuit change(s) is (are) made to accommodate this unit? (c) What is the function of R_b? (d) What is the function of R_2? (e) What is the function of R_E? (f) What is the function of C_E?

28. Is R-C coupling suitable for a transistor common-base configuration? Explain.

29. Refer to Fig. 11-13a: (a) What type of bias circuit is this? (b) What should the polarity of V_{CC} be if the transistor is a *PNP* type? (c) Knowing the component values, how can we calculate the quiescent value of base current? (d) How can we find I_{CQ}? (e) How can we find V_{CEQ}?

30. In step 1 of the solution to Example 11-3, where does the value 0.6 come from?

31. Refer to Fig. 11-13b: (a) How would you alter the diagram to establish the transistor as a *PNP* unit? (b) What polarity should V_{CC} then be? (c) If R_1 is four times larger than R_2, how would the base potential (V_B) compare with V_{CC}? (d) Is the emitter potential (V_E) greater or smaller than V_B? Give the relation. (e) What is the *exact* relation between I_E and I_C? (f) In a practical situation, why can we consider these two currents as equal?

32. What are the quiescent values for the transistor in Example 11-4?

33. Refer to Fig. 11-14: (a) What diagram change would establish the transistor as an *NPN*? (b) Specify the correct polarity for V_{CC}. (c) What currents flow through resistor R_c? (d) In a practical case, can we neglect either of these currents? Explain. (e) When calculating for the value of the base current I_B from V_{CC}, R_c, R_b, and V_{BE}, how is the value of R_c modified? Why?

34. Is Fig. 11-15 an exact equivalent diagram? Explain.

35. In the FET equivalent circuit (Fig. 10-6) no input side is shown. Yet an input side is shown in Fig. 11-15. Why the difference?

36. Refer to Fig. 11-15: (a) What does r_i represent? (b) What does R_b' represent? (c) What does $r_{T(in)}$ represent? (d) What does r_o represent? (e) How does r_o differ from r_{oe}? From h_{oe}? (f) How is R_e obtained? (g) Under what conditions may R_L be considered the same as R_e?

37. In Equation 11-15, what is I_B and in what units is it expressed?

38. Does Equation 11-15 give the correct value for the r_i of all units? Explain.

39. Refer to Fig. 11-14b: **(a)** As far as the ac signal input is concerned, why is R_1 in parallel with R_2? **(b)** What is this resistance combination called in Fig. 11-15 on the input side? **(c)** How is it included in the output side?

40. Refer to the input side of Fig. 11-15: **(a)** How can the total resistance of this circuit be evaluated? **(b)** Under what condition can we consider this total resistance as R_b' in parallel with r_i? **(c)** Under what condition can it be considered as R_s in series with r_i? **(d)** Under what condition can it be considered as r_i alone? **(e)** When R_s is negligible compared to $R_b' \parallel r_i$, give the equation for calculating i_b. **(f)** When R_b' is very large compared to r_i, give the equation for calculating i_b.

41. From the equivalent circuit (Fig. 11-15) give the equation for the voltage gain of the transistor circuit alone—i.e., neglecting the source resistance.

42. A BJT is operated from a 12-V dc power supply. **(a)** What is the (approximate) maximum possible peak-to-peak value of the output voltage? **(b)** What is the maximum safe value for V_{CE} to prevent thermal runaway?

43. Refer to the solution to Example 11-8: **(a)** In step 4, what fixes the operating point at $V_{CE} = 12$ V and $I_C = 10$ mA? **(b)** In step 8, where does the 23.3 V for the numerator of the proportion come from?

PROBLEMS AND DIAGRAMS

1. Draw the circuit diagram for a common-emitter R-C coupled *NPN* transistor amplifier using two supplies. Show supply polarities.

2. An R-C coupled common-emitter amplifier (as in Fig. 11-1) has the following circuit values: $V_{CC} = 16$ V, $V_{BB} = 6$ V, $R_b = 60$ kΩ, $R_c = 3300$ Ω, and $r_i = 1500$ Ω. The collector characteristic curves for this transistor, CE connection, are shown in Fig. 11-17. **(a)** Draw the dc load line and locate the operating point. **(b)** Find the quiescent values of collector current and voltage. **(c)** Calculate the value of the ac load resistance and draw the ac load line. (Keep this solution for use in Problems 6 and 7.)

3. Repeat Problem 2, using $V_{CC} = 22$ V, $V_{BB} = 3$ V, $R_c = 5000$ Ω, $R_b = 30$ kΩ, and $r_i = 2000$ Ω.

4. Refer to Fig. 11-2. Calculate the peak-to-peak value of the collector current swing, and calculate the current gain for a base current swing of ± 20 μA.

5. In Fig. 11-2, using a Q point at 80 μA, calculate the collector current swing and current gain for base current swings of **(a)** ± 20 μA and **(b)** ± 40 μA.

6. Using the data (given and calculated) from Problem 2 and a peak signal swing of ± 50 μA, find **(a)** the peak-to-peak value of the collector current swing; **(b)** the rms value of the ac component of the collector current; **(c)** the collector voltage swing and the rms value of this ac component; **(d)** the current gain A_i; **(e)** the voltage gain A_r.

7. Repeat Problem 6, using an input signal swing of ± 100 μA.

8. Using the data of Example 11-1, and its load line of Fig. 11-2, but for an operating point of 60 μA, and a peak signal swing of ± 20 μA, find **(a)** the collector current swing and the rms value of the ac component; **(b)** the collector voltage swing and the rms value of the ac component; **(c)** current gain A_i and voltage gain A_v.

9. Draw the circuit diagram for an R-C coupled *NPN* transistor amplifier using only one dc supply source and fixed bias.

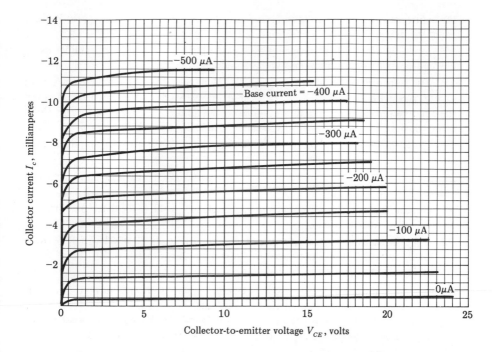

Figure 11-17 Collector characteristic curves for Problems 2 and 3 (*Raytheon*).

10. In the circuit of Fig. 11-4, it is desired to obtain a base-bias current of 60 μA. The supply voltage is 32 V. **(a)** What components determine the bias? **(b)** What values are needed? **(c)** How is this affected if a *PNP* transistor is used?

11. The transistor of Fig. 11-6 has a current gain of $h_{FE} = 50$. If the collector current is 6 mA, find **(a)** the base current; **(b)** the emitter current; **(c)** the value of each resistor required to produce the potentials shown.

12. Repeat Problem 11 for $h_{FE} = 90$ and a collector current of 5 mA.

13. Draw the circuit diagram for an R-C coupled *NPN* transistor amplifier using current-feedback stabilization and a voltage-divider base-bias circuit.

14. A 2N5133 silicon transistor has an h_{FE} range of 60 to 1000 at $I_C = 1.0$ mA. It is used in a simple fixed bias circuit with $V_{CC} = 21$ V. **(a)** Using the minimum value for h_{FE}, calculate the value of base current needed for a collector current of 1 mA. **(b)** Find the value of the base-bias resistor required. **(c)** If another transistor of the same type is used, but with $h_{FE} = 1000$, what is the collector current now? **(d)** What is the change (ratio) in collector current?

15. The transistor of Problem 14 is used in a voltage-divider circuit as in Fig. 11-7, with $R_E = 2.2$ kΩ; $R_1 = 68$ kΩ; and $R_2 = 10$ kΩ. **(a)** Find the emitter current using Equation 11-10a and the minimum value for h_{FE}. **(b)** Repeat, using the maximum value for h_{FE}. **(c)** Find I_E from Equation 11-10b.

16. A 2N4125 silicon transistor (h_{FE} 50–150 at $I_C = 2.0$ mA) is used in the circuit of Fig. 11-7, with $R_E = 2.2$ kΩ; $R_1 = 68$ kΩ; $R_2 = 16$ kΩ; and $V_{CC} = 24$ V. **(a)** Find the emitter current using Equation 11-10a and the minimum value for h_{FE}. **(b)** Repeat, using the maximum h_{FE} value. **(c)** Find I_E from Equation 11-10b.

17. Draw the circuit diagram for a CE R-C coupled *PNP* transistor amplifier using current-feedback stabilization and voltage-divider base bias.

18. The transistor of Problem 16 (h_{FE} = 50–150 at I_C = 2.0 mA) is used in the circuit of Fig. 11-9, with V_{CC} = +20 V; V_{EE} = −20 V; R_b = 16 kΩ; and R_E = 10 kΩ. **(a)** Find the emitter current using Equation 11-12a and the minimum value of h_{FE}. **(b)** Repeat, using the maximum value of h_{FE}. **(c)** Find I_E using Equation 11-12c.

19. Draw the circuit diagram for a CE R-C coupled *PNP* transistor amplifier using a separate emitter-bias supply.

20. Draw the circuit diagram for a CE R-C coupled *PNP* transistor amplifier using voltage-feedback stabilization.

21. Repeat Problem 20, using current-feedback and voltage-feedback stabilization and voltage-divider base bias.

22. Find the quiescent values in a fixed bias circuit (Fig. 11-13a) if V_{CC} is 30 V, R_b is 680 kΩ, R_c is 5100 Ω, and the transistor is a silicon unit with h_{FE} = 70.

23. Repeat Problem 22 for V_{CC} = 48 V, R_b = 1.2 MΩ, R_c = 6200 Ω, and h_{FE} = 120.

24. A voltage-divider base-bias circuit with emitter stabilization (Fig. 11-13b) has the following circuit values: R_1 = 27 kΩ, R_2 = 10 kΩ, R_c = 6800 Ω, R_E = 4700 Ω, and V_{CC} = 30 V. The transistor is a silicon unit with h_{FE} = 100. Find the quiescent values.

25. Repeat Problem 21 for R_1 = 33 kΩ, R_2 = 15 kΩ, R_c = 5600 Ω, R_E = 3300 Ω, V_{CC} = 9.0 V, and h_{FE} = 80.

26. A voltage-feedback bias circuit (Fig. 11-14) has the following values: R_b = 470 kΩ, R_c = 7500 Ω, V_{CC} = 40 V. The transistor is a silicon unit with h_{FE} = 120. Find the quiescent values.

27. Repeat Problem 26, using a 24-V supply, R_b = 330 kΩ, R_c = 6800 Ω, and a silicon transistor with h_{FE} = 150.

28. Two 2N2483 silicon transistors are used in the circuit of Fig. 11-16. The transistor parameters are h_{fe} = 80–450 and h_{oe} = 30 μmho. The circuit values are V_{CC} = 12 V, R_1 = 100 kΩ, R_2 = 47 kΩ, R_c = 5600 Ω, R_E = 3300 Ω. The source voltage is 0.02 V and has an internal resistance of 30 Ω. Using the minimum value of h_{fe}, find **(a)** the quiescent values; **(b)** the input resistance of the transistors; **(c)** the transistor gain; **(d)** the output voltage and the overall gain (from the source to the output of Q_1).

29. Repeat Problem 28, using the maximum value for h_{fe}.

30. Repeat Problem 28, but with a source resistance of 1500 Ω.

31. Two 2N3251 *PNP* silicon transistors are used in the circuit of Fig. 11-16. The transistors have an h_{fe} spread of 100 to 400 and an h_{oe} spread of 10 to 60 μmho. The circuit values are V_{CC} = 30 V, R_1 = 56 kΩ, R_2 = 22 kΩ, R_c = 7500 Ω, R_E = 3900 Ω. The source has an output of 0.1 V and an internal resistance of 1200 Ω. Using the minimum parameter values, find **(a)** the quiescent values; **(b)** the input resistance of the transistors; **(c)** the transistor gain; **(d)** the overall gain and the output voltage.

32. Repeat Problem 31, using the maximum parameter values but with the source voltage reduced to 0.025 V.

12

Bipolar Transistor Equivalent Circuits

For the purpose of detailed mathematical analysis, any electrical element or device can be replaced by an equivalent electrical network. The network can then be solved by using ac network circuit theory. Following this technique, equivalent circuits were used in Chapter 10 to analyze the gain and phase characteristics of FET and vacuum-tube circuits. Again, in Chapter 11, a simplified equivalent circuit was used for the BJT. Now we will see how a more rigorous analysis can be applied to this transistor.

The equivalent-circuit technique is not a cure-all; it has its limitations. The constants used to represent the device apply only for the given set of conditions for which these constants are valid. With FETs, for example, an equivalent circuit applies only for the operating point at which the FET parameters (g_{fs}, g_{os}) were evaluated. Any change in operation that would affect any of these parameters would impair the validity of the equivalent-circuit analysis. The same limitation applies also to equivalent circuits when used to analyze any transistor circuits. Specifically, then, a given equivalent circuit can be used to represent operation only for that given operating point and *only for a small swing to either side of that operating point*. (For this reason, some texts separate their amplifier discussions under headings of "Small-Signal Amplifiers" and "Large-Signal Amplifiers.")

This corollary limitation is made necessary by the fact that the dynamic (mutual) characteristic is not perfectly straight, and therefore the parameters of the unit will vary depending on the instantaneous point of operation. For a small signal swing, the curve can be considered linear (the tangential value) with negligible error. However, with a large signal swing, excessive curvature can be included and the error introduced would be too great. Furthermore, equivalent-circuit analysis cannot be used to evaluate the amount of harmonic or amplitude distortion produced by an amplifier circuit. Such distortion is caused by curvature of the dynamic characteristic, whereas the equivalent-circuit technique is valid only for operation on the linear portion. Evaluation of nonlinear distortion requires the use of characteristic curves and load lines for a graphic analysis.

With FETs and vacuum tubes, the equivalent circuits were simple circuits representing only the *output* side of the device. It was possible to make this simplification because the input circuit (class A, small signal) has an infinite impedance (no current) and also because there is negligible interaction between the output and input circuits. This simplification is not possible with BJTs, and so their equivalent diagrams must duplicate input and output conditions and the interaction between them. Many types of equivalent diagrams are possible to represent this transistor. Several are discussed in this chapter.

T EQUIVALENT CIRCUIT.   The equivalent circuits of Fig. 12-1 are quite easy to visualize because they resemble the actual physical construction and connection of the transistor. It is probably for this reason that they were the first type of equivalent circuit used to represent the transistor. The three diagrams (*a*), (*b*), and (*c*) are for the common-emitter, common-

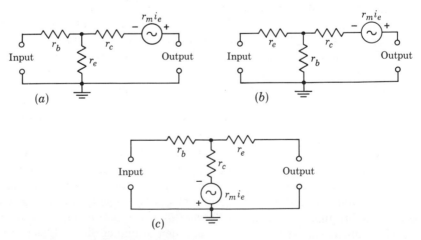

Figure 12-1   The *T*-type transistor equivalent circuit.

base, and common-collector configurations, respectively. In these diagrams each resistance represents the resistance of the corresponding element, including the junction resistance. The voltage source $r_m i_e$ represents the amplified voltage that appears on the output side due to the current i_e in the input, and the term r_m is regarded as the mutual resistance (or forward transfer function) between the input and output circuits. Typical values of the equivalent-circuit T parameters are $r_e = 25\ \Omega$, $r_b = 800\ \Omega$, $r_c = 2\ M\Omega$, $r_m = 1.96\ M\Omega$ ($r_m \cong \alpha r_c$), $\alpha = 0.98$. At one time these parameters could be found listed in the manufacturers' data sheets. However, since about 1956 they have been superseded by other parameters pertaining to another type of equivalent circuit. The reason for this will become apparent later in this discussion.

The equivalent diagram can be used to calculate current, voltage, and power gain. In addition, since impedance matching is a desirable goal in BJT amplifiers, it is also necessary to know the input and output impedance of a transistor stage. But because of the interaction between input and output, these values Z_i and Z_o are affected by the source and load impedances. The equivalent circuit *with the source and load connected* can be used to evaluate these input and output impedances. As an illustration, Fig. 12-2 shows the T equivalent circuit for a transistor, common-emitter

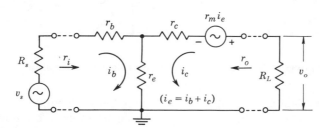

Figure 12-2 T equivalent circuit for a common-emitter connection, including source and load.

connected, including the source and load. For simplicity, the circuit is considered at some mid-frequency condition and all impedances are taken as resistances. By inspection of this diagram it is obvious that the value of the input resistance r_i will be affected by the load value R_L and, similarly, that the value of the output resistance r_o will be affected by the source resistance R_s. The exact relationships can be determined mathematically by setting up appropriate mesh equations and solving for the desired values. The mathematical details will not be shown here. The results for this common-emitter equivalent circuit are

$$r_i = r_b + r_e - \frac{r_e(r_e - r_m)}{R_L + r_c + r_e - r_m} \tag{12-1}$$

$$r_o = r_c + r_e - r_m - \frac{r_e(r_e - r_m)}{R_s + r_b + r_e} \qquad \text{(12-2)}$$

OPEN-CIRCUIT r PARAMETERS.   One drawback of the T equivalent circuit is that its parameters cannot be measured directly. This should be obvious, since a measurement cannot be made inside the transistor at the fictitious junction of r_b, r_e, and r_c. The only measurements that can be made are of currents and/or voltages at the input and/or output sides. This led to the development of the open-circuit r parameters. The input and output sides of the previous T network are converted into constant-voltage equivalent circuits with open-circuit generators and series resistances.* The resulting circuit is shown in Fig. 12-3. Notice that each parameter is identified by a double subscript. In this notation, the first digit refers to a voltage and the second digit to a current measurement, while the numerical value of the digit (1 or 2) specifies the terminals to

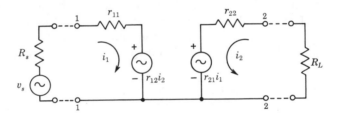

Figure 12-3   The open-circuit r-parameter equivalent diagram.

which the measurements apply. A number 1 indicates input side, whereas a number 2 refers to the output side. As an example, the parameter r_{21} signifies a resistance value obtained from a voltage measurement at the output side (the first digit is 2) and a current measurement at the input side (the second digit is 1). On this basis, each parameter shown has a specific meaning as follows:

1. r_{11} = input resistance with output terminals ac open:

$$r_{11} = \frac{e_1}{i_1} \qquad (i_2 = 0)$$

2. r_{22} = output resistance with input terminals ac open:

$$r_{22} = \frac{e_2}{i_2} \qquad (i_1 = 0)$$

*This is done by use of Thévenin's network theorem.

3. r_{12} = reverse *transfer* resistance with input terminals ac open:

$$r_{12} = \frac{e_1}{i_2} \qquad (i_1 = 0)$$

4. r_{21} = forward transfer resistance with output terminals open:

$$r_{21} = \frac{e_2}{i_1} \qquad (i_2 = 0)$$

Notice that in each of the above relations one side or the other is ac open-circuited. However, the circuit is open only insofar as ac signals are concerned. The dc operating bias currents and potentials must be maintained to establish the proper operating point.

Table 12-1
COMPARISON OF r AND T PARAMETERS

| r PARAMETERS | EQUIVALENT T PARAMETERS | | |
	CE	CB	CC
r_{11}	$r_b + r_c$	$r_e + r_b$	$r_b + r_c$
r_{22}	$r_c + r_e - r_m$	$r_c + r_b$	$r_c + r_e - r_m$
r_{12}	r_e	r_b	$r_c - r_m$
r_{21}	$r_e - r_m$	$r_m + r_b$	r_c

SHORT-CIRCUITED g PARAMETERS.   Evaluation of the above open-circuit parameters can impose some extremely difficult measurement limitations. For example, determination of r_{11} and r_{21} requires that the output side be ac open-circuited while the dc path is maintained normal, for bias. Such a condition can be approached by using a high inductance value or a high-Q parallel resonant circuit as the load through which to feed the dc supply. However, the output resistance (r_{22}) can run as high as 100,000 Ω for a common-emitter connection and over 1 MΩ for a common-base connection. The ac impedance required for an effective ac open circuit becomes so high that evaluation of this open-circuit r parameter becomes extremely impractical. This led to the development of the short-circuit g parameter equivalent diagram.

By use of appropriate network theory* the input and output sides of the T equivalent diagram can be converted into "constant-current" diagrams containing a constant-current generator and a shunt conductance. The current generator value is made equal to the current that would flow in the short-circuited terminals of the original T network.

This new equivalent diagram solves the problem of evaluating parameters requiring an ac open circuit on the output side but in turn creates

*Norton's network theorem.

impractical conditions requiring ac short circuit of the input side. For this reason, this technique never became popular and will not be discussed further.

HYBRID EQUIVALENT CIRCUIT.   This next step should be obvious. By combining the desirable elements of the open-circuit and the short-circuit equivalent diagrams, it was possible to arrive at a more practical equivalent circuit. Since this new diagram is the offspring of two different species, it has been called a *hybrid* (*h*) equivalent circuit and its parameters are known as *h* parameters.

A *general* hybrid equivalent circuit is shown in Fig. 12-4. It does not refer to any specific configuration. Therefore the currents and voltages are merely labeled with subscripts 1 and 2 to distinguish input from output, and no attempt is made to specify the transistor elements to which they

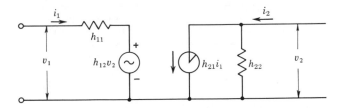

Figure 12-4   General form of the hybrid equivalent circuit.

apply. Notice that the constant-voltage or open-circuit type of diagram is used for the input side, while the constant-current or short-circuit type of diagram is used for the output side. Note also that the double subscript (numerical) notation is used to identify the parameters.

Now let us evaluate each of the *h* parameters:

1. h_{11} = input resistance with output terminals ac shorted:

$$h_{11} = \frac{v_1}{i_1} \qquad (v_2 = 0) \qquad (12\text{-}3)$$

2. h_{22} = output conductance with input terminals ac open:

$$h_{22} = \frac{i_2}{v_2} \qquad (i_1 = 0) \qquad (12\text{-}4)$$

3. h_{12} = input-to-output (*reverse transfer*) voltage ratio, or voltage-feedback ratio with input terminals *ac open:*

$$h_{12} = \frac{v_1}{v_2} \qquad (i_1 = 0) \qquad (12\text{-}5)$$

4. h_{21} = output-to-input (forward transfer) current ratio with output terminals short-circuited:

$$h_{21} = \frac{i_2}{i_1} \qquad (v_2 = 0) \qquad \text{(12-6)}$$

If a manufacturer's data sheet listed a specific value for the parameter h_{11}, it could refer to the common-base, common-emitter, or common-collector value. This presents a problem because the value of any one parameter is not the same for all configurations. In the early 1950s the common-base connection was considered as the *reference* configuration, and the value listed would probably have referred to the common-base value. However, because of the popularity of the common-emitter connection in practical circuit applications, manufacturers began to list the common-emitter parameters. To avoid confusion in identifying the parameters, the IEEE established supplementary standards to cover transistors (1956). These standards recommend the use of *letter* subscripts. In addition, they add that numerical subscripts may be used when convenient, but then they also include a third (letter) subscript (*e*, *b*, or *c*) to remove any ambiguity as to the configurations to which the parameters apply.

The letter subscript system is a dual-letter system. The *second* subscript is *b*, *e*, or *c*, referring to the configuration; the letters used for the *first* subscript have the following meanings: *i* = input, *o* = output, *f* = forward transfer (current function implied), *r* = reverse transfer (voltage function implied). Manufacturers began using this system almost immediately, and by 1960 the listing of *h* parameter values, using the double-letter subscript system, superseded the earlier parameter data and nomenclature. Sometimes both the common-emitter and the common-base parameter values are listed. When only one set is given, it is usually for the common-emitter configuration.

Figure 12-5 shows this system of notation used with the hybrid equivalent circuit for a common-emitter connection. Notice the replacement of the general terms with the specific values pertinent to the common-emitter circuit. Now h_{ie} for this circuit could not readily be misused as the value

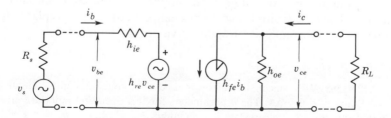

Figure 12-5 Hybrid equivalent circuit for a common-emitter configuration.

of its equivalent component h_{ib} in the common-base circuit. Notice also the use of i_b and i_c in place of the general terms i_1 and i_2. Now, if we re-examine the forward transfer ratio, it becomes

$$h_{fe} = \frac{i_c}{i_b} \qquad (v_2 = 0) \tag{12-7}$$

This should be recognized as numerically equal to the transistor current gain β, and you are more likely to see the symbol h_{fe} in the manufacturers' data in place of β. Similarly, h_{fb} is used in place of α for the common-base circuit.

HYBRID CIRCUIT EQUATIONS.   As mentioned before, the equivalent circuit is useful in calculating the input and output resistance of the transistor as well as the current, voltage, and power gain. The relation between these quantities and the h parameters can be developed by setting up suitable equations from the equivalent circuit and solving for the desired quantities. The mathematical manipulations are not given here (see later reference). The results are as follows:*

1. $r_i = \dfrac{v_{be}}{i_b} = \dfrac{h_{ie} + R_L\Delta_{he}}{1 + h_{oe}R_L}$ \hfill (12-8)

2. $r_o = \dfrac{v_{ce}}{i_c} = \dfrac{h_{ie} + R_s}{\Delta_{he} + h_{oe}R_s}$ \hfill (12-9)

3. $A_i = \dfrac{i_c}{i_b} = \dfrac{h_{fe}}{1 + h_{oe}R_L}$ \hfill (12-10)

4. $A_v = \dfrac{v_{ce}}{v_{be}} = \dfrac{-h_{fe}R_L}{h_{ie} + R_L\Delta_{he}}$ \hfill (12-11)

5. $A_p = A_iA_e$ \hfill (12-12)

6. $V_o = v_{ce} = \dfrac{-h_{fe}R_LV_s}{h_{oe}R_sR_L + R_L\Delta_{he} + h_{ie} + R_s}$ \hfill (12-13)

Although the above equations are expressed in terms of the common-emitter configuration, the same form applies for the common-base and common-collector circuitry. The only change needed is to replace the h_e parameters by their equivalent h_b or h_c values, as required.

EXAMPLE 12-1

A 2N1565 NPN silicon transistor is used in a common-emitter circuit with $V_{CE} = 5$ V and $I_b = 0.1$ mA. The h parameters for this operating point are given by the manufacturer as $h_{ie} = 660\ \Omega$, $h_{oe} = 65\ \mu$mho, $h_{fe} = 70$, and $h_{re} = 1.1 \times 10^{-4}$.

---

*The symbol Δ_{he} is used to replace the expression $h_{ie}h_{oe} - h_{re}h_{fe}$. Since this expression occurs in many formulas, this substitution simplifies the equations and their solution.

The transistor circuit has an ac load resistance of 10,000 Ω and a source resistance (including the effect of the bias network) of 800 Ω. Find the input and output resistances, current, voltage, and power gain of the circuit.

Solution

1. $\Delta_{he} = h_{ie}h_{oe} - h_{re}h_{fe} = 660 \times 65 \times 10^{-6} - 1.1 \times 10^{-4} \times 70$

   $= 3.52 \times 10^{-2}$

2. $r_i = \dfrac{h_{ie} + R_L\Delta_{he}}{1 + h_{oe}R_L}$

   $= \dfrac{660 + 1 \times 10^4 \,(3.52 \times 10^{-2})}{1 + 65 \times 10^{-6} \times 1 \times 10^4}$

   $= 614 \, \Omega$

3. $r_o = \dfrac{h_{ie} + R_s}{\Delta_{he} + h_{oe}R_s}$

   $= \dfrac{660 + 800}{(3.52 \times 10^{-2}) + 65 \times 10^{-6} \times 800}$

   $= 16,750 \, \Omega$

4. $A_i = \dfrac{h_{fe}}{1 + h_{oe}R_L} = \dfrac{70}{1 + 65 \times 10^{-6} \times 1 \times 10^4}$

   $= 42.4$

   [Notice that whereas the current gain of the transistor itself with the output shorted (β or h_{fe}) is 70, the current gain of the *complete circuit* has dropped to 42.4.]

5. $A_v = \dfrac{-h_{fe}R_L}{h_{ie} + R_L\Delta_{he}}$

   $= \dfrac{-70 \times 1 \times 10^4}{660 + 1 \times 10^4 \,(3.52 \times 10^{-2})}$

   $= -692$

   (The minus sign is a vestige of the mathematical derivation and merely indicates that there is a 180° phase reversal between input and output voltage.)

6. $A_p = A_iA_v = 42.4 \times -692$

   $= -29,200 \qquad$ or 44.7 dB

POWER GAIN COMPARISONS

Example 12-1 was concluded by calculating the power gain (A_p). This is the *actual* power gain obtained with this transistor operating from a specific source and into a specific load. However, we might have obtained a higher gain by changing (if we could) the effective load resistance, the effective source resistance, or both. Since the transistor draws input current and is a "current-operated device," power gain and efficiency of operation are important in evaluating transistor merits and circuit comparisons. Toward this end, several power gain terms have been developed.

TRANSDUCER GAIN.   To get a maximum power transfer from the source (or previous stage), the transistor circuit input resistance (r_i) should match the source resistance (R_s). If this match is not achieved, we will not obtain as an input all the power that the source is capable of delivering, and the power output to the load will not be as high as it could have been. In other words, the source is not being used efficiently. As a measure of the efficiency with which the source is utilized, the term *transducer gain* (A_{pt}) is used. It is the ratio of the actual power output *delivered to the load* compared to the maximum power input that would be available from the source if the input impedance matched the source impedance:

$$A_{pt} = \frac{\text{power output}}{\text{available source power (matched input)}} \tag{12-14}$$

Remembering that the maximum power delivered by the source occurs when $r_i = R_s$, and that under such a condition half the source power is dissipated internally, we can readily see that the power available from the source is limited to

$$P_{\text{avail}} = \frac{V_s^2}{4R_s} \tag{12-15}$$

If we calculate power output from the equivalent circuit and use the above relation for "available source power," we will get the transducer gain as

$$A_{pt} = \frac{4R_s}{R_L}\left[\frac{h_{fe}}{R_s h_{oe} + \Delta_{he} + (h_{ie} + R_s)/R_L}\right]^2 \tag{12-16}$$

AVAILABLE POWER GAIN.   For a fixed power input, the available power output from any given transistor and circuit would increase if we could adjust the load (R_L) to match the output resistance (r_o) of the transistor circuit. All the *available* power output and power gain is realized when this impedance match is effected. Now assume that we have several circuits, all of which match their respective loads, but the load values differ for each circuit. If we apply equal input signal levels to each of these circuits, the power outputs need not be the same and yet each circuit is delivering its available power output for that given input. The ratio of their output to input, or *available power gain* (A_{pa}), is therefore a measure of the merit of each circuit. Remember that the input power must be alike.

But each circuit probably has a different input resistance (r_i), and even when connected to identical sources, the power inputs will not be equal. To eliminate this variable from the measure of available power gain, the equal input value is taken as the *available source power* $V_s^2/4R_s$. The available power gain is therefore defined as the ratio of the power output available from the transistor (with matched load) to the available source power (regardless of whether any input match exists):

$$A_{pa} = \frac{\text{available transistor power output}}{\text{available source power}} \qquad (12\text{-}17)$$

Calculating the power output and gain with these conditions, from the equivalent diagram

$$A_{pa} = \frac{h_{fe}^2 R_s}{(\Delta_{he} + R_s h_{oe})(R_s + h_{ie})} \qquad (12\text{-}18)$$

MAXIMUM AVAILABLE GAIN (MAG).   To achieve the maximum possible gain it is necessary both that the maximum power be extracted from the source feeding the transistor circuit and that the maximum power be delivered by the transistor circuit to its load. This of course requires that the source resistance match the transistor input resistance $(R_s = r_i)$ and that the load resistance match the transistor output resistance $(R_L = r_o)$. Solving for these matching values, in terms of the transistor parameters

$$R_s = \sqrt{\frac{h_{ie}}{h_{oe}} \Delta_{he}} \qquad (12\text{-}19)$$

and

$$R_L = \sqrt{\frac{h_{ie}}{h_{oe}}} \times \frac{1}{\Delta_{he}} \qquad (12\text{-}20)$$

Also, solving for the power gain and using these matched impedance relations, the maximum available gain is found as

$$\text{MAG} = \frac{h_{fe}^2}{(\sqrt{\Delta_{he}} + \sqrt{h_{ie}h_{oe}})^2} \qquad (12\text{-}21)$$

Notice that neither the source nor load resistance $(R_s$ or $R_L)$ appears in the equation of maximum available gain. This quantity is therefore a function of the transistor parameters only and can be used as a figure of merit to compare transistors. In actual practice, this condition is seldom realized with untuned amplifier circuits since it would require use of transformer coupling.

In this discussion of power gain comparisons, we have again used the common-emitter parameters. These equations apply equally well to the other two configurations as long as the corresponding h parameters are used in place of the common-emitter symbols.

Y AND S PARAMETERS.   By the early 1960s, because of improved manufacturing techniques, transistors became available that could operate into the VHF and the lower UHF bands. At these high frequencies, measurement of the open circuit h parameters (h_{oe} and h_{re}) presented serious difficulties. Consequently, for RF applications into the VHF band (and higher), y or *admittance parameters* became a more logical choice. These are based on short-circuit conditions for both the input and output sides. They are:

1. y_{ie} = Input admittance, with output short-circuited:

$$y_{ie} = \frac{I_1}{V_1} \qquad (v_2 = 0)$$

2. y_{re} = Reverse transfer admittance, input short-circuited:

$$y_{re} = \frac{I_1}{V_2} \qquad (v_1 = 0)$$

3. y_{fe} = Forward transfer admittance, output short-circuited:

$$y_{fe} = \frac{I_2}{V_1} \qquad (v_2 = 0)$$

4. y_{oe} = Output admittance, input short-circuited:

$$y_{oe} = \frac{I_2}{V_2} \qquad (v_1 = 0)$$

Unfortunately, use of y parameters is not a cure-all. At high frequencies, lead inductances and capacitances make short circuits (and open circuits) difficult to obtain. Tuning stubs, separately adjusted at each measurement frequency, would be required to reflect a short-circuit condition to the transistor terminals. Even with specialized test equipment, inaccuracies of the instrument and their residual errors can cause wide variations in the parameter evaluations. Errors of 30 to 100 percent are possible.

For these reasons, in the past few years there has been a shift from y parameters to s or *scattering parameters*. These values are easier to measure because they are made with matched-load conditions at the input and output, instead of with open or short circuits. The parameter designations are:

1. s_{11} — Input voltage-reflection coefficient, with the output terminated by a matched load.
2. s_{22} — Output voltage-reflection coefficient, with the input terminated by a matched load.

3. s_{21} — Forward voltage (transmission-insertion) gain, with matched source and load.

4. s_{12} — Reverse voltage (transmission-insertion) gain, with matched source and load.

On data sheets, these parameter values are plotted on generalized Smith charts to show they vary with frequency.

The s-parameter design is considered easier, quicker, and more accurate than the y-parameter technique. However, when an exact design is desired, the labor for either design is excessive, and computer-aided design becomes very helpful. A number of computer-aided designs are available for s-parameter equivalent circuits. One of these is SPEEDY,* developed by Fairchild, and available worldwide on the General Electric Timeshare Computer System. Since both s- and y-parameter designs are used for RF circuitry and involve computer aid for accuracy, they will not be discussed further in this text.

REVIEW QUESTIONS

1. Explain why use of equivalent-circuit analysis is limited to small-signal amplifiers.

2. Explain why amplitude distortion cannot be evaluated from an equivalent-circuit analysis.

3. Why does a BJT equivalent circuit show input and output conditions, whereas the FET equivalent circuits generally show only the effective output circuit?

4. (a) What type of equivalent circuit was first used to represent a transistor? (b) What feature made this circuit seem desirable?

5. In Fig. 12-1, which transistor configuration is represented by (a) diagram (a)? (b) Diagram (b)? (c) Diagram (c)?

6. With reference to Fig. 12-1, explain briefly the significance of each of the following: (a) r_b; (b) r_e; (c) r_c; (d) r_m; (e) $r_m i_e$.

7. (a) With reference to the T equivalent circuit, give a typical value for r_b, r_c, r_e, r_m. (b) How is each of these values affected by a change from common-emitter to common-base connection? (c) How are the values affected by a change to common-collector connection?

8. Refer to Fig. 12-2: (a) What does R_s represent? (b) What does R_L represent? (c) What do the symbols r_i and r_o refer to? (d) Under what condition can these quantities be considered resistive? (e) Are their values affected by the transistor configuration? Explain. (f) For a given configuration, are their values affected by factors outside the transistor itself? Explain.

9. State a disadvantage to the use of the T equivalent circuit.

10. In a four-terminal circuit network, measurement values are identified by double numerical subscripts. Give the significance of each of the following: (a) the number 1

---

*SPEEDY, a computer-aided design program for RF and microwave circuits. Fairchild Semiconductor, May 1973.

as the first subscript; **(b)** the number 2 as the first subscript; **(c)** the number 2 as the second subscript; **(d)** the number 1 as the second subscript.

11. Refer to Fig. 12-3: **(a)** What does r_{11} represent? **(b)** Give the equation for this quantity. **(c)** What conditions must be met for a practical evaluation of this quantity? **(d)** What does r_{22} represent? **(e)** Give the equation for this quantity. **(f)** What conditions must be met for a practical evaluation?

12. State a disadvantage to the use of the open-circuit r parameter equivalent diagram.

13. **(a)** Why is the term *hybrid* used in connection with the equivalent diagram of Fig. 12-4? **(b)** Why has this type of equivalent diagram been generally adopted? **(c)** To which transistor configuration does this circuit (Fig. 12-4) apply? **(d)** How is each of these values affected by a change in transistor configuration?

14. Refer to Fig. 12-4: **(a)** What does h_{11} represent? **(b)** How is it evaluated? **(c)** Give the letter subscript symbol for h_{11} for the common-emitter connection.

15. Repeat Question 14 for h_{12}.

16. Repeat Question 14 for h_{21}.

17. Repeat Question 14 for h_{22}.

18. What is the advantage of the letter subscript system?

19. **(a)** In the letter subscript notation, what four letters are used as first subscripts? **(b)** What does each signify? **(c)** What three letters are used for second subscripts? **(d)** What does each signify?

20. Refer to Fig. 12-5: **(a)** What does R_S signify? **(b)** What does R_L signify? **(c)** What does h_{ie} signify? **(d)** How is it evaluated? **(e)** What does h_{re} signify? **(f)** How is it evaluated? **(g)** What does h_{fe} signify? **(h)** How is it evaluated? **(i)** Give another symbol often used for this quantity. **(j)** What does h_{oe} signify? **(k)** How is it evaluated?

21. With reference to Fig. 12-5, what changes are necessary in *circuitry* to convert this diagram **(a)** for common-collector configuration? **(b)** For common-base configuration?

22. With reference to Fig. 12-5, enumerate item by item the changes in symbols that are necessary to convert this diagram into a common-base circuit.

23. Repeat Question 22 for a common-collector circuit.

24. Refer to Equation 12-8: **(a)** What is r_i? **(b)** For which configuration does this equation apply? Explain. **(c)** Rewrite this equation for a common-base circuit.

25. Refer to Equation 12-10: **(a)** What is A_i? **(b)** For what configuration does it apply? **(c)** How does it differ from β? **(d)** Rewrite this equation for a common-base circuit.

26. **(a)** Under what condition will maximum power be delivered from a source to a transistor? **(b)** Give the equation for the avilable power from any source in terms of its open-circuit voltage and resistance. **(c)** Define *transducer gain*. **(d)** Of what significance is this term?

27. **(a)** Under what condition will an amplifier develop all its available power output? **(b)** Define *available power gain*. **(c)** Does the available power gain depend on the load resistance value? Explain. **(d)** Does the available power gain depend on the source resistance? Explain.

28. **(a)** What does the abbreviation MAG stand for? **(b)** Define this quantity. **(c)** Does the value depend on the source resistance? Explain. **(d)** Does the value depend on the load impedance? Explain. **(e)** What factors determine this value? **(f)** If matching is achieved at the input and output of a transistor circuit, what relation exists between power gain, transducer gain, available gain, and MAG?

29. **(a)** When are y parameters used in place of h parameters to characterize a BJT? **(b)** Why is the change made? **(c)** Give the symbol and name for each of the four y parameters. **(d)** Give the equation for evaluating each quantity.

30. **(a)** Why are s parameters preferable to y parameters? **(b)** Give the symbol and significance of each of the s parameters. **(c)** How does evaluation of s-parameter values differ from evaluation of the y parameters?

PROBLEMS AND DIAGRAMS

1. A manufacturer's data sheet for the 2N334A transistor lists the following characteristics: forward current transfer ratio $h_{fe} = 38$; input impedance $h_{ie} = 1700\,\Omega$; output admittance $h_{oe} = 6.0\,\mu\text{mho}$; feedback voltage ratio $h_{re} = 1.3 \times 10^{-4}$; input impedance $h_{ib} = 40\,\Omega$; output admittance $h_{ob} = 0.18\,\mu\text{mho}$; reverse voltage transfer ratio $h_{rb} = 1.2 \times 10^{-4}$. The transistor is to be used in a common-emitter circuit with an ac load of $8000\,\Omega$ and a source resistance of $1000\,\Omega$. Calculate **(a)** the input and output resistance and **(b)** the current, voltage, and power gains. (Save these answers for use in later problems.)

2. In Problem 1, if the source voltage is 0.3 V, find the output voltage across the load.

3. Calculate the forward current transfer ratio (h_{fb}) for the common-base connection of the transistor in Problem 1. (*Note:* This is a negative quantity.)

4. The transistor in Problem 1 is used with the same source and load in the common-base connection. Calculate parts (a) and (b) as in Problem 1. (Save the answers for use in later problems.)

5. In Problem 4, find the output voltage across the load if the source voltage is 2.5 V.

6. A 2N118 transistor has the following parameters: $h_{ib} = 42\,\Omega$, $h_{ob} = 0.4\,\mu\text{mho}$, $h_{rb} = 250 \times 10^{-6}$, and $h_{fb} = -0.96$. It is used in a common-base circuit with an ac load of $60,000\,\Omega$ and a source resistance of $300\,\Omega$. Calculate parts (a) and (b) as in Problem 1. (Save the answers for use in later problems.)

7. At an operating point of $V_{CE} = 5$ V and $I_C = 5$ mA, a 2N1565 transistor has the following parameter values: $h_{fe} = 70$, $h_{ie} = 660\,\Omega$, $h_{oe} = 65\,\mu\text{mho}$, and $h_{re} = 1.1 \times 10^{-4}$. It is used with an ac load of $10,000\,\Omega$ and a 0.02-V source having an internal resistance of $800\,\Omega$. Calculate **(a)** the input and output resistance; **(b)** the current, voltage, and power gain; **(c)** the voltage across the load. (Save the answers for use in later problems.)

8. In Problem 7, the 2N1565 is replaced by a 2N1566 with the following parameter values: $h_{fe} = 160$, $h_{ie} = 1000\,\Omega$, $h_{oe} = 95\,\mu\text{mho}$, and $h_{re} = 5 \times 10^{-4}$. Calculate parts (a), (b), and (c) as in Problem 7.

9. Using the data of Problem 1, calculate **(a)** the transducer gain, **(b)** the available power gain, and **(c)** the maximum available gain.

10. **(a)** In Problem 1, what value of source and load resistances would produce the maximum available gain? **(b)** Using these values, calculate the voltage, current, and power gain.

11. Using the data of Problem 4, calculate **(a)** the transducer gain, **(b)** the available power gain, and **(c)** the maximum available gain.

12. Using the data of Problem 7, calculate **(a)** the transducer gain, **(b)** the available power gain, and **(c)** the maximum available gain.

13

Distortion in
R-C Amplifiers

The purpose of any voltage amplifier is to produce in its output an exact replica of the input signal, but with higher proportionate amplitude. Any deviation in the output waveshape is considered distortion. As stated in Chapter 9, distortion can be classified under one of three general headings —amplitude, frequency, or phase distortion. In this chapter each of these three areas is discussed in more detail, including the causes of distortion and how it can be avoided or reduced. Although this discussion on distortion is with specific reference to the R-C coupled amplifier of the previous chapters, much of it applies equally well to amplifiers using other means of coupling. (These will be discussed in Chapter 14.)

AMPLITUDE DISTORTION

It was pointed out in Chapter 9 that amplitude distortion occurs when the gain of the amplifier varies depending on the amplitude of the input signal, and that this effect is due to nonlinearity of the *dynamic mutual characteristic curve*. This effect was shown in Fig. 9-6. However, the curvature of that dynamic characteristic was deliberately exaggerated to bring out the point. Now let us see how to obtain a dynamic mutual characteristic, after which a realistic condition will be examined.

DYNAMIC MUTUAL CHARACTERISTIC. This curve shows how the output current varies with changes in the control value—i_d versus v_{gs} for a

FET; i_c versus i_b for a BJT; and i_p versus e_g for a vacuum tube. Dynamic curves are not given in a manufacturer's data sheets because this relationship is affected by the value of the power supply voltage, the dc load resistance, and the ac load resistance. Obviously, then, they apply only to a given circuit. Once the circuit values are fixed, we can obtain a dynamic mutual (or transfer) characteristic from the output (static) family of the device—*after we plot the ac load line.* Then all we need do is pick off the current values where the load line crosses each curve of the family and then plot current versus control value. Let us illustrate this technique, using the *NPN* transistor from Example 11-1.

EXAMPLE 13-1

Using the circuit data from Example 11-1 and the characteristic curves of Fig. 11-2, take data and plot the curve for the dynamic transfer characteristic for this transistor and circuit.

Solution

The dc and ac load lines are already drawn from Example 11-1. Now, from the intersection of the ac load line and each collector characteristic curve, pick off the data for the dynamic transfer curve:

I_B (μA)	I_C (mA)	I_B (μA)	I_C (mA)
160	7.2	60	3.2
140	6.6	40	2.2
120	5.9	20	1.2
100	5.1	0	0.2
80	4.2		

The transfer curve is shown in Fig. 13-1.

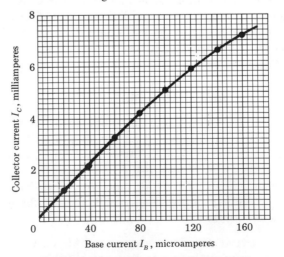

Figure 13-1   Curve from Example 13-1.

Now let us do a complete solution, using a field-effect transistor.

EXAMPLE 13-2

Figure 13-2 shows the common-source drain characteristics of an N-channel FET. It is used in an R-C coupled amplifier circuit with $V_{DD} = 24$ V; dc load resistance $R_c = 8000\ \Omega$; and an ac load resistance R_L of 3300 Ω. The bias is set at $V_{GS} = 1.0$ V.

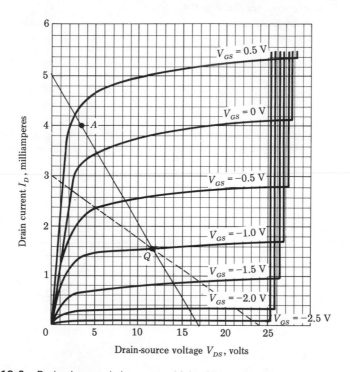

Figure 13-2   Drain characteristic curves with load lines added (*Texas Instruments*).

1. Draw the dc load line and locate the operating point.
2. Draw the ac load line.
3. Tabulate values for the dynamic transfer curve and plot the curve.

Solution

1. To draw the dc load line, two points are

$$I_D = 0 \quad \text{at } V_{DS} = 24 \text{ V}$$

$$V_{DS} = 0 \quad \text{at } I_D = \frac{V_{DD}}{R_c} = \frac{24}{8000} = 3.0 \text{ mA}$$

The dc load line is shown as the broken line in Fig. 13-2.

2. To draw the ac load line:
   (a) One point is Q, at $I_D = 1.54$ mA and $V_{DS} = 11.5$ V.

(b) For a second point, select a ΔI_D of 2.46 mA, making the total $I_D = 4.0$ mA. Then the added voltage drop is

$$\Delta V_{DS} = \Delta I R_L = (2.46 \times 10^{-3}) \, 3300 = 8.1 \text{ V}$$

The second point is therefore at $I_C = 4.0$ mA and $V_{DS} = 3.4$ V $(11.5 - 8.1)$. Locate this point (A) and draw the ac load line. See the solid line in Fig. 13-2.

3. Pick off values for the dynamic curve:

V_{GS} (V)	I_D (mA)	V_{GS} (V)	I_D (mA)
+0.5	4.03	−1.5	0.85
0	3.46	−2.0	0.34
−0.5	2.53	−2.5	0.09
−1.0	1.54		

The plot for this dynamic transfer curve is shown in Fig. 13-3 (upper left). The curvature, particularly at low values of I_D, is readily evident.

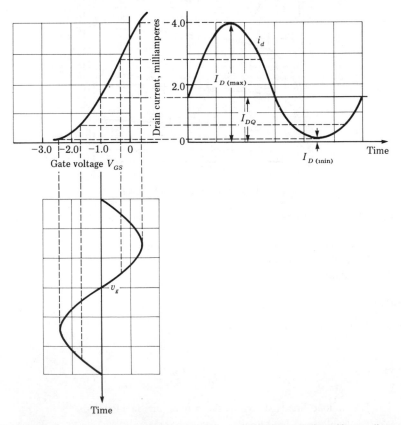

Figure 13-3 A dynamic transfer characteristic and distortion produced by nonlinear operation.

EVALUATION OF DISTORTION LEVEL. The dynamic characteristic curve can be used to evaluate the extent of the amplitude distortion produced by an amplifier. For this purpose, let us use the FET of Example 13-2 and apply a sine-wave input signal of 1.0 V (± 1.41 V peak) between gate and source. Since the FET is biased at -1.0 V, the instantaneous gate voltage will swing from -1.0 V to $+0.41$ V and to -2.41 V. This operating point and gate signal swing are shown in Fig. 13-3.

Also shown in Fig. 13-3 is the resulting drain current waveshape i_d as obtained from the dynamic curve and the given gate signal v_g. Notice the flattening of the negative current peak. This is an indication that harmonic components have been created. Now, using the quiescent value I_{DQ} as reference, notice that the positive current swing—from 1.54 to 4.13 mA —is 2.59 mA, whereas the negative current swing is only 1.44 mA (from 1.54 to 0.10). This unequal swing creates a dc component. Therefore, instead of being a pure sine wave, the drain current curve is a complex wave* containing a dc component and harmonic components. The general equation for such a wave is

$$i_D = I_{D0} + A_0 + A_1 \sin \omega t - A_2 \cos 2\omega t - A_3 \sin 3\omega t \cdots \qquad (13\text{-}1)$$

where I_{D0} is the quiescent or no-signal dc component; A_0 is the *increase* in dc component due to unequal positive and negative loops; A_1 is the amplitude (peak value) of the fundamental; A_2 is the amplitude of the second harmonic; and A_3 is the amplitude of the third harmonic.

In Fig. 13-2, the distortion is not too great and the only distortion component (ac) of any consequence is the second harmonic. (This is generally true with amplifiers when only one alternation is flattened.) The equation for this curve reduces to

$$i_D = I_{D0} + A_0 + A_1 \sin \omega t - A_2 \cos 2\omega t \qquad (13\text{-}2)$$

The composition of such a wave is shown in Fig. 13-4. Notice that the heavy curve, the resultant, is similar to the drain current waveshape of Fig. 13-3. Notice also that the values A_0, A_1, and A_2 are each measured from I_{D0} as reference. To evaluate each of these constants, we will examine the instantaneous values at three points in the wave (three-point schedule).

First examine the values at time $= 0$. Notice that $i_D = I_{D0}$. But this means that the sum of the other three components must be zero. Notice also that the fundamental component is at zero amplitude, while the second harmonic is at maximum (negative) amplitude. The amplitude of the added dc component A_0 must therefore be equal to the amplitude of the second harmonic component:

$$A_0 = A_2 \qquad (13\text{-}3)$$

*J. J. DeFrance, *Electrical Fundamentals*, Englewood Cliffs, N.J.: Prentice-Hall, Inc., 1969, part 2, chap. 4.

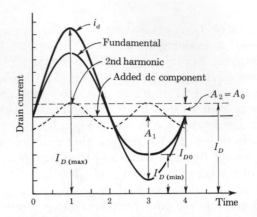

Figure 13-4   Effect of operating point and signal level on amplitude distortion.

Now let us turn out attention to time $= 1$ ($\omega t = 90°$). Here the instantaneous drain current is at its maximum value and can be expressed as

$$I_{D(\text{max})} = I_{D0} + A_0 + A_1 + A_2 \qquad \text{(13-4)}$$

Examine time instant 3 ($\omega t = 270°$). The instantaneous drain current is at its minimum value. Starting with I_{D0} as the reference, the only negative component value is A_1. Therefore

$$I_{D(\text{min})} = I_{D0} + A_0 - A_1 + A_2 \qquad \text{(13-5)}$$

Here we have three equations and three unknowns (A_0, A_1, and A_2). Solving simultaneously we get

$$A_1 = \tfrac{1}{2}(I_{D(\text{max})} - I_{D(\text{min})}) \qquad \text{(13-6)}$$

$$A_0 = A_2 = \frac{\tfrac{1}{2}(I_{D(\text{max})} + I_{D(\text{min})}) - I_{D0}}{2} \qquad \text{(13-7)}$$

Distortion levels are generally expressed in percent, as the ratio of the amplitude of the harmonic(s) to the fundamental. In the above case,

$$\text{Percent distortion (second harmonic)} = \frac{A_2}{A_1} \times 100 \qquad \text{(13-8)}$$

Or, if you prefer combination formulas,

$$\text{Percent distortion} = \frac{\tfrac{1}{2}(I_{D(\text{max})} + I_{D(\text{min})}) - I_{D0}}{I_{D(\text{max})} - I_{D(\text{min})}} \times 100 \qquad \text{(13-9)}$$

EXAMPLE 13-3

Find the distortion level produced by the FET circuit of Example 13-2.

Solution

1. Since the distortion level is not too great, it can be assumed that only the second harmonic component is of any consequence.
2. From Fig. 13-3:

$$I_{D(max)} = 4.13 \text{ mA}$$

$$I_{D(min)} = 0.10 \text{ mA}$$

$$I_{DO} = 1.54 \text{ mA}$$

3. $A_1 = \frac{1}{2}(I_{D(max)} - I_{D(min)}) = \frac{1}{2}(4.13 - 0.10) = 2.02 \text{ mA}$

$$A_2 = \frac{\frac{1}{2}(I_{D(max)} + I_{D(min)}) - I_{DO}}{2}$$

$$= \frac{\frac{1}{2}(4.13 + 0.10) - 1.54}{2} = 0.29 \text{ mA}$$

4. Percent distortion $= \dfrac{A_2}{A_1} \times 100 = \dfrac{0.29}{2.02} \times 100 = 14.3\%$

In Example 13-3 an additional dc component of 0.29 mA ($A_0 = A_2$) was produced due to the distortion. The total dc component—or the average value of the *complex* wave—is $1.54 + 0.29 = 1.83$ mA. Therefore a dc milliammeter in the drain (or source) circuit of this FET will read 1.54 mA under no-signal conditions, but the drain current will rise to 1.83 mA when this signal voltage is applied. *Any change in the quiescent value of drain current when a signal is applied is an indication of distortion.* (In fact, since $A_0 = A_2$, the change in dc milliammeter reading is equal to the peak value of the second harmonic.) Unfortunately there are cases where the converse is not true. It *may* happen that both alternations are distorted equally and the average value as read by the dc milliammeter would not be affected.

Under conditions of greater distortion levels, such as the severe flattening of the negative swing of the current waveshape (such as occurs with class AB or class B operation), the third harmonic component amplitude should be considered. This is also necessary when both the positive and the negative swings of the resultant current waveshape are distorted. For example, power amplifiers generally have more third harmonic distortion than second harmonic. Under such circumstances it would be necessary to find the amplitude A_3 of the third harmonic and possibly also of A_4, the fourth harmonic. This would give us five unknowns (A_0, A_1, A_2, A_3, and A_4) and we would need to develop five equations for evaluation. Following a technique similar to that given above, a five-point schedule is used. Three "instants" for examination of the current curve could be as before ($\omega t = 0°$,

90°, and 270°). One technique* for obtaining the other two points or instants uses I_x corresponding to $\omega t = 45°$ and I_y corresponding to $\omega t = 135°$. On this basis the distortion levels are evaluated as†

$$\text{Percent 2nd harmonic} = \frac{I_{max} + I_{min} - 2I_0}{I_{max} - I_{min} + 1.41(I_x - I_y)} \times 100 \qquad \textbf{(13-10)}$$

$$\text{Percent 3rd harmonic} = \frac{I_{max} - I_{min} - 1.41(I_x - I_y)}{I_{max} - I_{min} + 1.41(I_x - I_y)} \times 100 \qquad \textbf{(13-11)}$$

$$\text{Total percent distortion} = \sqrt{(\text{percent 2nd})^2 + (\text{percent 3rd})^2} \qquad \textbf{(13-12)}$$

(In the above equations, the references to I_{max}, I_{min}, and I_0 obviously refer to the drain or collector current even though the subscript D or C has been left out.)

DISTORTION LEVEL FROM OUTPUT CHARACTERISTICS.   Our evaluations, so far, have been based on a current curve obtained from the dynamic (mutual) characteristic. But remember that the data for this dynamic curve is often obtained from the output family of curves. Naturally, then, distortion data can also be obtained directly from the drain or collector (or plate) characteristic curves. Figure 13-5 shows a sketch of the original drain characteristic curves and ac load line from the FET of Example 13-2 and Fig. 13-2. The operating point for this circuit was set at $V_{GS} = -1.0$ V. Now let us apply an ac input signal $V_{g(max)} = 1.5$ V. The instantaneous gate voltage will swing to $+0.5$ V and to -2.5 V. Notice in Fig. 13-5 how this input signal is shown: on an axis perpendicular to the load line. This input signal looks distorted—but it is not. It only looks distorted because the voltage scale (the ordinate) is uneven.‡ Because of this uneven V_{GS} voltage scale, the accuracy of the three waveshapes is questionable. For accurate waveshapes, the dynamic transfer-curve method is recommended. However, the limits of the waveshapes in Fig. 13-5 are correct, and accurate

---

*A. V. Eastman, *Fundamentals of Vacuum Tubes* (3rd ed.), New York: McGraw-Hill Book Company, 1949.

†RCA *Receiving Tube Manual*.

‡Compare the distance representing 1 V from the -1.5 V line to the -2.5 V line, with the span from the -0.5 V line to the $+0.5$ V line! Note also that the amplitude of the positive half-cycle is numerically equal to the amplitude of the negative half-cycle. Because of this unequal voltage-axis scale, it is extremely difficult to draw the input waveshape accurately. In fact, if the peak value of the input signal did not cause the maximum and minimum V_{GS} values to coincide with a drain characteristic curve, even the overall accuracy could be poor. For example, if the input signal were ±1.2 V peak, V_{GS} would swing to $+0.2$ V on the positive swing. Since there is no $V_{GS} = +0.2$ V curve, we must interpolate. We could select a point two-fifths of the distance between $V_{GS} = 0$ and $V_{GS} = +0.5$. But this is only an approximation, since the scale is uneven.

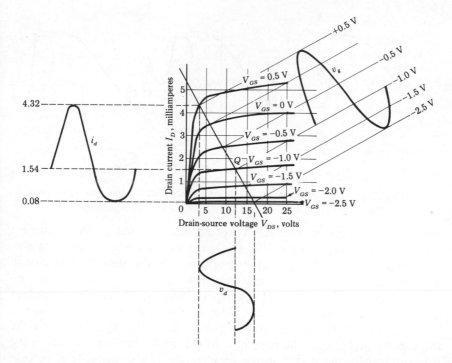

Figure 13-5   Waveshapes and distortion from drain family curves.

distortion calculations can be made. The calculations would be the same as those shown in Example 13-3 and need not be repeated here.

FACTORS AFFECTING DISTORTION.   By now it should be clear that amplitude distortion is caused by nonlinearity of the dynamic (mutual) characteristic. Let us study the situation more closely to see the reason for the distortion, the effect of the operating point on the distortion level, and how the distortion level can be minimized. We will use an N-channel FET for this discussion. In Fig. 13-6, all three diagrams show a dynamic characteristic with curvature at the low values of drain current (as was noted earlier) and with curvature also for operation in the positive gate region.

In Fig. 13-6a, notice that the bottom (negative) half-cycle of the drain *current* waveshape is flattened. This is caused by the gate signal driving the FET into the curved portion of the characteristic, near cutoff. This distortion could have been avoided by reducing the gate bias so that the operating point is at the center of the straight-line portion of the characteristic. Figure 13-6b shows the opposite effect. The bias is too low and the FET is being driven into curvature due to positive gate operation. This time the upper or positive half-cycle of the *current* waveshape is

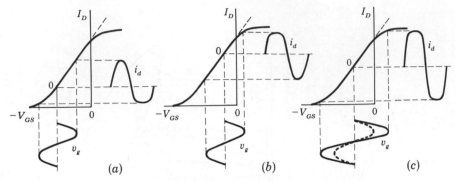

Figure 13-6   Effect of operating point and signal level on amplitude distortion.

flattened. Obviously, the remedy is to increase the bias. In the third case (Fig. 13-6c) the operating point is selected at the center of the straight-line portion of the dynamic characteristic. Yet the distortion is worse. The drain current waveform is flattened at both top and bottom. This time the amplitude of the gate signal V_g is appreciably larger than the value used in either of the two earlier cases. This is obviously a case of *overdrive*. This FET (for this value of load and power supply voltage) cannot handle this input signal level.* The remedy is to reduce the gate signal level to the value shown by the dotted wave. Incidentally, if the signal level in case (a) or (b) had been much lower, distortion would not have occurred. However, the proper bias point will allow a FET to handle a larger signal input (and develop a higher output voltage) before distortion results.

INTERMODULATION DISTORTION.   We have seen that if a sine wave is applied to an amplifier with a perfectly linear transfer characteristic, the output will contain only that one frequency; whereas if the dynamic transfer curve is not linear, the output will also contain harmonics of the original frequency. How is the output affected if several different frequencies are applied to the amplifier at the same time, as in music reproduction? In a perfectly linear amplifier, the output will contain only the original frequencies. Again, if there is nonlinearity, the output will contain harmonics of each of the original input frequencies—*but that is not all*. When there is more than one input frequency (and nonlinearity), each input frequency is "affected" or *modulated* by the others and additional frequencies are created in the output. These frequencies are the sums and differences of the original frequencies and the sums and differences of their harmonics.

*A given transistor can handle a somewhat higher input signal level, without distortion, if the supply voltage is raised and/or the load impedance is increased. For a moderate increase in signal level some other small-signal amplifier should be used. However, to prevent distortion with large input signals, power transistors are necessary.

These components are called *intermodulation* (IM) products. Since they were not present in the original input, they add to the distortion level. The more nonlinear the amplifier characteristic, the higher the harmonic distortion and the higher the intermodulation distortion.

As an illustration of IM distortion, if the input frequencies are 800 and 1500 Hz, the major IM products will be 2300 and 700 Hz (sum and difference of the original frequencies). Notice that they are not harmonically related to the input signals. In a music amplifier, such IM products would sound much more objectionable (discordant) than harmonic distortion. When testing amplifiers for IM distortion, it is common practice to use two input frequencies (60 Hz and 7 kHz) having a relative amplitude ratio of 4:1. With this as a "standard" input signal, we can compare the IM distortion level ratings of amplifiers.

FREQUENCY AND PHASE DISTORTION

In Chapter 10, it was shown that the gain of an amplifier depends on the load impedance value. It was further stated that *under ideal conditions* the load impedance is resistive and is presented by R_L. Unfortunately, this is not always the case; reactance effects cannot be ignored, and the load impedance must at times be considered in its phasor sense as Z_L. But if the load contains reactive elements, the impedance will vary with frequency, causing the gain of the amplifier to vary with frequency. This is *frequency distortion*. In general, the gain of an amplifier tends to decrease appreciably at very low and very high frequencies.

The load impedance could contain capacitive elements as in the R-C coupled amplifier. It could also contain inductive elements and combinations of both inductance and capacitance, as we will see in other coupling techniques. But R-L and R-C circuits are also phase-shift circuits,* and can cause *phase distortion*. In other words, phase and frequency distortion go hand in hand.

What is the physical significance of phase and frequency distortion? In audio amplifiers, frequency distortion would cause the loss of the brilliance of the piccolo, the crash of the cymbals, or the hearty depth of the tuba or bass viol. Phase distortion would not in itself be noticeable, but with modern feedback circuits,† phase shift can cause serious instability and oscillations. In television, phase-shift distortion is serious. Since a phase shift is a time delay, a shifting of the relationship of the picture components causes blurring of the fine details (high frequencies) and

*J. J. DeFrance, *Electrical Fundamentals,* Englewood Cliffs, N.J.: Prentice-Hall, Inc., 1969, part 2, chap. 9.

†Feedback circuits and effects are discussed in Chapter 16.

smearing of large uniform objects (low frequencies). An example of distortion due to phase shift is shown in Fig. 9-5. Notice that the input and output waveshapes have little (if any) resemblance to one another.

It was mentioned above that the reactive elements could be capacitive, inductive, or both—but where do they come from? To answer this question, let us examine *in detail* all the effective components in the R-C coupling network between two stages. Such a circuit is shown in Fig. 13-7 and applies equally well to FETs, BJTs, and vacuum tubes. Notice the extra capacitors C_o, C_1, and C_2 shown with dotted connecting lines. These *capacitances* are not real capacitors wired into the circuit; you could not reach in and touch or remove any of them. Yet their capacitance is very real and must be considered when studying circuit action.

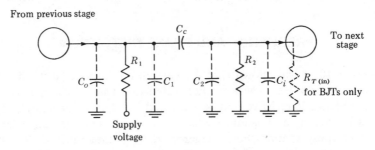

Figure 13-7   Effective coupling network showing shunt capacitances.

C_o represents the output capacitance of the previous stage and consists of the interelement capacitance (drain to source, collector to emitter, or plate to cathode) plus a capacitance due to the *Miller effect,* which will be described shortly. Depending on the specific transistor or tube, this capacitance value may range from 0.5 to about 12 pF. C_1 and C_2 are stray capacitances due to wiring, sockets (if used), and even to the components (R_1, R_2, and C_c) themselves. As an example, the connection from the output of the previous stage to resistor R_1 and capacitor C_c is a conductor. This and other conductors, separated by the insulation of the circuit board, form stray or unwanted capacitances. The longer these connections and the closer they are (particularly to ground conductors), the higher the stray capacitance. This effect cannot be eliminated, but with good layout, good quality of components, and good wiring techniques, this stray capacitance can be held to low values. An average value is about 10 pF. In Fig. 13-7, the stray capacitance is shown as two capacitors merely to indicate that part of this effect is caused by components pertaining to the output of one stage and another part is due to components associated with the input to the next stage. In calculations, these would be treated as a single capacitor of $C_1 + C_2$.

INPUT CAPACITANCE—MILLER EFFECT.   The only other shunt capacitance in Fig. 13-7 is produced by C_i, the input capacitance of the next
stage. What does it consist of? Obviously, one factor is the capacitance
between the input elements (gate to source, base to emitter, or grid to
cathode). But there is another, even more important component: the capacitance between input and output elements (drain to gate, collector to base,
or grid to plate). The effect of this latter component is magnified by the
gain of the stage. This is the Miller effect mentioned earlier. *Any impedance
Z connected between input and output elements appears as*

1. An input impedance equal to $Z/(A_v + 1)$.
2. An output impedance equal approximately to Z.

In terms of capacitance and using a FET as an example, C_{dg} between drain
and gate acts as

1. An input capacitance—$C_{dg}(A_v + 1)$
2. An output capacitance—C_{dg}

The effective increase in input capacitance can be shown with the aid
of Fig. 13-8. In Fig. 13-8a the source, voltage V_g, is supplying two currents:
I_1 through the gate-source capacitance and I_2 through the gate-drain capacitance. Since both are pure capacitive paths, the two currents are in
phase and can be added directly. Let us evaluate these currents:

$$I_1 = \frac{V_g}{X_{C_{gs}}} = \frac{V_g}{2\pi f C_{gs}}$$

The net voltage drop across capacitor C_{dg} is the difference between V_g on
one side and V_o on the other side. But since $V_o = A_v V_g$, this net voltage
is $V_g + A_v V_g$ and

$$I_2 = \frac{V_g + A_v V_g}{X_{C_{dg}}} = \frac{V_g(1 + A_v)}{2\pi f C_{dg}}$$

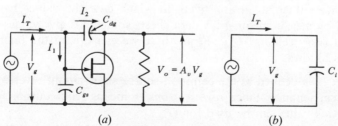

(a)   (b)

Figure 13-8   Evaluation of input capacitance of a FET.

Also, from Fig. 13-8*b*,

$$I_T = \frac{V_g}{X_{C_i}} = \frac{V_g}{2\pi f C_i}$$

Equating I_T with $I_1 + I_2$,

$$I_T = \frac{V_g}{2\pi f C_i} = \frac{V_g}{2\pi f C_{gs}} + \frac{V_g(1 + A_v)}{2\pi f C_{dg}}$$

Factoring and canceling like terms,

$$C_i = C_{gs} + C_{dg}(1 + A_v) \tag{13-13}$$

This also applies to BJTs and vacuum tubes.

EXAMPLE 13-4

A FET amplifier has a voltage gain of 55. The interelement capacitances for this FET are $C_{dg} = 2.8$ pF, $C_{gs} = 3.0$ pF, and $C_{ds} = 1.0$ pF. Find the input and output capacitances for this stage.

Solution

1. To find the input capacitance,

$$C_i = C_{gs} + C_{dg}(1 + A_v)$$
$$= 3.0 + 2.8(1 + 55) = 160 \text{ pF}$$

2. The output capacitance is

$$C_o = C_{ds} + C_{dg} = 1.0 + 2.8 = 3.8 \text{ pF}$$

Notice that the actual gate-to-source capacitance is practically negligible in comparison to the Miller-effect capacitance.

Example 13-4 shows how the input and output capacitances can be calculated by using the Miller effect—if we know the interelement capacitances. The data sheets for vacuum tubes invariably specify the interelectrode capacitances, but this is not done for transistors. Sometimes no capacitance values at all are given, in which case the values have to be obtained by measurement. FET data sheets sometimes give values for C_{iss}, C_{rss}, and (very seldom) C_{oss}. From these we can obtain the interelement values as follows:

1. C_{rss} is defined as the common-source short-circuit reverse transfer capacitance. Since *reverse transfer* means from output to input, this value can be used for C_{dg}.
2. C_{iss} is the input capacitance for the common-source connection—with the third element (in this case the drain) shorted to the source.

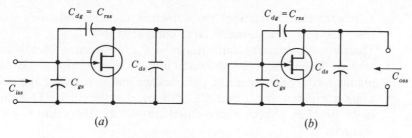

Figure 13-9 FET capacitances.

From Fig. 13-9 we can see that C_{iss} is effectively $C_{gs} + C_{dg}$ in parallel, and replacing C_{dg} by C_{rss},

$$C_{gs} = C_{iss} - C_{rss} \qquad (13\text{-}14)$$

3. C_{oss} is the common-source output capacitance—with gate shorted to source. From Fig. 13-9b, it is obvious that C_{oss} is equal to $C_{ds} + C_{rss}$ and that

$$C_{ds} = C_{oss} - C_{rss} \qquad (13\text{-}15)$$

This capacitance value is generally quite small (about 1 pF) and compared to the Miller-effect output value can even be neglected.

With bipolar junction transistors we have a somewhat similar problem. Sometimes no capacitance values are given in the data sheets. Measurements must be made. Sometimes (but seldom) the interelement values are given in direct form: that is, C_{cb} for the collector-to-base capacitance and C_{eb} for the emitter-to-base capacitance. At other times, capacitances C_{ob} and C_{ib} (or C_{obo} and C_{ibo}) are listed. These are, respectively, the *common-base* output capacitance *with the emitter open* and the *common-base* input capacitance *with the collector open*. A little thought will show the relationship between these CB values and the desired CE values:

1. Capacitance C_{ob} is from collector to base. Also from collector to base there are *in series* the capacitances of collector to emitter (C_{ce}) and emitter to base (C_{eb}). This series value is rather low; therefore, to a very close approximation $C_{ob} = C_{cb}$. This is the feedback capacitance in the CE configuration. Since this capacitance is due mainly to a reverse-biased junction, the depletion region is wide and this capacitance value is fairly low. However, it varies appreciably with the collector voltage.
2. Capacitance C_{ib} is between emitter and base. Again, neglecting the shunting effect of the series C_{ce} and C_{ob}, we can consider C_{ib} to be the same as C_{eb}. It should be realized that C_{eb} is the capacitance of

a forward-biased junction; the depletion region is relatively narrow; and this capacitance is large compared to C_{ob}.

3. The third interelement capacitance is C_{ce}, between collector and emitter. This value is not shown in any manner because C_{ce} is due mainly to the capacitance of the package and is usually negligibly small compared to C_{cb} and C_{eb}. A typical value for a TO-5 case is 0.6 pF. For plastic-encapsulated transistors, the value is even smaller.

Once the interelement capacitance values are known, the input and output capacitances can be found as explained for FET circuits.

EXAMPLE 13-5

A 2N4390 *NPN* transistor has $C_{ib} = 40$ pF and $C_{ob} = 6$ pF. When used as a CE amplifier, the voltage gain is 50. Find the input and output capacitances.

Solution

1. $C_{cb} = C_{ob} = 6$ pF; $C_{eb} = C_{ib} = 40$ pF
2. $C_i = C_{eb} + C_{cb}(1 + A_v)* = 40 + 6(1 + 50) = 346$ pF
3. $C_o = C_{ce} + C_{cb} = C_{cb} = 6$ pF

Now let us consider the effect of these reactive components on frequency and phase distortion. The technique generally employed is to consider the circuit of Fig. 13-7 at three different frequency ranges:

1. Mid-frequency range, where all reactance effects can be ignored (approximately 100 to 10,000 Hz)
2. High-frequency range, where only the shunt reactance effects need be considered (generally above 10,000 Hz)
3. Low-frequency range, where only the series reactance effects are detrimental (generally below 100 Hz)

MID-FREQUENCY RANGE.   In a typical coupling network (as shown in Fig. 13-7) the series (or coupling) capacitor value is relatively high. Consequently, even at some rather low frequency such as 400 Hz, the reactance of any coupling capacitor is generally negligible compared to the associated circuit resistances. The coupling capacitor can be considered as a short circuit. Meanwhile, since the shunt capacitance is only a few picofarads, the reactance of the shunt capacitance is so high that its shunting action can also be neglected. Depending on circuit values, reactance effects can be ignored down to about 100 Hz. Unfortunately, at still lower frequencies the series reactance of the coupling capacitor can become high enough to produce a

---

* This is the BJT equivalent of Equation 13-13.

noticeable voltage drop. Its effect now cannot be neglected. This is the low-frequency limit of the amplifier.

Conversely, at higher signal frequencies the reactance due to shunt capacitance keeps decreasing. A frequency will be reached at which this effect impairs circuit operation. This occurs when the shunt reactance value approaches the equivalent load resistance value R_e. Again, the specific frequency at which we can no longer neglect the shunt capacitance effect depends on the value of the ac load resistance, on the transistors used (C_o and C_i), and also on the layout and wiring design (stray C). A rounded-off approximation of this upper frequency limit is 10,000 Hz.

Within these two frequency limits, the equivalent diagram becomes purely resistive, the gain is constant, and there is no frequency distortion. This ideal condition was used in Chapters 10 and 11, and the mid-frequency gain was therefore given as

$$A_{v(\text{mid})} = g_{fs}R_e \qquad \text{(for FETs)} \tag{13-16}$$

or

$$A_{v(\text{mid})} = h_{fe}\frac{R_e}{r_i} \qquad \text{(for BJTs—neglecting source resistance)} \tag{13-17}$$

As for phase distortion within this frequency band, again we have no problem. Since the circuit acts as a purely resistive network, there is no phase shift produced by the coupling network and consequently no phase distortion. However, remember that due to the transistor action the input and output voltages are 180° apart.

HIGH-FREQUENCY RANGE—GAIN.   As we pass the upper frequency limit of this ideal condition, the reactance of the coupling capacitor is still negligible (in fact, more so), but the shunt capacitance effect can no longer be ignored. Because of this capacitance effect, the load must be treated as an impedance Z_L and the equivalent circuit must now include this capacitance effect. Figure 13-10 shows equivalent diagrams for a FET (Fig. 13-10a) and for a BJT (Fig. 13-10b) for the high-frequency condition. We are considering the action from the gate or base of one stage to the gate or base of the next stage.* The gain equations now become

$$A_{v(\text{high})} = g_{fs}\dot{Z}_e \qquad \text{(for the FET)} \tag{13-18}$$

$$A_{v(\text{high})} = h_{fe}\frac{\dot{Z}_e}{r_i} \qquad \text{(for the BJT)} \tag{13-19}$$

The rest is merely a problem in ac circuit theory to calculate the impedance Z_e. These calculations will not be covered here. However, one point should be obvious: the addition of the capacitive branch reactance will lower the

---

*A vacuum-tube analysis is similar to the FET case and so is not discussed here.

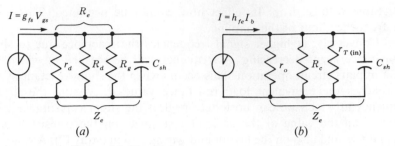

Figure 13-10   High-frequency equivalent diagrams.

impedance so that the gain at high frequencies will decrease. The higher the frequency, the lower the value of X_C, the lower the total load impedance, and the less the gain.

However, rather than calculating the actual gain at any frequency, it is more common in practice to use the mid-frequency gain as a reference and then to find the relative gain at the higher frequency:

$$\frac{A_{v(high)}}{A_{v(mid)}} = \frac{X_{C_{sh}}}{\sqrt{R^2 + X^2_{C_{sh}}}} \qquad (13\text{-}20)$$

Alternatively, a standard for design or comparison of amplifiers is to specify the frequency at which the gain drops 3 dB below the mid-frequency gain. A loss of 3 dB corresponds to a drop in output voltage to 0.707 of the mid-frequency value. This in turn means that the impedance Z_e must have dropped to 0.707 of the mid-frequency value R_e. Therefore, and this is an important point to remember, *the high-frequency gain will drop by 3 dB at that frequency which makes the shunt reactance $X_{C_{sh}}$ equal to R_e, or*

$$f_H = \frac{1}{2\pi R_e C_{sh}} \qquad (13\text{-}21)$$

This is known as the high-end *cutoff frequency.*

EXAMPLE 13-6

A FET with parameters $r_d = 95$ kΩ and $g_{fs} = 1000$ μmho is used in an R-C coupled amplifier with $R_d = 270$ kΩ and $R_g = 470$ kΩ. The total shunt capacitance is 80 pF. The gate input signal is 0.5 V. Find:

1. The gain and output voltage at mid-frequency
2. The upper frequency limit at which the gain will drop by 3 dB
3. The gain and output voltage at this frequency

Solution

1(*a*). Solving for the equivalent resistance R_e,

$$\frac{1}{R_e} = \frac{1}{r_d} + \frac{1}{R_d} + \frac{1}{R_g}$$

$$= \frac{1}{95 \text{ k}\Omega} + \frac{1}{270 \text{ k}\Omega} + \frac{1}{470 \text{ k}\Omega}$$

$$R_e = 61.2 \text{ k}\Omega$$

1(*b*). Mid-frequency gain and output voltage:

(i) $A_{v(\text{mid})} = g_{fs}R_e$

$$= 1000 \times 10^{-6} \times 61.2 \times 10^3$$

$$= 61.2$$

(ii) $V_o = V_i \times \text{gain}$

$$= 0.5 \times 61.2$$

$$= 30.6 \text{ V}$$

2. The upper-frequency 3-dB loss occurs when X_C of the shunt capacitances equals R_e. For $X_C = R_e$,

$$f = \frac{1}{2\pi R_e C_{sh}} = \frac{0.159 \times 10^{12}}{(61.2 \times 10^3) \times 80}$$

$$= 32.5 \text{ kHz}$$

3. Since there is a 3-dB loss, the gain and output voltage have dropped to 0.707 of their mid-frequency values, or

$A_{v(\text{high})} = 0.707 A_{v(\text{mid})}$

$$= 0.707 \times 61.2$$

$$= 43.2$$

$V_o = 30.6 \times 0.707$

$$= 21.6 \text{ V}$$

PHASE SHIFT AT HIGH FREQUENCIES.   To analyze the phase distortion at the higher frequencies, we can use the equivalent diagrams of Fig. 13-9. However, let us simplify them by using only one resistor, R_e, in place of the three parallel resistors. This modification is shown in Fig. 13-11*a*. This diagram can represent both the FET and the BJT circuit. The current, $I = g_{fs}V_{gs}$ or $I = h_{fe}I_b$, is now labeled I_T. In either case, an input voltage V_i must be applied between gate and source, or between base and emitter, to produce this current. At mid-frequencies, the output voltage is in phase

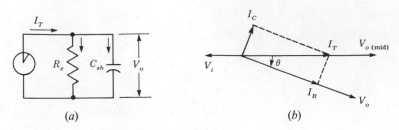

Figure 13-11 Phase distortion of output voltage at higher frequencies.

with the current I_T and is 180° out of phase with the input voltage. These phase relations are shown in Fig. 13-11b, with I_T and the mid-frequency output voltage as reference.

At the higher frequencies, because of the capacitor C_{sh}, the current I_T divides between the two branches I_R and I_C. The output voltage must be in phase with the current I_R ($V_o = IR_e$), but it will lag the line current I_T by some angle θ. The capacitive branch current must lead the resistive branch current by 90°, and their sum ($I_R + I_C$) must equal I_T. These current relations are also shown in Fig. 13-11b. From these current relations we can see that the tangent of angle θ is equal to the current ratio I_C/I_R. Then, since currents are inversely proportional to oppositions,

$$\theta = \text{arc tan} \frac{R_e}{X_{C_{sh}}} \qquad (13\text{-}22)$$

Notice that the output voltage at high frequencies *lags* the mid-frequency output voltage. Also, as the frequency rises, X_C decreases and the lag angle (θ) increases. At the cutoff frequency, when $X_C = R_e$, the angle is 45° and will increase to 89° as X_C drops to about one-fiftieth of R_e.

LOW-FREQUENCY RANGE — GAIN.   When we analyze the circuit action at the low frequencies, it should be obvious that the reactance of the shunt capacitance is now very high, compared to R_L or R_e, and does not affect the load value. Meanwhile, the reactance of the coupling capacitor C_c has also increased, but this capacitor is in *series* with the output, and so its action cannot be neglected. Suitable low-frequency equivalent diagrams are shown in Fig. 13-12. In both these diagrams it should be noted that the reactance X_C is part of one branch in determining the load impedance Z_e. This gives us a series-parallel circuit which is rather laborious to analyze. We can instead, looking in at x-x' in Fig. 13-12, use Thévenin's theorem to convert the transistors and their dc load resistors into constant-voltage-source equivalent circuits.

The converted diagrams are shown in Fig. 13-13. Resistor R' is the parallel combination of $r_d \| R_d$, or $r_o \| R_c$. Let us use Fig. 13-13a for the

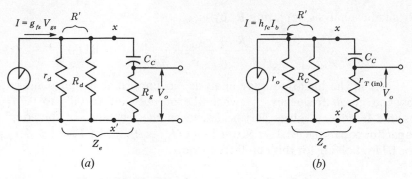

Figure 13-12   Low-frequency equivalent diagrams.

FET to develop suitable gain equations. At mid-frequencies, the reactance of the coupling capacitor is negligible and

$$V_{o(\text{mid})} = V_s \frac{R_g}{R' + R_g}$$

At low frequencies, due to the voltage-divider action of C_C and R_g, the output voltage drops to

$$V_{o(\text{low})} = V_s \frac{R_g}{\dot{R}' + \dot{X}_C + \dot{R}_g}$$

Notice that the denominator is a phasor addition. Comparing this low-frequency output with the mid-frequency value, we get

$$\frac{V_{o(\text{low})}}{V_{o(\text{mid})}} = \frac{R' + R_g}{\dot{R}' + \dot{X}_C + \dot{R}_g}$$

Since the output voltages are dependent on gain, and taking into consideration the 90° phase relation between R and X values,

$$\frac{A_{v(\text{low})}}{A_{v(\text{mid})}} = \frac{R' + R_g}{\sqrt{(R' + R_g)^2 + X_C^2}} \qquad \text{(for FETs)} \qquad \text{(13-23)}$$

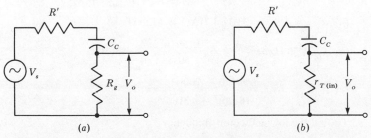

Figure 13-13   Constant-voltage low-frequency equivalent diagrams.

By similar analysis, replacing R_g by $r_{T(in)}$,

$$\frac{A_{v(low)}}{A_{v(mid)}} = \frac{R' + r_{T(in)}}{\sqrt{(R' + r_{T(in)})^2 + X_C^2}} \qquad \text{(for BJTs)} \qquad \text{(13-24)}$$

As in the high-frequency analysis, an important design point is the low-end cutoff frequency, f_L, when the output voltage falls to 0.707 of its mid-frequency value. This occurs when the reactance of the coupling capacitor equals the total series resistance $(R' + R_g)$ for FETs or $(R' + r_{T(in)})$ for BJTs. Solving for this cutoff frequency,

$$f_L = \frac{1}{2\pi(R' + R_g)C_C} \qquad \text{(for FETs)} \qquad \text{(13-25)}$$

$$f_L = \frac{1}{2\pi(R' + r_{T(in)})C_C} \qquad \text{(for BJTs)} \qquad \text{(13-26)}$$

EXAMPLE 13-7

The BJT of Example 11-6 is coupled to the next stage through a 1-μF capacitor. In Example 11-6 we were given (or found that) $h_{fe} = 175$, $r_o = 33.3$ kΩ, $R_c = 4.7$ kΩ, $r_i = 2640$ Ω, and $r_{T(in)}$ for the next stage = 2000 Ω. Calculate:

1. The mid-frequency gain
2. The gain at 50 Hz
3. The low-end cutoff frequency

Solution

1. To find the mid-frequency gain:

   (a) $R_e = r_o \| R_c \| r_{T(in)} = (33.3 \| 4.7 \| 2.0)$ k$\Omega = 1350$ Ω

   (b) $A_{v(mid)} = h_{fe}\dfrac{R_e}{r_i} = 175\dfrac{1350}{2640} = 89.5$

2. To find the gain at 50 Hz:

   $$A_{v(50)} = A_{v(mid)}\frac{R' + r_{T(in)}}{\sqrt{(R' + r_{T(in)})^2 + X_C^2}}$$

   (a) $R' = r_o \| R_c = (33.3 \| 4.7)$ k$\Omega = 4120$ Ω

   (b) X_C at 50 Hz $= \dfrac{1}{2\pi fC} = \dfrac{0.159 \times 10^6}{50 \times 1} = 3180$ Ω

   (c) $A_{v(50)} = 89.5\dfrac{4120 + 2000}{\sqrt{(6120)^2 + (3180)^2}} = 79.4$

3. To find the low-end cutoff frequency:

   $$f_L = \frac{1}{2\pi(R' + r_{T(in)})C_C} = \frac{0.159 \times 10^6}{6120 \times 1} = 26.0 \text{ Hz}$$

Another possible cause for poor low-frequency response could be the R-C network used for source bias in FET circuits or for emitter bias-stabilization in BJT circuits. These are simple circuits—a capacitor in parallel with a resistor. When we discussed bias circuits, it was pointed out that the reactance of the bypass capacitor should not be more than one-tenth the resistor value *at the lowest operating frequency*. For example, if the bias resistor is 1000 Ω and the lowest signal frequency of interest is 20 Hz, then X_C of the bypass capacitor should not exceed 100 Ω at 20 Hz. If the capacitor is so chosen, there will be no loss in the low-frequency gain at 20 Hz. In this illustration, the capacitor should not be smaller than

$$C = \frac{1}{2\pi f X_C} = \frac{0.159 \times 10^6}{20 \times 100} = 80 \ \mu F$$

Using this capacitance value and the basic cutoff frequency equation, we find that a 3-dB loss in gain would occur at

$$f_L = \frac{1}{2\pi RC} = \frac{0.159 \times 10^6}{1000 \times 80} = 2.0 \ \text{Hz}$$

Consequently, at 20 Hz (a frequency 10 times greater than the cutoff frequency) the loss in gain is zero.

PHASE SHIFT AT LOW FREQUENCIES.   To investigate the phase shift of the output voltage at low frequencies, we will use the constant-voltage equivalent diagrams of Fig. 13-13. For reference we will use the mid-frequency output voltage. Within the mid-frequency range, the reactance of X_C is negligible; the equivalent circuit is purely resistive; the current I_{mid} is in phase with the constant-voltage source V_s; and the output voltage $V_{o(mid)}$ is in phase with I_{mid}. The phase relations, at low frequencies, are shown in Fig. 13-14. The increasing value of X_C causes the circuit current I_{low} to *lead* the source voltage by some angle θ. The output voltage $V_{o(low)}$ developed across a resistor (R_g or $r_{t(in)}$) is in phase with this current. Therefore, the output voltage at low frequencies *leads* the mid-frequency output voltage. The lower the frequency, the greater the lead angle θ. At the cutoff frequency, when X_C equals the total series resistance, the lead angle is 45°. This angle increases to 89° at a still lower frequency when X_C rises to about 50 times the series resistance.

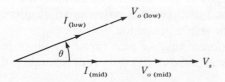

Figure 13-14   Phase distortion of the output voltage at low frequencies.

FREQUENCY RESPONSE CURVES

Quite often, in giving the specifications of an amplifier (single-stage or complete amplifier), a manufacturer or an engineer will state "frequency response 20 to 80,000 Hz." or "flat from 5 Hz to 3.6 MHz." They are referring to the lower and upper frequency points at which the loss in gain is 3 dB (or the output voltage drops to 70.7 percent of its mid-frequency value). Sometimes even more precise data is given, such as "40 to 22,000 Hz $\pm$ 0.5 dB." Of course, the ultimate in specification of frequency response is when the information is given by a curve. Such a curve, showing the effect of frequency on gain and on phase shift, is shown in Fig. 13-15.

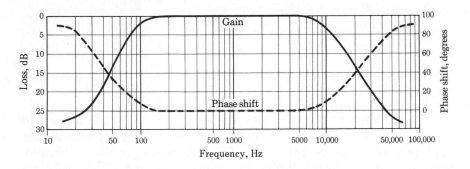

Figure 13-15   Typical amplifier frequency-response curve.

These curves are plotted on semilogarithmic paper with frequency on the logarithmic scale. This technique is always used when investigating a wide range of frequencies, as in untuned or nonresonant amplifiers. (A 100-Hz variation at a frequency of around 200 Hz is a much more drastic change than the same 100 Hz considered at a frequency of 10,000 Hz. Use of the semilogarithmic scale gives better weighting to these increments.)

In a design situation, the data for the response curves would be calculated from equivalent diagram analysis as shown above. However, in a finished unit this data would be obtained by test. A typical test setup is shown in Fig. 13-16. For phase-shift evaluations, the oscilloscope is used (see Appendix B) and the Lissajous patterns are resolved for several input frequencies from the signal generator.

In the evaluation of relative gain (decibel gain or loss), the basic idea is to measure the input and output voltage for a number of input frequencies

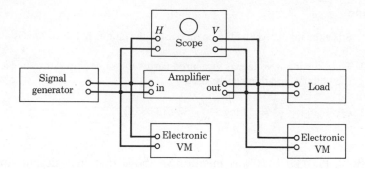

Figure 13-16   Test setup for frequency-response run.

over the range desired. However, several precautions should be observed:

1. The input level should not be so high as to produce *amplitude* distortion. (The oscilloscope, using internal sweep for waveshape analysis, can be connected across the output to monitor for possible amplitude distortion.)
2. Electronic voltmeters (solid state or VTVM) are desirable to prevent loading of the input or output circuits. Be sure the frequency response of the voltmeters is adequate.
3. The test is preferably made by varying the frequency and at each test value adjusting the *input* voltage so as to maintain a *constant output* voltage. The output level at which the response curve was taken should be noted. Response curves—particularly for power output stages—may vary appreciably at different levels. (Usually, the higher the level, the poorer the frequency response curve.) With this technique, the voltmeter at the output could be omitted (once the level is noted) and the oscilloscope could then be used to monitor level as well as waveshape.

Notice in Fig. 13-15 that the relative gain of the amplifier is given in decibels and that in the mid-frequency range the level is 0 dB. That is because a frequency in the mid-frequency range is used as the reference level. The frequency selected is usually 400 or 1000 Hz.

EXAMPLE 13-8

In order to maintain constant output in a frequency test run, it was necessary to increase the input voltage from 0.08 V at 400 Hz to 0.36 V at 1.5 MHz. Calculate the decibel gain or loss at the higher frequency.

Solution

1. Since the input level had to be increased, the gain is reduced. This is a decibel loss.
2. Since the input voltage is measured across the same impedance in each case, the voltage equation can be used.
3. $\text{dB} = 20 \log \dfrac{V_1}{V_2} = 20 \log \dfrac{0.36}{0.08} = -13.08 \text{ dB}$

BODE PLOTS.   It was mentioned above that in a design situation the data for a frequency response curve can be calculated point by point for various frequencies by using the equations developed earlier in the chapter. This procedure can be quite laborious. There is a much simpler method—a *Bode plot* (named after its originator, Hendrik W. Bode). Only three calculations are required:

1. The mid-frequency gain
2. The frequency for a 3-dB loss at the high end
3. The frequency for a 3-dB loss at the low end

From these three values we can draw the complete response curve—over any frequency range—using the Bode plot technique. This method is based on the fact that for any simple R-C circuit, the roll-off (slope or fall in output with frequency) is at the same rate—regardless of the values of R and C. The R and C values determine the frequency at which the roll-off starts, but not how fast it drops. This fixed roll-off is at the rate of 6 dB per octave and 20 dB per decade.* According to the Bode technique, the frequency response is considered constant from the low-end cutoff frequency to the high-end cutoff frequency. Then it drops (at each end) at the rate of 20 dB per decade. If this "curve" is drawn on a logarithmic frequency scale, the roll-off is a straight line.

EXAMPLE 13-9

An R-C coupled amplifier stage has its 3-dB (cutoff) frequencies at 150 Hz and at 8000 Hz. Draw the full Bode plot frequency response.

Solution

1. We will use four-section semilogarithmic paper and plot frequency on the log scale.
2. The curve will be flat at 0 dB (no loss) between 150 and 8000 Hz.

*An octave is a 2:1 frequency change; a decade is a 10:1 frequency change.

3. The curve drops 20 dB at 15 Hz ($\frac{1}{10}f_L$) and at 80 kHz ($10f_H$).
4. This curve is shown as Fig. 13-17.

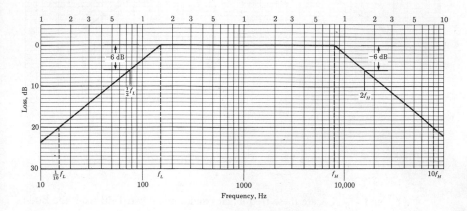

Figure 13-17   Bode plot for Example 13-9.

Notice in Fig. 13-17 the 6-dB drops at $\frac{1}{2}f_L$ and at $2f_H$. These are the octave points. Notice also the sharp *break*—or *corners*—in the curve at the low and high cutoff frequencies. Because of this, these frequencies are also known as the *break frequencies* or *corner frequencies*.

The curve shown in Fig. 13-17 could be called a *basic* or *idealized* Bode plot. Since it uses a straight-line segment to the cutoff frequency, we know it is not an exact curve. However, if a more accurate curve is desired, it can be obtained readily without further calculations. We apply some simple corrections at each end:

1. At the corner frequencies, the true response should be 3 dB down.
2. At one octave to each side of the corner frequencies, the true response is down 1 dB from the Bode plot.
3. The true curve and the Bode plot are tangent at about two octaves to each side of the corner frequencies.

EXAMPLE 13-10

Figure 13-18 shows a basic Bode plot for an R-C coupled amplifier. Draw the corrected response curve.

Solution

1. At the corner frequencies (f_L and f_H) locate points *A* 3 dB down.
2. The points one octave to each side of the corner frequencies are $\frac{1}{2}f_L$, $2f_L$,

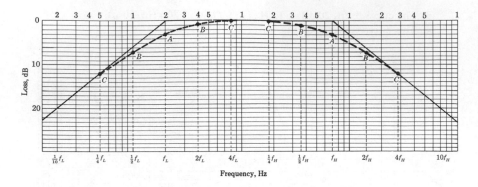

Figure 13-18   Correcting a Bode plot.

$\frac{1}{2}f_H$, and $2f_H$. Locate these frequencies and come down 1 dB from the Bode plot line. Mark these points *B*.

3. The tangency points are two octaves away at $\frac{1}{4}f_L$, $4f_L$, $\frac{1}{4}f_H$, and $4f_H$. Locate these points and mark them *C*.

4. Draw the corrected response through these new points. This curve is shown by the broken line in Fig. 13-18.

MULTISTAGE RESPONSE CURVES.   So far we have considered a single stage, wherein only one R-C network affected the high-frequency response and only one R-C network affected the low-frequency response. This is generally the case in a single-stage amplifier. How do we handle a multistage amplifier? Obviously, we should be able to calculate the response of each stage—individually—as described above. Then, since the response data is in decibels, we merely add the decibel loss of each stage at any given frequency to determine the overall frequency response. If the stages are alike, we need calculate only the response of one stage and multiply the decibel loss by the number of stages. For example, with two identical stages the response will be down by 6 dB at the corner frequencies; the response is down 2 dB below the Bode plot at one octave to each side of the corner frequencies, and the roll-off for the basic Bode plot is at the rate of 40 dB per decade. Since this is rather simple, no numerical example will be given. However, let us examine in more detail a case where the stages are not alike.

EXAMPLE 13-11

A two-stage amplifier has first-stage cutoff frequencies of 150 and 18,000 Hz and second-stage cutoff frequencies of 60 and 8000 Hz. Draw the overall response curve.

Solution

1. Since neither stage has any loss between 150 and 8000 Hz, this is our mid-frequency range. Start by plotting this portion at 0 dB. (See line A to D in Fig. 13-19.)

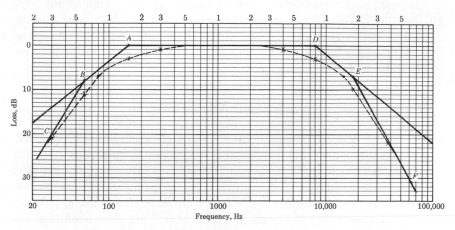

Figure 13-19   Development of overall response curve for a two-stage amplifier.

2. Now the response starts to drop off at the rate of 20 dB per decade—at the low end because of the first-stage response and at the high end because of the second stage. Draw these 20-dB roll-off lines at each end (from A and D).
3. At the low-frequency end, the second stage comes into play at 60 Hz. Mark this point B. From 60 Hz down, since two R-C networks are now active, the roll-off is at 12 dB per octave. Since point B at 60 Hz is already down 8 dB, at one octave below (30 Hz), the total loss will be $8 + 12 = 20$ dB. Mark this point (30 Hz, -20 dB) with a C. Draw the new roll-off through B and C.
4. At the high-frequency end, the first-stage cutoff becomes effective at 18 kHz. Locate this point (E) on the original curve. Now the slope increases at the rate of 12 dB per octave. At point D (18 kHz) the loss is 7 dB. Two octaves higher brings us to 64 kHz, and the loss must be $12 + 12 + 7 = 31$ dB. Mark this point F, and draw the new roll-off through E and F.
5. This gives us the overall *basic* Bode plot. Now, if we wish, we can make corrections for the more accurate curve. For example:
   (*a*) At A and D the true response is down 3 dB.
   (*b*) At 75, 300, 4000, and 16,000 Hz (one octave to each side), the true response is down 1 dB.
   (*c*) At B and E, the second corner frequencies, the response should be down an additional 3 dB.
   (*d*) At one octave below B (30 Hz) and at one octave above E (36 kHz), the correction for the second slope is 1 dB below the line.
   We can now sketch in the corrected frequency response (broken line).

SINGLE STAGE WITH TWO R-C NETWORKS.   Sometimes, even in a single stage, the frequency response can be affected by two R-C networks. This can happen at the low- and/or high-frequency ends. For example, if a stage uses source (or emitter) bias and the bypass capacitor is not large enough, the low-frequency response is affected by this bias R-C network as well as by the coupling R-C network. The cutoff frequency of each of these networks is found, and the treatment is then the same as for the two-stage circuit described above.

WIDEBAND AMPLIFIERS

In an average R-C amplifier for other than high-fidelity audio work, frequency response does not present any problem. A frequency range of approximately 80 to 8000 Hz is quite adequate and easily obtained. The design of such amplifiers is generally predicated on maximum gain. However, for other applications, much wider frequency response is needed. Specifications such as "flat from 5 Hz to 5 MHz" are not uncommon for video or pulse amplifiers. Two techniques are used to achieve wideband response. The first is to remove, or reduce as far as practicable, all causes for loss of gain. The second step is to add compensating circuit components. Let us examine both these techniques as they apply to the low- and high-frequency ranges.

IMPROVEMENT OF LOW-FREQUENCY RESPONSE.   You will recall that poor low-frequency response is due to the voltage-divider action between X_C of the coupling capacitor and the following resistor (R_g in a FET, $r_{T(in)}$ in a BJT circuit). As long as X_C is low compared to this resistor value, there is no problem. This would suggest making the capacitor and resistor values larger and larger, so that X_C would be negligible—compared to R—at any frequency. But this is not always possible. In a BJT circuit, the value of R is primarily limited by the input resistance of the next stage (r_i). Even with FETs, there is a limit. Too large a gate resistor can cause breakdown of the gate insulation in MOSFETs or can upset the bias in a JFET because of leakage current. Increasing the capacitance value also has limitations. No capacitor has a perfect dielectric. The larger the capacitance value, the greater the possible leakage, and the bias to the next stage can be upset. Also, the larger the capacitor, the greater the stray capacitance (to ground) and the poorer the high-frequency response.

The next step is compensation. The basic idea is quite simple. Refer to Fig. 13-20a.* We know that some of the voltage V_o' will be lost because

*Although a BJT circuit is shown in Fig. 13-20a, this diagram and discussion apply equally well to a FET—merely change the resistor designations to fit the FET circuitry.

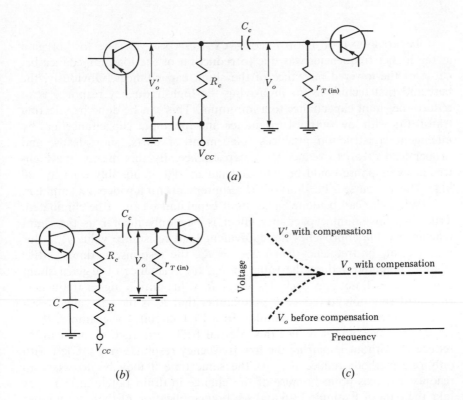

Figure 13-20   Low-frequency compensation.

of the voltage drop across C_c. However, if we could deliberately increase voltage V_o'—as the frequency decreased—we could afford this drop across C_c. In fact, if the increase in V_o' were just equal to the drop across C_c, the voltage output to the next stage (V_o) would remain constant. This means that the gain—at low frequencies only—must be increased. We can do this by adding a compensating R-C network *in series* with R_c as shown in Fig. 13-20*b*.

The addition of resistor R will reduce the collector potential and will probably necessitate some readjustment of the bias point (or an increase in the supply voltage V_{CC}). As far as ac operation at mid- and high frequencies is concerned, resistor R is effectively shorted to ground by the compensating capacitor C. It therefore does not alter the load impedance. At low frequencies, the reactance of the compensating capacitor increases, the impedance of the compensating circuit (C and R) increases, the voltage V_o' increases, and this will offset the voltage drop across the coupling capacitor (Fig. 13-20*c*). For best compensation, the product CR_d on the collector side should equal $C_c r_{T(\text{in})}$ on the base side of the coupling network.

IMPROVEMENT OF HIGH-FREQUENCY RESPONSE.   The loss of gain at the higher frequencies was due to reduction of the load impedance because of the lowered reactance of the shunt capacitances. Obviously, the best and first technique for improving the high-frequency response is to reduce the shunt capacitance to a minimum. This can be done by selecting transistors with low output capacitance and low input capacitance and by intelligent construction practices (placement of parts, short leads, and proper lead dress) to reduce stray capacitance. By this means, the high-frequency response could be extended flat to 100 or possibly even to 200 kHz. This of course is far short of the requirements for a wideband amplifier.

What else can be done? The shunt capacitance cannot be eliminated. True, but this shunt capacitance effect is bad only insofar as the shunt reactance value approaches the equivalent load resistance value R_e. (For example, at the frequency when $X_{C_{sh}} = R_e$, the response is down 3 dB.) If we deliberately reduce the value of R_e, the reactance of a given shunt capacitance will not approach this lower R_e value until a higher frequency. In considering how to reduce R_e, remember that R_e is the equivalent resistance of three resistances in parallel. In a FET circuit, for example, $R_e = r_d \| R_d \| R_g$. We cannot reduce the internal FET resistance r_d; nor can we reduce R_g without impairing the low-frequency response. We are left with only one method: reduce R_d. (At the same time it may be necessary to readjust the bias point because of the change in drain potential.)* Let us take the data of Example 13-6 and see how application of these techniques will raise the upper cutoff frequency limitation.

EXAMPLE 13-12

The FET of Example 13-6—$r_d = 95$ kΩ; $g_{fs} = 1000$ μmho—is used in an R-C coupled amplifier with $R_d = 270$ kΩ and $R_g = 470$ kΩ. The total shunt capacitance is 80 pF.

1. Find (a) the mid-frequency gain and (b) the high-end cutoff frequency.
2. By improved design and layout, the shunt capacitance is reduced to 40 pF. Repeat (a) and (b) above.
3. The drain resistor is reduced to 27 kΩ. Repeat (a) and (b) above.

Solution

1. From Example 13-6 we have the following answers:

(a) $A_{v(mid)} = g_{fs}R_e = 61.2$

(b) f_H (for 3-dB loss) $= 32.5$ kHz

---

*If low-frequency compensation is also used, these two effects tend to cancel each other, and it is possible that little or no change in bias or supply voltage will be required.

2. With shunt capacitance reduced to 40 pF:

(a) $A_{r(mid)} = g_{fs}R_e = 61.2$

(b) f_H (cutoff frequency) $= \dfrac{1}{2\pi R_e C_{sh}} = \dfrac{0.159 \times 10^{12}}{(61.2 \times 10^3) \times 40} = 65 \text{ kHz}$

Notice that since R_e was unchanged (61.2 kΩ), the mid-frequency gain does not change. However, the reduction of C_{sh} from 80 to 40 pF raised the high-end cutoff frequency from 32.5 to 65.0 kHz. (This should have been obvious, since halving the capacitance requires doubling the frequency to obtain the same reactance.)

3. With reduced drain resistor R_d:

$$\frac{1}{R_e} = \frac{1}{r_d} + \frac{1}{R_d} + \frac{1}{R_g} = \frac{1}{95 \text{ k}\Omega} + \frac{1}{27 \text{ k}\Omega} + \frac{1}{470 \text{ k}\Omega}$$

$$R_e = 20.1 \text{ k}\Omega$$

(a) $A_{r(mid)} = g_{fs}R_e = 1000 \times 10^{-6} \times 20.1 \times 10^3 = 20.1$

(b) $f_H = \dfrac{1}{2\pi R_e C_{sh}} = \dfrac{0.159 \times 10^{12}}{(20.1 \times 10^3) \times 40} = 198 \text{ kHz}$

Example 13-12 shows the improvement in high-frequency response. Reducing shunt capacitance increased the upper frequency limit to 65 kHz, and reducing the value of the drain resistor further increased this upper frequency limit to almost 200 kHz.* But there is a penalty to this last frequency improvement. Notice that the mid-frequency gain (step 3a) dropped from 61.2 to only 20.1. The price that must be paid for improved high-frequency response is lower overall gain. This effect is shown in Fig. 13-21 for three values of plate resistor R_d.

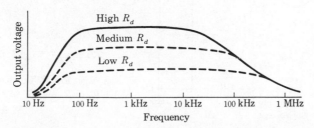

Figure 13-21   Effect of drain resistor (R_d) on frequency response.

Can the frequency response of the above amplifier be extended to about 400 kHz? This is double the value obtained in step 3b. Therefore merely reduce R_e to half, or approximately 10,000 Ω. To extend the fre-

*The reduction in gain in turn reduces the Miller effect, which reduces the input capacitances, reduces the shunt reactances, and further improves the frequency response.

quency further—for example, to 4 MHz or 10 times higher than 400 kHz —it would be necessary to reduce R_e to one-tenth, or 1000 Ω. This should give us the desired 4-MHz frequency limit. But before we accept this step, let us check for gain:

$$A_v = g_{fs}R_e = 1000 \times 10^{-6} \times 1000 = 1$$

We have increased the frequency response, but the gain is only 1. In other words, if we apply an input signal voltage of 1 V, we will get an output of 1 V. There is no value in extending the frequency response to such a limit. In practice, reduction of R_d is seldom carried to the point where the gain drops below 8 to 10. In this respect, FETs with a high mutual conductance g_{fs} and low input and output capacitance are desirable for wideband amplifiers. The ratio $g_{fs}/(C_i + C_o)$ is taken as the "figure of merit" of a FET.*

HIGH-FREQUENCY COMPENSATION.   These techniques (reducing shunt capacitance and drain, or collector, resistor) still leave us far short of the upper frequency limits needed by pulse amplifiers or other types of wideband amplifiers. Any further extension of the upper frequency limit requires compensation. If an impedance were added to the coupling network so as to increase Z with frequency, this could offset the decrease in X_C of the shunt reactance. The total impedance Z_e would remain constant, and the gain would remain constant. Impedance increasing with frequency means that an inductance is needed. One technique is to add inductance in series with the drain (or collector) resistor. Such a circuit, shown in Fig. 13-22, is known as *shunt peaking*. The capacitor C_s in this circuit represents

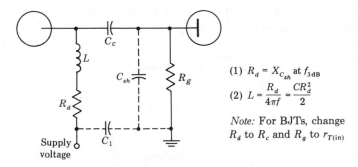

(1) $R_d = X_{C_{sh}}$ at f_{3dB}

(2) $L = \dfrac{R_d}{4\pi f} = \dfrac{CR_d^2}{2}$

Note: For BJTs, change R_d to R_c and R_g to $r_{T(in)}$

Figure 13-22   High-frequency compensation—shunt peaking.

the total shunt capacitance $(C_o + C_1 + C_{\text{stray}})$. At these high frequencies, the coupling capacitor C_C and the bypass capacitor C_1 can be considered as direct connections (zero impedance). Picturing the circuit this way, components L and C_{sh} form a parallel tuned circuit. This circuit should

*It should be realized that the entire discussion of improvement of frequency response applies equally well to BJT and FET circuits.

resonate at a frequency close to the desired maximum frequency limit. Since the impedance of a parallel resonant circuit rises to a maximum at resonance, it is possible to overcompensate and produce a "hump" or rise in response at the high-frequency end. This is counteracted by the series resistance R_d (or R_c), which lowers the Q of the circuit. For the flattest response curve the optimum values for R_d and L are shown in Fig. 13-22. Higher inductance values can produce overcompensation.

Another compensating circuit is shown in Fig. 13-23. Again an inductance is used, but this time the inductance is in series with the *signal path*. It is therefore called *series peaking*. By this technique the shunt capacitance is split into two separate values. C_1 represents the output capacitance of the previous stage plus that portion of the stray capacitance to the left of the peaking coil; C_2 represents the remainder of the stray capacitance plus

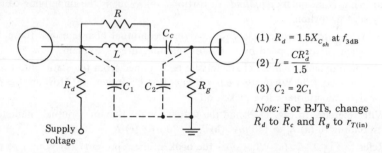

(1) $R_d = 1.5 X_{C_{sh}}$ at f_{3dB}

(2) $L = \dfrac{CR_d^2}{1.5}$

(3) $C_2 = 2C_1$

Note: For BJTs, change R_d to R_c and R_g to $r_{T(in)}$

Figure 13-23   High-frequency compensation—series peaking.

the input capacitance of the next stage. This combination of capacitance and inductance forms what is known as a π-type low-pass filter.* Such filters pass all frequencies up to a certain value known as the *cutoff frequency*. By choice of L and C values this cutoff frequency can be made equal to or higher than the highest desired frequency. For proper design of this type of filter, it is necessary for C_2 to have twice the capacitance of C_1. Here is an ironical twist: whereas poor high-frequency response is due to shunt capacitance, now it may sometimes be necessary to deliberately *add* capacitance to get the proper value for C_2.

Since the peaking coil has distributed capacitance, it is possible for this coil to become self-resonant at some frequency within the desired frequency range. This could cause a sharp rise in impedance and a hump in the response curve. The resistor R, shunted across the coil, lowers the Q of the circuit and reduces this resonant rise. The advantage of this circuit as compared to shunt peaking is that a higher value of R_d (or R_c) can be used and higher circuit gain is achieved.

*J. J. DeFrance, *Communications Electronics Circuits* (2nd ed.), New York: Holt, Rinehart and Winston, 1972, chap. 2.

In some cases, both series and shunt peaking are incorporated into the same circuit. This allows use of even higher values for R_d (or R_c)—up to $2X_{C_{sh}}$—and therefore further increases the gain. Regardless of which specific form of high-frequency compensation is used, in addition to reducing frequency distortion these circuit components also reduce phase distortion. Consequently a compensated amplifier will maintain constant phase shift as well as constant gain over a much wider range of frequencies.

REVIEW QUESTIONS

1. **(a)** Give a general definition for the term *distortion* as applicable to an amplifier. **(b)** Name the three basic types of distortion that can occur in an amplifier.

2. **(a)** What is meant by *amplitude* distortion? **(b)** What is the underlying cause for this type of distortion?

3. **(a)** What are the coordinates for a dynamic mutual characteristic for a FET? **(b)** Are such curves supplied by transistor manufacturers? Explain.

4. **(a)** Can a dynamic *mutual* characteristic be obtained from the static output family of characteristics? **(b)** What other curves must first be drawn? **(c)** How are the points for the dynamic mutual curve picked off?

5. With reference to Fig. 11-2, find the values for the dynamic mutual characteristic corresponding to **(a)** $I_B = 140\ \mu A$; **(b)** $I_B = 80\ \mu A$; **(c)** $I_B = 20\ \mu A$.

6. Refer to Fig. 13-2: **(a)** What does the broken line represent? **(b)** What two points are used to draw this line? **(c)** How is point Q obtained? **(d)** What does the solid straight line represent? **(e)** How is point A for this line obtained? **(f)** Pick off values for a dynamic transfer curve at $V_{gs} = 0$ and at $V_{gs} = -2.0$ V.

7. Refer to Fig. 13-3: **(a)** What does the plot in the upper left represent? **(b)** How were the values for this curve obtained? **(c)** For what value of load resistance is this curve valid? **(d)** For what value of supply voltage is this curve valid? **(e)** What does the waveform in the lower left represent? **(f)** What is the peak value of this wave? **(g)** What are the minimum and maximum values of gate voltage produced by this input signal? **(h)** What does the waveshape in the upper right represent? **(i)** What is the quiescent value of the drain current? **(j)** Is this i_d curve a pure sine wave? How can you tell? **(k)** What caused this distortion? **(l)** Does this i_d wave have a dc component? How can you tell? **(m)** Since the distortion is not too great, what other component is present in this wave?

8. With reference to Fig. 13-4, what do the following symbols mean: **(a)** i_d; **(b)** I_{d0}; **(c)** $I_{d(max)}$; **(d)** $I_{d(min)}$; **(e)** A_1; **(f)** A_2; **(g)** A_0?

9. When a signal is applied to an amplifier stage, its collector current increases from 1.1 to 1.4 mA. **(a)** What is the probable reason for this change? **(b)** What added information can be obtained from the values themselves?

10. In evaluating distortion effects from a waveform (using the three-point method), give the equation for **(a)** the amplitude of the fundamental component; **(b)** the amplitude of the second harmonic; **(c)** the percent distortion.

11. If an accurate plot of output *voltage* (versus time) were available, give an equation showing how we could evaluate the percent distortion in this wave.

12. Can the presence of distortion always be detected by the change in dc drain or collector current reading? Explain.

13. (a) In evaluating distortion, when is the use of a five-point schedule necessary? **(b)** Compared to the three-point schedule, what are the other two evaluation instants called? **(c)** To what bias values do these other two instants correspond? **(d)** Using the five-point schedule, give the equations for second harmonic distortion (in percent), third harmonic distortion (in percent), total distortion (in percent).

14. (a) Can distortion effects be evaluated directly from the output characteristic curves? **(b)** What additional curve(s) must be added?

15. Refer to Fig. 13-5: **(a)** For what value of supply voltage do these curves apply? **(b)** For what value of bias? **(c)** What value of gate signal is applied? **(d)** What is the minimum instantaneous gate voltage? **(e)** What drain current—symbol and value—corresponds to this gate voltage? **(f)** What is the maximum instantaneous gate voltage? **(g)** What drain potential—symbol and value—corresponds to this gate voltage? **(h)** Is the input voltage waveshape (v_g) a sine wave or is it distorted? **(i)** Why does it *look* distorted?

16. Refer to Fig. 13-6: **(a)** What is the *specific* cause of distortion in diagram (a)? Explain. **(b)** In diagram (b)? Explain. **(c)** In diagram (c)? Explain.

17. From output *voltage* waveshapes, how can you distinguish a low bias condition from too high a bias?

18. (a) A complex wave is applied to an amplifier with a nonlinear transfer characteristic. Name the type of distortion this produces. **(b)** Will this type of distortion be produced with sine-wave input? **(c)** Which produces poorer sound quality: 3 percent harmonic distortion or 3 percent IM distortion? **(d)** What standard input signal is used for IM distortion testing?

19. (a) What is meant by frequency distortion? **(b)** What is the basic cause for this type of distortion?

20. (a) What is meant by phase distortion? **(b)** Why is this form of distortion undesirable in oscilloscope amplifier circuits?

21. Why do frequency and phase distortion generally occur at the same time?

22. Refer to Fig. 13-7: **(a)** Why are four of the capacitors shown with dotted connections? **(b)** What do each of these four capacitors represent?

23. (a) What is meant by *stray capacitance*? **(b)** What factors contribute to this quantity? **(c)** What is an average value for this quantity?

24. (a) What is meant by the Miller effect? **(b)** Name two quantities that contribute to this effect. **(c)** Which of these two quantities has the greater effect on the input capacitance? Why?

25. (a) What are the approximate frequency limts of the "mid-frequency" range? **(b)** What circuit analysis simplification is valid within this range?

26. Repeat Question 25 for the "low-frequency" range.

27. Repeat Question 25 for the "high-frequency" range.

28. Draw the constant-current equivalent circuit for the mid-frequency range.

29. **(a)** Give the equation for mid-frequency gain for a FET circuit. **(b)** What does the quantity R_L represent? **(c)** Repeat part (a) for a BJT circuit. **(d)** How is the quantity R_e evaluated?

30. What is a typical value for the phase distortion within the mid-frequency range? Explain.

31. Refer to Fig. 13-10: **(a)** What does diagram (a) represent? **(b)** What does diagram (b) represent? **(c)** What does C_{sh} represent? **(d)** Knowing the individual component values, how is Z_e evaluated?

32. Give the equation for the gain at high frequencies as obtained from **(a)** the FET equivalent circuit and **(b)** the BJT equivalent circuit.

33. **(a)** What is the significance of the "3-dB point" with regard to frequency response? **(b)** Compared to the mid-frequency load, at what load impedance value does it occur? **(c)** At what value of shunt reactance does this occur? **(d)** Give a name for this 3-dB frequency point.

34. Refer to Fig. 13-11b: **(a)** What circuit is the basis for this phasor diagram? **(b)** Does this diagram apply to a FET or to a BJT? **(c)** Why is the mid-frequency output voltage in phase with I_T? **(d)** Why does I_R lag I_T? **(e)** What is the phase angle between I_R and I_C? Why? **(f)** How is the value of angle θ determined?

35. **(a)** At the cutoff frequency, what is the phase of the output voltage compared to the output voltage at mid-frequency? **(b)** What happens to this phasing if the frequency of operation is made still higher? **(c)** What is the maximum value that this phase angle can reach?

36. Refer to Fig. 13-12: **(a)** What is the distinction between diagrams (a) and (b)? **(b)** Give the steps in calculating Z_e. **(c)** Give the steps in calculating V_o.

37. Refer to Fig. 13-13: **(a)** How are these diagrams obtained? **(b)** Why were these conversions made? **(c)** How is the value of R' obtained? **(d)** What values of the components will cause the output voltage to drop to 0.707 of its mid-frequency value? **(e)** What is this loss in decibels? **(f)** What is the term for the frequency at which this occurs? **(g)** Give the equation for the high-end cutoff frequency in an R-C coupled BJT amplifier.

38. **(a)** Give a possible cause for poor low-frequency response other than the coupling capacitor. **(b)** How can this problem be avoided?

39. Refer to Fig. 13-14: **(a)** What circuit is the basis for this phasor diagram? **(b)** What quantity is used as the reference? **(c)** Why is the mid-frequency current in phase with the mid-frequency output voltage? **(d)** At low frequencies, why does the current lead V_s? **(e)** Why is the output voltage in phase with this current? **(f)** What happens to the phase of the output voltage as the signal frequency is lowered further? **(f)** Give an equation for calculating this phase angle.

40. **(a)** What is meant by the frequency response curve of an amplifier? **(b)** What are the coordinates of such a curve? **(c)** What type of graph paper is used for this curve? **(d)** Which coordinate is plotted on which scale? **(e)** Why is this technique used?

41. **(a)** Unless otherwise stated, what limits are generally used when giving frequency response data of an amplifier? **(b)** If more stringent limits are desired, how is the response stated?

42. In getting data for phase-shift and gain measurements of an amplifier, what test equipment would be needed?

43. Refer to Fig. 13-16: **(a)** What is the purpose of the two voltmeters? **(b)** State two purposes of the oscilloscope. **(c)** Describe briefly how the test is made.

44. What are the coordinates for a Bode plot?

45. **(a)** In discussing frequency response, what is an octave? **(b)** What is a decade?

46. If the frequency response of an amplifier is affected by only one R-C network, **(a)** what is the rate of fall in the response at the low frequencies? **(b)** At the high frequencies? **(c)** What is the term used to denote "rate of fall"?

47. Refer to Fig. 13-17: **(a)** Give two names for the frequencies marked f_L and f_H. **(b)** At the points marked $\frac{1}{2}f_L$ and $2f_H$, why is the response down by 6 dB? **(c)** At the points marked $\frac{1}{10}f_L$ and $10f_H$, why is the response down 20 dB? **(d)** In drawing this response curve, which values had to be calculated? **(e)** Is this response curve completely accurate? Explain.

48. Refer to Fig. 13-18: **(a)** Specify the corner frequencies by letter designation. **(b)** What correction is applied at the corner frequencies? **(c)** What is the frequency relation of the points marked B to the points marked A? **(d)** What correction is applied to the B points? **(e)** What is the frequency relation of the points marked C to the A points? **(f)** What is the relation of the corrected curve to the basic Bode plot at the C points?

49. In a three-stage amplifier, the first stage has a 3-dB loss at 12 kHz. The second stage has a 4-dB loss at the same frequency, and the third stage has a 2-dB loss at this frequency. How does the overall response at this frequency compare to the mid-frequency output voltage?

50. Refer to Fig. 13-19: **(a)** What is the roll-off of the line through A-B? **(b)** What is the roll-off of the line through B-C? **(c)** Under what condition would this increased roll-off take place? **(d)** Since this plot is for a two-stage amplifier, give the possible frequency response (within 3 dB) for each stage.

51. State two *general* techniques used to achieve wideband response.

52. **(a)** Explain how the low-frequency response of an R-C amplifier could be improved without resorting to compensation. **(b)** State two disadvantages of increasing the value of the coupling capacitor. **(c)** State a limitation on increasing the R_g value in a FET circuit. **(d)** Can the $r_{T(in)}$ value in a BJT circuit be materially increased? Explain.

53. Refer to Fig. 13-20: **(a)** Which components are responsible for the improved response? **(b)** What is the effect of this compensating network at the higher frequencies? Explain. **(c)** In diagram (c), why does the V_o curve drop off at low frequencies before compensation? **(e)** For optimum compensation, what relative values should be used?

54. State two ways by which the shunt capacitance in an amplifier circuit can be kept to a minimum.

55. **(a)** Why does reduction of the collector resistor value R_c improve the high-frequency response? **(b)** What is the disadvantage of this technique?

56. **(a)** What is the purpose of "shunt peaking" in an R-C amplifier? **(b)** What component is added? **(c)** How is it connected? **(d)** Explain its action.

57. Refer to Fig. 13-23: **(a)** What type of compensating circuit is this? **(b)** Which component is primarily responsible for the compensation? **(c)** What does C_1 represent?

(d) What does C_2 represent? **(e)** What basic type of circuit do these components form? **(f)** Are the inherent values of C_1 and C_2 always correct? Explain. **(g)** Explain why resistor R is used.

58. (a) Which type of high-frequency compensation is better? Why? **(b)** Are both types ever used in one circuit? Explain.

59. Draw the circuit diagram of a complete R-C coupled FET amplifier incorporating full compensation (high and low). Include the input circuit of the next stage.

60. Repeat Question 59, using BJTs.

PROBLEMS

1. Figure 13-24 shows the output characteristics for a 2N364 *NPN* transistor CE connection. Also shown is an ac load line. Using this load line, tabulate values for a dynamic transfer curve. Plot the curve.

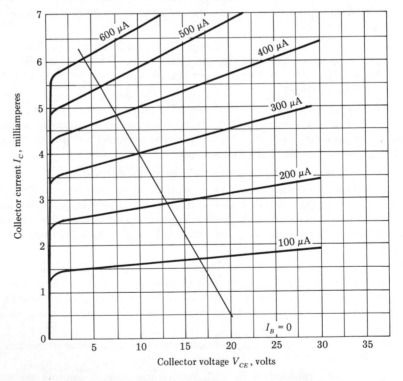

Figure 13-24 Common-emitter output characteristics.

2. Using the FET characteristics and ac load line from Fig. 13-25, tabulate values for a transfer curve and plot the curve. (Save your answer for a later problem.)

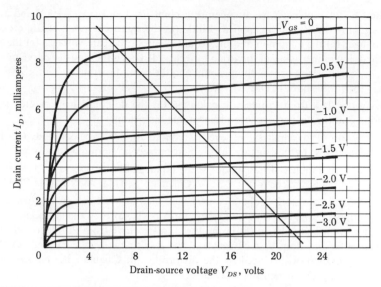

Figure 13-25   Output characteristics—FET.

3. The *PNP* transistor (characteristics in Fig. 13-26) is operated from a 10-V supply. The collector resistor is 200 Ω and the ac load resistance is 130 Ω. The base bias is adjusted for a base current of 0.4 mA. **(a)** Draw the dc load line. At what point does it intercept the *y* axis? **(b)** Draw the ac load line. At what point does it intersect the *y* axis? **(c)** Using this ac load line, tabulate points for a dynamic transfer characteristic and plot the curve. (Save this curve for a later problem.)

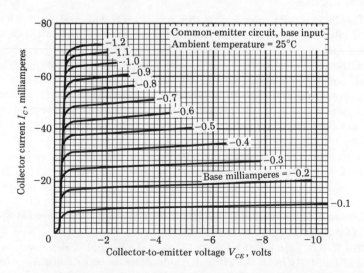

Figure 13-26   Average collector characteristics of type 2N408.

4. The FET of Problem 2 has its operating point fixed at $V_{GS} = -1.5$ V and a sine-wave signal of 1.5 V peak is applied. Find: **(a)** the quiescent value of drain current and voltage; **(b)** the minimum and maximum swing of the gate voltage; **(c)** the minimum and maximum swing of the drain current. **(d)** Using the transfer curve of Problem 2, draw waveshapes of the input signal and the resulting drain current. (Save these answers for Problems 5, 6, and 7.)

5. Using the data from Problem 4, calculate **(a)** the dc component created by the distortion and **(b)** the percent second harmonic distortion.

6. In Problem 4, the input signal level is reduced to 1.25 V peak. Find: **(a)** the dc component created by this signal; **(b)** the percent second harmonic distortion.

7. The FET in Problem 4 has its quiescent point shifted to $V_{GS} = -1.0$ V and the input signal level is reduced to 1.0 V peak. Calculate parts (a) and (b) as in Problem 6.

8. The BJT of Problem 3 is operated at a base bias of -0.4 mA. A sine-wave input signal produces a base current change of 0.3 mA peak. **(a)** Find the quiescent value of collector current and voltage. **(b)** Using the transfer curve from Problem 3, draw the waveshapes for the input base current and the resulting collector current. **(c)** Calculate the dc component produced by the distortion. **(d)** Calculate the percent second harmonic in the collector current wave.

9. Using the curves and load line in Fig. 13-24, a basic bias of 300 μA, and an input signal swing of ± 200 μA (peak), find **(a)** the quiescent values of collector current and voltage; **(b)** the peak value of the output voltage *on each alternation;* **(c)** the maximum and minimum values of the collector current; **(d)** the dc component added to this current because of distortion; **(e)** the percent second harmonic distortion in this current waveshape.

10. The MFE4010 *P*-channel FET has $C_{iss} = 5.5$ pF, $C_{rss} = 0.8$ pF, and $C_{oss} = 1.5$ pF. It is used as an amplifier with a voltage gain of 30. Find the input and output capacitances of this stage.

11. The 2N3904 *NPN* silicon transistor has $C_{ob} = 4.0$ pF and $C_{ib} = 8.0$ pF. It is used as an amplifier with a voltage gain of 40. Find the input and output capacitances of this stage.

12. A 2N4867 *N*-channel FET with $g_{os} = 1.5$ μmho and $g_{fs} = 2000$ μmho is used in an R-C coupled amplifier. The coupling network has a drain resistor of 47 kΩ, a coupling capacitor of 0.1 μF, and a following gate resistor of 470 kΩ. The circuit has a total shunt capacitance of 70 pF. Find: **(a)** the gain at mid-frequency; **(b)** the upper-frequency 3-dB point; **(c)** the gain at this frequency. (Save your answers for later problems.)

13. A 2N4339 ($g_{os} = 15$ μmho; $g_{fs} = 2400$ μmho) is used in an R-C coupled amplifier with $R_d = 75$ kΩ, $C_c = 0.2$ μF, and the next stage $R_g = 1$ MΩ. The circuit has a total shunt capacitance of 80 pF. An input signal of 0.1 V is applied to the gate. Find: **(a)** at mid-frequency, the gain and the voltage fed to the next gate; **(b)** the high-end cutoff frequency; **(c)** the gain at this cutoff frequency and the voltage fed to the next stage. (Save your answers for later problems.)

14. The BJT of Example 11-6 ($h_{fe} = 175$; $h_{oe} = 30$ μmho) is used (as before) in the circuit of Fig. 11-16 ($R_c = 4.7$ kΩ; voltage-divider base-bias resistor $R_1 = 68$ kΩ and $R_2 = 33$ kΩ; $r_i = 2640$ Ω). The total shunt capacitance in the coupling circuit is

200 pF. Find: **(a)** the mid-frequency gain; **(b)** the high-end cutoff frequency; **(c)** the gain at this frequency. (Save your answers for later problems.)

15. A 2N4074 *NPN* silicon transistor ($h_{fe} = 175$; $h_{oe} = 75\ \mu$mho) is used in an R-C amplifier circuit with $R_c = 1200\ \Omega$, $R_E = 600\ \Omega$, and base-bias resistors $R_1 = 56\ k\Omega$ and $R_2 = 16\ k\Omega$. (See Fig. 11-16.) The total shunt capacitance in the coupling network is 250 pF. The supply voltage is 30 V. Find: **(a)** the mid-frequency gain; **(b)** the high-end cutoff frequency and the gain at this frequency. (Save your answers for later problems.)

16. In Problem 12, compared to the output voltage at mid-frequency, what is the phase of the output voltage **(a)** at the high-end cutoff frequency? **(b)** At the frequency that the shunt reactance drops to 20 kΩ? **(c)** What is this frequency?

17. Repeat Problem 16 for the circuit of Problem 14.

18. The BJT of Problem 14 is coupled to the next stage through a 0.5-μF capacitor. Find: **(a)** the gain at 80 Hz; **(b)** the low-end cutoff frequency.

19. The BJT circuit of Problem 15 uses a 2.0-μF coupling capacitor. Find: **(a)** the gain at 50 Hz; **(b)** the low-end cutoff frequency.

20. In the FET circuit of Problem 12, find **(a)** the gain at 20 Hz and **(b)** the low-end cutoff frequency.

21. Repeat Problem 20 for the FET of Problem 13.

22. Using the mid-frequency output voltage as reference, calculate the phase of the output voltage **(a)** at 20 Hz for the circuit in Problem 13 (and 21); **(b)** at 20 Hz for the circuit in Problem 12 (and 20); **(c)** at 50 Hz for the circuit in Problem 15 (and 19); **(d)** at 80 Hz for the circuit in Problem 14 (and 18).

23. In making a frequency response run, the output voltage was held constant at 10 V. However, the input voltage had to be varied from 0.45 V at 20 Hz, to 0.12 V at 400 Hz, and to 0.31 V at 2 MHz. Calculate the decibel gain or loss at the extreme frequencies.

24. Repeat Problem 23, using 5 V for the constant output voltage and inputs of 0.1 V at 400 Hz, 0.54 V at 100 kHz, and 0.25 V at 10 Hz.

25. An R-C coupled amplifier stage has cutoff frequencies of 120 Hz and 10 kHz. **(a)** Draw the full basic Bode plot frequency-response curve. **(b)** What is the relative response at 40 Hz and at 30 kHz? **(c)** Draw the corrected response curve. **(d)** Find the response at 80 Hz and at 20 kHz.

26. The 3-dB points in the frequency response of an R-C amplifier stage occur at 80 Hz and at 30 kHz. Repeat parts (a), (b), and (d) as in Problem 25. (*Note:* This plot can be fitted on one sheet of three- or four-section semilog paper by drawing the low end and high end separately.)

27. Each stage of a two-stage R-C amplifier has a frequency response (3-dB limits) of 80 Hz to 20 kHz. Draw the overall basic Bode response curve.

28. A two-stage R-C amplifier has a first-stage response of 80 Hz to 20 kHz and a second-stage response of 40 Hz to 40 kHz. Draw the overall basic Bode response curve.

29. The original design of a FET R-C coupled amplifier had a shunt capacitance of 70 pF, a drain resistor of 47 kΩ, and a following gate resistor of 470 kΩ. The FET parameters are $g_{fs} = 3000\ \mu$mho, and $g_{os} = 10\ \mu$mho. **(a)** Calculate its high-end cutoff frequency. **(b)** This response was not satisfactory, and by wiring changes the shunt capacitance was reduced to 40 pF. Find the new cutoff frequency. **(c)** This response was still not satisfactory, and R_c was reduced to 22 kΩ. Calculate the original gain, the new gain, the loss in decibels, and the new cutoff frequency.

30. A BJT R-C coupled amplifier was designed with $h_{fe} = 150$, $h_{oe} = 30 \,\mu$mho, $R_c = 4.7$ kΩ, $r_i = 2640 \,\Omega$, and $r_{T(in)}$ of the next stage $= 2000 \,\Omega$. The total shunt capacitance was 300 pF. **(a)** Calculate its high-end cutoff frequency. **(b)** A new layout reduced the shunt capacitance to 120 pF. Calculate the new cutoff frequency. **(c)** The collector resistor was reduced to 3.3 kΩ. Find the cutoff frequency. **(d)** What value of collector resistor would increase the frequency limit to 2.0 MHz? **(e)** What was the original mid-frequency gain? What is the mid-frequency gain now? What is the loss in decibels?

14

Other Coupling Methods

Thus far the discussion of nonresonant amplifiers has centered mainly on R-C coupled circuits. This much time has been devoted to one type of circuit because the R-C type of coupling is the most widely used in discrete-component circuits. However, transistors and tubes are also used with other types of coupling. In this chapter these other types of untuned voltage or small-signal amplifiers are discussed.

IMPEDANCE COUPLING

With regard to untuned voltage amplifiers, impedance coupling is now seldom used. However, it can still be found in power amplifier circuits, particularly of the resonant type (RF amplifiers). This method of coupling is introduced here for its historical value and as a basis for the next topic: transformer coupling.

In the early days of radio, triode tubes were the only types of active devices, and batteries (or motor-generator sets) were the only sources of dc power. The triode required a fairly high dc voltage (at least 90 V) for reasonable gain. If R-C coupling had been used—because of the high voltage drop in the dc load resistor R_b—the supply voltage would have had to be around 180 V. Using a battery supply, this meant many cells in series (four 45-V units in series) and a bulky overall package. On the other hand, if the plate resistor R_b was replaced by an inductor X_b having negligibly low dc resistance, the plate potential would approach the full E_{bb} value.

347

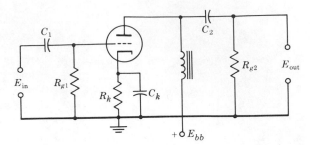

Figure 14-1 Typical impedance-coupled untuned triode voltage amplifier.

A high inductance value can be used to produce the high load impedance needed for high output voltage and high gain. Such a circuit is shown in Fig. 14-1.

A comparison of this circuit with the triode circuit of Fig. 10-16 reveals only one change: R_b is replaced by X_b. The inductor is generally referred to as a *choke coil*. The choice of choke inductance value depends on the tube with which it is used (more specifically on its internal plate resistance). Because of the phasor relationship, gains of from 90 to 98 percent of the tube's mu can be achieved if the reactance of the choke coil falls between two to five times the plate resistance of the tube. Unfortunately, at low frequencies the inductance of the choke coil decreases, resulting in poor low-frequency gain. Also, hysteresis and saturation effects in the iron cause amplitude distortion. Obviously, this type of coupling—for untuned voltage amplifiers—was never too popular.

TRANSFORMER COUPLING

In those early days of radio, transformer coupling was the common type of coupling used in the audio amplifier section of receivers and transmitters. With a low-mu triode, this type of coupling gave the highest overall stage gain. The transformers used had more turns in the secondary winding than in the primary, and the gain of the tube circuit was increased by the step-up ratio of the transformer windings.* In addition, just as with impedance coupling, the voltage drop in the plate circuit was low and transformer coupling could be used efficiently with moderately-low plate potentials. However, the early transformer introduced all the disadvantages of the choke coil: the distortion level was high and the fidelity was poor. But no one cared about fidelity; the very idea that music or speech could be received at all was satisfaction enough. (Furthermore, the other components

*The gain with transformer coupling is $\mu(N_2/N_1)$, where N_2 and N_1 are the number of turns in the secondary and primary windings, respectively.

such as loudspeakers or microphones were also poor in their quality of reproduction.)

Improved know-how in the science and art of manufacturing transformers, specifically for amplifier coupling, has made units available that can handle any desired signal level without amplitude distortion and with flat frequency response, within better than 0.5 dB, for frequencies well beyond the audio range. However, such transformers are expensive, so it is not surprising that with the advent of rectifier power supplies to replace batteries and the development of high-mu triodes and pentodes, transformer coupling for *untuned voltage* amplifiers fell into disuse. R-C coupling became the favorite technique for this type of amplifier. This new technique did not relegate transformers to limbo. They are still used whenever impedance matching is necessary, such as for coupling between a *power* amplifier and its load. They may also be seen in BJT circuits. In fact, transformer coupling *must* be used to get any voltage gain if the transistors are connected in the common-base configuration.

IMPEDANCE MATCHING.   If the maximum available gain is to be obtained from a bipolar transistor amplifier, proper matching to the next stage (or load) is necessary. As we have seen, such matching cannot be achieved with R-C coupling. However, it can be obtained with transformer coupling. When a transformer is used as the coupling between two BJTs, it acts as an impedance-matching device. Current flows in its secondary winding. This secondary current, I_2, produces its own flux ϕ_2. Because of this flux, a voltage $E_{1,2}$ is induced in the primary and a "load component" of current flows in the primary. This is exactly like the action of a power transformer as explained in Chapter 5, and the current ratio is equal to the inverse turns ratio:

$$\frac{I_1}{I_2} = \frac{N_2}{N_1}$$

As far as the BJT looking into the transformer primary winding is concerned, it is delivering a voltage E_1 to the winding and a current I_1 is flowing. The BJT "sees" an impedance Z_1 equal to the ratio of E_1/I_1 or

$$Z_1 = \frac{E_1}{I_1}$$

This is shown in Fig. 14-2a. The transformer, *because of the load, acts* as if it were an impedance or resistance equal to E_1/I_1. Although this term is often referred to as the "primary impedance," it is not the phasor sum of the X_L and R of the primary winding itself. This impedance Z_1 depends on I_1, the ac component of primary current, which in turn *depends on the load*

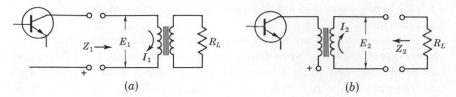

Figure 14-2 Transformer impedance.

resistance and the turns ratio. Any change in the load resistance will change the impedance Z_1 as seen by the BJT.

In similar fashion the term $Z_2 = E_2/I_2$ is often referred to as the "secondary impedance." Again, it is not the X_L and R of the secondary winding but rather the impedance "seen" by the load looking into the secondary winding *with the primary connected to the BJT.* It must be emphasized that when the transformer is loaded, it has no primary or secondary impedance of its own but rather an *impedance ratio,* and the actual value seen on one side then depends on the load value connected to the other side. The specific relationship can be developed mathematically by dividing the equation for Z_1 by the equation for Z_2:

$$\frac{Z_1}{Z_2} = \frac{E_1/I_1}{E_2/I_2} = \frac{E_1}{E_2} \times \frac{I_2}{I_1}$$

But since E_1/E_2 and I_2/I_1 each equals N_1/N_2, then

$$\frac{Z_1}{Z_2} = \left(\frac{N_1}{N_2}\right)^2 \tag{14-1}$$

Expressed in words, the impedance ratio is equal to the square of the turns ratio.

The following example shows how to convert the low 310-Ω input impedance of one transistor circuit to match the 5000-Ω output impedance of the previous stage.

EXAMPLE 14-1

Find the transformer turns ratio needed to convert a 310-Ω load into an equivalent 5000-Ω load.

Solution

1. $\dfrac{Z_1}{Z_2} = \left(\dfrac{N_1}{N_2}\right)^2$

2. $\dfrac{N_1}{N_2} = \sqrt{\dfrac{Z_1}{Z_2}} = \sqrt{\dfrac{5000}{310}} = 4$

The turns ratio needed is 4:1.

A transformer for the above requirement must have four times more turns on the primary as compared to the secondary. The rest of the design (actual number of turns, size of wire, size of core, and so forth) will depend on the power level at which the transformer will be used as well as on the frequency range and fidelity desired.

Since the output impedance of a transistor amplifier is higher than its input impedance, it is obvious that a step-down transformer must be used. Considering the small size of a transistor, the bulky transformers used with vacuum tubes are hardly desirable. Special miniature transformers have been made for interstage coupling between transistors. One such unit is shown in Fig. 14-3*a*.

The base bias for a transformer-coupled common-emitter amplifier can be fed either in shunt or in series with the ac signal. Figure 14-3*b* shows a *PNP* transistor with the bias applied between base and ground, in parallel with the signal voltage from the transformer. This is *shunt-fed* bias. Since the resistance of the transformer secondary winding is quite low, capacitor C_1 must have a fairly high capacitance—at least 10 μF—in order to main-

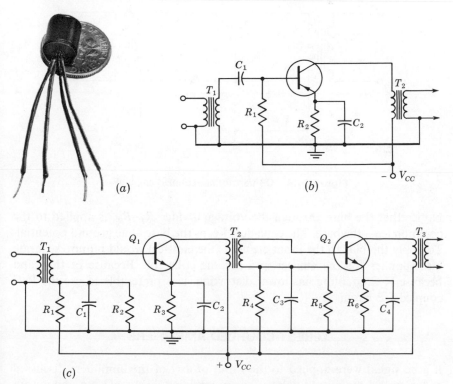

(a) (b)

(c)

Figure 14-3 Transformer and transformer coupling for transistor circuits (*photograph: United Transformer Co.*)

tain good low-frequency response. Resistor R_1 sets the base-bias current, and the R_2C_2 network is used for dc feedback bias stabilization.

The circuit of Fig. 14-3c shows two *NPN* transistors used with transformer coupling and with *series-fed* base bias. (Notice that this time the bias is applied in series with the signal voltage from the transformer.) The voltage-divider bias circuits R_1–R_2 and R_4–R_5 increase the bias stability. Because of the improved impedance match obtained with transformer coupling, this two-stage circuit will, in general, provide as much voltage gain as a three-stage R-C coupled amplifier. When calculating the gain of a transformer-coupled BJT amplifier, the equivalent-diagram technique shown in Chapter 11 can still be used. This time, however, the $r_{T(in)}$ of the next stage must be increased by the square of the transformer turns ratio.

COMMON-BASE AMPLIFIER.   By using suitable interstage transformers, it is possible to employ the common-base connection and get maximum gain and good stability. One such circuit is shown in Fig. 14-4.

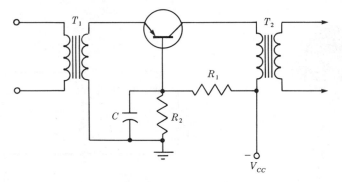

Figure 14-4   CB transformer-coupled amplifier.

Notice that the bias, through the voltage divider R_1–R_2, is applied to the common leg—the base. The capacitor keeps the base at ac ground potential. To apply the bias to the input element (the emitter) would require separate bias supplies for the emitter and for the collector. Because of this, and because R-C coupling has lower distortion, it is preferable to use the R-C coupled CE circuit.

DIRECT-COUPLED AMPLIFIERS

If a dc signal were applied to the input of any of the amplifiers discussed so far, would the output voltage be an amplified dc signal? No, the output would be zero. With R-C coupling, the capacitor blocks out the dc component regardless of whether it was the dc supply voltage or the output signal. For discussion, let us consider a FET circuit. When a dc signal is applied, the drain current of the FET will change, but it will be *constant* at this new value. When the dc signal is removed, the drain current will change back to its *steady* no-signal value. While the current is changing— increasing or decreasing to its new steady value—the drain potential must also be changing. These changing potentials are passed through the capacitor and appear as two pulses, one when the dc signal is applied and the other when the dc signal is removed. In between, the output is zero. With transformer coupling, exactly the same effect is produced. While the drain current is changing—signal just applied or just removed—the magnetic flux is changing and a pulse of voltage is induced in the secondary. Again, while the dc signal is applied, because of the steady drain current there is no change in flux and no voltage is induced in the output.

When it is necessary to amplify dc signals (or very-low-frequency signals), *direct-coupled amplifiers* must be used. (These circuits are also known as *dc amplifiers*.) From the above discussion it should be obvious that neither coupling capacitors nor transformers can be used. One type of direct-coupled amplifier is shown in Fig. 14-5. The circuit represents a two-stage voltage amplifier, with each stage having identical operating voltages and component values. Let us analyze how this is done. Because of the voltage-divider action (R_1, R_2, R_3, R_4), points a, b, c, and d have the potentials shown. The gate of Q_1 is at zero potential with respect to the grounded power supply negative terminal and is 3 V negative with respect to its own source. This sets the bias for this FET. In the drain circuit, the load resistor is connected to point c on the voltage divider ($+53$),

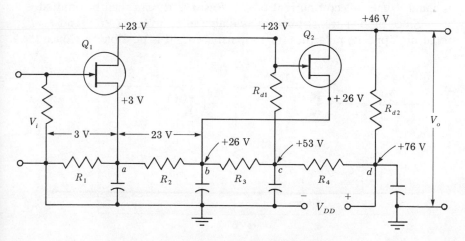

Figure 14-5 A direct-coupled FET amplifier.

but because of a 30-V drop in the load resistor R_{d1}, the drain potential is +23 V. The drain voltage V_{DS}—between drain and source—is 20 V.

Now examine FET Q_2. Its gate potential (with respect to ground) is +23 V. Such a high positive value would ruin the FET if it were the gate bias. However, notice that the source of Q_2 is connected to point b on the voltage divider. The source potential is +26 V. Therefore V_{GS}—the gate-to-source voltage—is actually −3 V. In the drain circuit of Q_2 notice that the drain supply voltage is 76 V. Because of the voltage drop (30 V) across its drain resistor R_{d2}, the drain potential is +46 V. What is the drain-source voltage V_{DS}? Since its source is at +26 V, the drain-source voltage is 46 − 26, or 20 V.

Compare the two stages. In each case we have a gate bias of −3 V and a drain-source voltage of 20 V. Identical operating voltages were obtained without the use of blocking capacitors or isolating transformers.

A direct-coupled amplifier using BJTs is shown in Fig. 14-6. A slightly different technique is used here. Instead of increasing the supply voltage to the second stage, the resistor values are changed. (R_{c2} is made smaller than R_{c1}, and R_{E2} larger than R_{E1}.) Again we get the right potentials in each stage. Notice that in each stage the base is positive (compared to the emitter) by 0.2 V and that the collector is positive (compared to base) by 10 V. Yet the collector potential of the second stage (compared to ground) is almost double that of the first stage.

Because bipolar transistors draw current on the input side, one simplification can be made in BJT direct-coupled circuits. In Fig. 14-6, resistor R_1 serves in common as a path for collector current for Q_1 and base current for Q_2. If Q_2 is a transistor of higher power rating than Q_1 it is possible to adjust the operating point of each transistor so that the base current of Q_2 is equal to the collector current of Q_1. Resistor R_1 can then be omitted.

Since bipolar transistors are available in two types (*NPN* and *PNP*) requiring opposite polarities for proper biasing, it is possible to reduce the

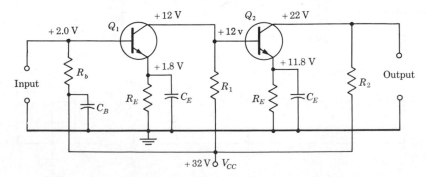

Figure 14-6   A direct-coupled bipolar junction transistor amplifier.

power supply voltage requirements of the direct-coupled amplifier. Figure 14-7 shows a two-stage amplifier using one *NPN* and one *PNP* transistor. The first-stage circuitry using the *NPN* transistor is identical to that of Fig. 14-6. However, component values have been altered to readjust the element potentials. Notice that each transistor still maintains a 10 V collector-to-base voltage with a total power-supply voltage of only 22 V. In the second transistor, instead of having to raise the collector voltage to a higher positive value, we reverse and return toward ground potential by using the *PNP* unit.

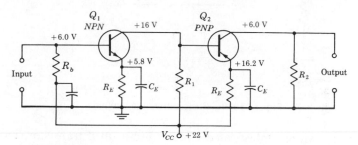

Figure 14-7   Direct coupling using alternate transistor types.

At first thought it might seem that direct coupling should be used in preference to the other types previously discussed. Unfortunately, however, there are drawbacks which limit the application of dc amplifier circuits to only those cases where dc signals or very-low-frequency signals are to be handled. One objection is the higher power-supply voltage requirement. Unless complementary BJTs (*NPN* and *PNP*) are being used, the required supply voltage doubles for each added stage. An even more serious problem is stability. Direct-coupled amplifiers have a tendency to drift; that is, the operating points could shift drastically because of temperature effects, power supply voltage variations, or changes in transistor characteristics (ageing or replacement). Such drift could cause heavy distortion or even ruin the transistor. Stabilizing circuitry makes these amplifiers too expensive for common usage.

DIFFERENTIAL AMPLIFIER

The poor stability of the dc amplifiers discussed above can be overcome by using two (or more) transistors—in each stage—connected to form a *differential amplifier*. The basic circuit for such an amplifier is shown in Fig. 14-8. Notice that the emitters are biased from a voltage supply V_{EE} through a *constant-current source*. In its simplest form, and if V_{EE} is large enough, this could be a large resistance value, so that I_E will depend only on the

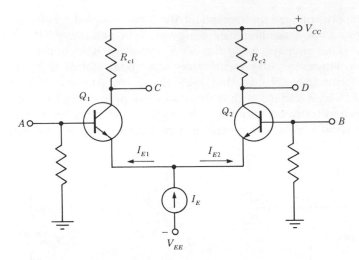

Figure 14-8 Basic differential amplifier.

value of R_E. More sophisticated circuits use a third transistor, base-bias stabilized, and the collector output is the constant-current source. In these circuits, it is also important that the amplifying transistors Q_1 and Q_2 be (ideally) perfectly matched. Therefore, with no input signals at A and B current I_E splits equally between I_{E1} and I_{E2}. Collector currents I_{C1} and I_{C2} are equal; the voltage drops across R_{c1} and R_{c2} are also equal; and the collector potentials at the output points, C and D, are also equal. What is the output voltage between points D and C? Obviously zero. We have effectively removed the dc component in the output (the collector potentials) without using capacitors or transformers.

Let us apply a positive input signal to Q_1. Since it is an *NPN* transistor, this constitutes forward bias. I_{E1} increases, I_{C1} increases, and the potential at C decreases. This is the normal action in a CE amplifier. Similarly, an ac signal applied at the base of Q_1 will·appear amplified but inverted at the output of Q_1 (point C). But there is another effect: Q_1 is "emitter-coupled"* to Q_2; that is, signal can travel from the emitter of Q_1 to the emitter of Q_2. Q_2 now acts as a CB amplifier and an amplified output appears at the collector of Q_2 (point D). The "output" from the emitter of Q_1 is in phase with its own base input. Also, in a CB amplifier there is no phase inversion. Therefore the output from the collector of Q_2 (point D) is in phase with the signal at its own input (the emitter)—and with the input signal at the base of Q_1. So an ac signal applied at the base of Q_1 appears amplified and in phase at the collector of Q_2. One more point: at first thought, it might seem that the output at D would be greater than at C

---

*Emitter-coupled circuits (emitter followers) will be discussed in detail in Chapter 16.

because it passes through both transistors. This is not so; the outputs are equal. Q_1, as an emitter follower, has a maximum (ideal) gain of 1; the only gain in this signal path is from Q_2.

Now let us consider an ac signal applied at the external input (the emitter) of Q_2. By symmetry, it can be seen that this action follows the above discussion. Outputs will appear at C and D—the output at C in phase and the output at D inverted compared to the input signal. With two inputs (at A and B), the outputs (with respect to ground) are

$$\text{Output } D = A_d(v_1 - v_2) \tag{14-2}$$

$$\text{Output } C = A_d(v_2 - v_1) \tag{14-3}$$

where v_1 and v_2 are the input signals to Q_1 and Q_2, respectively, and A_d is the voltage gain of the differential amplifier.

Notice in Fig. 14-8 that we can use either, or both, of two input terminals (A and B) and that we can also use either or both of the two output terminals (C and D). This gives us a variety of operating conditions:

1. *Single-ended input.* One input signal is fed to one input terminal (A or B) and ground; the other input terminal is grounded.
2. *Dual input.* We have two input signals (with respect to ground). One input signal is fed between each input terminal (A and B) and ground.
3. *Differential input.* We have one input signal. It is applied between terminals A and B. Obviously, in this connection neither line of the input signal can be grounded. Otherwise this would revert to condition 1—single-ended input.
4. *Common-mode input.* We have only one input signal. It is applied simultaneously between terminal A and ground and between terminal B and ground. In other words, terminals A and B are tied together and the same signal is fed to both.
5. *Single-ended output.* The output is taken from across point D and grounded.
6. *Double-ended output.* Two output voltages—180° apart—are obtained: one from point C and ground, the other from point D and ground. This is suitable for driving a balanced line or a push-pull amplifier (discussed in later chapters).
7. *Differential output.* A single output is taken from across points D and C. The output is then $A_d(v_1 - v_2) - A_d(v_2 - v_1)$. Combining and simplifying, this becomes

$$\text{Output } D - C = 2A_d(v_1 - v_2)$$

Notice that whether we use single-ended or dual inputs, the differential output is twice as much as we would get from single-ended output. However, neither terminal of the output signal can be grounded. In certain applications this is desirable.

INVERTING AND NONINVERTING AMPLIFIER.   In this case, the differential amplifier is used with single-ended input and single-ended output. The output is always taken from terminal D. Notice from Equation 14-2 that the output will be $A_d(v_1 - v_2)$. But for single-ended input we will use only one input signal. Let us feed this input signal to terminal A and ground terminal B. Obviously, v_2 will be zero and the output becomes $A_d(v_1)$. This output voltage is in phase with the input. Our amplifier does not invert the input signal and is therefore called a *noninverting amplifier*. More specifically, terminal A of this circuit is called the *noninverting input terminal*.

What will happen if we connect the input signal to terminal B? This means that we ground terminal A and v_1 becomes zero. The output—still from point D (and ground)—becomes $A_d(-v_2)$. The minus sign on v_2 shows that the output is inverted (or 180° from the input signal). Terminal B is marked as the *inverting input terminal*. Therefore, when using single-ended input and output we have a choice in the polarity of the output voltage by selecting either terminal A or B as the input terminal.

COMMON-MODE REJECTION RATIO.   If we tie terminals A and B together and feed the input signal between this common junction and ground, we are operating in the common-mode input connection. Obviously, $v_1 = v_2$, and the output, a function of $(v_1 - v_2)$, is zero—*if both halves of the circuit are identical*, i.e., perfect matching. Naturally, such a connection would not be used for a desired signal amplification, but it can be used as a figure of merit for the differential amplifier. For example, let us designate the voltage gain of the common-mode input circuit by A_c. For a perfectly matched pair (Q_1 and Q_2), this gain should ideally be zero ($A_c = 0$). In a practical situation, however, any slight mismatch in the two halves of the circuit will result in some small output voltage and A_c will have some finite value. The lower this value A_c—compared to the normal differential amplifier voltage gain A_d—the better the quality of the amplifier. The ratio of these two gain values serves as a figure of merit and is called the *common-mode rejection ratio* (CMRR):

$$\text{CMRR} = \frac{A_d}{A_c} \tag{14-4}$$

On manufacturers' data sheets, this rejection ratio is generally expressed in decibels. Since gains are voltage ratios, we can convert to decibels by using

$$\text{CMRR (in decibels)} = 20 \log \frac{A_d}{A_c} \tag{14-5}$$

EXAMPLE 14-2

A differential amplifier produces an output voltage of 15 V when fed with a 0.1-V single-ended input and produces only 0.005 V when this voltage is applied to both inputs. Find:

1. The normal differential amplifier gain
2. The common-mode gain
3. The common-mode rejection ratio (both as a number and in decibels)

Solution

1. Normal gain:

$$A_d = \frac{V_o \text{ (normal)}}{V_i} = \frac{15}{0.1} = 150$$

2. Common-mode gain:

$$A_c = \frac{V_o \text{ (CM)}}{V_i} = \frac{0.005}{0.1} = 0.05$$

3. Common-mode rejection ratio:

$$\text{CMRR} = \frac{A_d}{A_c} = \frac{150}{0.05} = 3000$$

$$= 20 \log 3000 = 69.6 \text{ dB}$$

Multistage commercial amplifiers—using integrated circuitry—are available with rejection ratios as high as 1 million, or 120 dB. Such CMRR ratios represent high-quality amplifiers. However, as mentioned above, we would obviously not connect a desired signal to both inputs. Then of what value is this mode of operation? It eliminates undesired signals. Quite often an undesired signal is picked up by the amplifier. This is particularly bad when the desired signal level is very low (low signal-to-noise ratio). For example, the signal from a velocity microphone, tape recorder, or magnetic phono pickup is very low (a few millivolts). The first stage of an amplifier fed from these sources has a very weak desired signal. Yet it may also pick up an even stronger undesired interfering signal—such as static from a nearby motor or even ripple from its own power supply. These interfering signals are picked up equally by both inputs, i.e., in the common-mode operation. If the rejection ratio of the amplifier is high, the interference will be minimized while the desired signal is amplified.

EXAMPLE 14-3

A differential amplifier with a gain $A_d = 150$ is fed with an input signal of 0.01 V. It also picks up, as a common-mode input, a 5-mV ripple voltage.
1. Find the signal-to-noise ratio at the input.
2. This amplifier has a CMRR of 60 dB. Find the signal-to-noise ratio at the output.

Solution

1. Input signal-to-noise ratio:

$$\frac{S_i}{N_i} = \frac{0.01}{0.005} = 2:1$$

2. Output signal-to-noise ratio:

   (a) With CMRR of 60 dB,

$$20 \log \frac{A_d}{A_c} = 60$$

$$\log \frac{A_d}{A_c} = 3$$

$$\frac{A_d}{A_c} = 10^3$$

$$A_c = \frac{A_d}{10^3} = \frac{150}{10^3} = 0.15$$

   (b) Output signal $= V_i \times A_d = 0.01 \times 150 = 1.5$ V

   (c) Output noise $= V_{i(N)} \times A_c = 0.005 \times 0.15 = 0.75$ mV

   (d) $\dfrac{S_o}{N_o} = \dfrac{1.5}{0.75 \times 10^{-3}} = 2000:1$

Notice in this example how the rejection quality of the differential amplifier has minimized the hum level of the power supply ripple voltage from an extremely annoying level (half as strong as the signal) to a negligible value.

FET DIFFERENTIAL AMPLIFIER.   Although the above discussion has used bipolar transistors, it applies equally well to field-effect transistors. A FET circuit has the added advantage of higher input impedance. A differential amplifier using FETs as the "amplifying" transistors is shown in Fig. 14-9. This circuit also incorporates a BJT with zener diode bias as the constant-current source.

MULTISTAGE DIFFERENTIAL AMPLIFIERS.   A multistage dc amplifier can be obtained by cascading two (or more) differential amplifiers. Depending on the type of input signal, the first stage could have single-ended or differential input. (Most often it is single-ended.) Similarly, depending on what the last stage feeds, the output could be single-ended or push-pull. A dual-stage differential amplifier is shown in Fig. 14-10. Each stage uses three transistors and a zener diode. The third transistor and the zener diode form the constant-current source. This circuit can be used with single-ended, differential, or balanced (dual) input voltages. For single-

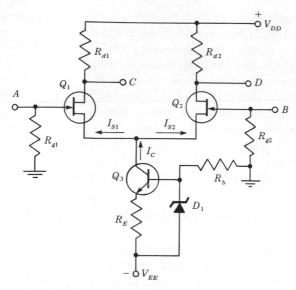

Figure 14-9   A differential amplifier using FETs.

ended operation, terminal B is grounded. Notice that the first stage uses double-ended output to feed a differential input signal to the second stage. The output from the second stage is single-ended. Note also that the collector resistor for Q_4 has been omitted. This is permissible, since no output is taken from this transistor.

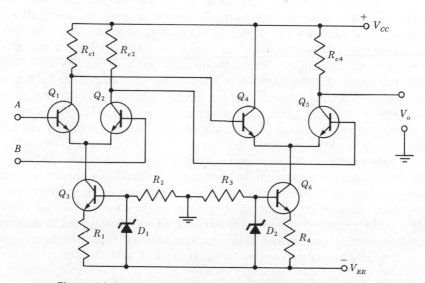

Figure 14-10   A two-stage direct-coupled differential amplifier.

Although such circuits were known, and were possible, many years ago—with tubes as well as with discrete semiconductor components—they were never popular. Close-to-perfect matching was impossible with discrete semiconductors. Adding balancing components increased the cost, increased the bulk, and lowered the gain. However, with the advance in the art of integrated circuitry it has become possible to achieve excellent matching, low cost, small bulk, and multistage circuits. Since about 1970 these ICs have been used in greater and greater quantities, in both industrial and consumer products, as analog or linear amplifiers. A name given to these multistage integrated circuits is *operational amplifier* (*op amp*).* The basic circuit in the op amp is still the differential amplifier.

CURRENT-DIFFERENCING AMPLIFIER

In the above differential amplifier (and in the op amps using this basic circuit), the output is proportional to the difference in the two input voltages. They could have been called voltage-differencing amplifiers. Notice also that these circuits have always been shown with two supply voltages, $+V_{CC}$ and $-V_{EE}$. In other words, they require a dual or split power supply, such as ± 18 V. Yet many electronic circuits are designed to operate from a single power source. We can use the "conventional" differential amplifier (and op amp) from a single power source by grounding the V_{EE} feed, but this impairs the action of the constant-current source and results in reduced performance characteristics.

The current-differencing (CD) amplifier was designed to meet the needs of low-cost, single-power-supply control systems. Basically, it is a high-gain CE amplifier with a variety of circuitry added to give it stability and flexibility. The schematic of "one stage" of a CD amplifier is shown in Fig. 14-11. Q_1 is the main CE amplifying unit. For high gain, its equivalent load resistance (R_e) should be high. First of all, this requires that the normally used dc load resistor (R_c) should have a high value. Notice that there is no physical resistor here. In its place it uses another BJT, Q_2, acting as a constant-current source. (The bias applied to the base terminal sets the effective value of this resistance.)

Now notice that Q_1 is directly coupled to the base of Q_3. Therefore, the input resistance of Q_3 is the ac load on Q_1. To make this input resistance a high value, Q_3 is connected as an emitter follower. Also, for high input resistance the normally used emitter resistor for Q_3 should be a large value. Here again, instead of a physical resistor we see a BJT, Q_4, used as another constant-current source; and again the base bias sets this resistance value.

An input signal current fed to the base of Q_1 will be inverted, passed

*These circuits will be discussed in more detail in Chapter 19.

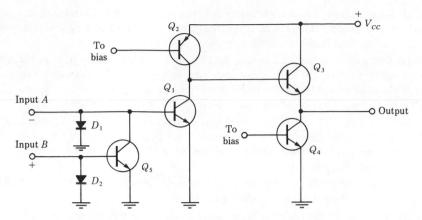

Figure 14-11   Current-differencing amplifier.

on to the emitter follower Q_3, and will reach the output 180° out of phase with the original input signal. Therefore, terminal A is the inverting input terminal. The CE amplifier has only this one input. But a general-purpose amplifier requires two inputs: one for inverting (as above) and another as the noninverting input. This second function is obtained by adding transistor Q_5 and diode D_2. This combination is known as a *current-mirror circuit*. The gain of Q_5 is fixed at unity, so that a (+) input current from terminal B will reach Q_1 (base) with the same polarity and gain as a (−) input fed directly into Q_1 from input A. Hence Q_5 gets the name *current mirror*.

Because this amplifier operates on the difference between two *currents*, it is also called a *Norton amplifier*. A commercial application of this circuit is the LM3900. On a single chip it contains four of these CD stages plus an additional 12 transistors, 4 diodes, and 10 resistors to provide bias, clamping, referencing voltages, and other regulating and stabilizing features.

REVIEW QUESTIONS

1. Refer to Fig. 14-1: **(a)** How does this circuit differ from an R-C coupled circuit? **(b)** What name is given to the inductor in the plate circuit? **(c)** Explain the advantage of using impedance coupling, as far as the power supply requirements are concerned.

2. **(a)** In an impedance-coupled triode amplifier, what range of choke reactance values should be used? **(b)** What is the approximate gain available when using such values? **(c)** Why do such circuits tend to have poor low-frequency response?

3. **(a)** Is impedance coupling popular in modern untuned voltage amplifiers? Give two reasons. **(b)** Is this type of coupling used at all? Explain.

4. **(a)** What advantage does transformer coupling have over R-C coupling when used with low-mu triodes? **(b)** What disadvantage does it have? **(c)** Is this type of coupling popular in modern untuned voltage amplifiers? Explain.

5. **(a)** Can gain be obtained from an R-C coupled amplifier using BJTs in the CB connection? Explain. **(b)** Can transformer coupling correct this situation? Explain. **(c)** When used with BJTs, is the transformer used as a step-up device? Explain. **(d)** What is the basic function of the transformer in BJT interstage coupling?

6. Refer to Fig. 14-2: **(a)** What component determines the value of the impedance Z_1? **(b)** What other quantity affects this value? **(c)** In addition to the turns ratio of the transformer, what other parameter sets the value of Z_2? **(d)** What does the transformer impedance ratio depend on?

7. Refer to Example 14-1: **(a)** What is the impedance seen by the 310-Ω load? **(b)** By the 5000-Ω load? **(c)** What is this effect called?

8. Refer to Fig. 14-3b: **(a)** What is the name given to the way the bias is applied to the base of this transistor? **(b)** What is the function of C_1? **(c)** Should it be a large or a small capacitor? Explain.

9. Refer to Fig. 14-3c: **(a)** Does this circuit use series-fed or shunt-fed bias? Explain. **(b)** Why does this circuit use a $+V_{CC}$ instead of a $-V_{CC}$ as in diagram (b)?

10. Refer to Fig. 14-4: **(a)** What is the emitter potential (voltage with respect to ground)? **(b)** Is the emitter positive or negative with respect to the base? **(c)** Is this forward or reverse bias? **(d)** Explain why the bias voltage—from the voltage divider—is not fed directly to the emitter. **(e)** How can we obtain forward bias and also feed the bias directly to the emitter?

11. **(a)** When are direct-coupled amplifiers necessary? **(b)** Why?

12. Refer to Fig. 14-5: **(a)** What is the gate bias on FET Q_1? Explain. **(b)** What makes the drain potential of Q_1 +23 V? **(c)** What is the value of V_{DS} for Q_1? **(d)** What is the value of the gate bias for Q_2? Explain. **(e)** What is the drain potential for Q_2? Explain. **(f)** What is V_{DS} for Q_2? Explain. **(g)** Compare the gate bias and drain-source voltage for each transistor.

13. In Fig. 14-5, the drain current of FET Q_1 for some reason decreases, so that the voltage drop across R_{d1} decreases from 30 V to 26 V. **(a)** What is the drain potential of Q_1 now? **(b)** What is the gate bias of Q_2? **(c)** What may happen to Q_2? **(d)** Why are the applications of this type of circuit limited?

14. Refer to Fig. 14-6: **(a)** With the potentials shown, what are the values of V_{EE} for Q_1 and Q_2? **(b)** What are the values of V_{CE} for Q_1 and Q_2? **(c)** How will their collector current values compare? Why? **(d)** Since Q_1 and Q_2 are fed from a common power supply, account for the difference in their collector potentials. **(e)** Account for the difference in their emitter potentials.

15. Refer to Fig. 14-7: **(a)** What is the collector-base voltage V_{CB} for Q_1? **(b)** In Q_2, to maintain the same V_{CB} as for Q_1, shouldn't the collector potential be +26 V? Explain. **(c)** Compare the operating voltages (V_{BE} and V_{CE}) for each stage.

16. Why is stability such an important factor with all of the above direct-coupled amplifiers?

17. Refer to Fig. 14-8: **(a)** What does the small circle marked I_E represent? **(b)** In its simplest form, what might this be? **(c)** What would constitute a better constant-current source? **(d)** With no input to either Q_1 or Q_2, what happens to current I_E? **(e)** When signals are applied to the inputs of Q_1 and/or Q_2, what happens to the total value of I_E? **(f)** If the input at A causes I_{E1} to increase, what happens to I_{E2}?

18. Refer to Fig. 14-8 again: **(a)** An input signal is applied between point A and ground. Terminal B is grounded. What is this input condition called? **(b)** Reverse the setup. Apply the signal to B and ground. Ground terminal A. What is this condition called? **(c)** An input signal is applied between A and B. What is this called? **(d)** Explain common-mode input. **(e)** If we want a single-ended output, specify the output connections. **(f)** Repeat part (e) for differential output. **(g)** Which connection (single-ended or differential) provides the larger output voltage? **(h)** What is a limitation to this connection? **(i)** If we want two output voltages, 180° apart, can this circuit supply them? Explain.

19. We wish to use the circuit of Fig. 14-8 as a noninverting amplifier. **(a)** Where do we connect the input? **(b)** From where do we take the output?

20. Repeat Question 19 for an inverting amplifier.

21. **(a)** In an ideal case, how much gain do we get from a differential amplier in the common-mode connection? **(b)** What term is used to express the nearness of a practical amplifier to the ideal case? **(c)** Give an equation for this quantity.

22. **(a)** When do we connect a differential amplifier in the common-mode connection? **(b)** Of what value is this mode of operation? **(c)** In Example 14-3, what is the effect of this mode on the sound that would have been heard from a loudspeaker connected at the output of this amplifier? **(d)** If a circuit uses this type of amplifier, is excellent power-supply filtering necessary? Explain.

23. **(a)** Can FETs be used in differential amplifier circuits? **(b)** What is an advantage of using FETs over BJTs?

24. Refer to Fig. 14-9: **(a)** What is the function of Q_3? **(b)** What is D_1? **(c)** What is its function?

25. Refer to Fig. 14-10: **(a)** How many stages does this circuit have? **(b)** Which transistors are in which stage? **(c)** The collector resistor for Q_4 is missing. Is this a design error? Explain. **(d)** How would we use this circuit for single-ended input? **(e)** Repeat part (d) for balanced inputs. **(f)** Can we use this circuit for double-ended output? Explain.

26. What is an *operational amplifier?*

27. Describe the type of power supply generally required for optimum efficiency of an operational amplifier.

28. Refer to Fig. 14-11: **(a)** Give an advantage of this circuit over a "conventional" operational amplifier. **(b)** How many "amplifier stages" are shown in this diagram? **(c)** Which transistor serves mainly as the amplifier? **(d)** What is the function of Q_2? **(e)** Is the output taken from Q_3 or from Q_4? **(f)** What type of circuit connection applies to Q_3? **(g)** What is the function of Q_5?

PROBLEMS AND DIAGRAMS

1. Draw the circuit diagram for an impedance-coupled triode amplifier.

2. What transformer turns ratio will be needed to match an 8000-Ω source to a 200-Ω load?

3. The circuit of Fig. 14-4 is used with a ceramic phono pickup cartridge (output impedance 500 kΩ). The input and output impedances of this stage are 120 Ω and 100 kΩ, respectively. The next stage has an input impedance (r_{i2}) of 95 Ω. Find the turns ratio required in each transformer to achieve the maximum transducer available gain (MAG).

4. A transistor with an ouput resistance of 4000 Ω is coupled to its load through a 3.5:1 step-down transformer. What value of load should be used as a match?

5. Draw the circuit diagram for an *NPN* transformer-coupled CE amplifier.

6. Repeat Problem 5 for a CB amplifier.

7. The characteristic curves for the FETs used in Fig. 14-5 are given in Fig. 10-19 (page 249). Find the values needed for R_{d1} and R_{d2} in the direct-coupled amplifier.

8. Draw the circuit diagram for a two-stage dc amplifier using FETs.

9. Repeat Problem 8, using *PNP* transistors.

10. Repeat Problem 8, using a *PNP* transistor in the first stage and an *NPN* transistor in the second stage.

11. Draw the circuit diagram for a differential amplifier using *PNP* transistors as the amplifying units and as the constant-current source.

12. A differential amplifier has a normal gain of 78, but when the input is connected in common mode the gain drops to 0.08. Calculate the common-mode rejection ratio as a number and in decibels.

13. A differential amplifier is fed an input signal of 0.08 V. The output is 9.3 V. When the same signal is applied in the common-mode connection, the output is 6.4 mV. Find: **(a)** the normal gain; **(b)** the common-mode gain; **(c)** the common-mode rejection ratio in decibels.

14. Because of severe static conditions, a tape amplifier with a desired input signal of 12 mV develops a 1.5:1 signal-to-noise ratio at its input. The amplifier gain is 120 and its CMRR is 40 dB. Calculate the signal-to-noise ratio in the output.

15. In Problem 14, what value of common-mode rejection ratio would improve the output signal-to-noise ratio to 3000:1?

16. Draw the schematic for a basic differential amplifier using *PNP* transistors.

15

Untuned Power Amplifiers

In the circuit discussions so far, the interest was mainly on gain—current, voltage, and even power gain—but the power *level* in these circuits was relatively low. A complete amplifier requires several such stages in cascade to build up the signal level. However, the ultimate goal is to deliver sufficient power to "operate" its load. Toward this end, the most important criterion is the *amount* of power delivered, not the gain. This function is accomplished in the last stage of the amplifier;* this stage is called a *power amplifier* since it handles high power levels. At this time a question may arise: How much power is "sufficient power"? That depends on the type of load. In an audio amplifier with a loudspeaker as the load, the power needed may range from less than 1 W for a personal radio to hundreds of watts for a theater or public address system. In modulation amplifiers for broadcast transmitters and in industrial electronic applications, the power level in the final amplifier may be in the thousands of watts.

POWER AMPLIFIER DEVICES

Theoretically, any of the active devices previously discussed could be used as power amplifiers. However, to achieve high power-ratings, high currents

*In very high power amplifiers, the previous stage (*the driver*) would be an intermediate power amplifier.

are required. High currents in turn mean higher I^2R losses and more heat to be dissipated by the device. In this respect, vacuum tubes are best, with bipolar transistors second and field-effect transistors in very poor last place.* Tubes for power amplifier service are designed with large heat-dissipating surfaces. Several such tubes were shown in Fig. 1-13. Note their physical size and the added cooling features: forced air for the 2.7-kW unit and water cooling for the 110-kW tube.

At moderate power levels, bipolar junction transistors have been replacing tubes. Unfortunately, BJTs have serious temperature limitations: if their temperature rise exceeds the maximum rating, transistor action will be ruined. Since most of the heat is developed at the collector junction (because of its higher resistance), commercial designs use special techniques to remove heat from this junction as quickly as possible. These techniques include attaching more massive metal structures (with fins) to the body of the transistor and provisions for bolting the transistor case directly to a larger piece of metal which acts as a *heat sink*. The heat sink then carries heat quickly away from the collector junction by conduction. Several power transistors are shown in Fig. 15-1. Figure 15-1a shows two plastic-encapsulated units. They are available with power ratings up to 90 W when the flanges are bolted to suitable heat sinks. The unit in Fig. 15-1b is in a TO-3 metal case and is available with power ratings to 150 W. Figure 15-1c shows two units with studs for direct bolting to the heat sink. Power ratings for this type of unit are as high as 300 W. Figure 15-1d to g shows transistors used with various types of heat sinks.

BJT DISSIPATION RATING.   Manufacturers' data sheets list the maximum power than can be dissipated by a transistor at some specified temperature. If the unit is operated at a higher temperature, the power rating should be decreased to compensate for the higher initial temperature. The amount of derating will vary with the unit's construction and method of mounting. Derating coefficients are also listed in the data sheets. For example, the 2N1050 has a 40-W rating with heat sink and a 1-W rating in free air, for a *case* temperature of 25°C. The derating coefficients are given as 228 mW per degree Celsius with heat sink and 5.7 mW per degree Celsius in free air.

EXAMPLE 15-1

The 2N1050 is used with heat sink in a circuit wherein it dissipates 31 W. The case temperature is found to be 75°C. Is the unit operating within its rating?

*Most FETs have power ratings under 1 W, although a few are available with ratings as high as 10 W. One manufacturer announced in mid-1975 a new development, the "vertical field effect transistor (V-FET)," with a maximum power rating of 63 W. However, since FETs are not generally used as power amplifiers, they will not be considered further in this chapter.

Figure 15-1   Bipolar junction power transistors [(d), (e), (f), (g): *International Electronic Research Corp.*]

Solution

1. The increase in temperature is $75 - 25 = 50°C$.
2. The derating at 228 mW per degree Celsius is $228 \times 10^{-3} \times 50 = 11.4$ W.
3. The maximum dissipation at 75°C should be $40 - 11.4 = 28.6$ W.

The unit is operating *above* its rating.

It was mentioned above that the dissipation rating is given for some specified temperature. With respect to the 2N1050 (Example 15-1), the specified temperature referred to the temperature of the transistor case (T_c). Sometimes the reference temperature is the *ambient* temperature (T_a); other manufacturers refer to the temperature of the junction itself (T_j). Obviously, if the transistor circuit is not energized all three temperatures will be alike, but if the transistor is operating at near its maximum rating there may be a wide difference in the three values, with the junction having the highest temperature and the ambient the lowest. For power transistors the case temperature is most generally specified. Of course, the actual limiting factor is the junction temperature. Data sheets therefore generally show the maximum allowable junction temperature. The junction temperature cannot be measured, but it can be calculated from the case temperature. The difference in their temperatures depends on the actual power dissipated and on the *thermal resistance* (θ_{j-c}) of the unit. This factor is listed in the data sheets and is expressed in terms of the rise in temperature per watt of dissipation. An example will show how this is used.

EXAMPLE 15-2

A 2N1157 has a thermal resistance θ_{j-c} of 0.7°C per watt. When used in a circuit wherein it dissipates 20 W, the case temperature is found to be 70°C. Calculate the actual junction temperature.

Solution

1. The temperature difference is

$$\Delta T = \theta_{j-c} \times P_d = 0.7 \times 20 = 14°C$$

2. The junction temperature is therefore

$$T_j = T_c + \Delta T = 70 + 14 = 84°C$$

POWER DISSIPATION CURVE. The power dissipation rating of a transistor is an *instantaneous* maximum value. This means that at no instant during a cycle should the product of the instantaneous values of collector current and voltage $(i_c v_c)$ exceed the dissipation rating. In the design or analysis of a power amplifier, such instantaneous overloads can be avoided with the help of a power dissipation curve plotted directly on the collector characteristic curves. Data sheets for power transistors generally include this curve. Figure 15-2 shows the collector characteristics for the 2N376A power transistor and includes the power dissipation curve. This transistor has a power dissipation rating of 40 W. If the dissipation curve is not given, it can be added quite readily. The coordinates of any point on the curve are such that the product of the V_{CE} and I_C values must equal the dissipation rating. Merely locate several such points, and draw a smooth curve con-

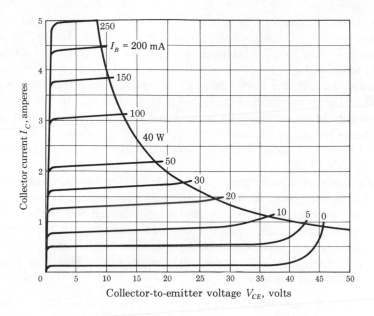

Figure 15-2 Power transistor collector characteristics showing power dissipation curve (*Motorola Semiconductors*).

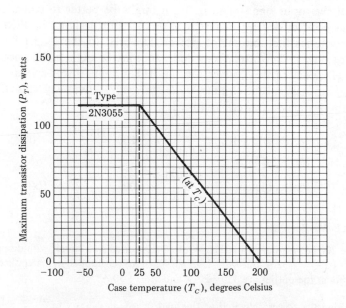

Figure 15-3 Power dissipation derating curve (*RCA*).

necting them. Notice in Fig. 15-2, for example, at $V_{CE} = 40$ V on the dissipation curve, the corresponding value of $I_C = 1$ A; also, at $V_{CE} = 20$ the corresponding I_C is 2 A. The product of the coordinate values in all cases is 40 W.

Unfortunately, the power dissipation capability of a transistor is also affected by the temperature of the unit—decreasing with a rise in operating temperature. Manufacturers' data therefore also includes derating constants (watts per degree Celsius above a referenced temperature) or derating curves. Figure 15-3 shows a typical derating curve for a 2N3055 rated at 115 W at 25°C.

Vacuum-tube power amplifiers are not limited by the instantaneous power dissipation. Instead, they can withstand instantaneous overloads as long as the average power per cycle does not exceed the rating. Consequently, power dissipation curves are not used. Furthermore, tubes operate at considerably higher temperatures than semiconductors, and their power dissipation ratings are based on these high temperatures. Therefore derating curves (or data) are not required.

COLLECTOR AND PLATE EFFICIENCY.　The title of this chapter— Untuned Power Amplifiers—is somewhat of a misnomer. The circuits to be discussed do not "amplify power" but rather convert dc power from the dc power supply into ac power delivered to the load—under the influence of the input signal. (Actually, it might be better to consider these circuits as control devices, wherein a small control signal applied to the input determines the amount of power delivered to the load.) Since this amount of power, converted from dc to ac, may be high, we are concerned with the efficiency of this conversion. In BJT circuits this is called *collector efficiency;* in vacuum-tube circuits, *plate efficiency.* Obviously, this efficiency is the ratio of the two powers involved:

$$\text{Collector efficiency} = \frac{P_o \text{ (ac)}}{P_i \text{ (dc)}} \times 100 = \frac{V_c I_c}{V_{CC} I_C} \times 100 \qquad \textbf{(15-1a)}$$

$$\text{Plate efficiency} = \frac{P_o \text{ (ac)}}{P_i \text{ (dc)}} \times 100 = \frac{E_p I_p}{E_{bb} I_b} \times 100 \qquad \textbf{(15-1b)}$$

EXAMPLE 15-3

The 2N3055 (power rating 115 W at 25°C) is used in a circuit with dc operating conditions of $V_{CC} = 32$ V and $I_C = 1.33$ A. When an ac input signal is applied, the ac components in the output are $V_c = 20$ V and $I_c = 1.30$ A.

1. Find the power delivered to the load.
2. Find the collector efficiency.

3. Find the power dissipated by the transistor.
4. Is this transistor operating within safe limits?

Solution

1. Power delivered to the load: since the ac components are rms values,

$$P_o = V_c I_c - 20 \times 1.30 = 26.0 \text{ W}$$

2. To find the collector efficiency:

   (a) The dc power input is

   $$P_{dc} = V_{CC} \times I_C = 32 \times 3.5 = 112 \text{ W}$$

   (b) Collector eff $= \dfrac{P_o \text{ (ac)}}{P_i \text{ (dc)}} \times 100 = \dfrac{26.0}{112} \times 100 = 23.2\%$

3. The power dissipated (internally) in the transistor is

$$P_{diss} = P_i - P_o = 112 - 26.0 = 86.0 \text{ W}$$

4. At a case temperature of 60°C, the transistor rating is reduced to 85 W (see Fig. 15-3). The transistor is operating in a slightly overloaded condition. This can be dangerous.

CLASSES OF OPERATION

When the power level is low, as in voltage amplifier stages, efficiency of operation is relatively unimportant and fidelity or minimum distortion level is the prime consideration. For this reason, voltage amplifiers are always operated in class A. Since output current flows for a full cycle, distortion can be held to extremely low values. For similar reasons, low-power amplifiers are also operated in class A. Efficiency, however, is poor, generally below 25 percent. As the required power level increases, higher efficiencies become more desirable and operation is shifted to class AB and B. With class B operation, efficiencies up to 60 percent and higher are obtained.

CLASS A OPERATION.   For class A operation, output current must flow for the full cycle of input. Obviously the input circuit, the base in a CE amplifier (or the grid in a tube amplifier), should not be driven below the cutoff value nor beyond the saturation value.* The solution is to bias the transistor (or tube) in the center of the straight-line portion of its dynamic characteristic curve. Then, by maintaining the signal swing within the straight-line portion, distortion is avoided. Class A operation is shown

---

*With vacuum tubes, the grid should not be driven positive. In other words, grid current does not flow. This operating condition is called class A_1.

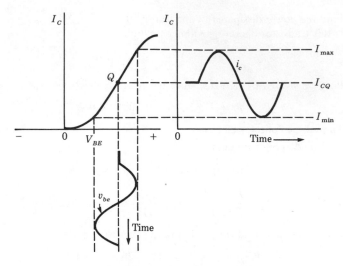

Figure 15-4   Class A operation—*NPN* transistor.

in Fig. 15-4 for an *NPN* transistor. Notice that collector current flows for the full input cycle, or 360°. Note also the high value of collector current I_{CQ}, which flows even during the no-signal condition. This means that the full dc power input is supplied even when there is no power output. This accounts for the low efficiency of class A operation. What is even worse is that since there is no output, this full dc power input must be dissipated by the transistor itself. In other words, the transistor power rating must equal the full dc power input.

CLASS AB OPERATION.   Using an *NPN* transistor and setting the dc base bias at a somewhat less positive value (nearer to cutoff), the no-signal collector current I_{CQ} can be reduced materially. This improves the efficiency of operation. Unfortunately, a full signal swing, such as to drive the base up toward saturation, will also drive the base past the low-current area and into cutoff. Collector current will cease for some portion of the input cycle. This operation is called class AB and is shown in Fig. 15-5. The following characteristics should be noted for class AB operation:

1. The base is never driven into saturation.
2. The base is driven beyond cutoff for a portion of the cycle.
3. Collector current flows for more than half but less than one full cycle of input (between 180 and 360°).
4. A lower forward bias is used, compared to class A.
5. The no-signal collector current is lower than for class A, resulting in higher operating efficiency.

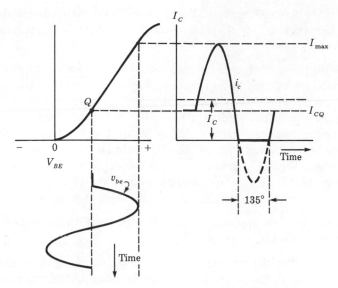

Figure 15-5 Class AB operation—*NPN* transistor.

6. A greater signal swing (V_{be} amplitude) can be used, resulting in a higher collector current swing (I_c) and greater power output ($I_c^2 R_L$).
7. The unequal collector current swings, and the cutoff of part of the negative half-cycle, produce a high distortion level.

The lack of symmetry in the collector current waveshape indicates that this distortion is mainly second (and other even) harmonics. This type of distortion can be eliminated by using two transistors in what is known as a *push-pull* circuit. (These circuits will be discussed later in this chapter.) Class AB amplifiers should be used only in push-pull circuitry.*

CLASS B OPERATION.   In *untuned* power amplifiers, maximum efficiency and power output are obtained when the transistor or tube is operated in class B. *Theoretically,* the transistor forward bias is set at zero and collector current flows for exactly one half-cycle of input. (Similarly, using

---

*With vacuum tubes, class AB is further divided into two subclassifications. Class AB_1 designates a class of AB operation wherein the grid signal input does not drive the tube into the positive-grid region. On the other hand, if the grid signal amplitude is increased, the grid will be driven positive and grid current will flow. This operating condition is specified as AB_2. The greater signal swing with class AB_2 results in higher power output, so that a given tube operated in class AB_2 can deliver more power output than in class AB_1. However, the distortion level will increase. Again, push-pull operation is required to reduce distortion. Another problem in class AB_2 is that because grid current flows, power is dissipated in the grid circuit. This power must be supplied by the previous stage, and to minimize distortion this "driver" stage must have good regulation (i.e., low resistance).

tubes, they are biased—theoretically—at cutoff and plate current flows for only one half-cycle of input.) This class B condition is a theoretical one for two reasons:

1. With transistors, even at zero bias, or supposedly at cutoff, some collector current flows because of minority carriers (I_{CEO} or I_{CBO}).
2. Near cutoff—in the low-current region—the dynamic transfer characteristic of the transistor (or tube) is curved. Operation in this region produces high distortion. On the other hand, if the transistor is operated with a slight forward bias (or the tube is operated with a slightly less negative bias), operation is taken out of the high-curvature area, eliminating what is known as *crossover distortion*. In practical situations, the bias is set at *projected cutoff*.

Figure 15-6 shows the operating conditions for an *NPN* transistor in class B. Several facts should be noted:

1. Notice that the dc bias is set at the point where the *continuation* of the straight-line portion of the dynamic transfer characteristic intersects the x axis. This is projected cutoff.
2. The quiescent (no-signal) value of collector current is very low— practically zero. This makes the quiescent dc power input very low and greatly improves efficiency.
3. Collector current flow is reduced to approximately one half-cycle. The distortion level is high, and again push-pull operation is required.

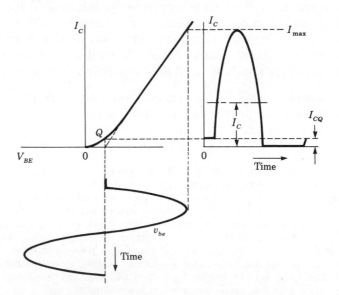

Figure 15-6 Class B operation—*NPN* transistor.

4. The base is driven deeper toward saturation, and higher input driving power is required from the previous stage.
5. Notice the drastic change between the no-signal dc collector current (I_{CQ}) and the dc component of collector current (I_C) when the input signal is applied. To maintain a constant V_{CE} with this heavy current increase, good power supply regulation is essential.

These five conditions also apply to vacuum-tube circuits.

If we compare the three classes of operation, we will find that class A operation has the least distortion but the poorest collector efficiency and the lowest power-handling ability. At the other extreme, a given unit in class B operation produces the most distortion but has the highest efficiency and the greatest power-handling ability. For portable equipment, or whenever economy of operation is a prime consideration, class B operation is preferable.

CLASS C OPERATION.   To complete the discussion on classes of operation, class C is included here. With this class of operation, it is possible to obtain still greater efficiency and power output. However, it must be clearly understood that class C operation is *not* used for nonresonant or untuned power amplifiers. The transistor (or tube) is now operated at a bias that puts the device deep into cutoff. Current flow is for less than half-cycle—around 120°. The distortion level is much greater and includes a high percentage of third and other odd harmonics. Push-pull operation will not reduce distortion to acceptable limits. This class of operation can be used only with resonant or tuned power amplifiers.

CLASS A POWER AMPLIFIER

Two class A vacuum-tube power amplifiers are shown in Fig. 15-7. The triode amplifier stage is shown with transformer coupling to the previous stage; in Fig. 15-7b, R-C coupling is used between the beam-power tube and its previous stage. It should not be inferred that either tube is limited

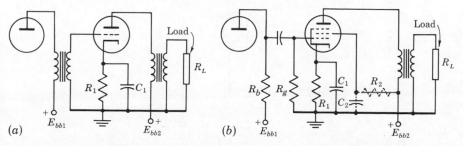

Figure 15-7   Class A power amplifier. **(a)** Triode tube with transformer-coupled input. **(b)** Beam-power tube with R-C coupled input.

to a specific type of input coupling. The diagrams could just as well have shown R-C coupling with the triode or transformer-coupled input to the beam-power tube.

Notice the series resistor and bypass capacitor shown in the screen grid circuit of the beam-power tube. These components are shown dotted because they may or may not be found in an actual circuit. Their presence or absence depends on the relative plate and screen potentials desired. As for the other circuit components, their function has been covered in previous sections and no further discussion is needed at this time. In vacuum-tube circuits, the power amplifier is generally transformer-coupled to the load. You will recall that this provides impedance matching between tube and load for maximum power output.

On the other hand, transistor power amplifiers may be transformer-coupled or direct-coupled to their loads. These two variations are shown in Fig. 15-8. Figure 15-8a shows a complete circuit with transformer-coupled load; the partial schematic in Fig. 15-8b shows the direct-coupled load. Since the power supply, load, and transistor are all in series, this type of circuit is also called a series load.

There are pros and cons regarding both types of load connections. Let us first mention the more obvious considerations. To avoid saturation, the iron core of the transformer must be of generous size. For good low-frequency response, the transformer primary winding must have a high X_L

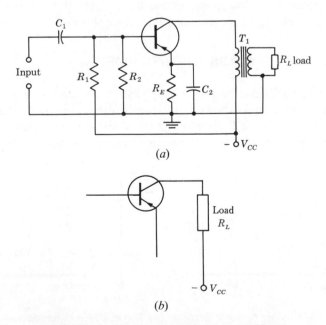

Figure 15-8 Class A transistor power amplifiers.

at the lowest desired signal frequency. This means many primary turns. For good high-frequency response, the transformer windings must avoid distributed shunt capacitance. All of this means that for high-fidelity applications the transformer is heavy, bulky, and expensive. For example, a high-fidelity 15-W vacuum-tube audio amplifier—vintage 1940—could have an output transformer $3\frac{1}{2} \times 4\frac{1}{2} \times 4\frac{1}{2}$ in. and weighing close to 10 lb.* With modern IC technology, a complete 15-W amplifier (less power supply) would be smaller and lighter than this output transformer. Elimination of the output transformer is therefore very desirable. However, there is a penalty: proper impedance matching becomes virtually impossible without a transformer. A sacrifice must be made in the power output obtainable from a given transistor. In addition, what may not be so obvious is that the direct-coupled load requires a higher power-supply voltage for the same operating condition—and, even more detrimental, it causes higher power dissipation in the circuit. Consequently, although most consumer high-fidelity audio amplifiers use transformerless circuits, commercial and industrial audio amplifiers and other high-power applications use matching transformers for coupling to their loads.

The last points—the need for higher power-supply voltage and the higher power dissipated in the circuit—may need clarification. Refer to Fig. 15-9. This shows a series-connected load (partial schematic) and a graphic analysis of the operating conditions. The load line is drawn in the usual manner as explained in Chapters 10 and 11. This time, the one line is both the dc and the ac load line. (There are no other paralleling resistors.) The two "operating" ends of this load line are labeled A and B. Limit A is set by the characteristic curves. Operation should not extend to the left of these curves since this is in the saturation area and would cause heavy distortion. Limit B is set by the supply voltage; that is, $V_{CE} = V_{CC}$. The operating point Q is selected halfway between A and B, since this will allow maximum signal swing and will produce maximum undistorted power output. Notice that—neglecting the small saturation voltage—the peak-to-peak value of the output voltage is from zero to V_{CC} and that the peak value $V_{cem} \cong \frac{1}{2}V_{CC}$. In rms terms, the maximum output voltage that can be obtained is $V_{ce} = 0.707V_{CC}/2$. Similarly, the peak value of collector current is from zero to I_{CQ}, or in rms value $I_c = 0.707I_{CQ}$. From these relations,

$$\text{dc power input} = P_{dc} = V_{CC} \times I_{CQ}$$

$$\text{ac power output} = P_o = V_{ce} \times I_c = \left(\frac{0.707V_{CC}}{2}\right)(0.707I_{CQ})$$

$$= \frac{V_{CC}I_{CQ}}{4} = \frac{1}{4}P_{dc}$$

---

*If this does not impress you, consider that the audio power amplifier matching transformer for a 100-kW AM broadcast transmitter weighs 2100 lb.

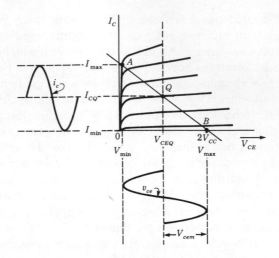

Figure 15-9 Operating conditions with a series-connected load (class A).

The remainder of the dc power input, $\frac{3}{4}P_{dc}$ is dissipated within the circuit. However, these relations apply only at maximum power output. If the input signal level is reduced, the collector current and output voltage swings will decrease and the power output will decrease; but since the dc power input remains constant, the power dissipated must increase. The worst-case condition occurs at no-signal input. The output is zero and the circuit must dissipate the full dc power input. If the load is matched to the transistor, this dissipation is split equally between transistor and load. More often, however, the load resistance is lower than the transistor output resistance, and the transistor dissipation can be appreciably more than half the dc power input. For safety, then, if the input signal level varies widely (as in audio applications), some designers recommend that the dissipation rating of the transistor should equal (approximately) the dc power input. Remember that this applies to class A amplifiers with direct-coupled loads.

EXAMPLE 15-4

A class A power amplifier with direct-coupled load has a collector efficiency of 20 percent and delivers a power output of 3.0 W. Find:
1. The dc power input
2. The power dissipation at full output
3. The power dissipated in the circuit with no input signal
4. The desirable power dissipation rating for the transistor

Solution

$$1. \ P_{dc} = \frac{P_o}{\text{Eff}} = \frac{3}{20} \times 100 = 15 \text{ W}$$

2. P_{diss} (at full output) $= P_{\text{dc}} - P_o = 15 - 3 = 12$ W
3. P_{diss} (at no input) $= P_{\text{dc}} = 15$ W
4. Transistor power dissipation rating $\cong P_{\text{dc}} \cong 15$ W

Now let us turn our attention to the transformer-coupled load. Figure 15-10 shows a partial schematic of this circuit and a graphic analysis of the operating conditions. For a direct comparison with the direct-coupled load, we will use the same collector characteristics, the same value of ac load resistance, and the same operating point (V_{CEQ} and I_{CQ}). This time we have both a dc load resistance and an ac load resistance. The direct-current path is from ground, through the transistor (emitter to collector), and through the primary winding of the output transformer to the positive power supply terminal. The dc load is the dc resistance of the primary winding. This is some rather small value, and for convenience we will consider it negligible. Therefore, the IR drop in this winding is also negligible and the collector-emitter voltage, V_{CEQ}, is equal to V_{CC}. The dc load line ($R_{\text{dc}} = 0$) will be a vertical line at the value of $V_{CE} = V_{CC}$. This is shown in Fig. 15-10b as the broken line through the operating point Q. The ac load line is drawn through Q at the same slope as in Fig. 15-9, since we are using the same value of ac load. As before, the input signal swing will cause operation to swing from Q to A and to B. The collector current swing (the i_c curve) and the output voltage swing (the v_{ce} curve) are the same as in Fig. 15-9. Notice that since the quiescent collector voltage $V_{CEQ} = V_{CC}$, the maximum positive swing is to $2V_{CC}$. Therefore the peak value of the output voltage, V_{cem}, is measured from V_{CEQ} (which is equal to V_{CC}) to $2V_{CC}$ (or $V_{cem} = V_{CC}$) and its rms value is $0.707V_{CC}$.

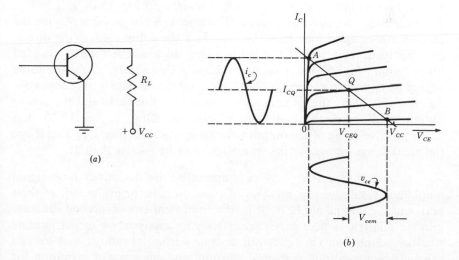

(a)

(b)

Figure 15-10 Operating conditions with a transformer-coupled load (class A).

Now let us analyze the maximum-swing power relations with a transformer-coupled load:

1. dc power input is

$$P_{dc} = V_{CC} \times I_{CQ}$$

2. ac power output is

$$P_o = V_{ce} \times I_c = 0.707V_{CC} \times 0.707I_{CQ}$$
$$= \tfrac{1}{2}P_{dc}$$

3. Power dissipated in the transistor is

$$P_{diss} = P_{dc} - P_o = P_{dc} - \tfrac{1}{2}P_{dc} = \tfrac{1}{2}P_{dc}$$

4. Collector efficiency is

$$\frac{P_o}{P_i} = \frac{\tfrac{1}{2}P_{dc}}{P_{dc}} \times 100 = 50\%$$

At the condition of maximum signal swing, we see that the power output is $\tfrac{1}{2}P_{dc}$ and the power dissipation in the transistor is also $\tfrac{1}{2}P_{dc}$. (Remember that with direct-coupled load the power output was $\tfrac{1}{4}P_{dc}$ and the dissipation $\tfrac{3}{4}P_{dc}$.) This proves the point made earlier: direct-coupled loads cause higher power dissipation.

When we started this discussion on the transformer-coupled load, we stated that we would use the same operating point for Figs. 15-9 and 15-10. This means that the quiescent values of V_{CEQ} were the same for both cases. But notice that in Fig. 15-9, $V_{CEQ} = \tfrac{1}{2}V_{CC}$; whereas in Fig. 15-10, $V_{CEQ} = V_{CC}$. To make these two operating points the same, it is obvious that V_{CC} for the transformer-coupled load must be only half the value used in the direct-coupled load. This proves the other point made earlier: direct-coupled loads require higher power-supply voltages. Conversely, using the same supply voltage the transformer-coupled circuit would have a quiescent voltage twice as high; the ac component would be doubled; and the power output would be doubled. A word of caution: notice in Fig. 15-10, that at the positive swing of the output voltage, v_{ce} rises to twice V_{CC}. Therefore, the breakdown voltage for the transistor must be greater than $2V_{CC}$.

GRAPHIC ANALYSIS.   Power amplifiers are essentially large-signal amplifiers. Consequently, analysis of these circuits, or evaluation of their performance, should not be done by the equivalent-circuit method. Instead, graphic analysis is necessary. From such an analysis we can determine whether a transistor (or tube) will operate within its ratings, and we can calculate power output, percent distortion, and efficiency of operation for any given operating conditions. The technique is best illustrated by examples. Since (except for very high power levels) transistors are generally used in new designs, we will use a transistor for our illustrations.

EXAMPLE 15-5

A 2N1068 is to be used as a class A power amplifier with the circuit of Fig. 15-8. The operating conditions are set at $V_{CC} = 20$ V, base bias $I_B = 8.0$ mA, and input signal current swing $= \pm 7.0$ mA. The load is an 8-Ω loudspeaker coupled through a 3.06:1 output transformer. The maximum ratings for this transistor are power dissipation 10 W, peak collector-to-emitter voltage 60 V, and peak collector current 1.5 A.

1. Will the transistor operate within its maximum ratings?
2. Find the percent distortion, assuming it to be mainly second harmonic.

Solution

The collector characteristics for this transistor are shown in Fig. 15-11. Since the collector is transformer-coupled to its load, we can neglect the low dc resistance of the primary winding and consider the dc load resistance as zero.

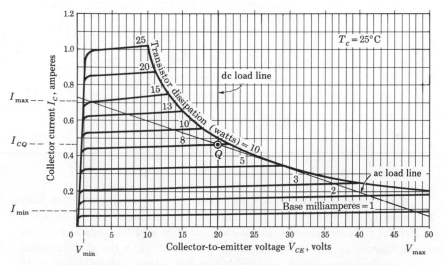

Figure 15-11 Graphic analysis—transistor power amplifier (*RCA*).

(a) For zero dc load resistance, the collector voltage V_{CE} is equal to the supply voltage V_{CC}. We can therefore draw the dc load line as a vertical line along the 20-V V_{CE} ordinate.

(b) Now locate the operating point Q at the intersection of this dc load line and the base current curve for 8.0 mA. The coordinates for this point give us the quiescent values of $I_{CQ} = 0.46$ A and $V_{CEQ} = 20$ V.

(c) The ac load line must go through this point. For a second point we can use $V_{CE} = 0$ when I_C increases such that $\Delta I_C R'_L = 20$ V, or $\Delta I_C = V_{CC}/R'_L$. The first step is to find the load value R'_L—as seen by the collector:

$$R'_L \text{ (collector side)} = R_L \left(\frac{N_1}{N_2}\right)^2 = 8\left(\frac{3.06}{1}\right)^2 = 75 \, \Omega$$

$$\Delta I_C = \frac{V_{CC}}{R'_L} = \frac{20}{75} = 0.276 \text{ A}$$

and

$$I'_C \text{ (for } V_{CE} = 0) = I_{CQ} + \Delta I_C = 0.46 + 0.267 = 0.73 \text{ A}$$

We can now draw the ac load line through these two points.

(d) For a signal swing of ± 7.0 mA, the operation extends *along the load line*, from Q (at $I_B = 8.0$ mA) up to $I_B = 15$ mA and down to $I_B = 1.0$ mA. The corresponding collector current and voltage values are

At $I_B = 15$ mA: $\qquad I_{max} = 0.71$ A $\qquad V_{min} = 2.0$ V

At $I_B = 1.0$ mA: $\qquad I_{min} = 0.09$ A $\qquad V_{max} = 48$ V

1. (a) No point along the load line falls to the *right* of the maximum dissipation curve. Therefore the dissipation rating is not exceeded.
   (b) The maximum voltage swing is 48.0 V and the maximum current swing is 0.71 A. Neither the current nor the voltage rating is exceeded.
2. The percent distortion can be found from the basic equation (13-9):

$$\text{Percent second} = \frac{\frac{1}{2}(I_{max} + I_{min}) - I_Q}{I_{max} - I_{min}} \times 100$$

Applying this equation to the transistor circuit,

$$\text{Percent second} = \frac{\frac{1}{2}(0.71 + 0.09) - 0.46}{0.71 - 0.09} \times 100$$

$$= 9.68\%$$

POWER OUTPUT FROM GRAPHIC ANALYSIS. In general, the greater the amplitude of the input signal, the greater the base current swing, the greater the collector current and voltage swings, and the greater the power output. However for class A operation, and to keep the distortion level down, the transistor should not be driven into cutoff or into the saturation region. Notice in Example 15-5 and Fig. 15-11 that we are just about at these limits. Once we have drawn the ac load line we can calculate power output. The swings from the minimum to the maximum values of current and voltage give us the peak-to-peak values of these quantities. (These are shown graphically in Figs. 15-9 and 15-10.) Also, assuming that these waveshapes are essentially sinusoidal (distortion level is low), these total swings correspond to twice the peak values of the ac components; that is, the peak values are

$$I_{c(max)} = \frac{I_{max} - I_{min}}{2}$$

$$V_{ce(max)} = \frac{V_{max} - V_{min}}{2}$$

Converting to rms values,

$$I_c = \frac{I_{c(max)}}{\sqrt{2}}$$

$$V_{ce} = \frac{V_{ce(max)}}{\sqrt{2}}$$

Replacing these maximum values with peak-to-peak swings,

$$I_c = \frac{I_{max} - I_{min}}{2\sqrt{2}} \tag{15-2}$$

$$V_{ce} = \frac{V_{max} - V_{min}}{2\sqrt{2}} \tag{15-3}$$

Then, since $P_o = V_{ce} \times I_c$ (rms values),

$$P_o = \frac{(V_{max} - V_{min})(I_{max} - I_{min})}{8} \tag{15-4}$$

EXAMPLE 15-6

Using the transistor power amplifier of Example 15-5, calculate:

1. The power output
2. The collector efficiency

Solution

1. The power output can be found from

$$P_o = \frac{(V_{max} - V_{min})(I_{max} - I_{min})}{8}$$

Substituting the transistor values,

$$P_o = \frac{(48.0 - 2.0)(0.71 - 0.09)}{8} = 3.56 \text{ W}$$

2. The collector efficiency can be found from

$$\text{Collector efficiency} = \frac{\text{ac power output}}{\text{dc power input}} \times 100$$

$$= \frac{3.56}{20 \times 0.46}$$

$$= 38.8\%$$

The collector efficiency in the above case could be improved by using a lower forward bias. This would reduce the dc value of collector current and the dc power input. Of course, if the signal swing drives the transistor into cutoff, class AB operation would result.

SELECTION OF OPERATING CONDITIONS.   In the above power amplifier examples we were given the transistor, the operating point, and the load value and were then asked to find the power output. Often, the converse is true. We know the power output we want, but we have to select the device and the operating conditions so as to obtain this output.

With vacuum tubes, the "design" is rather simple. Tube manufacturers, in their manuals and data sheets, not only give maximum ratings but also include "typical operation" or "characteristics." Under such headings they list dc plate and grid voltages, plate and grid currents, plate resistance, transconductance, load resistance, total harmonic distortion, and maximum-signal power output. Then, all the designer has to do is select a tube that will provide the desired power output within the tolerable percent distortion. The operating values—supply voltage, grid bias, load resistance—are given. Of course, if designers wish to experiment with other operating conditions, they can use a graphic analysis—such as we used for the transistor in Examples 15-5 and 15-6. However, they already have a good starting point: the manufacturer's recommended "typical operation" data. In addition, two design guides are generally used:

1. For triodes:
   (a) Maximum safe undistorted power output is obtained with load resistances of from two to four times the plate resistance.
   (b) A good dc grid bias is approximately $0.68E_b/\mu$.
2. With pentodes and beam-power tubes:
   (a) Since their r_p is so much higher than in triodes, optimum R_L values range from $\frac{1}{8}$ to $\frac{1}{10}r_p$.
   (b) The load line should pass close to the knee (but to the right) of the zero-bias plate characteristic curve, as the I_{max} point, and through an operating point so that $I_{BQ} \cong I_{max}/2$.

The question now is: Where do we start with transistors? Manufacturers do not provide "typical operation" data. All we have are the maximum ratings. A good starting point is the power dissipation rating. Remember that in class A the maximum theoretical collector efficiency is 25 percent for direct-coupled loads and 50 percent for transformer-coupled loads. In practice, however, these efficiencies are not realized. Efficiency values of 20 percent and 40 percent, respectively, are the most we can expect.

EXAMPLE 15-7

What power dissipation rating should a transistor have for class A operation at 4 W output if transformer-coupled to its load?

Solution

$$P_{diss} = \frac{P_o}{Eff} = \frac{4}{0.40} = 10 \text{ W or more}$$

With regard to power output, let us take another look at Fig. 15-11 and Equation 15-4. Notice that the load line forms a triangle with the collector current and voltage axes. Note also that to a very close approximation the power output is proportional to the product of the height of this triangle (I_{max}) and the base of the triangle (V_{max}). The greater the area under this triangle, the greater the power output. Figure 15-12a shows three load lines plotted on an I_C–V_{CE} set. Since the lines are parallel, they represent loads of equal resistance. (The broken-line curve shows the maximum power-dissipation limit of some given transistor.) Obviously, load line ① will produce the lowest power output, whereas load line ③ will

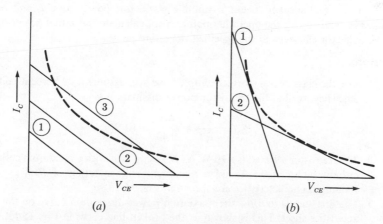

(a)   (b)

Figure 15-12   Load lines—power output comparisons.

yield the highest output. Notice, however, that load line ③ is at times *to the right* of the power dissipation curve. This means that for some portion of the input signal cycle, the instantaneous maximum power rating of the transistor is exceeded. This cannot be allowed. Maximum safe power output would be obtained by some line between load lines ② and ③—and tangent to the power dissipation curve.

Now examine Fig. 15-12*b*. Here we see two load lines, both tangent to the power dissipation curve but with different slopes. Load line ① is steep and is therefore a low-resistance load. Load line ②, being much flatter, represents a much higher resistance load. Obviously, any number of tangent lines can be drawn. Furthermore, a careful check will reveal that all tangent load lines produce the same triangle areas, or power output. Which load line should we use? Two additional limits should now be obvious. A line that is too steep will result in an I_{max} greater than the maximum collector current rating. Conversely, a line that is too flat will result in a V_{max} that exceeds the BV_{CEO} rating. This still leaves a rather wide margin for selection. With direct coupling we have no choice. The load line is fixed by the resistance of the load itself. With transformer coupling —allowing for a safe margin between V_{max} and BV_{CEO}—the flattest or highest resistance load line is preferred, since this results in lower currents and lower I^2R losses. (Also, operation at high currents requires higher base current drive, and the current-gain factor h_{fe} can decrease appreciably.) Impedance matching to the actual load is then made by selection of the proper output transformer turns ratio. Let us apply these ideas to a problem.

EXAMPLE 15-8

It is desired to construct a 4-W class A audio power amplifier transformer coupled to an 8-Ω speaker. Select a suitable transistor, power supply, and output transformer, and specify the operating point. Also, calculate the actual power output and the collector efficiency at the specified operating point.

Solution

1. For the class A transformer coupled, we will assume a 40 percent collector efficiency, so that the transistor power dissipation is

$$P_{diss} = \frac{P_o}{\text{Eff}} = \frac{4.0}{0.40} = 10 \text{ W}$$

2. The transistor selected is a 10-W *NPN* silicon unit with a maximum collector-current rating of 0.5 A and a maximum voltage rating of $BV_{CEO} = 30$ V. Its output characteristics are shown in Fig. 15-13.
3. The first step is to draw the maximum power-dissipation curve on the characteristics sheet. This is shown as the broken-line curve in Fig. 15-13.
4. To allow some safety margin, let us draw the ac load line starting at $V_{CE} = 25$ V and $I_C = 0$, passing just below tangency to the dissipation curve, and cutting

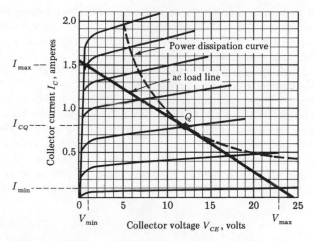

Figure 15-13   Graphic design—class A power amplifier.

the I_C axis at $I_C = 1.55$ A. We can calculate the resistance of this load line by using the full voltage swing from 0 to 25 V and the corresponding full current swing from 1.55 A to 0:

$$R_L' = \frac{\Delta V_{CE}}{\Delta I_C} = \frac{25}{1.55} = 16.1 \, \Omega$$

5. To minimize distortion, we will limit the maximum current swing to $I_{max} = 1.5$ A and $V_{min} = 1.0$ V. This keeps the operation out of the saturation area. (Notice that this also follows the pentode guideline: close to the knee, but to the right.) On the other end, to keep out of cutoff, we will limit the swing to an I_{min} of 0.1 A. Notice that V_{max} is now 23 V.
6. The operating point Q should be selected at about the center of this swing, or $V_{CEQ} = \frac{1}{2}(V_{max} + V_{min}) = \frac{1}{2}(23 + 1) = 12$ V. Mark point Q at the intersection of the $V_{CE} = 12$ V ordinate and the load line. The corresponding quiescent current value is $I_{CQ} = 0.8$ A.
7. Now let us check what power output we can get from the above operation:

$$P_o = \frac{(V_{max} - V_{min})(I_{max} - I_{min})}{8} = \frac{(23 - 1)(1.5 - 0.1)}{8} = 3.86 \text{ W}$$

Notice that we have missed our goal of 4 W, but only by 0.16 W (4 percent or 0.16 dB). This is negligible.*
8. Now for the supply voltage:
   (a) If we assume that the dc resistance of the output transistor primary winding is zero, then the supply voltage is the same as V_{CEQ}, or 12 V. The dc load line is the vertical ordinate line through $V_{CE} = 12$ V.

---

*If the full 4.0 W output is still desired, the operation can be redesigned either by using a transistor with a higher power-dissipation rating (such as 12 or 15 W) or in step 4 above by starting at a V_{CE} slightly higher than 25 V (for example, 28 V).

(b) If we wish to consider the dc resistance of the transformer primary, we proceed as follows. The transformer must "raise" the 8-Ω impedance of the speaker to 16.1 Ω. The impedance ratio is approximately 2:1, and the turns ratio is approximately 1.4:1. Looking up such a transformer, we find, for example, a primary resistance of 2.1 Ω. The IR drop in this winding due to the quiescent current is $0.8 \times 2.1 = 1.68$ V. The power supply voltage must compensate for this drop to maintain V_{CEQ} at 12 V. Therefore $V_{CC} = 12 + 1.68 = 13.68$ V. The dc load line passes through the quiescent point, Q, and intersects the abcissa at $V_{CE} = 13.68$ V.

9. To find the true turns ratio for the output transformer, consider that the collector should "see" a load of 16.1 Ω. But 2.1 Ω of this is due to the dc resistance of the transformer winding. Therefore, the reflected resistance from the secondary is only $16.1 - 2.1 = 14 \Omega$. The turns ratio is

$$\frac{N_p}{N_s} = \sqrt{\frac{R'_L}{R_L}} = \sqrt{\frac{14}{8}} = 1.32:1$$

10. The actual collector efficiency is

$$\text{Eff} = \frac{P_o \,(\text{ac})}{P_i \,(\text{dc})} = \frac{P_o \,(\text{ac})}{V_{CC} I_{CQ}} = \frac{3.86}{12 \times 0.8} = 40.2\%$$

Did you notice in Fig. 15-13 that the individual collector-characteristic curves are not identified by their corresponding base current values? In fact, in Example 15-8 we really did not use the collector output characteristic curves—except to fix the I_{max} value so as to avoid saturation. But this can also be done without these curves. Data sheets generally list $V_{CE(sat)}$ for several values of I_C. For the transistor in Example 15-8 at an I_C of 1.5 A, $V_{CE(sat)}$ is given as 0.8 V. So, had we worked from the data sheet, we probably would have added an extra margin for safety and still used 1.0 V as the limit point. Actually, it is just as well that this type of design can be handled without the characteristic curves, for two reasons:

1. Data sheets generally do not include such curves. In fact, short of obtaining your own curves by direct measurement, it is very difficult to obtain such published data.
2. Any such published set of curves is an "average" or "typical" set. It would probably not apply to any given transistor, and the deviation could be appreciable from one unit to another of the same type and from the same manufacturer. Finally, even if the curves did fit, a change in temperature would again cause deviations.*

From the data of Example 15-8, we cannot calculate the base current drive (swing) required for the full 4-W power output. For this, each of the

---

*These limitations do not apply to vacuum-tube circuits. Plate characteristic curves are readily available and apply to any tube of that type number. Graphic analyses are therefore accurate and valid.

collector characteristic curves in Fig. 15-13 must be identified with its base current value. Furthermore, if we wanted to know the input signal voltage requirement, we would also need *input* characteristic curves (I_B versus V_{BE}). Notice that no distortion calculation was made in Example 15-8. Again, the given data is not sufficient. Calculations for output distortion are based on collector current swings above and below the quiescent value—due to balanced swings of the base current (or input voltage). Instead we *selected* the amount of collector current we desired. Again, for such a calculation we would need collector characteristics with base current identification or transfer characteristic curves.

PUSH-PULL CLASS A OPERATION.   If the power output obtainable from one transistor (or tube) is not enough, it is often possible to select another device (usually larger in size and more expensive) with a higher power-dissipation rating. However, there are certain advantages in using two identical units in a *push-pull circuit*. Such a circuit, using transformer coupling, is shown in Fig. 15-14. Notice that the input and output transformers both have center taps. To analyze the circuit action, consider this circuit as two separate power amplifiers connected back to back and using a common base-bias voltage divider—R_1 and R_2. (You should recognize this circuit as a series-fed bias.) The $R_3 C_1$ and $R_4 C_2$ networks in the emitter legs provide emitter bias stabilization.

Let us trace the path of collector current flow under a condition of no-signal input. The collector current is dc. Starting from ground, current for Q_1 will flow through R_3, through Q_1, and *down* through the upper half of the transformer T_2 primary winding to the positive terminal of the power supply. In like manner current will simultaneously flow from ground through R_4, through Q_2, and *up* through the lower half of the transformer primary to the power supply. These dc components of collector current are shown in Fig. 15-14 by solid arrows. In the transformer primary, the collector current of each transistor tends to produce a magnetic field. But with matched transistors, since these two currents are equal and flowing in opposite directions, their effects cancel and the total magnetic field is zero. In other words, the dc components of collector current produce no magnetization. This is one advantage of push-pull operation: there is no danger of saturation due to a high dc component, and the size of the iron core of the transformer can be reduced. Meanwhile, even if not balanced, the net current flow is still dc and the output is zero.

Now consider the ac aspects. In the input circuit, notice that the center tap of the secondary winding of transformer T_1 is effectively grounded to ac. Therefore, with respect to ground the voltages produced at each end of the winding are 180° out of phase. This phase relation is shown by the sine-wave symbols alongside each base wire in Fig. 15-14. At some instant t_1, when the signal voltage at the base of Q_1 is maximum positive,

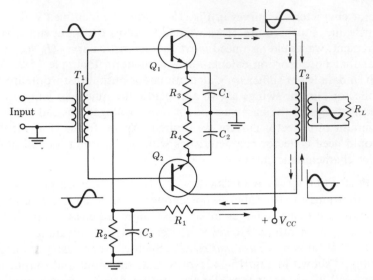

Figure 15-14 A transformer-coupled push-pull transistor power amplifier.

the signal applied to the base of Q_2 will be maximum negative. The collector current of Q_1 will increase. Since this increase is due to the ac component, it follows that this ac component must be flowing in the same direction as the dc component, or *down* through the upper half of the primary winding of transformer T_2. This ac component is shown in Fig. 15-14 as a dashed arrow. *At this same instant,* the signal at the base of Q_2 is at negative maximum and the collector current of Q_2 must decrease. This happens only if the instantaneous value of the ac component is *opposite* to the dc component; that is, the ac component of collector current must be flowing *down* through the lower half of the T_2 primary winding. (This current is also shown in Fig. 15-14 by a dashed arrow.) Notice that the ac components of collector current for both transistors flow in the same direction (down) through the output transformer primary winding. Their magnetic fields are additive. Since the flux is doubled as compared to either BJT alone, the secondary output voltage is doubled and twice as much power is delivered to the load.

The complete signal phase relations can be seen from Fig. 15-15. The base signals are 180° out of phase, because they were obtained from opposite ends of the center-tapped input coupling transformer. The resulting collector currents (i_c) are in phase with their own base voltages and are therefore also 180° out of phase with each other. Since the collector potential is a minimum when the collector current is a maximum (due to the $i_c R_L$ drop), each collector voltage (v_{ce}) must be 180° out of phase with its own collector current and also 180° out of phase with each other. The

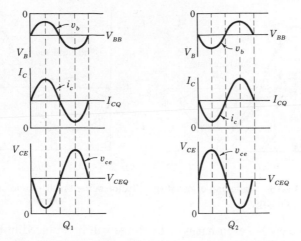

Figure 15-15 Phase relations in a push-pull amplifier.

output voltage across the load is shown in Fig. 15-14 as in phase with base signal to Q_1. This is not necessarily so. The polarity of this voltage depends on which end of the secondary winding is grounded. Opposite phasing can be obtained by reversing the secondary connections (grounding the other terminal of the winding).

CANCELLATION OF EVEN-HARMONIC DISTORTION.   In Fig. 15-15, all ac components are shown as perfect sine waves. This is true only if the dynamic (mutual) characteristic is linear. We know this is not the case. In fact, from Fig. 15-11 we calculated a second-harmonic distortion of over 9 percent. Second- and other even-harmonic distortion causes flattening of one of the current peaks of the ac component of collector current. This effect is shown in Fig. 15-16 for each BJT of a push-pull amplifier. The collector currents are 180° out of phase with each other because of the push-pull operation. Naturally, it follows that the fundamental components of each of these collector currents (shown by dashed lines) must also be 180° out of phase. But notice carefully: in order for the negative loops of each of these collector currents to become flattened, the second-harmonic components of each current (Q_1 and Q_2) are in phase with each other. (This would also be true for any other even-harmonic components.)* Since these harmonic current components are in phase and are flowing from opposite ends of the output transformer primary winding, their magnetic

---

*In Fig. 15-16, both second-harmonic components are negative cosine waves and produce flattening of the negative peaks of the collector current waveshape. When these second-harmonic components are positive cosine waves, the positive peaks of the collector current waveshape are flattened.

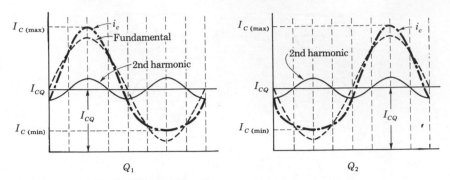

Figure 15-16 Flattening of negative collector-current peak due to second harmonic distortion.

fields will cancel. No even-harmonic voltage will be induced in the output.

This effect can be seen in another way. In Fig. 15-16, examine the first half-cycle of collector current through Q_1. Specifically consider the conditions at the quarter-cycle point. At this instant, the current value is a maximum and

$$I_{C(max)} = I_{CQ} + I_{1(max)} + I_{2(max)}$$

where $I_{1(max)}$ and $I_{2(max)}$ are the maximum values of the fundamental and the second harmonic, respectively. The dc component and these two ac components—at this instant—are all flowing through the output transformer primary *in the same direction*. This is shown in Fig. 15-17, the solid arrow representing the dc component; the dashed arrow, the fundamental; and the dotted arrow, the second harmonic.

Now let us examine this same time instant for Q_2. This time the collector current value is $I_{c(min)}$ and the instantaneous component values are

$$I_{c(min)} = I_{CQ} - I_{1(max)} + I_{2(max)}$$

Again, we end up with the fact that the second harmonic component is additive with the dc component, or flowing in the same direction as the dc component. These instantaneous current values are also shown in Fig.

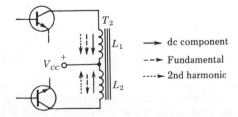

Figure 15-17 Instantaneous current values in a push-pull transformer.

15-17. Since (as was previously shown) the magnetic fields of the dc components canceled out, it follows that the second harmonic fluxes also cancel.

Cancellation of even-harmonic distortion is another advantage of push-pull operation. Two BJTs, in push-pull, can be used to deliver twice the power output as a single BJT—but at a much lower distortion level. Or, as is often done in practical circuits, by using a little more input driving power (or by changing the load resistance value), two BJTs can be made to deliver appreciably more than twice the power output for the same level of distortion as for a single unit.

In a similar fashion, any hum voltage introduced into this stage due to insufficient filtering of the power supply ripple components will be canceled out by the push-pull action. When the ripple content is at its maximum positive value, the collector supply voltage is a maximum and the collector currents of both BJTs will increase. Since both collector currents increase, it means that the ripple component of collector current for each BJT is in the same direction as its respective dc component. Their fluxes are in opposition and no hum voltage will be produced in the output winding.

BIAS COMPENSATION.   The above discussion implies complete cancellation of even-harmonic distortion, complete cancellation of hum voltages, and zero dc magnetization. This situation would be obtained only under theoretically ideal conditions of perfect balance of the input transformer secondary windings, perfect balance of the output transformer primary windings, and BJTs (Q_1 and Q_2) with identical characteristics. Such perfection would be difficult (and costly) to obtain. Fortunately, most applications do not require exact balance. However, when better matching is desired some manufacturers compensate for small inequalities by providing for separate adjustment of the individual base biases.

CLASS B AND AB POWER AMPLIFIERS

We have seen that a class A power amplifier has a maximum theoretical efficiency of 50 percent (with transformer coupling), but a more realistic maximum value is closer to 35 percent. (Direct coupling would yield only half these efficiencies.) On the other hand, it can be proved that with class B operation efficiencies as high as 78.5 percent can be achieved. Using 35 percent for class A and 70 percent for class B as an illustration, we can immediately see a 2:1 advantage for operation in class B. At low power-output levels, this is of little consequence, but consider instead a 200-W high-fidelity, quadraphonic audio amplifier. If we use class A, the required dc power input will be 200/0.35 or 572 W; with class B, a 286-W power supply will suffice. This can be quite a saving in the cost and bulk of the dc power unit.

This is not the only advantage. There is an even more important consideration. In class A power amplifiers, the full collector current (dc) flows at all times and the transistor power dissipation is highest when there is no input signal. This power dissipation can be drastically reduced by operating the transistors at zero bias. Neglecting the low value of leakage current (I_{CEO}), the collector current and the dc power input are zero. This would be true class B operation. Obviously, the power dissipation in the transistor with no signal input is also zero. Instead, maximum power dissipation in a class B amplifier occurs when the output is 42 percent of maximum and *is only 40.6 percent of the maximum power output*.

Let us see what this means with regard to a 200-W audio amplifier. For class A operation, the dc power input—even using the maximum theoretical efficiency of 50 percent—is 400 W; and using push-pull the minimum power-dissipation rating for each transistor is half the dc power input or 200 W. Now examine class B conditions. The total power dissipation is only 40.6 percent of the power output (0.046 × 200), or 81.2 W, and each transistor needs a power rating of just over 40 W. This is a tremendous saving in transistor size, in transistor cost, and in heat-sink provisions. For this reason, class A operation is generally not used for power output levels above (about) 5 W.

For true class B operation, the transistor (or tube) is biased at cutoff. With a BJT and neglecting leakage current, this would be at zero bias. Obviously, such a BJT will conduct for one half-cycle of input, when the signal drives the BJT into forward bias. This is a very serious case of even-harmonic distortion. However, by using two BJTs in push-pull most of the distortion is eliminated. This push-pull action can be explained with the aid of Fig. 15-18. Consider *NPN* transistors. During one half-cycle, when the base of Q_1 has a positive signal input, it conducts. At the same time Q_2 has a negative input signal and it is cut off. On the next half-cycle, the input signal polarities are reversed. Therefore Q_1 is cut off and Q_2 conducts. The collector currents for Q_1 and Q_2 are shown in Fig. 15-18b. Notice that these currents are distorted for low signal values due to the nonlinear characteristic. Since the collector currents for each transistor flow in the opposite directions through the output transformer primary, the combined collector-current effect is equivalent to a full-wave pattern as shown in Fig. 15-18c. Again notice the distortion in the low collector-current area as conduction crosses from one transistor to the other.

This *crossover distortion* can be greatly reduced by operating the BJTs at a slightly forward bias—or at *projected cutoff*. The improvement can be readily seen by graphic analysis using a back-to-back "composite" characteristic curve. Such an analysis is shown in Fig. 15-19. For zero bias, the back-to-back curves would have been aligned (vertically) at their zero-bias points. Instead, by using a small forward bias we align the curves along

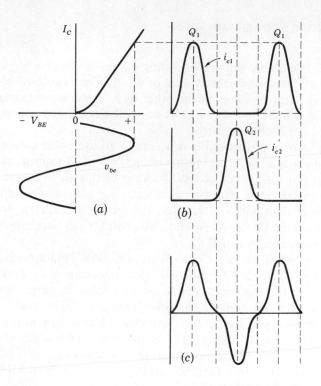

Figure 15-18 Collector current "crossover" distortion (class B) with operating point at cutoff.

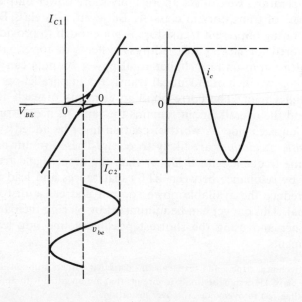

Figure 15-19 Push-pull class B with operating point at projected cutoff.

their bias lines. This shift in operating point produces a slight overlap; the low and distorted current values of one BJT are canceled by the equal and opposite current values of the other BJT. The composite characteristic becomes a straight line. The distortion in the combined collector current waveshape has been eliminated.

The amount of forward bias necessary to overcome crossover distortion is just about equal to the cut-in voltage of the BJT (approximately 0.6 V for silicon transistors). Of course, because of this small forward bias the quiescent current is no longer zero; moreover, each transistor conducts for slightly more than half a cycle, even though the conduction from the dc bias point to cutoff point is low and distorted. Strictly speaking, this operation could be called class AB.*

As for circuit diagrams, class B and class AB cannot be distinguished from class A. The only difference is in their operating point bias. So, referring to Fig. 15-14, if we want class A, we must make R_2 large enough to bias the BJTs to the center of their transfer characteristic curves. For class B, R_2 is made much smaller, so that the forward bias is decreased just enough to overcome crossover distortion (approximately 0.6 V). For class AB, R_2 is some in-between value. Since a change in a resistor value is not apparent from a schematic, a transformer-coupled class B or class AB retains the same circuit shown in Fig. 15-14.

PARALLEL POWER AMPLIFIERS.   We have seen that with class B push-pull operation, we can get appreciably more power output than from the same pair of transistors in class A. But even with class B, the power dissipation rating of present transistors is not enough to provide the power output required for large public address systems or for modulation of a large broadcast transmitter. Higher total power outputs can be obtained by connecting two or more identical transistors in parallel on each side of the push-pull circuit. The corresponding elements of each BJT are tied together, and the overall circuit remains the same as if only one BJT were being used on each side. A word of caution must be added: with parallel BJTs *parasitic oscillations* are likely to occur. These are unwanted oscillations at some frequency much higher than the input signal frequency and are caused by resonance between BJT capacitances and lead inductances. Parasitics reduce the available power output and cause distortion of the output signal. This danger can be minimized by placing the paralleled units close together and using the shortest possible connecting leads between similar elements.

*Some technical papers and texts call this class AB. Others still consider this class B and apply the class AB designation only to circuits that are operated with appreciably higher forward bias than just to overcome crossover distortion.

TRANSFORMERLESS POWER AMPLIFIERS

For high-fidelity applications, where low distortion and wide frequency responses are prime considerations, coupling transformers are the weakest link. Yet in all the push-pull amplifiers discussed so far we have used both input and output coupling transformers. Elimination of these components would be "a consummation devoutly to be wished." Not only would this improve fidelity, but it would also reduce the cost and bulk of the amplifier. Let us consider, first, techniques for eliminating the output transformer.

SERIES-OUTPUT AMPLIFIERS.   In discussing single-ended class A power amplifiers, we saw that with a direct-coupled (or series-output) load the maximum (theoretical) collector efficiency is 25 percent, whereas 50 percent efficiencies are possible with transformer coupling. With a direct-coupled load, the full dc component of collector current flows through the load. The poor efficiency is a direct result of the dc power loss ($I_{dc}^2 R_L$) across the load. Furthermore, in audio amplifiers any direct current flowing through the loudspeaker load pulls the voice coil out of its neutral position and causes heavy distortion, if not actual damage. And so we have used transformer coupling, and then kept on using transformer coupling even with push-pull class A and with class B amplifiers. Yet, let us stop and think for a moment. With push-pull operation the dc components of each transistor should be equal. It should therefore be possible to connect the two transistors in series across the supply voltage and still keep the dc out of the load. Two such circuits are shown in the partial schematics of Fig. 15-20.

In Fig. 15-20a, a single dc power supply is used, but the total voltage ($+V_{CC}$) should be twice the value that would have been used with transformer coupling, so that the V_{CE} for each transistor is the same as before. The dc potential of point A will obviously be at the center, or one-half this total voltage. For example, if the power supply voltage is 40 V, the potential of point A is $+20$ V and each transistor has an effective V_{CE} of 20 V. When the input signal is applied, during the half-cycle that Q_1 conducts, current (ac component) flows *up* through the load as shown by the solid arrows. This produces the positive half-cycle of output voltage. On the alternate half-cycles, Q_2 conducts and current (ac component) flows *down* through R_L (dashed arrows), giving us the negative half-cycles of output. Capacitor C is necessary to keep dc out of the load. (Remember that the dc potential of point A is $\frac{1}{2}V_{CC}$.) For good low-frequency response, the reactance of this capacitor must be low—compared to R_L—at the lowest desired signal frequency. This is a drawback. For example, if R_L is an 8-Ω loudspeaker

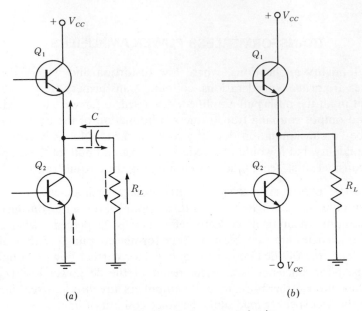

(a)　　　　　　　　　(b)

Figure 15-20　Series output circuits.

and we want X_C to be appreciably less than 8 Ω at 50 Hz, the capacitance value should be at least 2000 μF.

This coupling capacitor can be eliminated by using the circuit shown in Fig. 15-20b. Again this is a partial schematic that shows only the output connections. Notice that we are now using a dual (or split) power supply. The two supply voltages ($+V_{CC}$ and $-V_{CC}$) are equal, and since the potential of point A is at the center of these two voltages, point A is effectively at dc ground. Therefore, both ends of the load are at dc ground; no dc can flow through R_L and no blocking capacitor is necessary.

COMPLEMENTARY SYMMETRY.　Let us turn our attention to the input coupling transformer. Its job is to provide two input signals of equal amplitude, but opposite in phase, as needed by the two identical push-pull transistors (or tubes). One technique that eliminates the input transformer is to use some form of *paraphase amplifier*. These circuits (discussed in the next chapter) can be used with tubes as well as with transistors. Another technique became possible when improved semiconductor technology made transistors available as matched *NPN* and *PNP* units. By combining one of each type in a push-pull stage, the need for input push-pull transformers or for paraphase amplifiers is eliminated. Such an arrangement is known as a *complementary symmetry* circuit, and it can be used with any class of operation.

The operation of a push-pull complementary circuit can be shown readily from the simplified circuit of Fig. 15-21. We will assume that these transistors are properly biased for class A operation. (The bias circuit is not shown.) Notice that a series-output load and a dual power-supply source is being used. For class A operation, both transistors will conduct, even when no input signal is applied. These quiescent (dc) values of collector current are shown as the solid arrows above and below the load resistor. Notice that the collector current of Q_1 flows *to the left* while the collector current of Q_2 flows *to the right*. With matched transistors these currents are equal and opposite, and no direct current flows through the load. However, a direct current does flow through the overall series circuit containing the two transistors and the two power supplies.

Now let us add a positive signal at the input. Q_1 is an *NPN* unit. A positive signal—making its base more positive with respect to ground (or emitter)—increases the forward bias. This, in turn, increases its collector current. The *signal component* of the collector current is therefore "in phase" or in the same direction as the dc component. This signal component is indicated in Fig. 15-21 by the dashed arrow above the load resistor.

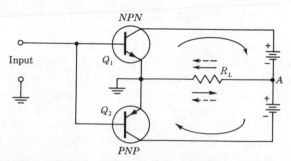

Figure 15-21   Simplified diagram—push-pull complementary symmetry (bias not shown).

At the same time, this positive signal is also applied to Q_2, a *PNP* transistor. In this case, the positive signal is in opposition to the normal forward bias and will cause a decrease in collector current. The signal component of collector current is therefore "out of phase" with the quiescent (dc) component. In Fig. 15-21 the signal component (dotted arrow) for Q_2 is drawn in the opposite direction to the quiescent value (solid arrow). Notice, however, that with respect to each other the two signal components are in phase or additive, and the output voltage, as developed across the load, will increase. The push-pull effect is obtained from a single-ended input signal.

Figure 15-22 shows a commonly used arrangement of a CE amplifier direct-coupled to a class B complementary symmetry push-pull amplifier.

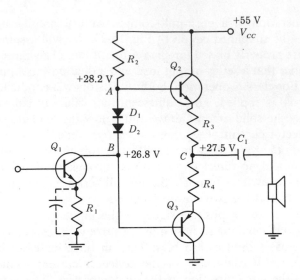

Figure 15-22 Class B complementary symmetry audio amplifier.

Q_1 is a CE class A driver stage. (Its input circuit is not shown.) Emitter resistor R_1 provides bias stabilization, and may or may not be bypassed.* The dc collector load for Q_1 is resistor R_2. The output from Q_1 is fed to the base inputs of the power amplifiers, Q_2 and Q_3. Notice that the collector of Q_3 is grounded directly, that the collector of Q_2 is at ac ground (through the power supply), and that the output is taken from the emitters. This is the *common-collector,* or *emitter-follower,* connection. Since a single power supply is used, the speaker load is capacitively coupled to the power amplifiers. (The capacitor can be eliminated if the collector of Q_3 is connected to a $-V_{CC}$ supply source instead of to ground.) Emitter resistors R_3 and R_4 are very small resistors (usually less than 0.5 Ω) and provide stabilization.

Now let us examine the dc biasing. For clarity, the power supply voltage and the corresponding potentials at key points are given. With a power supply voltage of 55 V, the output load point (C) is at the half-value, or $+27.5$ V. Since the emitter-stabilizing resistors are of very low value, let us also consider this $+27.5$ V as the potential of each emitter (Q_2 and Q_3). For zero bias, the base of these transistors should also be at $+27.5$ V. But we want a small forward bias (of about 0.6 to 0.7 V for silicon units) to prevent crossover distortion. Q_2 is an *NPN* transistor. Therefore its base should be more positive than its emitter. Notice the base potential: $+28.2$ V. Similarly Q_3, being a *PNP* unit, should have a base potential slightly less positive than its emitter. Notice that it is $+26.8$ V. Resistor R_2, in com-

---

*The unbypassed resistor produces inverse feedback for ac stabilization and improved frequency response—but reduces the gain. This will be discussed further in the next chapter.

bination with the collector current of Q_1 and diodes D_1 and D_2, sets this bias. Sometimes a resistor is used in place of the two diodes to provide this bias. However, the bias adjustment is critical, and is complicated by the need for compensation for temperature changes and for power supply voltage variations. This is achieved by using diodes that will match the V_{BE} changes in Q_3 and Q_4. A resistor may be added in series with the diodes to adjust the quiescent point to some specified value. A more sophisticated technique is to use the collector-to-emitter circuit of a transistor in place of the diodes. Then the bias point (V_{CE}) can be set to any desired value by changing the base-emitter voltage to this control transistor.

DARLINGTON CIRCUITS

In a high-power class B amplifier, the base current drive required to obtain full power output can also be quite high. This means that the driver stage must be a class B power amplifier—and must also be a push-pull stage. The need for two cascaded push-pull stages can be avoided by using a compound transistor connection known as a *Darlington amplifier* or *Darlington pair*. Two transistors (both *NPN* or both *PNP*) are connected together as shown in Fig. 15-23. Notice that the input sides are connected in series whereas the collectors are in parallel. The terminals B, E, and C of the compound unit are then connected to the rest of the circuit in the same manner as the base, emitter, and collector of a single transistor. This Darlington pair can be used in any single-ended, any standard push-pull, or any complementary push-pull circuit.

One advantage of the Darlington pair (over a single transistor) is its much higher current gain h'_{fe}. For example, assuming (for simplicity) identical units, if the base current applied to Q_1 is i_b then the collector current

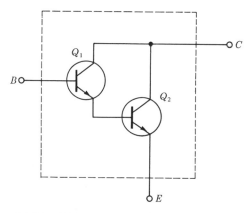

Figure 15-23 *NPN* Darlington pair (emitter-follower type).

for Q_1 is $h_{fe}i_b$. To a very close approximation, this is also the emitter current for Q_1—and the base current fed to Q_2. The action of transistor Q_2 would multiply this $h_{fe}i_b$ value by the second h_{fe} value, so that the collector current of Q_2 would be $(h_{fe})^2i_b$ and the overall current gain is

$$h'_{fe} \text{ (Darlington pair)} = (h_{fe})^2 \qquad \text{(for identical units)} \qquad \text{(15-5)}$$

or more accurately,

$$h'_{fe} = (1 + h_{fe1})(1 + h_{fe2}) \qquad \text{(15-6)}$$

In high-power amplifier applications, the two units would not be identical. The input BJT would probably be a transistor of lower power rating. For example, a 120-W amplifier* uses a 40883 250-W *NPN* transistor as the output half of one Darlington pair and a 40871 40-W transistor as the input unit.

Darlington pairs are also used at low power levels, not only for their high current gains but also because they provide very high input resistances. Notice that Q_1 in Fig. 15-23 is already connected as a common-collector or emitter-follower circuit, with the input resistance of Q_2 as its load. Now if Q_2 is also used in the same manner, the input resistance of Q_2 is (approximately) $h_{fe2}R_L$† and the overall input resistance is $(h_{fe1} \times h_{fe2}R_L)$ or

$$R_i \text{ (Darlington pair)} = h'_{fe}R_L \qquad \text{(15-7)}$$

Darlington pairs are available commercially as IC units with current gains as high as 300,000.

QUASI-COMPLEMENTARY POWER AMPLIFIERS.   In a full complementary symmetry circuit, as shown in Fig. 15-22, one of the output transistors must be a *PNP* type. Even using Darlington pairs, one of the output transistors must still be a *PNP* type. But for many years there was difficulty in manufacturing *PNP* silicon transistors of high power rating. This led to the use of a quasi-complementary circuit at high power levels. With this variation, *both* the output transistors are of the *NPN* type, yet the complementary feature is preserved, so that a push-pull input is not required. The basic circuit features are shown in the partial schematic of Fig. 15-24a. The upper half of this push-pull circuit (Q_1 and Q_2) is a standard (common-emitter) *NPN* Darlington pair, as described earlier.

Now examine the lower half of this circuit. Transistors Q_3 and Q_4 are connected with the inputs in series and the outputs in parallel, as in the Darlington circuit. Notice, however, that Q_3 is a *PNP* unit whereas Q_4 is an *NPN* unit—they are not both of the same type. Note also that the emitter-collector connections of Q_3 are reversed (compared to the standard

*From RCA data sheet VAQ-120S.
†See the next chapter for details on emitter followers.

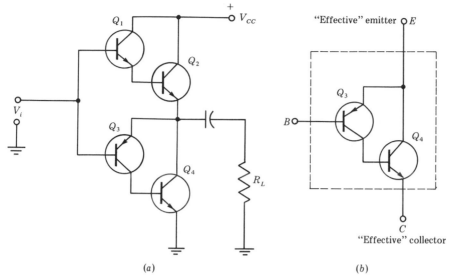

Figure 15-24   A basic quasi-complementary circuit with an inverted Darlington.

Darlington). This combination is therefore sometimes called an *inverted Darlington*. This portion of the circuit is shown by itself in Fig. 15-24*b*. Notice that the collector current of the input *NPN* unit, Q_3, becomes the base current of the output *PNP* unit, Q_4. This output transistor, in turn, operates as an emitter follower and provides additional current gain, but without phase inversion. If the emitter of this *NPN* output transistor is considered as the "effective" collector of the composite unit, it becomes apparent that the compound unit is equivalent to a high-gain, high-power *PNP* transistor. In other words, this compound connection effectively converts a low-power *PNP* unit and a high-power *NPN* unit into a high-power *PNP* unit. Then, since the upper Darlington pair functions as a compound *NPN* unit while the lower inverted Darlington functions as a matched compound *PNP* unit the combined circuit of Fig. 15-24*a* is a complementary symmetry circuit and does not need a push-pull input.

FULL COMPLEMENTARY SYMMETRY—INVERTED DARLINGTONS.   Notice in the "regular" Darlington pair of Fig. 15-23 that between the compound unit base (*B*) and its emitter (*E*) there are two base-emitter junctions and two V_{BE} voltages in series. On the other hand, in the inverted Darlington pair of Fig. 14-24*b* there is only one junction between the composite base (*B*) and the composite emitter (*E*). Therefore, the cut-in voltage for the inverted pair is (about) 0.6 V, but in the regular Darlington it is twice as much, or 1.2 V. Also, the transfer curve (I_C versus V_{BE}) for the inverted pair is steeper. This means higher gain and less crossover distortion for the inverted pair. Obviously, this circuitry is preferable. Consequently, when good high-power *PNP* silicon units became available the inverted Darlington was used on both sides of the push-pull circuit. Such a combination is shown in Fig. 15-25 together with a typical bias circuit and a CE driver stage.

Transistor Q_1 is a CE amplifier, with R_1 for bias stabilization and R_4 as its dc collector load resistance. The output from Q_1 is fed to the bases of Q_3 and Q_5, the input BJTs of the inverted Darlington pairs. On the positive half-cycles of the input signal, Q_3 conducts but Q_5 is driven into cutoff. Conduction in the Darlington amplifiers is reversed on alternate half-cycles. Push-pull operation is obtained. The quiescent bias point for the Darlingtons is fixed by the voltage V_{CE} across Q_2. This value, in turn, is set by resistors R_2 and R_3 to just overcome crossover distortion. The

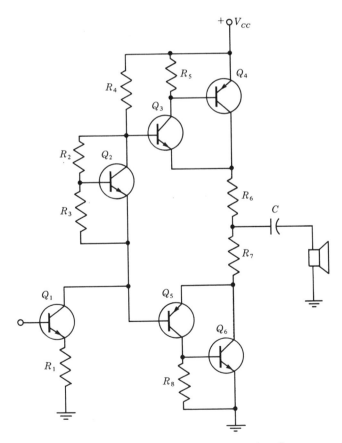

Figure 15-25 A push-pull complementary symmetry class B power amplifier using compound-connected output transistors.

output is taken from the "effective" emitter terminals of the compound units, making this an emitter-follower type of Darlington connection. Since we are using a single power supply, the speaker load is capacitively coupled. As mentioned before, the coupling capacitor can be eliminated if a dual dc power supply is used.

REVIEW QUESTIONS

1. **(a)** What is the basic function of a power amplifier? **(b)** In a power amplifier, which is more important: power gain, voltage gain, or power output? Explain.

2. **(a)** With reference to tubes, FETs, and BJTs, which are suitable as power amplifiers? **(b)** Which can deliver the highest power outputs?

3. **(a)** What is a heat sink? **(b)** What is its effect on the rating of a transistor?

4. **(a)** Explain the effect of temperature on the power dissipation rating of a transistor. **(b)** How is this effect taken into consideration in the manufacturer's data sheets?

5. **(a)** Give three references that may be used in specifying dissipation ratings. **(b)** Under what condition would these three reference values be the same? **(c)** In general, which reference would have the highest temperature? The lowest temperature?

6. **(a)** What is meant by the thermal resistance of a transistor? **(b)** Give an example of this quantity and show how it is applied.

7. **(a)** What are the coordinates for a power dissipation curve? **(b)** With what other data would you expect to find such a curve? **(c)** Explain the purpose of such a curve.

8. Refer to Fig. 15-2: **(a)** What is the power dissipation rating for this transistor? **(b)** For $I_C = 3$ A, what is V_{CE}? What is the power dissipation?

9. Refer to Fig. 15-3: **(a)** What is the listed power dissipation rating for this transistor? **(b)** For what temperatures does this rating apply? **(c)** What is the maximum power it can dissipate (safely) at $100°C$?

10. **(a)** What is the significance of the term *collector efficiency?* **(b)** Give the equation for this efficiency. **(c)** Is there an equivalent vacuum-tube term? If so, what is it?

11. What is the approximate value for the efficiency obtainable in a practical circuit operating in **(a)** class A? **(b)** Class B?

12. In view of the relative efficiencies of classes A and B, why are voltage amplifiers operated in class A?

13. Are power amplifiers ever used in class A? Explain.

14. State four characteristics typical of class A operation.

15. Refer to Fig. 15-4: **(a)** How is the operating point Q selected with respect to the dynamic characteristic? **(b)** Could a smaller input signal voltage be used and still maintain class A operation? Explain. **(c)** Could a larger input signal be applied and still maintain class A operation? Explain.

16. Compare class AB operation to class A with respect to **(a)** bias value; **(b)** angle (duration) of output current flow; **(c)** no-signal (quiescent) current (I_{co}) value; **(d)** distortion level.

17. If a given transistor or tube can be operated in class A or AB, under which condition would it give **(a)** higher efficiency? Explain. **(b)** Higher power output? Explain.

18. Refer to Fig. 15-5: **(a)** Why is this called class AB operation? **(b)** If a larger signal were applied, would it still be class AB? Explain. **(c)** If the signal level were reduced to one-half the given value, would it still be class AB operation? Explain.

19. Refer to the collector current waveshape of Fig. 15-5: **(a)** Is this a sine wave? Explain. **(b)** From an examination of the waveshape, what is the main distortion component? Explain.

20. Will single transistor circuits provide a "clean" output **(a)** with class A operation? **(b)** With class AB operation?

21. **(a)** What type of circuit is used to eliminate even-harmonic distortion? **(b)** How many transistors are needed in such a circuit?

22. What class of operation is capable of maximum power output and maximum efficiency for an untuned amplifier?

23. In an ideal condition, what value of bias would be used for transistor class B operation?

24. Refer to Fig. 15-6: **(a)** Is this transistor biased at cutoff? **(b)** What is this bias point called? **(c)** How is this bias value obtained? **(d)** Why is this slight forward bias used?

25. For a given transistor, which class of operation has **(a)** the lowest distortion level? **(b)** The highest distortion level? **(c)** The lowest collector circuit efficiency? **(d)** The highest collector circuit efficiency? **(e)** The lowest power-handling ability? **(f)** The highest power-handling ability?

26. With respect to class C operation, **(a)** what relative value of bias is used? **(b)** What is the duration (angle) of output current flow? **(c)** What types of distortion are produced? **(d)** Is it used for nonresonant amplifier service?

27. Refer to Fig. 15-7: **(a)** Can the triode circuit be used with R-C coupled input? **(b)** Explain what circuit changes would be necessary. **(c)** In the beam-power tube circuit, why is resistor R_2 shown dotted? **(d)** When is this resistor needed? **(e)** When the resistor is used, why is capacitor C_2 also used? **(f)** How would the circuit for a pentode power tube compare with either of these two circuits? **(g)** Could the beam-power tube be used with transformer-coupled input? **(h)** What circuit changes would be necessary?

28. Refer to Fig. 15-8: **(a)** What type of transistor is this? **(b)** What components set the base-bias value? **(c)** Is any form of bias stabilization used? If so, what components are involved? **(d)** Give two names for the circuit variation shown in diagram (*b*).

29. Comparing the load-connection variations in Fig. 15-8, which circuit would **(a)** have a lower cost? **(b)** Be lighter in weight and smaller in size? **(c)** Have better frequency response? **(d)** Have higher power-output capability? **(e)** Require a higher power-dissipation rating for the transistor? **(f)** Require a higher dc power-supply voltage?

30. Refer to Fig. 15-9: **(a)** Is the load line shown in diagram (*b*) the dc or the ac load line? Explain. **(b)** Can a larger input signal be used? Explain. **(c)** Can a smaller input signal be used? **(d)** For maximum undistorted power output, how is point Q selected? **(e)** How does the quiescent value, V_{CEQ}, compare to the supply voltage value? **(f)** At maximum power output, how does the amplitude of the output voltage compare with

V_{CC}? **(g)** With V_{CEQ}? **(h)** How does the amplitude of the ac component of collector current compare with the quiescent current value?

31. Compared to the dc power input, what is the maximum power output obtainable from a class A power amplifier with a direct-coupled load?

32. In a class A power amplifier with direct-coupled load, compared to the dc power input, what is the power dissipated by the transistor **(a)** at maximum power output? **(b)** Under quiescent conditions?

33. Refer to Fig. 15-10: **(a)** Why is the dc load line drawn vertically? **(b)** What correlation—if any—exists between V_{CEQ} and V_{CC}? **(c)** Why is this so? **(d)** Compared to V_{CC}, what should the maximum voltage rating of the transistor be? **(e)** How do the amplitudes of the collector current waveshapes (i_c) in Figs. 15-9 and 15-10 compare? **(f)** How do the v_{ce} amplitudes compare?

34. Repeat Question 31 for a transformer-coupled load.

35. Repeat Question 32 for a transformer-coupled load.

36. Refer to Fig. 15-11: **(a)** What is the supply voltage value V_{CC}? **(b)** What is the quiescent value of base bias? **(c)** How was this value obtained? **(d)** Why is the dc load line vertical? **(e)** How was the operating point Q obtained? **(f)** What is the quiescent value of collector current? **(g)** What is the value of the ac load line? **(h)** For this value of load, is the power rating of this transistor exceeded? Explain. **(i)** For a signal swing of ± 7 mA, is the collector voltage rating of 60 V exceeded? Explain. **(j)** Is the collector current rating of 1.5 A exceeded? Explain.

37. Refer to Fig. 15-11: **(a)** How are the values labeled V_{max} and V_{min} obtained? **(b)** How is the peak-to-peak output voltage related to these values? **(c)** How are the values of I_{max} and I_{min} obtained? **(d)** How is the peak-to-peak ac component of collector current related to these values?

38. Account for the denominator 8 in Equation 15-4.

39. In the design of vacuum-tube power amplifiers, **(a)** what range of load values is suitable with triodes? **(b)** With pentodes? **(c)** Compared to the plate potential, give a suitable value for the grid bias, using triodes. **(d)** What manufacturers' data simplifies such designs?

40. Refer to Fig. 15-12a: **(a)** How do these three load lines compare in resistance value? Explain. **(b)** Which load line will produce the highest power output? **(c)** Will any of these load lines cause the transistor to exceed any of its ratings? Explain. **(d)** Under what condition would the maximum safe power output be obtained?

41. Refer to Fig. 15-12b: **(a)** Which load line will produce the higher output? **(b)** Which load line is preferable? Why?

42. Refer to Fig. 15-13: **(a)** For what value of power dissipation does the power curve apply? **(b)** From the curve itself, how can you tell? **(c)** Does this ac load line exceed any rating of this transistor? Explain. **(d)** Could any greater power output be obtained with some other load-line value? Explain. **(e)** Explain why the V_{min} swing is not taken all the way down to zero. **(f)** What is the value of I_{min}? **(g)** Explain why the I_{min} swing is not taken all the way down to zero. **(h)** How is the Q point value obtained? **(i)** How does the dc power supply voltage compare to the quiescent value of V_{CE} if we assume that the output transformer windings have negligible resistance? **(j)** Repeat (i) if we wish to consider the resistance of the windings.

43. Refer to Fig. 15-14: **(a)** What components determine the base bias of transistor Q_2? **(b)** Repeat for Q_1. **(c)** What is the function of R_3–C_1? **(d)** What class of operation is this? **(e)** What determines the class of operation? **(f)** If the values were set for class A, what changes would be needed to convert to class AB? To Class B?

44. Refer to Fig. 15-14: **(a)** How does the amplitude of the input signal applied to the base of Q_1 compare with that applied to Q_2? **(b)** How do these input signals compare in phase? **(c)** How is this phase relation obtained? **(d)** How does I_{CQ} for each transistor compare in magnitude? **(e)** What does the solid arrow in this diagram represent? **(f)** What is the total magnetic effect produced by the two collector currents under no-signal conditions? **(g)** What do the dashed arrows in this diagram represent? **(h)** What is their phase relation? Why? **(i)** How do these components combine to produce an output? **(j)** What do the waveshapes drawn at the top and bottom of the transformer T_2 primary winding represent? **(k)** How are these phase relations obtained?

45. Refer to Fig. 15-16: **(a)** What operating condition causes flattening of the negative portion of the i_c waveform? **(b)** Why are the fundamental components of Q_1 and Q_2 collector currents 180° out of phase? **(c)** What are the phase relations of the second-harmonic components to each other? **(d)** Why must this relation be so? **(e)** Draw the resultant i_c waveform of Q_2 if the second-harmonic component in this BJT were shifted 180° with respect to Q_1. **(f)** At any instant what components are effective in producing the resultant current i_c? **(g)** When the resultant current for Q_1 is at its maximum positive value, what are the values and polarities of each component? **(h)** Repeat part (g) for Q_2 *at this same time instant.*

46. Refer to Fig. 15-17: **(a)** At what instant of time do these current relations apply? **(b)** Why are all the components for Q_1 shown as flowing in the same direction? **(c)** Why is the fundamental component of Q_2 shown as flowing opposite to the other components? **(d)** Why do the second-harmonic components produce no output?

47. State three conditions that must be met for ideal push-pull operation.

48. **(a)** At high power levels, which class of operation is preferable: class A or class B? **(b)** Give two reasons for this preference.

49. **(a)** Can single-ended operation be used with class B? With class AB? **(b)** Why is this so?

50. Refer to Fig. 15-18: **(a)** What class of operation is shown in diagram (a)? **(b)** What bias value is used? **(c)** What does the waveshape in diagram (c) represent? **(d)** What is the cause for the distortion shown? **(e)** What is this distortion called? **(f)** How is this condition remedied?

51. **(a)** What class of operation is indicated in Fig. 15-19? **(b)** What value of bias is used? **(c)** Explain how crossover distortion is eliminated.

52. How does a circuit diagram for push-pull class AB differ from that for class A?

53. **(a)** Can BJTs (or tubes) be operated in parallel? **(b)** When would such operation be necessary? **(c)** Give a disadvantage of paralleling.

54. Refer to Fig. 15-20: **(a)** Compared to previously shown push-pull circuits, what advantage does this type of load connection have? **(b)** Does any direct current flow through the load in diagram (a)? Explain. **(c)** Repeat for diagram (b). **(d)** What advantage does the circuit in diagram (b) have? **(e)** What is its disadvantage?

55. (a) What advantage does a complementary symmetry push-pull circuit have over a "conventional" push-pull circuit? (b) How many transistors are needed? (c) What types of transistors are used? (d) Can this technique be applied to vacuum-tube circuits? Explain.

56. Refer to Fig. 15-21: (a) Is this an actual working circuit? Explain. (b) What class of operation is intended? (c) What do the solid arrows represent? (d) Trace the complete current path for transistor Q_1 alone. (e) Trace the current path for transistor Q_2 alone. (f) Why are the directions of the two emitter currents opposite? (g) For matched transistors, what is the total current through the load? (h) Trace the actual no-signal current path for the combined Q_1–Q_2 circuit. (i) What do the dashed arrows signify? (j) For what polarity of input signal do these dashed arrows apply? (k) Explain how this direction of the signal component of collector current for Q_1 is obtained. (l) Repeat for transistor Q_2. (m) Do two transistors in such a circuit produce a higher output than either one alone? (n) Do two transistors in such a circuit tend to cancel even-harmonic distortion? Explain.

57. Refer to Fig. 15-22: (a) What is the function of Q_1? (b) Does it produce the two 180° out-of-phase input voltages for the push-pull stage? Explain. (c) What type of circuit is used with Q_2–Q_3? (d) What is the function of resistor R_2? Resistor R_3? Resistor R_4? (e) Why is capacitor C_1 used? (f) What is the function of diodes D_1 and D_2? (g) Explain why a resistor is not used in place of those diodes. (h) Why are *two* diodes used? (i) On the positive swing of the input to Q_1, which output transistor will conduct? Explain.

58. Refer to Fig. 15-23: (a) How does the current gain of this compound unit compare with the individual transistor current gains? (b) Why would they be used in power amplifiers? (c) What other advantage of this pair makes them also useful at low power levels?

59. Why was the quasi-complementary symmetry circuit developed?

60. Refer to Fig. 15-24: (a) Which of the transistors form an inverted Darlington pair? (b) Which form a regular Darlington pair? (c) On the positive swing of the input signal, which transistors will conduct? (d) What is the polarity of the output voltage at this instant? Explain.

61. In class B push-pull power amplifier service, what is the advantage of the inverted Darlington over the regular Darlington pair?

62. Refer to Fig. 15-25: (a) What is the function of Q_1? (b) Of Q_2? (c) What type of circuit is used with transistors Q_5 and Q_6? (d) With transistors Q_3 and Q_4? (e) What is the advantage of using Q_2 instead of the diodes shown in Fig. 15-22? (f) How can capacitor C be eliminated?

PROBLEMS AND DIAGRAMS

1. The 2N511 transistor has a maximum power rating of 80 W at a mounting base temperature of 25°C. Its derating coefficient is 1.067 W per degree Celsius. What is the maximum power it can dissipate at a case temperature of 40°C?

2. Repeat Problem 1 for a case temperature of 75°C.

3. The transistor of Problem 1 is used in a circuit so that it dissipates 55 W. What is the maximum allowable mounting base temperature?

4. The 2N296 has a dissipation rating of 20 W at a base temperature of 25°C. It has a maximum allowable junction temperature of 85°C and a temperature gradient (base to collector) of 3°C per watt. Find the junction temperature if the unit is operated at maximum dissipation and the base temperature is 25°C.

5. The 2N511 of Problem 1 has a maximum allowable junction temperature of 100°C. Find the temperature gradient or thermal resistance for this unit.

6. The 2N296 of Problem 4 is operated with a power dissipation of 16 W, but at a case temperature of 32°C. Is it operating above its rating?

7. (a) Using the characteristics shown in Fig. 15-2, draw the power dissipation curve for this unit, assuming that the maximum dissipation is 30 W. **(b)** For this rating, what are the maximum values of collector current and voltage that could be used at a base bias of 30 mA?

8. A transistor (common-emitter connection) delivers 40 W of signal power to its load. Its operating point is set at $V_{CE} = 40$ V and collector current $I_C = 3$ A. The ac power input from the previous stage is 0.8 W. Calculate: **(a)** the collector efficiency; **(b)** the power dissipated in the transistor.

9. A class A triode power amplifier draws 60 mA from a 300-V dc power supply. The ac component of plate current is 0.035 A, and the ac output voltage is 140 V. Find: **(a)** the plate efficiency; **(b)** the plate dissipation.

10. Sketch a typical dynamic characteristic curve (I_C versus V_{BE}). Select a suitable operating point for class A. Show the maximum input signal swing that can be used and the resulting collector current waveshape. (Save these curves for comparison with Problems 11 and 12.)

11. Using the same scale, repeat Problem 10 for class AB operation.

12. Using the same scale, repeat Problem 10 for class B operation.

13. Draw the circuit diagram for a triode class A power amplifier, using R-C coupling to its previous stage.

14. Draw the circuit diagram for a pentode class A power amplifier, using R-C coupling to its previous stage and with its screen grid operating at a lower potential than its plate.

15. Draw the circuit diagram for an *NPN* transistor used in the common-emitter configuration as a class A power amplifier. Show transformer coupling for input and output. Include some form of bias stabilization.

16. A class A transistor power amplifier with direct-coupled load has a quiescent point of $V_{CEQ} = 20$ V and $I_{CQ} = 2.0$ A. Assuming that this is an optimum operating point, calculate **(a)** the maximum power output obtainable; **(b)** the power dissipated at maximum power output; **(c)** the power dissipated at no-signal input.

17. Repeat Problem 16 for a transformer-coupled load.

18. A power output of 3.5 W is desired. It is planned to use a class A circuit with direct-coupled load. Assuming that a collector efficiency of 18 percent will be obtained, calculate **(a)** the power that must be supplied by the dc power supply; **(b)** the power dissipated in the transistor at full power output; **(c)** the power dissipated at no-signal input; **(d)** the minimum power rating for the transistor.

19. Repeat Problem 18 for a transformer-coupled load and a collector efficiency of 36 percent.

20. Using the collector characteristic curves and operating point for the 2N1068 of Fig. 15-11, can this transistor be used with an ac load of 120 Ω? If it cannot, explain why. If it can, calculate the power output.

21. Using the collector characteristic curves of Fig. 15-11, an operating point of $V_{CE} = 15$ V and $I_B = 8.0$ mA, and an ac load of 100 Ω, calculate **(a)** power output for a signal swing of ± 7.0 mA; **(b)** percent distortion (assuming only second harmonic); **(c)** collector efficiency at full power output and with a transformer-coupled load. (Neglect the resistance of the transformer windings.)

22. A *PNP* transistor (characteristics shown in Fig. 15-26) is to be operated as a transformer-coupled class A power amplifier from a 28-V power supply. The base bias is to be set at 5 mA, the ac load will be 150 Ω, and a signal swing of ± 4.0 mA will be applied as input. The dc resistance of the transformer primary is 27.5 Ω. Calculate: **(a)** power output; **(b)** percent second-harmonic distortion; **(c)** collector efficiency.

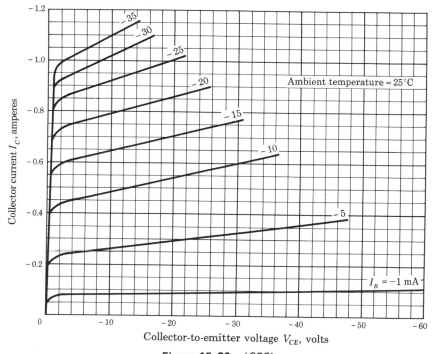

Figure 15-26   (*CBS*).

23. It is desired to construct a class A power amplifier with a power output of 2.5 W. What should the minimum power rating of the transistor be **(a)** if direct-coupled load with 20 percent efficiency is used? **(b)** If transformer-coupled load with 40 percent efficiency is used?

24. It is desired to construct a 5.0-W class A transformer-coupled audio power amplifier to feed a 4-Ω speaker load. The transistor of Fig. 15-26 has a 15-W dissipation rating, $BV_{CEO} = 60$ V, and $I_{C(max)} = 2.0$ A. **(a)** Can this transistor be used? If so, neglect the dc resistance of the transformer windings and **(b)** select a suitable $V_{CE(max)}$ voltage, and a resistance for the ac load line. Specify these values. **(c)** Select maximum and minimum collector current swings, and pick off corresponding values for V_{min} and V_{max}. **(d)** Select the operating point Q and give the quiescent values. **(e)** Calculate the maximum power output. Does it meet (or exceed) the original requirement? If not, repeat the process, using another ac load line. **(f)** Select the output transformer turns ratio, assuming 1.5 Ω as the dc resistance of the primary winding. **(g)** Calculate the required power-supply voltage.

25. The transistor whose ratings and characteristics are shown in Fig. 15-27 is to be used as a class A power amplifier with a direct-coupled load of 8 Ω. **(a)** Draw the load line for maximum safe power output, and specify where it cuts the I_C and V_{CE} axes. **(b)** Select minimum and maximum swings for I_C and V_{CE}, and calculate the power output. **(c)** Select a suitable operation point and specify the quiescent values of collector current and voltage. **(d)** Calculate the required power-supply voltage.

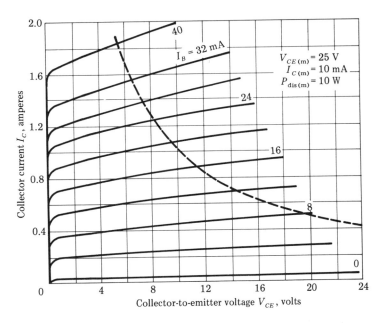

Figure 15-27   Transistor characteristics for Problem 25 (*Sylvania*).

26. Draw the circuit diagram for a transformer-coupled class A push-pull power amplifier using *PNP* transistors.

27. Repeat Problem 26 for class AB.

28. Repeat Problem 26 for class AB, but using two transistors in parallel for each side of the push-pull circuit.

29. Draw the circuit diagram for a class AB power amplifier using *PNP* transistors with transformer-coupled input, series-output load, and a single power supply.

30. Repeat Problem 29, using a dual power supply.

31. Draw the circuit diagram for a class B power amplifier using a CE-connected *PNP* driver stage, complementary symmetry output stage, and a single power supply.

32. Repeat Problem 31, using a dual power supply.

33. Repeat Problem 31, using Darlington pairs in the push-pull stage.

34. Repeat Problem 31, using inverted Darlingtons in the push-pull stage.

35. Repeat Problem 31, using inverted Darlingtons and a dual power supply.

16

Auxiliary Untuned
Amplifier Circuits

Several circuits commonly used with untuned amplifiers have not yet been discussed. They may be used in connection with voltage or power amplifiers. The delay in presentation was quite deliberate because these circuits employ, to some degree, a principle known as *negative or inverse feedback*. We could not explain this principle until we had explained the action of the basic voltage and power amplifier circuits. The topics covered in this chapter include inverse feedback, paraphase amplifiers, followers, and mixers.

INVERSE FEEDBACK

In our studies of active devices (tubes and semiconductors), we have seen that their static and dynamic characteristics are not completely linear. This is especially true of pentode power tubes and bipolar junction power transistors. The result of this nonlinearity could be serious harmonic and intermodulation distortion, were it not for the application of inverse (negative) feedback. In addition, such feedback circuitry also stabilizes gain and improves frequency response. Let us see how this is accomplished.

BASIC FEEDBACK PRINCIPLE.   The general idea in any feedback circuit is to take a portion of the output voltage and "feed it back" to some point in the input circuit. This feedback *loop* can be taken over one stage or over several stages. If the amplifier has no phase distortion, the feedback voltage (depending on the connections and points over which it is taken) can be either in phase or 180° out of phase with the input voltage. Figure 16-1 shows this basic feedback principle in block diagram form. The feedback network is some form of voltage divider whereby only a portion of the output voltage is returned to the input circuit. The Greek letter β (beta) is used to signify this feedback factor, or percent feedback, expressed as a

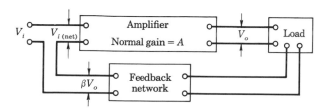

Figure 16-1   Block diagram—feedback amplifier.

decimal. If the voltage fed back is in phase with the input signal, it adds to the effective input level. The feedback factor is positive, and this type of feedback is called *positive* or *regenerative* feedback. The effect of such feedback is cumulative; that is, the increase in input level causes a higher output voltage which increases the feedback voltage and further increases the input level. If the percent feedback is high enough, this regenerative action will cause the circuit to break into oscillation. Since positive feedback is not the effect desired at this time, it is not discussed further in this chapter.

When the feedback voltage is 180° out of phase with the input voltage, we have an ideal condition for *inverse, negative,* or *degenerative* feedback. Since the feedback voltage opposes the input voltage, the feedback factor (β) is taken as *negative*. In Fig. 16-1, if V_o is the output voltage then for inverse feedback the voltage fed back is $-\beta V_o$ and the net input voltage $V_{i(net)}$ becomes

$$V_{i(net)} = V_i - \beta V_o \qquad (16\text{-}1)$$

To produce the output voltage V_o, this net input signal $V_{i(net)}$ when multiplied by the amplifier gain A must equal V_o:

$$V_o = A(V_i - \beta V_o) \qquad (16\text{-}2)$$

and

$$V_o = \frac{AV_i}{1 + A\beta} \qquad (16\text{-}3)$$

Since gain is the ratio of output to input, and using V_i as the actual input voltage, the overall gain with feedback, A',* becomes

$$A' = \frac{V_o}{V_i} = \frac{A}{1 + A\beta} \qquad (16\text{-}4)$$

EXAMPLE 16-1

An amplifier having a normal gain of 35 is used with 15 percent feedback. Find the overall gain with feedback.

Solution

1. The gain with feedback must be *less* than the original gain.

2. $A' = \dfrac{A}{1 + A\beta} = \dfrac{35}{1 + 35(0.15)} = 5.6$

Notice in Example 16-1 the drastic reduction in gain. From analysis of the equation, it can be seen that the greater the feedback factor, the lower the overall gain. If the feedback factor is 100 percent, the gain would be less than 1. The amount of feedback is also commonly expressed in decibel units as a ratio of the loss in amplification. For example, the expression "20-dB feedback" means that the gain with feedback is 20 dB lower, or the voltage gain is only one-tenth of the original value. As an equation, the amount of feedback expressed in decibels would be 20 times the logarithm of the gain ratio.

EXAMPLE 16-2

Express the feedback level of Example 16-1 in decibel units.

Solution

$$\text{Feedback in decibels} = 20 \log \frac{A}{A'} = 20 \log \frac{35}{5.6}$$

$$= 15.9 \text{ dB}$$

The gain is *decreased* by 15.9 dB.

Although inverse feedback reduces the gain, the importance of this type of feedback is that *it also stabilizes the gain;* that is, the gain tends to remain constant regardless of other circuit variations. For example, in an industrial electronic control application wherein a critical control function

---

*The gain with feedback, A', is often called the *closed-loop gain* because the feedback circuit forms a loop (from output to input and back through the amplifier to the output). If this loop is opened, feedback is removed, and so the gain without feedback, A, is called the *open-loop gain*. Also, the product $A\beta$ is called the *loop gain*.

depends on some specific signal voltage level, if the gain of the amplifier changes because of line voltage changes, aging of circuit components, or any other reason, the entire production can be ruined. In test equipment such as the electronic voltmeter, a change in gain could make the instrument readings worthless. Furthermore, since amplitude and frequency distortion are caused by changes in gain (with signal level or frequency) both these effects are reduced by inverse feedback.

EFFECT ON GAIN—EXAMPLES.   So far we have discussed the effect on gain from a mathematical derivation and equation. It is at times more meaningful to analyze by examples. This is done in Table 16-1. Several cases are shown to illustrate the stabilizing effect of feedback on gain. The term *nominal gain* refers to the gain of the amplifier without feedback. *Net input* is the effective voltage reaching the amplifier. Since the feedback voltage opposes the input signal, the *total input* voltage must be increased to maintain a net input voltage of 1 V.

Several interesting facts should be noted from the feedback tabulations of Table 16-1. In the first three cases, the normal amplifier gain varies over a 4:1 or 12-dB range. This could have been caused by amplitude or frequency distortion or any of the other factors mentioned above. Using only 10 percent feedback, this variation was reduced—in the worst case— to less than 2:1 or 4.08 dB for case (*a*) and to practically no change or 0.56 dB for case (*c*). Notice also in cases (*b*) and (*c*) that with 10 percent feedback the gain is approximately 10. The gain is being stabilized at this value. This brings out an important conclusion: *the maximum possible gain with feedback* is $1/\beta$. For $\beta = 10$ percent, the maximum gain is 10. This is

Table 16-1
STABILIZATION OF GAIN BY INVERSE FEEDBACK
(Net Input = 1 V)

CASE	NOMINAL GAIN	OUTPUT VOLTAGE (volts)	PERCENT FEEDBACK β	FEEDBACK VOLTAGE (volts)	TOTAL INPUT (volts)	OVERALL GAIN
(*a*)	40	40	10	4	5	8
	10	10	10	1	2	5
(*b*)	400	400	10	40	41	9.7
	100	100	10	10	11	9.1
(*c*)	800	800	10	80	81	9.87
	200	200	10	20	21	9.55
(*d*)	200	200	5	10	11	18.3
(*e*)	200	200	20	40	41	4.88
(*f*)	200	200	30	60	61	3.28

further verified by cases (d), (e), and (f) in the tabulation. Each case has a nominal gain of 200, but feedback factors vary from 5 to 30 percent. Notice that the overall gain A' with feedback, for each of these cases, approaches $1/\beta$. This can also be shown mathematically from the gain equation (16-4). When the product $A\beta$ is large compared to 1 (due to high circuit gain or high feedback factor), the denominator $(1 + A\beta)$ is approximately equal to $A\beta$ and the gain equation simplifies to

$$A' \cong \frac{1}{\beta} \qquad (16\text{-}5)$$

EFFECT ON DISTORTION.   In Chapter 9 it was shown that amplitude and frequency distortion are caused by variations in gain. Obviously, then, any technique that stabilizes gain must also correct such distortion. Since with inverse feedback the gain is reduced to $A' = A/(1 + \beta A)$, it follows that distortion will also be reduced by the same factor:*

$$D' = \frac{D}{1 + \beta A} \qquad (16\text{-}6)$$

where D' is the distortion with feedback and D is the distortion before feedback.

Just as amplitude and frequency distortion are minimized by use of negative feedback, so is phase distortion. However, showing this effect requires phasor representation. This is done in Fig. 16-2. We start with

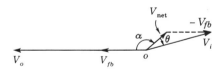

Figure 16-2   Phase-shift correction by inverse feedback.

the output voltage V_o as our reference phasor (drawn horizontally). The feedback voltage is in phase with the output voltage but is only a portion of this output voltage. It is shown as V_{fb}. Let us assume that this amplifier has a phase shift (at this frequency) of less than 180° as represented by α. In order to produce an output V_o, the net input voltage must be at this phase angle. It is shown by V_{net} and is the phasor sum of the actual input voltage V_i and the feedback voltage. Working backward to find the actual input voltage, we must *subtract* the feedback voltage from the net voltage. This is done in Fig. 16-2. Notice that the total phase shift between the actual

*F. E. Terman, *Electronic and Radio Engineering* (4th ed.), New York: McGraw-Hill Book Company, 1955, chap. 11.

input voltage and the output voltage is now much closer to 180°. Phase distortion has been reduced.

INVERSE FEEDBACK CIRCUITS.   There are two general types of circuits used for inverse feedback—*voltage-feedback* and *current-feedback* circuits. In either case, a *voltage* is fed back. The method of obtaining this voltage and the factor controlling the amount of this voltage will differ. With voltage feedback, the amount of feedback is some fixed percentage of the output voltage and nearly independent of the load current. Because the takeoff network is in parallel with the load, voltage-feedback circuits are also called *shunt-feedback* circuits. In current-feedback circuits, the amount of feedback is directly proportional to the load current and the takeoff is in series with the load. This type of feedback is therefore also referred to as *series feedback*. Another distinction is also made in the way the feedback voltage (whether obtained by voltage feedback or by current feedback) is applied to the input. If the feedback is applied in series with the input signal it is called *series-fed*. Conversely, in a *shunt-fed* circuit the feedback is applied in parallel or in shunt with the input signal. As we will see later, the takeoff method affects the output impedance of the circuit whereas the method of feed affects the input impedance.

Feedback circuits can also vary as to the number of stages in the feedback loop. This can span from the output to the input of the same stage, or further back to the input of the previous stage, or can even include three stages. Feedback for more than three stages is seldom attempted because of the possibility of instability (discussed later in this chapter).

A very simple feedback circuit is shown in Fig. 16-3. Basically this circuit is a single-stage R-C coupled FET amplifier with the source bypass capacitor omitted. This introduces feedback. Let us assume that the ac input signal is swinging so as to make the gate positive with respect to

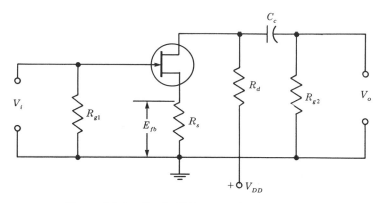

Figure 16-3   Simple FET current-feedback circuit.

source, causing a rise in drain current. The ac component of drain current will flow *up* through the source resistor R_s, making the source positive with respect to ground. This feedback voltage opposes the input signal. It is inverse feedback. Since the voltage is proportional to the load current, it is current feedback. The effective ac output load resistance R_L (at mid-frequencies) is equal to the parallel resistance of R_d and R_{g2}. The same load current flows through R_s as through R_L. Therefore, the ratio of feedback voltage to output voltage is the same as the resistance ratio, and the percent feedback is

$$\beta = \frac{R_s}{R_L + R_s} \times 100 \qquad (16\text{-}7)$$

In the circuit of Fig. 16-3, the resistor R_s also serves as the dc bias resistor. The same value of resistance may not be satisfactory for dc bias and for feedback. In such cases, the larger resistance value is used and correction is then made to reduce the feedback or the bias. These modifications are shown in Fig. 16-4. In the first circuit, the dc bias is produced across the total resistance $R_1 + R_2$. However, since R_2 is bypassed the feedback voltage is developed only across R_1, thereby reducing the percent feedback. When a high amount of feedback is desired, the source resistor could be so large as to produce excessive dc bias. One correcting technique is shown in Fig. 16-4b. By returning the gate resistor to the tap between R_1 and R_2, the dc bias is reduced to the voltage drop across R_1 alone. The combination of $R_1 + R_2$ is effective for producing the feedback voltage.

Current feedback can be applied to vacuum-tube circuits by using an unbypassed cathode-bias resistor. Since the circuits are identical to FET circuits, they need not be shown here. Again, as with FETs, the variations shown in Fig. 16-4 can be used if the desired cathode bias or percent feedback requires different-sized resistors.

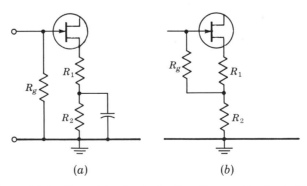

(a)   (b)

Figure 16-4   Current-feedback circuits with correction for dc bias.

The same techniques can also be used with BJTs. An unbypassed emitter resistor will provide inverse feedback in an R-C coupled CE amplifier. A typical circuit is shown in Fig. 16-5. Resistor R_E provides the inverse feedback in addition to bias stabilization. If more bias stabilization (than ac inverse feedback) is desired, the value of R_E is increased by using two resistors, but part of it (R_{E2}) is bypassed as shown in Fig. 16-5b. However, the proper amount of base bias is obtained from the voltage divider R_1–R_2. (Any of the other bias circuits discussed in Chapter 11 could have been used.)

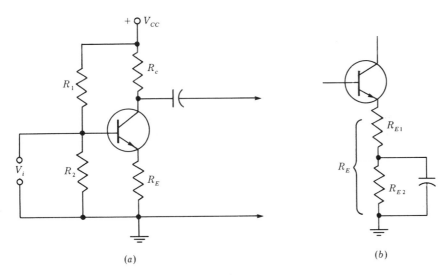

Figure 16-5   An *NPN* current-feedback circuit.

Two simple FET *voltage-feedback* circuits are shown in Fig. 16-6. A portion of the FET output voltage is fed back into the gate circuit. Since the input and output voltages from a single stage are 180° apart, proper phasing for negative feedback is obtained.* Either circuit, (a) or (b), could be used with R-C or transformer-input coupling. In Fig. 16-6a, the feedback voltage (developed across R_1) is shunt-fed, or in parallel with the input signal voltage. However, the effective feedback resistance is not just R_1 but some equivalent resistance R_1' which combines the shunting effect of the previous stage and coupling network (r_d and R_d). The percent feedback for this circuit is then

For shunt-fed circuit:
$$\beta = \frac{R_1'}{R_2 + R_1'} \times 100 \qquad \textbf{(16-8)}$$

*Capacitor C_1 is sufficiently large that its reactance at the signal frequency causes no phase shift.

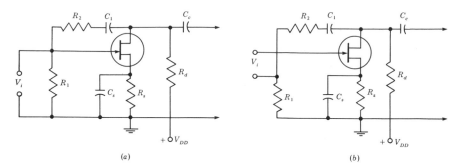

Figure 16-6 Single-stage FET voltage-feedback circuits.

The circuit of Fig. 16-6*b* is simpler. The feedback voltage is in series with the signal voltage (or series-fed). There is no shunting action across the feedback resistor, and the percent feedback is simply

For series-fed circuit:
$$\beta = \frac{R_1}{R_1 + R_2} \times 100 \qquad (16\text{-}9)$$

Voltage-feedback circuits using a BJT are shown in Fig. 16-7.* The feedback resistor, R_2, is also part of the base-bias voltage divider. Notice, however, that R_2 is connected to the collector side of the load instead of to the power supply. A positive-going input signal will forward-bias this transistor, increasing the collector current—but dropping the collector potential—and producing a negative-going output voltage. This output, 180° reversed from the input, is fed back to the base by the voltage divider R_2-R_1. In Fig. 16-7*a* shunt feed is used, and R_1 is shunted not only by the previous stage values (as for the FET circuit) but also by the BJT input resistance. The effective feedback resistance, R_1', is the parallel combination of all these components. The percent feedback is then determined using Equation 16-8. In the series-fed circuit (Fig. 16-7*b*) the feedback factor depends only on R_1 and R_2, and Equation 16-9 is used.

MULTISTAGE FEEDBACK.   It has been found in multistage amplifiers that better gain stability and better corrective action are obtained by using an overall feedback loop (from output to input) instead of an individual feedback loop for each stage. However, one point must be carefully observed: for negative feedback the feedback voltage must always "oppose" the input signal voltage. Let us consider a BJT circuit. If the feedback voltage is 180° out of phase with the base input signal, the feedback can be injected into the base circuit (series-fed or shunt-fed as shown in Fig. 16-7).

*Again no vacuum-tube circuits need be shown, because these are identical to the FET circuits.

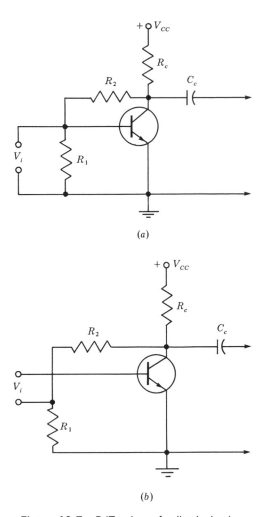

(a)

(b)

Figure 16-7 BJT voltage-feedback circuits.

On the other hand, if the feedback voltage is in phase with the base input signal and the feedback is again injected into the base circuit, it would aid the base signal. This is positive feedback and may cause oscillations. However, this same in-phase feedback voltage—*if fed to the emitter*—will oppose the base input signal producing inverse feedback. In other words, regardless of the phase of the feedback voltage (compared to the input signal), we can still get inverse feedback by injecting the feedback voltage either to the base or to the emitter, as appropriate. Two methods for injecting the feedback voltage into the emitter circuit are shown in the partial schematics of Fig. 16-8.

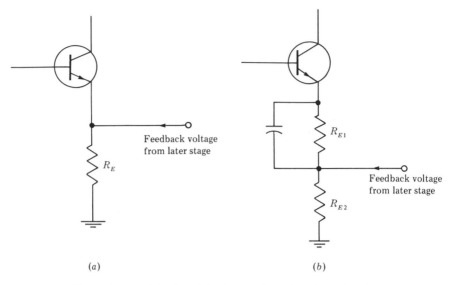

(a) (b)

Figure 16-8   Injection of feedback voltage into the emitter leg.

Refer to Fig. 16-8a. If the feedback voltage from the later stage is removed, resistor R_E still provides *current* feedback within its own stage. In reality, then, this circuit provides local feedback as well as overall multistage feedback. However, when designed for multistage feedback the resistor value (R_E) is quite small, and so the local feedback action can generally be neglected. Notice that resistor R_E also provides stabilization of the dc bias. If the resistor is too small, this stabilization may be inadequate. In that case, a larger value of resistance can be used—if it is split into two parts, R_{E1} and R_{E2}, as shown in Fig. 16-8b. The full value, $R_{E1} + R_{E2}$, will give dc stabilization; but because R_{E1} is bypassed, only R_{E2} will give ac multistage inverse feedback. A complete two-stage feedback circuit is shown in Fig. 16-9. Notice that a positive input signal applied to Q_1 would produce a negative-going signal at the base of Q_2 and a positive-going signal at the collector of Q_2. To obtain negative feedback, the feedback signal must therefore be applied to the emitter of Q_1 as shown. (If this were a three-stage circuit, the output from Q_3 would be negative and the feedback signal would be applied to the base of Q_1.)

The gain calculations for a multistage amplifier with overall feedback follow the technique shown earlier for single-stage feedback. Equation 16-4 is still valid, but this time the gain without feedback (or open-loop gain), A, is the overall gain of all the stages in the loop. The feedback factor β is also found as explained before. Similarly, distortion calculations use Equation 16-6. Let us apply these techniques to a problem.

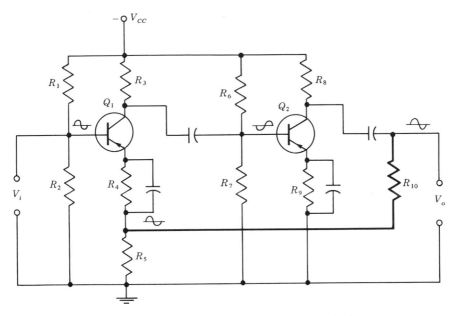

Figure 16-9 A two-stage negative feedback circuit.

EXAMPLE 16-3

Each stage of the two-stage amplifier in Fig. 16-9 has a normal (open-loop) gain of 80 and an overall harmonic distortion of 10 percent. In the feedback loop, R_{10} is 10,000 Ω and R_5 is 100 Ω. Calculate:

1. The gain with feedback
2. The distortion with feedback

Solution

1. Since the circuit uses voltage feedback,

$$\beta = \frac{R_5}{R_{10} + R_5} = \frac{100}{10,000} = 0.0099$$

(a) The overall gain without feedback is

$$A = A_1 \times A_2 = 80 \times 80 = 6400$$

(b) The gain with feedback is

$$A' = \frac{A}{1 + A\beta} = \frac{6400}{1 + 6400(0.0099)} = 99.3$$

2. Distortion with feedback is

$$D' = \frac{D}{1 + A\beta} = \frac{10}{1 + 6400(0.0099)} \times 100 = 0.155\%$$

With push-pull circuits, if the feedback is taken from the collector of the BJT as in Fig. 16-9 two feedback loops would be necessary—one from each side of the push-pull circuit. (This would also apply to a vacuum-tube circuit.) The circuitry can be simplified by taking the feedback voltage from the secondary of the output transformer and feeding it to the emitter of an earlier stage as in Figs. 16-8 and 16-9. This has another advantage in that it also tends to correct for any distortion originating in the output transformer itself. To obtain the proper amount of feedback, resistor voltage dividers are sometimes used. However, since most output transformers have tapped secondary windings, the proper tap is merely selected. The polarity of the feedback voltage from the secondary winding depends on the "phasing" of the winding. It can be reversed by grounding the opposite end of the winding. In this way, this circuit can be used to feed back over an odd or even number of stages.

EFFECT OF FEEDBACK ON OUTPUT IMPEDANCE.*   It was mentioned earlier that the way the feedback signal is obtained will affect the output impedance of the stage. For example, with a constant input signal level a voltage feedback circuit tends to maintain the output voltage constant, regardless of variations in the load. Obviously, if the load current increases, the output voltage will remain constant only if the internal resistance decreases. In other words: *voltage feedback reduces the output resistance* of the amplifier. This is of special importance in power amplifiers. It can be shown mathematically† that

For voltage feedback:
$$R_o' = \frac{R_o}{1 + A\beta}$$
(16-10)

where R_o is the output resistance without feedback, R_o' is the output resistance with voltage feedback, A is the gain without feedback, and β is the negative feedback factor.

Conversely, using current feedback the load current tends to remain constant—as from a constant-current source. This implies high output resistance: *current feedback increases the output resistance*. Mathematically,† if the feedback resistor is small compared to the output resistance,

For current feedback:
$$R_o'' = R_o(1 + A\beta)$$
(16-11)

Because of this increase in output resistance, current feedback should be avoided in power amplifiers, particularly if the output is direct-coupled to a low-resistance load.

---

*This discussion applies equally well to the collector resistance of a BJT, the drain resistance of a FET, and the plate resistance of a vacuum tube.

†J. D. Ryder, *Electronic Fundamentals and Applications* (4th ed.), Englewood Cliffs, N.J.: Prentice-Hall, Inc., 1970, chap. 12.

EFFECT OF FEEDBACK ON INPUT RESISTANCE.   The input resistance of an amplifier (whether BJT, FET, or vacuum-tube circuit) is affected by the application of a feedback voltage. In a series-fed circuit, the input resistance will increase. This is of great advantage in BJT small-signal amplifiers. You will recall that their input resistance is generally quite low and they load down the previous stage, reducing its possible gain. With series-fed inverse feedback, the increase in input resistance will reduce this loading effect. Mathematically, the increase in input resistance is given by*

For series feed: $$R_i' = R_i(1 + A\beta) \tag{16-12}$$

EXAMPLE 16-4

An R-C coupled *NPN* transistor amplifier has an input resistance of 1500 Ω and a gain of 60, without feedback. Twenty percent inverse feedback is applied, using an unbypassed emitter resistor. Calculate the input resistance with feedback.

Solution

Feedback into the emitter leg is a form of series-fed circuit. Therefore

For series feed: $R_i' = R_i(1 + A\beta) = 1500(1 + 60 \times 0.2) = 19,500 \ \Omega$

The gain of the amplifier in Example 16-4 is reduced by feedback from 60 to approximately 5 (that is, $1/\beta$). However, the gain of the previous stage is dependent on the total equivalent load resistance R_e, which is in turn dependent mainly on the input resistance of this stage. By increasing this input resistance from 1500 to 19,500 Ω, we increase the gain of the previous stage by about the same factor as the loss of gain due to feedback. The overall two-stage gain will remain about the same, but now we have reduced the distortion and have stabilized the operation.

If the feedback voltage is applied in shunt with the input signal (i.e., is shunt-fed), the input resistance of the amplifier will decrease:

With shunt feed: $$R_i'' = \frac{R_i}{1 + A\beta} \tag{16-13}$$

When an amplifier is driven from a low-impedance source (for example, a dynamic microphone or a magnetic phono cartridge), a low input impedance is necessary to achieve maximum power transfer from the source. Since transformer coupling is not desirable (because of bulk, cost, and frequency response), inverse feedback can be used to effect such a match.

---

*J. D. Ryder, *Electronic Fundamentals and Applications* (4th ed.), Englewood Cliffs, N.J.: Prentice-Hall, Inc., 1970, chap. 12.

EXAMPLE 16-5

The BJT amplifier of Example 16-4 is to be used as a microphone amplifier. The microphone impedance is 50 Ω. What change in the feedback circuit will be needed?

Solution

1. Since we wish to reduce the input impedance, a shunt-fed feedback circuit must be used.
2. To find the percent feedback needed to reduce the input resistance to 50 Ω, Equation 16-13 can be rewritten as

$$\beta = \frac{R_i/R_i'' - 1}{A}$$

$$= \frac{1500/50 - 1}{60} = 0.483$$

$$= 48.3\%$$

STABILITY OF FEEDBACK CIRCUITS.   So far, in the discussion of negative feedback we have taken an ideal view. The feedback voltage has been assumed as exactly reversed in phase (180°) with respect to the input signal. At mid-frequencies this may be so, but at very high or very low frequencies this may be far from true. Let us see how such a condition might affect a feedback circuit. Assume that the feedback loop covers two stages. At mid-frequencies, since each stage introduces a normal 180° phase reversal, the input and output voltages are in phase. To produce degeneration, the feedback voltage must be applied to the emitter, source, or cathode. But at some low frequency each stage may have a phase shift of only 115° as compared to the normal 180° reversal. Two such stages would result in an output voltage shift of 230° (115 + 115) lagging, or 130° leading. These phase relations can be seen in Fig. 16-10. The second-stage output voltage is essentially "out of phase" with the input voltage. *If this were fed back, for example, to the emitter of a BJT, positive* feedback or *regeneration* would result. The overall gain would tend to increase, and if the feedback effect were enough, the circuit would break into oscillations. The same effect can also occur at high frequencies.

Stability can also be analyzed mathematically from the gain equation. In our earlier discussion, the gain A' with feedback was given as

$$A' = \frac{A}{1 + A\beta}$$

But if the phase of the feedback voltage is reversed, positive feedback results and the denominator becomes $(1 - A\beta)$. Now if the nominal gain (A) is high enough or the percent feedback (β) is high, such that their

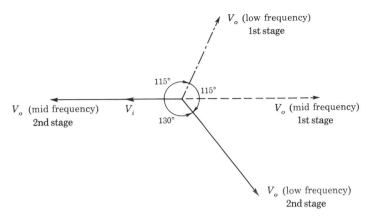

Figure 16-10   Effect of low-frequency phase distortion in a two-stage amplifier.

product $(A\beta)$ is 1 (or greater), the numerator becomes zero (or negative) and oscillations will occur.* This possibility increases as more stages are included in the feedback loop.

PARAPHASE AMPLIFIERS

One of the problems encountered with push-pull amplifiers is that these circuits require a "balanced input." Or, more specifically, two input signals are required, one for each base of the push-pull pair, and these input signals should be equal in magnitude but opposite in phase. The output from a normal amplifier stage is only a single voltage. In Chapter 15, to convert from such a "single-ended" output to a balanced output, a transformer with a center-tapped or split secondary winding was used (see Fig. 15-14). Such a transformer is known as a push-pull input transformer. However, transformers are not desirable. They are bulky, heavy, and costly, and their frequency response is rather limited. Because of these drawbacks of transformer coupling, a number of R-C coupled circuits have been devised that will take a single-ended signal and produce two output voltages that are essentially equal in magnitude and in phase opposition. These circuits are known as *paraphase amplifiers*. They are sometimes also referred to as *phase inverters*. (It should be realized that any R-C coupled amplifier is in reality a phase inverter, since its input and output are 180° out of phase. Yet an R-C amplifier does not produce a balanced output.)

PHASE-SPLITTER CIRCUIT.   One of the simplest paraphase circuits is the circuit shown in Fig. 16-11. This circuit is often called a *phase-splitter.* (It is also variously known as the *split-load* circuit. The term *split-load* is

*H. Nyquist, "Regeneration Theory," *Bell System Technical Journal*, Jan. 1932.

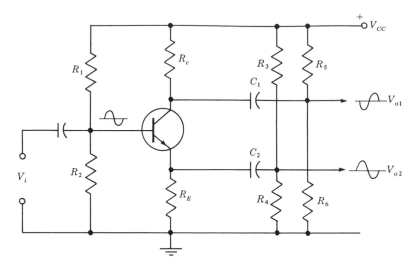

Figure 16-11 The "phase-splitter" paraphase amplifier.

well taken, in that what would have been the normal load value for the "standard" R-C circuit is here split equally between the collector resistor R_c and the unbypassed emitter resistor R_E. An output voltage is then developed across each of these resistors.

Let us consider the phase relations in this circuit. During the interval when the input signal V_i is rising from zero to maximum positive, the collector current must also increase from its quiescent (I_{CQ}) to its maximum value. The voltage drop in the collector resistor is increasing, the collector potential is dropping, and the voltage V_{o1} fed to the next stage will rise from zero to its negative maximum. This is the normal action that takes place in any R-C amplifier. Meanwhile, since this rising collector current also flows *up* through R_E, the emitter potential will rise to a positive maximum. The coupling capacitor C_2 will remove the dc component due to the quiescent current, and an ac output will be developed across the resistor R_4. This voltage, V_{o2}, is in phase with the collector current and in phase opposition to V_{o1}.

We have obtained two output voltages 180° out of phase. Also, their amplitudes are equal because the same collector current flows through each resistor R_c and R_E, and the resistors are equal in value. The common 20 percent tolerance resistors should not be used for R_c and R_E. The degree of balance depends on the accuracy of their values. Five percent tolerance would therefore be better.* In fact, matching the two resistor values is often

---

*The imbalance, caused by the emitter current being slightly higher than the collector current, is generally negligible.

resorted to. Another advantage of this circuit is that aging or replacement of the BJT does not unbalance the outputs. Any change in the BJT characteristics may affect the collector current value, but the effect will be the same in R_c as in R_E. The balance of the output voltages is not affected. Imbalance may arise at some high frequency because the shunting capacitance effects from collector to ground are not necessarily equal to the emitter-to-ground value. Capacitance can of course be added to the lower of these two values to restore balance.

If the BJT in a normal R-C circuit (Fig. 16-11) had a gain of 40 and the input voltage were 1 V, we might expect to get 40 V for V_{o1}. But this ignores the effect of inverse feedback due to the unbypassed emitter resistor. Since this is current feedback, the percent feedback can be found by using Equation 16-7. Neglecting the shunting effect of the resistors R_4 and R_6, this reduces to

$$\beta = \frac{R_E}{R_c + R_E} \times 100$$

and since $R_c = R_E$,

$$\beta = 50\%$$

The maximum possible gain is

$$A' \cong \frac{100}{\beta} = \frac{100}{50} = 2$$

With a 1-V input, the *total* output voltage approaches 2 V as a maximum. This 2-V output is divided equally across R_c and R_E. The paraphase output voltages V_{o1} and V_{o2} approach but do not quite equal 1 V each.

If we consider the overall output, the gain of the phase-splitter circuit is less than 2; or if we consider *each* output voltage, less than 1. At first thought, it might seem that this circuit has no value. But consider our goal. It is not to provide amplification, but to produce a balanced output suitable for feeding a push-pull circuit. Compared to a transformer, the phase-splitter circuit is much cheaper, less bulky, and much lighter, and it has a far wider frequency response range. Similar circuits are used with FETs and vacuum tubes. The drawing of such circuits is left to the student (see Problems 30 and 31 at the end of this chapter).

DUAL-FET VOLTAGE-DIVIDER CIRCUIT.   To overcome the lack of gain of the previous circuit, several dual-unit circuits are in use. One of these is shown in Fig. 16-12. Q_1 is connected in standard manner as an R-C coupled amplifier. Its input and output voltages are self-explanatory. There is only one slight departure from the conventional circuit: the normal gate-return resistor is in two parts, R_1 and R_2. Let us assume that the actual gain of this stage (Q_1) is 40. This means that for a 40-V output (V_{o1})

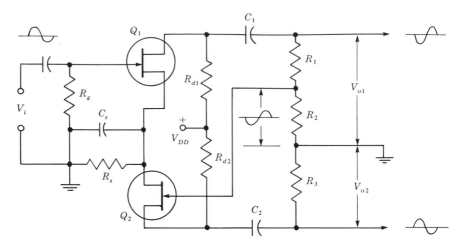

Figure 16-12 Dual-FET voltage-divider type, paraphase amplifier.

the input V_i is 1 V. Suppose the resistors R_1 and R_2 were deliberately and carefully chosen so that R_2 is exactly one-fortieth of the total gate-return resistance $(R_1 + R_2)$. Then the voltage across R_2 would be 1 V, or exactly the same magnitude as the input voltage V_i. But notice that the phasing of this voltage V_{R2} is in phase opposition to V_i. If we feed this voltage (1 V) to the gate of Q_2 and the Q_2 circuit values are otherwise identical with Q_1, the output voltage V_{o2} will be equal and opposite to V_{o1}. We have produced a balanced output. In addition this circuit provides voltage gain.

For proper balance with this circuit, the drain resistors R_{d1} and R_{d2} should have equal values, the gate-return circuits R_3 and $R_1 + R_2$ should have equal resistances, and the voltage divider ratio $R_2/(R_1 + R_2)$ must be equal to the reciprocal of the stage gain. In cases where exact balance is desired, R_1 is made somewhat lower in value and R_2 is replaced with a potentiometer of correspondingly higher resistance value. The gate of Q_2 is connected to the potentiometer arm. The setting is then adjusted to equalize V_{o1} and V_{o2}. This also corrects for small inequalities in the drain resistors and in the FET characteristics. Unfortunately this circuit is not self-balancing. Unequal aging of the FETs or their replacement with unmatched FETs can cause serious imbalance. With a potentiometer divider, the outputs can be rebalanced. At low frequencies the circuit tends to produce unbalanced outputs because the output V_{o2} is attenuated by an additional coupling capacitor. To compensate, capacitor C_2 is sometimes made larger than C_1. Somewhat better balance, particularly at the lower frequencies, is obtained by removing the bypass capacitor and introducing current feedback. This will of course also reduce the gain. Similar circuits are used with vacuum tubes and with BJTs. The drawing of such circuits is left to the student. (See Problems 34 and 35.)

FLOATING PARAPHASE CIRCUIT.   A disadvantage of the above circuit is that, even after it has been carefully balanced, it can be thrown out of balance by changes during operation. By a slight modification in wiring and in component values, a self-balancing circuit is produced. Such a circuit, shown in Fig. 16-13, is known as the *floating paraphase*. The change in wiring is the rearrangement of the resistors R_1, R_2, and R_3. The change in component values also affects the same resistors. All three resistors are equal in value, and the value used is in the range generally used for gate-return circuits.

To analyze the action of the circuit, let us first completely neglect the effect of FET Q_2. Q_1 then acts as a normal amplifier. Its output voltage is developed across $R_1 + R_2$ and the output polarity is in phase opposition to the input signal. Part of this output voltage is developed across R_2. Therefore the potential at point P will also be in phase opposition to the input signal. If we start with an input signal V_i that is going positive, the *ac component* of drain current I_{d1} will flow down through R_1 and *to the right* in R_2. (This is indicated on the diagram by a solid arrow.) Since the input signal is going positive, point P will swing negative. But this negative-going voltage is applied to the gate of FET Q_2. Its drain current will decrease, which in turn means that the ac component I_{d2} is negative, or flowing from ground through resistors R_2 and R_3 toward the drain of the FET. This current is shown by the dashed arrow. Notice that the drain currents of both FETs flow through the resistor R_2, but these currents are in opposition

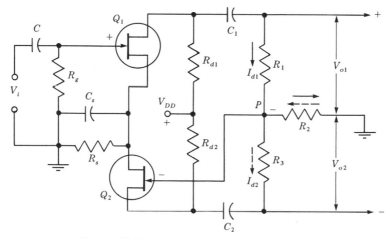

Figure 16-13   The floating paraphase circuit.

and the "net" current through R_2 decreases. Obviously, these two ac components cannot be equal; otherwise V_{R2} fed to Q_2 would be zero and current I_{d2} could not exist. Actually I_{d1} is slightly greater than I_{d2}, so as to make V_{R2} just enough smaller than V_i for Q_1 to cause this slight unbalance in collector currents. If for any reason an imbalance were to occur, such as V_{o2} increasing (compared to V_{o1}), then I_{d2} would increase, the net current through R_2 would decrease, and the signal voltage applied to Q_2 would decrease, offsetting the tendency for output voltage to increase. Balance is restored. Similarly, any change in V_{o1} (compared to V_{o2}) is counteracted in like manner. In other words, the circuit is self-balancing. Actually, the circuit is never in perfect balance, but is always slightly off-balance. This off-balance is reduced by using higher-gain FETs. Furthermore, if perfect balance of the two output voltages is desired, R_1 should be made slightly larger than R_3. In this way, V_{o2} can equal V_{o1} even though I_{d1} is slightly greater than I_{d2}.

Another commonly used paraphase amplifier—particularly with integrated circuitry—is the differential amplifier with single-ended input and double-ended output. (See Figs. 14-8 and 14-9.) Since this circuit was analyzed in Chapter 14, no further discussion is needed at this time.

FOLLOWER CIRCUITS

Figure 16-14 shows an *NPN* transistor circuit with the output taken from across the emitter resistor R_E. On the positive half of the input signal, the forward bias is increased, the emitter current increases, and the potential of the emitter rises. Obviously, the output voltage is in phase with the input signal. From this the circuit gets the name *emitter follower*. Notice

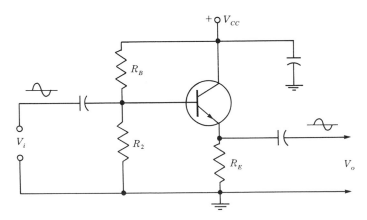

Figure 16-14 Emitter follower.

also that the collector is at ac ground potential because of the bypass action of capacitor C_1. Since the collector is effectively grounded (as far as the input signal is concerned), it can be considered as common to both the input and the output side of the circuit. This connection is called a *common-collector* circuit.

Note also, in Fig. 16-14, that the emitter resistor is not bypassed. Therefore, inverse or negative feedback is present. Both the output voltage and the feedback voltage are developed across this resistor. In fact, they are one and the same voltage. Since all the voltage is fed back, the feedback is 100 percent. Also, regardless of the load current, the feedback level remains the same: 100 percent. Therefore, this is a *voltage*-feedback circuit.

Using the approximate gain equation (16-5), the maximum possible *voltage* gain from this circuit is

For emitter followers:
$$A_v' \cong \frac{100}{\beta} \cong \frac{100}{100} \cong 1$$

The exact formula will show that the gain is actually somewhat less than 1.* However, the importance of this circuit is not in its gain, but in its impedance relationships.

Similar circuits are used with FETs and with vacuum tubes. They are shown in Fig. 16-15. Again the maximum possible voltage gain is only 1, and the usefulness of such circuits is due to the input-output impedance relationships.

EFFECTIVE INPUT AND OUTPUT IMPEDANCES.   From the schematic diagrams of the follower circuits discussed above, it is obvious that the

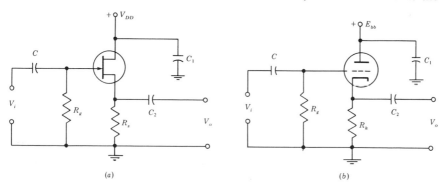

(a)   (b)

Figure 16-15   Source and cathode followers.

*The exact equation (from Equation 16-4) would be

For follower circuits:
$$A_v' = \frac{A}{1 + A}$$

feedback voltage is in series with the input signal. Therefore, as in any series-fed feedback circuit, the input impedance is increased. With FETs (Fig. 16-15c) it can be seen that the source is positive with respect to ground. This biases the gate negative with respect to source, and gate current cannot flow. The input *resistance* of the FET is infinite. Even if a strong positive signal were applied, the drain current would increase, making the source potential more positive and maintaining the gate negative compared to source. The input resistance remains infinite. In addition, because the voltage gain is less than 1 the Miller effect capacitance and the total input capacitance are very much lower than in a conventional amplifier. As a result, follower circuits have appreciably higher input impedance than standard amplifier circuits. Consequently they produce negligible loading on the circuits to which they are connected.

Now let us consider the output impedance aspects. The capacitive portion is unaffected and remains as previously discussed. (Also, at low to moderate frequencies their effect is generally negligible.) On the other hand, the drain resistance is reduced because of voltage feedback to

$$1. \quad r_d' = \frac{r_d}{1 + A_v}$$

In turn the voltage gain for this FET amplifier, but without feedback, is

$$2. \quad A_v = g_{fs}r_d$$

and, substituting step 2 into step 1, the effective drain resistance is

$$r_d' = \frac{r_d}{1 + g_{fs}r_d} \tag{16-14}$$

Using this value in the Norton equivalent diagram for a source follower, we get the circuit of Fig. 16-16. The FET is replaced with a constant-voltage source and a series resistance r_d'. Since the source is considered as having

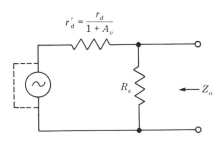

Figure 16-16   Equivalent circuit for the source follower.

zero resistance (and again neglecting capacitance effects), the output impedance looking into this circuit consists of r_d' in parallel with R_s:

$$Z_o = \frac{[r_d/(1 + g_{fs}r_d)] \times R_s}{[r_d/(1 + g_{fs}r_d)] + R_s}$$

Simplifying by combining the denominator, inverting, and multiplying, we get

$$Z_o = \frac{r_d R_s}{r_d + R_s(g_{fs}r_d + 1)} \tag{16-15}$$

Then, since the product $g_{fs}r_d$ is generally appreciably greater than 1, the term $(g_{fs}r_d + 1)$ can be reduced to $g_{fs}r_d$, and dividing both numerator and denominator by r_d,

$$Z_o = \frac{R_s}{1 + g_{fs}R_s} \tag{16-16}$$

Finally, if the transconductance g_{fs} and/or the source resistor are high the output impedance approaches $1/g_{fs}$ as a maximum value:

$$Z_o \cong \frac{1}{g_{fs}} \tag{16-17}$$

EXAMPLE 16-6

A 2N4868 JFET has a transconductance g_{fs} of 3000 μmho and an output conductance g_{os} of 4 μmho. It is used as a source follower with $R_s = 3300 \ \Omega$. Find:

1. The gain of the stage (using both the accurate and the approximate method)
2. The output resistance (accurate and approximate)

Solution

1. To calculate the gains:

$$A_v \text{ (without feedback)} = g_{fs}r_d = \frac{g_{fs}}{g_{os}} = \frac{3000 \times 10^{-6}}{4 \times 10^{-6}} = 750$$

$$A_v' \text{ (with feedback)} = \frac{A_v}{1 + A_v} = \frac{750}{1 + 750} = 0.987$$

$$A_v' \text{ (approx)} \cong \frac{1}{\beta} = \frac{100}{100} = 1.0$$

2. To calculate the output impedance:

$$Z_o = \frac{R_s}{1 + g_{fs}R_s} = \frac{3300}{1 + 3000 \times 10^{-6} \times 3300} = 303 \ \Omega$$

$$Z_o \cong \frac{1}{g_{fs}} = \frac{1}{3000 \times 10^{-6}} \cong 333 \ \Omega$$

With BJTs we have similar, but not identical, relationships. For example, although the input resistance of an emitter follower is much higher than with a CE connection, it is not infinite. Furthermore, the increase in input resistance depends on the current gain A_i. Summarizing the relations for the emitter follower in terms of its hybrid parameters,* we have

Current gain: $$A_i = \frac{1 + h_{fe}}{1 + h_{oe}R_E} \tag{16-18}$$

Input resistance: $$R_i' = h_{ie} + A_i R_E \tag{16-19}$$

Output admittance: $$Y_o' = h_{oe} + \frac{1 + h_{fe}}{h_{ie} + R_E} \tag{16-20}$$

Output resistance: $$R_o' = \frac{1}{Y_o'} \tag{16-21}$$

EXAMPLE 16-7

The 2N2483 *NPN* silicon transistor has the following parameter values: $h_{ie} = 1.5$ to $13\ \Omega$, $h_{fe} = 80$ to 450, $h_{oe} = 30\ \mu$mho. It is used in an emitter-follower circuit with an emitter resistor of $2700\ \Omega$. Find the range of input and output resistances for this transistor, using both minimum and maximum parameter values.

Solution

1. To find the input resistance:

$$(a)\ A_i = \frac{1 + h_{fe}}{1 + h_{oe}R_E} = \frac{1 + 80}{1 + (30 \times 10^{-6} \times 2700)} = 100\ \text{(min)}$$

$$= \frac{1 + 450}{1 + (30 \times 10^{-6} \times 2700)} = 556\ \text{(max)}$$

$$(b)\ R_i' = h_{ie} + A_i R_E = A_i R_E = 100 \times 2700 = 270\ \text{k}\Omega\ \text{(min)}$$
$$= 450 \times 2700 = 1.21\ \text{M}\Omega\ \text{(max)}$$

2. To find the output resistance:

$$(a)\ Y_o' = h_{oe} + \frac{1 + h_{fe}}{h_{ie} + R_E} = 30 \times 10^{-6} + \frac{1 + 80}{1.5 + 2700} = 30.0\ \text{mmho (min)}$$

$$= 30 \times 10^{-6} + \frac{1 + 450}{13 + 2700} = 166\ \text{mmho (max)}$$

$$(b)\ R_o' = \frac{1}{Y_o'} = \frac{10^3}{30} = 33.3\ \Omega$$

$$= \frac{10^3}{166} = 6.02\ \Omega$$

*J. Millman and C. Halkias, *Integrated Electronics,* New York: McGraw-Hill Book Company, 1972, chap. 8.

From the above discussion and examples, it can be seen that follower circuits have a very high input impedance and a very low output impedance, compared to conventional amplifier circuits.

APPLICATIONS.   We have already seen how an iron-core transformer can be used for impedance matching. We also saw the disadvantages of the transformer—limited frequency range, harmonic distortion, phase shift, and high cost. Now we see that because of their input-output impedance relation, follower circuits can be used to match a high-impedance source to a low-impedance load. Matching the output impedance of the follower to the load can be effected by selecting a device with suitable parameters and by choosing the proper value for the source or the emitter resistor. Because this circuit uses 100 percent feedback, distortion of any type is very low and frequency response range is excellent. Furthermore, from Figs. 16-14 and 16-15 it is obvious that the cost of these circuits should not be a problem.

The low output impedance of these follower circuits is also used to advantage when it becomes necessary to interconnect two pieces of equipment that are some distance apart. The distributed capacitance of a connecting cable would represent a serious loading effect with a conventional amplifier. By feeding the cable from a follower circuit, the loading effect is greatly reduced; higher frequencies can be carried by the line and over longer distances. In such applications, the follower *isolates* the cable and its loading effect from the last stage of the "source" equipment. Even without long interconnecting cables, follower circuits are often used between two stages of a piece of equipment whenever the loading effect of the input impedance of the second stage would seriously affect the output of the first stage. The follower is a natural coupling device. Its high input impedance produces negligible loading on the previous stage, while its low output impedance makes it insensitive to loading by the next stage. Although it contributes no gain itself, use of a follower circuit makes it possible (because of reduced loading) to increase the gain and to extend the frequency response of the "regular" amplifier stages.

BOOTSTRAP AMPLIFIER.   We have seen that the input resistance of the follower circuits is very high (infinite for the FET and 270 kΩ or more for the BJT of Example 16-7). But this is the input resistance of the *device alone*. Unfortunately, bias resistors shunt the transistor and seriously reduce this input resistance. (See R_B and R_2 for the BJT of Fig. 16-14 and see R_g for the FET and tube of Fig. 16-15.) To overcome the effect of the bias resistors, a circuit modification called *bootstrapping* is used. This modification is shown in Fig. 16-17a for the BJT. Notice the addition of resistor R_1 and capacitor C_1 as compared to Fig. 16-14. Capacitor C_1 should have negligible reactance at the lowest signal frequency, so that as far as the ac signal is concerned, the bottom end of R_1 is tied to the emitter. We know

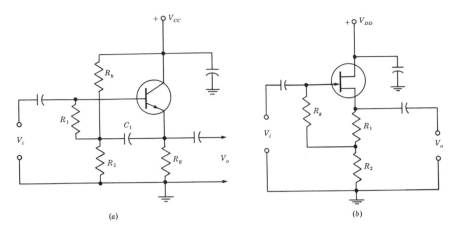

Figure 16-17 Bootstrapped follower circuits.

that the voltage gain of this circuit is slightly less than 1. Let us use a value of 0.99. (See Example 16-6.) Then the output voltage V_o is 99 percent of the input voltage and the signal voltage across R_1 is 1 percent of V_i. On the other hand, if R_1 were connected to ground (as is R_2 in Fig. 16-14), the full input voltage would be across it. Obviously, the ac signal current through R_1 is only 1 percent (or one-hundredth) of the signal current in the grounded circuit. The resistor acts as if it were 100 times larger. Mathematically, the effective resistance of the R_1, R_B, R_2 combination becomes*

$$R_{\text{eff}} \text{ (bootstrap bias)} = \frac{R_1}{1 - A_v} \tag{16-22}$$

EXAMPLE 16-8

A BJT used as a follower (Fig. 16-17a) has a voltage gain of 0.99 and a device input resistance of 400 kΩ. The bootstrap resistors are $R_1 = R_2 = 56$ kΩ and $R_B = 33$ kΩ. Find:

1. The input resistance of the stage without bootstrap components R_1 and C_1
2. The input resistance with bootstrapping

Solution

1. Without bootstrapping, R_B in parallel with R_2 shunts the transistor input. Therefore

$$R_i \text{ (without bootstrapping)} = R_B \parallel R_2 \parallel \text{device input resistance}$$

$$= (33 \parallel 56 \parallel 400) \text{ k}\Omega = 19.8 \text{ k}\Omega$$

---

*J. Millman and C. Halkias, *Integrated Electronics*, New York: McGraw-Hill Book Company, 1972, chap. 8.

2. With bootstrapping:

(a) R_{eff} (bootstrap bias) $= \dfrac{R_1}{1 - A_v} = \dfrac{56 \text{ k}\Omega}{1 - 0.99} = 5.6 \text{ M}\Omega$

(b) $\quad\quad R_i$ (bootstrap) $= R_{\text{eff}} \parallel$ device input resistance

$\quad\quad\quad\quad\quad\quad = 5.6 \text{ M}\Omega \parallel 400 \text{ k}\Omega = 374 \text{ k}\Omega$

Notice the vast improvement due to bootstrapping. The bias resistors in a normal emitter-follower circuit reduce the input resistance from $400 \text{ k}\Omega$ to $19.8 \text{ k}\Omega$. The bootstrap circuit raises this input resistance back to $374 \text{ k}\Omega$.

Even with source followers, bootstrapping is used when a very high input resistance is required. Although the input resistance of the FET *alone* is infinite, the shunting action of the gate resistor may make the input resistance *of the circuit* lower than desired. In such cases, the circuit of Fig. 16-17b can be used. Notice that the source resistor is in two parts, R_1 and R_2. The R_1 portion sets the desired quiescent point bias. The input resistance of this bootstrap circuit is raised to

$$R_i \text{ (FET bootstrap)} = \frac{R_g}{1 - A_v} \tag{16-23}$$

EXAMPLE 16-9

A FET follower circuit, operating at high ambient temperature, uses a gate resistor of only $120 \text{ k}\Omega$ to avoid biasing effects from excessive gate leakage current. Calculate its input resistance using a bootstrap circuit as in Fig. 16-17b. (Assume a gain of 0.97.)

Solution

$$R_i \text{ (FET bootstrap)} = \frac{R_g}{1 - A_v} = \frac{120 \times 10^3}{1 - 0.97} = 4.0 \text{ M}\Omega$$

This bootstrap circuit has another advantage. If R_g were returned to ground as in Fig. 16-15a, the full $R_1 + R_2$ value could make the gate appreciably negative compared to source. A strong negative signal on top of this could readily drive the FET into cutoff. Instead, by returning the gate resistor to the tap between R_1 and R_2 the no-signal gate bias is decreased, reducing the danger of driving the FET into cutoff. This circuit can be used in pulse applications and can handle strong negative pulses without distortion from cutoff.

MIXING CIRCUITS

There are two types of circuits used in electronic applications that are known as *mixers*. Their function and principle of operation are radically different. It is unfortunate that both are called mixers. Yet because their functions are so different, there should be little doubt as to which one is specifically meant in any application. One of these circuits is part of a frequency-changing system. Two input signals are applied and an output is produced *at the difference frequency.* Since this application is generally in RF communications circuits, it will not be discussed at this time.

The other type of mixer again accepts two input signals, but the output contains both signals. A better name would have been *combining amplifier.* Such mixing circuits are used in a variety of applications. In a sound studio, the signals picked up by several microphones may have to be mixed to produce complete sound coverage; or it may be necessary to add commentary to some other sound pickup. In Loran timing equipment, marker signals from various timing oscillators need to be mixed to produce the overall time axis. In the Armstrong FM transmitter, sidebands and carrier must be mixed to produce phase modulation. Regardless of the application, the circuits used fall into several categories.

SERIES-RESISTANCE MIXING.   This type of mixing circuit, shown in Fig. 16-18, is the simplest of all. The two signal sources are connected across suitable potentiometers. The output from each potentiometer is then series-fed to the input of the amplifier. The level from each source is independently controlled by its own potentiometer. Except for its simplicity, there is nothing to recommend use of this circuit. Neither side of the channel *A* signal is at ground potential. Hum can be easily picked up by this signal source. Also, stray capacitances between channel *A* and ground will tend to bypass or short-circuit the high-frequency signals in channel *B*.

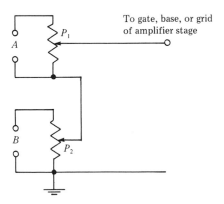

Figure 16-18   Series-resistance mixing.

PARALLEL-RESISTANCE MIXING.   The hum pickup danger and the bypassing action of the above circuit can be eliminated by connecting the individual source potentiometers in parallel. This is shown in Fig. 16-19a. The output from each channel is controlled by its own potentiometer. Resistors R_1 and R_2 are necessary to reduce the interaction between the two channels. For example, assume that channel B is set for zero output (see Fig. 16-19b). *If resistors R_1 and R_2 were not used,* the base of the BJT and any output from channel A would be shorted to ground through the potentiometer arm of channel B. Obviously, the larger the value of these isolating resistors, the less the interdependency. It is common practice to make these resistance values equal to or greater than the potentiometer resistance values.

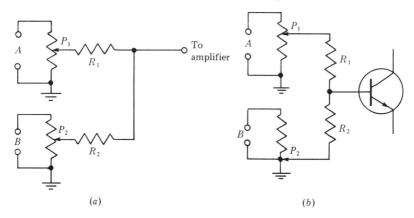

(a)   (b)

Figure 16-19   Parallel-resistance mixer.

Unfortunately, because of these isolating resistors this circuit (Fig. 16-19) has an "insertion loss"; that is, all of the signal from either channel can never reach the base of the BJT. For example, consider again channel A set for maximum signal level and channel B set for zero. The lower end of resistor R_2 is therefore at ground potential. Resistors R_1 and R_2 act as a second voltage divider across the output of potentiometer A. For equal-value resistors, only half of the input signal will reach the base. This is an insertion loss of 6 dB (20 log 2/1). Of course, as the setting of potentiometer B is raised, the resistance to ground increases and the insertion loss decreases, but only slightly.

ELECTRONIC MIXING — COMMON LOAD.   To reduce interaction without this insertion loss, each signal can be fed to a separate amplifier and the output of each device can be combined in a common load resistor. Such a circuit, using FETs, is shown in Fig. 16-20. Since the drain current of both FETs flows through the common drain load resistor R_d, the value of this resistor is only half of the value that would be used in a standard R-C FET amplifier. This tends to reduce the gain. However, a more serious

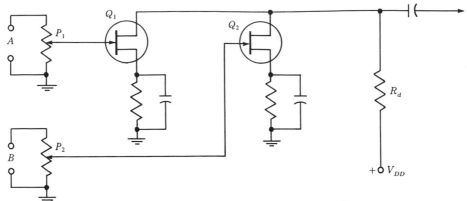

Figure 16-20 Electronic mixing—common R_d.

disadvantage is that the drain resistance of each FET acts as a shunting load on the other FET. For example, if r_d is 10,000 Ω and R_d is 30,000 Ω, the effective drain load resistance is $r_d \| R_d$ or only 7500 Ω. Because of this low effective drain load resistance, not only is there a loss in gain but, even more detrimental, there is a drastic limitation on the maximum output voltage swing obtainable without distortion. For this reason, the FETs used in this circuit should have high internal resistance (low g_{os}) to minimize the shunting action.

The identical circuitry could be used with vacuum tubes, and with normal bias-circuit modifications it applies also to BJTs. The drawing of such circuits is left to the student (see Problem 51).

ELECTRONIC MIXING—SEPARATE LOADS.   The shunting effect discussed above can also be avoided by using separate drain load resistors and adding *isolating resistors*. The circuit, using FETs, is shown in Fig. 16-21.

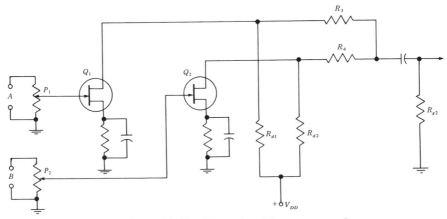

Figure 16-21 Electronic mixing—separate R_d.

The individual drain resistors are R_{d1} and R_{d2}. Notice that they are separated from each other by the isolating resistors R_3 and R_4. Each drain resistor now has the full value that would be used in any R-C amplifier, and the isolating resistors are made from one to four times the R_d values. These isolating resistors also cause some loss in gain, but not nearly so much as the shunting effect of the other FET. At the same time, this circuitry also prevents the distortion that would occur from too low a load resistance value. Again, with suitable modifications, this circuit can be used with BJTs and vacuum tubes.

An ideal electronic mixing circuit with separate loads—but without the insertion loss of the isolating resistors—is the differential amplifier with dual inputs and single-ended output. Again BJTs, FETs, or vacuum tubes could be used. Since these circuits were discussed in Chapter 14, they need not be repeated here. This type of mixer is generally used with integrated circuits.

REVIEW QUESTIONS

1. Refer to Fig. 16-1: **(a)** What does the "normal" gain refer to? **(b)** What is the purpose of the feedback network? **(c)** What does the letter β signify? **(d)** What is the value of β corresponding to feedbacks of 10 percent, 15 percent, 23 percent? **(e)** If β is a positive value, what does this mean with respect to the feedback voltage? **(f)** Give two names for this type of feedback. **(g)** What does a negative β value signify with respect to the feedback voltage? **(h)** Give three names for this type of feedback. **(i)** Using the voltage symbols shown in this figure, write the equation for $V_{i(net)}$ for negative feedback.

2. Explain a possible effect from positive feedback.

3. **(a)** What is the effect on the overall gain of an amplifier when using inverse feedback? **(b)** Give the equation for gain with inverse feedback. **(c)** Give another name for "gain with feedback." **(d)** Give another name for "gain without feedback." **(e)** Give a name used to represent the product $A\beta$.

4. What is the meaning of the term *decibel feedback*?

5. **(a)** What is the disadvantage of using inverse feedback? **(b)** State three advantages.

6. Give an illustration to show the importance of gain stability.

7. Refer to Table 16-1: **(a)** What is the net input voltage in each case? **(b)** How is this net input obtained? **(c)** What is meant by the nominal gain? **(d)** In row 1 of this table, how is the output voltage of 40 V obtained? Feedback of 10 percent obtained? Feedback voltage of 4 V obtained? Total input of 5 V obtained? Overall gain of 8 obtained? **(e)** In case (*a*) of the table, what is the gain variation without feedback? With 10 percent feedback? **(f)** In case (*c*), what is the gain variation without feedback? With 10 percent feedback? **(g)** Draw a conclusion from a comparison of cases (*a*) and (*c*). **(h)** Draw a conclusion from a comparison of cases (*d*), (*e*), and (*f*).

8. Give an equation that shows the maximum possible gain that can be obtained from a circuit using inverse feedback.

9. Refer to Fig. 16-2: **(a)** Which phasor is used as the reference? **(b)** What is V_{fb}? **(c)** If this amplifier were used without feedback, (1) what input would be needed to produce V_o? (2) What would the phase shift be between input and output? **(d)** What does V_i represent? **(e)** How are the value and angle for V_i obtained? **(f)** What is the phase shift between input and output, with feedback?

10. (a) Name two general types of feedback circuits. **(b)** In which type of circuit is a *current* fed back into the input? Explain. **(c)** What determines into which category a specific circuit belongs? **(d)** Name two further classifications of feedback circuits, based on how the feedback is injected into the input circuit.

11. Refer to Fig. 16-3: **(a)** How does this circuit differ from a conventional R-C amplifier stage? **(b)** Why is this a current feedback circuit? **(c)** Explain why this is *inverse* feedback. **(d)** What determines the percent feedback in this circuit? **(e)** Give the equation for the percent feedback.

12. Refer to Fig. 16-4a: **(a)** What basic type of feedback does this circuit use? **(b)** What component or components are effective in producing this feedback? **(c)** Why is R_2 bypassed? **(d)** Explain why R_2 is not omitted.

13. With reference to Fig. 16-4b: **(a)** Which component or components are effective in producing feedback? **(b)** Why is R_g connected to the junction of R_1 and R_2 rather than to ground?

14. Refer to Fig. 16-5: **(a)** What type of feedback is used in diagram (a)? **(b)** What component is responsible for the feedback? **(c)** How is the percent feedback evaluated? **(d)** What is the function of R_1 and R_2? **(e)** When is the modification shown in diagram (b) used?

15. Refer to Fig. 16-6a: **(a)** What basic type of feedback does this circuit use? **(b)** How does this circuit differ from a conventional R-C amplifier? **(c)** How does capacitor C_1 affect the feedback voltage? **(d)** What is the purpose of this capacitor? **(e)** What component, or components, determine the amount of feedback produced? **(f)** Give the equation for the percent feedback.

16. Refer to Fig. 16-6: **(a)** What is the essential difference between diagrams (a) and (b)? How can these circuits be distinguished by name? **(c)** Give the equation for the percent feedback in Fig. 16-6b.

17. Refer to Equation 16-8: **(a)** What does the symbol R_1' stand for? **(b)** Why must we use R_1' instead of just R_1 (as in Equation 16-9)?

18. Refer to Fig. 16-7: **(a)** Which of these circuits uses current feedback? **(b)** Which uses shunt-fed voltage feedback? **(c)** Which uses series-fed voltage feedback? **(d)** Do Equations 16-8 and 16-9 apply to these BJT circuits?

19. Comparing Figs. 16-6a and 16-7a, notice that capacitor C_1 is missing from the feedback network in the latter diagram. **(a)** Is this an error? **(b)** Why is it left out?

20. (a) Can a feedback voltage be fed from the output of stage 2 (or stage 3) to the input of stage 1? **(b)** What is such feedback called?

21. (a) What is the correct phasing of the feedback voltage for injection into the gate of a FET? **(b)** Into the base of a BJT? **(c)** If the feedback voltage is of opposite phase, can it still be used? Explain.

22. Refer to Fig. 16-8: **(a)** What should the phasing of the feedback voltage be compared to the input signal? **(b)** What is the effect of resistor R_{E1} on the feedback? Explain. **(c)** When is the circuit of Fig. 16-8b used in preference to that of 16-8a?

23. Refer to Fig. 16-9: **(a)** What is the phase of the output voltage V_o compared to the input voltage V_i? **(b)** Why is this so? **(c)** What is the phase of the feedback voltage applied to Q_1, compared to the input signal to Q_1? **(d)** Will this be positive or negative feedback? Explain. **(e)** What components determine the percent feedback in this circuit? **(f)** Give the equation.

24. With push-pull output circuits, give two reasons for taking the feedback voltage from the secondary of the output transformer instead of from the primary side.

25. It is desired to increase the output resistance of an amplifier. **(a)** Should we use current feedback or voltage feedback? **(b)** Should we inject it using a series-fed or a shunt-fed circuit? Explain.

26. What type of feedback should *not* be used in a power amplifier stage? Explain.

27. It is desired to increase the input resistance of an amplifier. **(a)** Should we use current feedback or voltage feedback? Explain. **(b)** Should we inject it using a series-fed or a shunt-fed circuit? **(c)** Give the equation for the input resistance of an amplifier with a series-fed feedback circuit.

28. **(a)** What is the effect of voltage feedback on the output resistance of an amplifier? **(b)** Give the equation.

29. **(a)** When might we want to reduce the input resistance of an amplifier circuit? **(b)** What type of feedback circuit would we use now? **(c)** Give the equation for the input resistance with shunt-fed feedback.

30. Refer to Fig. 16-10: **(a)** How many stages of amplification are being considered? **(b)** At mid-frequency, what is the phase relation between the input and output voltages? **(c)** If inverse feedback were desired (from output to input), to what point should the feedback connection be made? **(d)** At some low frequency, what is the indicated phase relation between input and output? **(e)** With the feedback connection as in part (c) above, what would happen? Explain.

31. **(a)** What is meant by the stability of a feedback circuit? **(b)** How does the number of stages in a feedback loop affect stability? **(c)** How does percent feedback affect stability?

32. **(a)** State two basic functions of a paraphase amplifier. **(b)** By what other names are these circuits also known? **(c)** Compare the desirability of these circuits versus transformer coupling.

33. Refer to Fig. 16-11: **(a)** Give two other names commonly used for this circuit. **(b)** How do the values of R_c and R_E compare with the values used in a conventional R-C amplifier? **(c)** Explain how the phasing shown for the output voltage V_{o2} is obtained. **(d)** Are any component values critical? Explain. **(e)** Is this circuit self-balancing? Explain. **(f)** How much gain can be obtained from this circuit? Explain. **(g)** If the bias developed by this circuit is too high, how can this be corrected?

34. Refer to Fig. 16-12: **(a)** Explain how the phasing shown for the output voltage V_{o2} is obtained. **(b)** What two components are most important in determining the equality of the output voltages? **(c)** Explain how the value of the input voltage for Q_2 is made equal to the input voltage for Q_1. **(d)** What determines the relative value of R_1 and R_2? **(e)** Are the drain resistor values critical? Explain. **(f)** Is this circuit self-balancing? Explain. **(g)** How much gain can be obtained from this circuit?

35. Refer to Fig. 16-13: **(a)** What do the two arrows above resistor R_2 represent? **(b)** Can these two currents be equal? Explain. **(c)** Can I_{d2} be greater than I_{d1}? Explain.

(d) Explain why output V_{o2} has a positive polarity. **(e)** What are the relative values of R_1, R_2, and R_3? **(f)** Give an approximate value for R_2. **(g)** Is this circuit self-balancing? Explain. **(h)** Can exact balance of the output voltages be obtained with identical components? Explain.

36. (a) Give another name for the circuit shown in Fig. 16-14. **(b)** Why is it so named?

37. Refer to Fig. 16-14: **(a)** If a negative-going signal is applied to the base of this circuit, what is the polarity of the output voltage? Why? **(b)** What is the phase relation between input and output voltage in an emitter follower? **(c)** Does this circuit have any feedback? **(d)** What kind and how much feedback? **(e)** What is the maximum possible gain obtainable?

38. (a) Is it possible to drive the gate of a source follower positive with respect to its source? Explain. **(b)** What is the input resistance of this circuit? **(c)** Compare the input capacitance of this circuit with that of the same FET used as an R-C amplifier. Explain. **(d)** How does the input impedance of a source follower compare with that of a standard amplifier? Explain.

39. (a) How does the output impedance of a source follower compare with that of a standard FET amplifier? Explain. **(b)** What is the major factor that determines this output impedance?

40. (a) Is the input resistance of an emitter follower higher or lower than in a standard CE amplifier? Explain. **(b)** Is it infinite, as in a source follower? Explain.

41. (a) Under what condition can a follower circuit be used for impedance matching? **(b)** What advantage does this circuit have compared to a transformer?

42. (a) Explain what is meant when a follower circuit is used for isolation. **(b)** Give two examples of this application.

43. Refer to Fig. 16-17a: **(a)** What two components are responsible for the bootstrapping effect? **(b)** What is the advantage of bootstrap circuits over standard follower circuits?

44. Give another advantage of the bootstrap circuit in Fig. 16-17b over the circuit in Fig. 16-15b.

45. Give three applications of a mixer circuit when used to combine several signals.

46. (a) What is the simplest type of mixing circuit? **(b)** Why is this circuit not desirable?

47. Refer to Fig. 16-19: **(a)** What is the function of resistors R_1 and R_2? **(b)** What are these resistors called? **(c)** What disadvantage is created by these resistors? **(d)** With the channel controls set as in Fig. 16-19b and assuming each channel voltage to be 10 V, how much of channel A voltage is applied to the base of the BJT (1) if resistors R_1 and R_2 are not used? (2) If these resistors are used?

48. Refer to Fig. 16-20: **(a)** Explain why isolating resistors are not used to reduce interaction between channels. **(b)** Does this circuit have an insertion loss? **(c)** How does this value of R_d compare with the values used in a standard R-C amplifier? **(d)** Does this R_d value constitute the effective drain load resistance? Explain. **(e)** State two disadvantages of this effect. **(f)** What type of FET minimizes this effect? Explain.

49. Refer to Fig. 16-21: **(a)** What is the purpose of the additional resistors used in this circuit as compared to Fig. 16-20? **(b)** How do the values of R_{d1} and R_{d2} compare to R_d of Fig. 16-20? **(c)** How do the values of R_3 and R_4 compare to R_{d1} and R_{d2}? **(d)** Does this circuit have an insertion loss? Explain.

50. (a) Name another circuit (discussed in an earlier chapter) that makes an ideal two-channel mixer. **(b)** Does this circuit have any insertion loss? Explain.

PROBLEMS AND DIAGRAMS

1. An amplifier has a normal gain of 50. If it is used with 5 percent negative feedback, **(a)** find the maximum possible gain. **(b)** Find the actual gain. (Save this answer for Problem 4.)

2. The amplifier of Problem 1 is modified to incorporate 30 percent feedback. Find **(a)** and **(b)** as above. (Save this answer for Problem 5.)

3. An amplifier with a normal gain of 80 is used with 20 percent feedback. Express the feedback in decibels.

4. Using the answer from Problem 1(b), express the feedback in decibels.

5. Using the answer from Problem 2(b), express the feedback in decibels.

6. The amplifier of Problem 1 has an amplitude distortion of 10 percent before feedback. Find the distortion level with feedback.

7. The amplifier of Problem 3 has an amplitude distortion of 8 percent before feedback. Find the distortion level with feedback.

8. An amplifier has a gain of 120 and a distortion level of 12 percent. It is desired to reduce the distortion to not more than 2 percent. **(a)** What is the minimum amount of feedback necessary? **(b)** What is the gain with this feedback?

9. Draw the circuit diagram of a simple current-feedback circuit, using a P-channel FET.

10. In the circuit of Fig. 16-3, R_s is 3300 Ω, R_d is 33.3 kΩ, $R_{g2} = 1$ MΩ, V_{DD} is 50 V, and the FET parameters are $g_{fs} = 3000$ μmho and $g_{os} = 4$ μmho. Find: **(a)** the gain without feedback; **(b)** the percent feedback; **(c)** the gain with feedback.

11. Draw a circuit diagram for a current-feedback circuit with relatively low percent feedback and a high dc gate bias. Use a P-channel FET.

12. Draw a circuit diagram showing a current-feedback circuit with high percent feedback and low dc gate bias. Use a P-channel FET.

13. Refer to Fig. 16-4a: If $R_1 = 1000$ Ω, $R_2 = 2000$ Ω, and R_L (not shown) is 40,000 Ω, find the percent feedback.

14. Find the percent feedback for the circuit of Fig. 16-4b, using the same resistance values as in Problem 13.

15. Draw the circuit diagram for a simple shunt-fed current-feedback circuit using a *PNP* transistor. The circuit should provide for more dc bias stabilization than ac feedback.

16. Draw the circuit diagram for a shunt-fed voltage-feedback circuit using an N-channel FET.

17. Repeat Problem 16 for a series-fed P-channel FET.

18. The circuit of Fig. 16-6a has the following component values: $R_1 = 0.27$ MΩ, $R_2 = 180$ kΩ, $R_d = 47$ kΩ. The FET parameters are $g_{fs} = 4000$ μmho, $g_{os} = 10$ μmho. This stage is R-C coupled, on each side, to identical stages. Find: **(a)** the gain without feedback; **(b)** the percent feedback; **(c)** the gain with feedback.

19. Repeat Problem 18 for the circuit of Fig. 16-6b, but with this stage transformer coupled to the previous stage and with $R_1 = 12$ kΩ.

20. Draw the circuit diagram for a shunt-fed voltage-feedback circuit using a *PNP* transistor.

21. Repeat Problem 20 for a series-fed *NPN* circuit.

22. Draw the circuit diagram for an R-C coupled FET amplifier. Show how a feedback voltage that is in phase with the gate signal can be used for inverse feedback.

23. In the circuit of Fig. 16-9, without feedback, the gain of each stage can vary from 45 to 120 when transistors are replaced. Also, the amplifier has a harmonic distortion level of 15 percent. The feedback loop used has $R_{10} = 8200$ Ω and $R_5 = 390$ Ω. Find: **(a)** the overall gain range without feedback; **(b)** the overall gain range with feedback; **(c)** the distortion level range with feedback.

24. Draw a circuit diagram showing push-pull *NPN* transistors fed from a transformer-coupled *NPN* stage and with feedback from the secondary of the output transformer to the input stage.

25. A FET ($g_{os} = 4$ μmho) is used in an amplifier circuit with a gain of 50. Find the effective drain resistance r'_d when 12 percent voltage feedback is added.

26. It is desired to increase the effective drain resistance of the FET in Problem 25 to 0.8 MΩ. How can this be done? Give values.

27. A two-stage *NPN* amplifier has an input resistance of 1200 Ω and a gain of 6400. Overall feedback is introduced as in Fig. 16-9, using 8200 Ω for R_{10} and 390 Ω for R_5. Find: **(a)** the percent feedback; **(b)** the input resistance with feedback.

28. A two-stage amplifier (as in Problem 27) is to be used as a phono amplifier from a 100-Ω magnetic pickup. How can the input impedance of the amplifier be matched to this source? Give values.

29. Draw the circuit diagram for a split-load paraphase amplifier using a *P*-channel FET. Show input and output phase relations.

30. Repeat Problem 29, using an *NPN* silicon transistor.

31. Repeat Problem 29, using a triode.

32. Draw the circuit diagram for a dual-FET voltage-divider paraphase amplifier using *P*-channel FETs. Indicate input and output phase relations.

33. In the circuit of Fig. 16-12, the FETs and components are chosen so as to produce a voltage gain of 56, with $R_3 = 470$ kΩ. Find the proper values for R_1 and R_2.

34. Repeat Problem 32, using *PNP* transistors.

35. Repeat Problem 32, using triodes.

36. Draw the circuit diagram for a floating paraphase amplifier, using *P*-channel FETs.

37. Draw the circuit diagram for a *PNP* transistor used as an emitter follower. Indicate input and output phase relations.

38. Repeat Problem 37, using a *P*-channel FET.

39. The FET of Fig. 16-15 has a g_{fs} of 4000 μmho and a g_{os} of 10 μmho. The circuit values are $R_g = 1$ MΩ and $R_s = 3900$ Ω. Find the output impedance of the stage **(a)** by using the approximate method and **(b)** by using the exact equation.

40. Repeat Problem 39 for a 2N5558 N-channel junction FET with $y_{fs} = 6500\ \mu\text{mho}$, $y_{os} = 20\ \mu\text{mho}$, and $R_s = 2700\ \Omega$. (Consider conductance values equal to admittance values at these frequencies.)

41. A 2N6007 with $h_{fe} = 235$, $h_{ie} = 3\ \text{k}\Omega$, and $h_{oe} = 30\ \mu\text{mho}$ is used as an emitter follower with an emitter resistor of 3300 Ω. Find the input and output resistances of this transistor. (Save these answers for Problem 45.)

42. Repeat Problem 41 for a 2N3241A with $h_{fe} = 175$, $h_{ie} = 400\ \Omega$, $h_{oe} = 100\ \mu\text{mho}$, and an emitter resistor of 2700 Ω.

43. Draw the circuit diagram for a bootstrap emitter-follower circuit using a *PNP* transistor.

44. Draw the circuit diagram for a bootstrap source-follower circuit using a *P*-channel FET.

45. The emitter-follower circuit of Problem 41 and Fig. 16-14 has bias resistors $R_B = R_2 = 47\ \text{k}\Omega$. **(a)** Find the input resistance of the *stage*. **(b)** A bootstrap arrangement as in Fig. 16-17a is used, adding $R_1 = 47\ \text{k}\Omega$. Find the new value of the *stage* input resistance. (Assume a gain of 0.98.)

46. A FET source follower uses a gate resistor of 0.1 MΩ. Assuming a gain of 0.98, calculate the input resistance of the *circuit,* using **(a)** the connections of Fig. 16-15a and **(b)** the connections of Fig. 16-17b.

47. Draw the circuit diagram for a simple two-channel series-resistance mixing circuit using an *NPN* transistor. Show bias and output connections.

48. Repeat Problem 47, using an *N*-channel FET.

49. Draw the circuit diagram for a three-channel parallel-mixing circuit using a *PNP* transistor. Show bias and output connections.

50. Repeat Problem 49, using a *P*-channel FET.

51. Draw the circuit diagram for a two-channel electronic mixing circuit, with common load, using *NPN* transistors. Show bias circuitry.

52. Repeat Problem 51, using *P*-channel FETs.

53. Repeat Problem 51, using separate collector load resistors and isolating resistors.

54. Repeat Problem 51, using *P*-channel FETs, separate drain load resistors, and isolating resistors.

55. Refer to Fig. 16-21: If R_{d1} and R_{d2} are each 33 kΩ and R_3 and R_4 are each 100 kΩ and R_{g2} is 1 MΩ, calculate the insertion loss due to these resistors (in decibels).

56. Draw the circuit diagram for a two-channel mixing circuit using a BJT differential amplifier.

57. Repeat Problem 56, using FETs.

17

Audio Devices

In the previous chapters on untuned amplifiers, it was mentioned that these amplifiers are used in a variety of applications. However, their greatest use by far is in the field of audio. Audio amplifiers are used in all types of home electronic sound equipment, from the small personal radio to the large AM/FM console, the color television receiver, the hi-fi stereo unit, and the electronic organ. They are used in all types of public address and paging systems, in all types of sound transmitting and receiving equipment. They are used by the telephone companies on long-line operations. If it is a form of electronics that involves sound, it incorporates an audio amplifier. It is therefore only fitting that some attention be given to the devices and special circuits used in the audio field. This chapter deals mainly with *transducers* —the devices that are used either to convert sounds into electrical signals or to convert electrical signals back into sounds. Special circuits used in the audio field will be treated in Chapter 18.

Let us first turn out attention to the various sources used to feed a signal to the amplifier itself. If the sound originates locally, a *microphone* is used to convert the sound waves to electrical waves. On the other hand, sounds can also be recorded on a record or on a magnetic tape. In such cases, the source is a *phono pickup* or a *playback head*. A brief description of each of these devices will be given in this chapter.

MICROPHONE RATINGS

A microphone is a device which converts sound waves into electrical variations. Depending on the method used to accomplish this, microphones can

454

be classified into various types such as *carbon, capacitor, dynamic, velocity, crystal,* or *ceramic.* Before we discuss the basic principle and characteristics of each of these types, let us study the terminology and systems generally used in rating microphones.

In evaluating or comparing the performance of microphones many properties can be investigated. The most important characteristics and the ones generally quoted in catalog listings are (1) directional properties, (2) frequency response, (3) electrical impedance, (4) output level or sensitivity.

DIRECTIONAL PROPERTIES.   The directional characteristics of a microphone may be indicated by a polar plot showing the output level variation of the microphone versus the angle of incidence of the sound wave (with respect to the microphone axis) for a sound of constant intensity and frequency. On this basis, microphones fall into three categories:

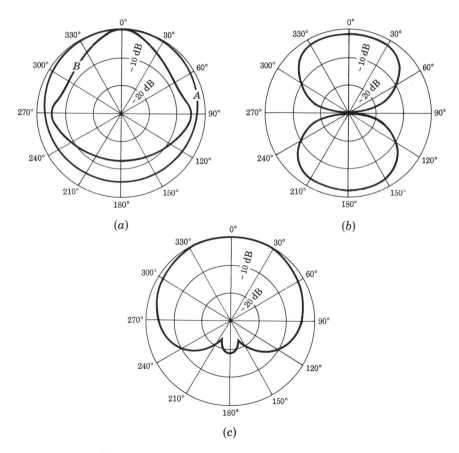

(a)

(b)

(c)

Figure 17-1 Microphone directional characteristics.

1. *Omnidirectional or nondirectional.*   Such a microphone picks up sound waves equally well from any direction. The polar plot for this type microphone is essentially circular, as shown in Fig. 17-1*a*.
2. *Bidirectional.*   This microphone is most sensitive to sounds along its axis, front or back. The output level decreases for sounds coming from an angle, and the pickup approaches zero for sounds at right angles to the microphone axis. See Fig. 17-1*b*.
3. *Unidirectional or cardioid.*   Maximum sensitivity for this microphone is for sounds along the axis *and in front.* Minimum pickup is obtained for sounds to the rear of the unit. This heart-shaped pattern is seen in Fig. 17-1*c*.

The selection of a microphone, with regard to its directional properties, depends on the specific application. For example, in a tête-à-tête or across-the-table interview, the bidirectional microphone is ideal. On the other hand, in a large hall with high audience noise level the cardioid microphone is necessary.

FREQUENCY RESPONSE.   Microphone frequency response is not nearly as good as is found or available in amplifiers. A microphone range of from 30 to 20,000 Hz is considered excellent and is obtainable only in the costliest models. However, wide range is not always needed. Microphones intended purely for paging (speech) may have a frequency response of only 200 to 4000 Hz. Notice that no decibel reference is given with these responses. Usually, in catalogs this rating is given as "uniform from 30 to 10,000 Hz" or "flat from 70 to 8000 Hz." In such cases you can expect at least a 3-dB drop at either end.

ELECTRICAL IMPEDANCE.   This refers to the impedance at the output terminals of the microphone. Since all microphones have reactive elements (L and/or C), the impedance will vary with frequency. The nominal impedance listed in catalogs is taken at a reference frequency of 1000 Hz. Depending on construction, microphone impedances may range from less than 1 Ω to over 1 MΩ. In many cases transformers are built into the microphone housing (particularly with low-impedance microphones) to provide a variety of output impedances, such as 50 Ω, 200 Ω, 500 Ω, or high impedance. The high impedance value is in the 30,000 to 50,000 Ω range.

OUTPUT LEVEL OR SENSITIVITY.   To provide adequate amplification to "fit" a microphone, it is necessary to know the level of output obtainable from the microphone. This knowledge is also desirable in comparing microphones or in selecting a microphone for a specific application. The sensitivity of microphones is expressed in decibels—gain or loss—*as compared to some reference level.* The basic idea would be simple enough except that many reference levels are used:

1. With high-impedance microphones, a *voltage* level of 1 V is generally used as the reference value.
2. With low-impedance units, a *power* level is preferred. The reference power level may be either 1 or 6 mW.
3. It should be obvious that the output level of any *one* microphone will depend on the intensity or loudness of the sound, or the sound pressure. To compare sensitivities, all measurements for output levels should be obtained at the same sound pressure level. However, three sound level values have been used—1 dyne per square centimeter, 10 dynes per square centimeter, and 0.0002 dyne per square centimeter.

The dyne per square centimeter is a unit of pressure used in the cgs (centimeter–gram–second) system, and it corresponds to a force of 1 dyne on an area of 1 sq cm. With relation to sound values:

1. The *peak* pressure value of normal speech is approximately 100 dynes per square centimeter up close to the lips.
2. At a distance of 1 ft, this peak pressure value drops to about 10 dynes per square cemtimeter.
3. The pressure decreases approximately 6 dB as the distance is doubled.

To complete the picture, one more term must be introduced—the *bar*. It is a unit of pressure equal to 1 million (10^6) dynes per square centimeter. The reference pressure levels of 10 dynes or 1 dyne per square centimeter would now be listed as 10 *microbars* or 1 *microbar,* respectively. Unfortunately, some catalog listings refer to these values as 10 bars or 1 bar, omitting the prefix *micro*. This should not cause confusion once it is realized that a sound pressure level of 10 bars is impossible and that 10 *micro*bars was the original intent.

To summarize, microphone sensitivity ratings could be listed as:

1. Decibels below 1 V per dyne per square centimeter (or per 10 dynes per square centimeter).
2. Decibels below 1 V per microbar (or per 10 microbars).
3. Decibels below 1 mW per dyne per square centimeter (or per microbar).
4. Decibels below 1 mW per 10 dynes per square centimeter (or per 10 microbars).
5. Decibels below 1 mW per 0.0002 dyne per square centimeter.*
6. Same as for (3) and (4) but with a 6 mW reference.

Some of this confusion is being clarified by using (1) and (2) only with high-impedance units and calling attention to voltage as the reference by

*This reference level is being advocated by EIA.

use of decibel volts (dBV) instead of just decibels. Similarly, system (3) or (4) is being used with low-impedance units and labeled "dBm." The use of 6 mW (system 6) as a reference level is becoming rare. One final point: a microphone that has been rated on a basis of decibels below 1 V per dyne per square centimeter can readily be converted to its 10 dynes per square centimeter value merely by adding 20 dB. The higher sound pressure (ten times) would cause a correspondingly higher *voltage* output. The decibel increase would be 20 log (pressure change) or 20 dB. Similarly, changing from a 1-dyne reference to 0.0002 dyne would *reduce* the output by 74 dB. Obviously, then a change from 10 dynes to 0.0002 dyne would correspond to a reduction by 94 dB. Be careful, however, not to try these conversions for comparing a voltage reference against a power reference.

EXAMPLE 17-1

A high-impedance microphone (25,000 Ω) has an output rating of 48 dB below 1 V per dyne per square centimeter. Find the open-circuit output voltage at this sound intensity.

Solution

Since the rating is *below* the reference level, the output voltage must be less than the 1-V reference level. Therefore:

$$20 \log \frac{V_r}{V_o} = 48$$

$$\frac{V_r}{V_o} = 2.51 \times 10^2$$

$$V_o = \frac{V_r}{2.51 \times 10^2} = \frac{1}{2.51 \times 10^2} = 3.98 \text{ mV}$$

EXAMPLE 17-2

A low-impedance microphone (200 Ω) has an output rating of 52 dB below 1 mW per 10 dynes per square centimeter. Find:

1. Its rated output in watts
2. The voltage output across 200 Ω
3. The average voltage developed by a pressure of 80 dynes per square centimeter due to close talking

Solution

1. Again, the output is less than the reference level, or:

$$10 \log \frac{P_r}{P_o} = 52$$

$$\frac{P_r}{P_o} = 1.59 \times 10^5$$

$$P_o = \frac{P_r}{1.59 \times 10^5} = \frac{1 \times 10^{-3}}{1.59 \times 10^5} = 6.3 \times 10^{-9} \text{ W}$$

2. $E = \sqrt{PR} = \sqrt{6.3 \times 10^{-9} \times 200} = 1.12 \text{ mV}$

3. With a sound pressure of 80 dynes per square centimer, the output power would be eight times higher and the voltage output would increase by the factor $\sqrt{8}$, or

$$E_{80} = E_{10}\sqrt{8} = 1.12\sqrt{8} = 3.17 \text{ mV}$$

The above microphone specifications were taken directly from manufacturers' literature. Notice what low output levels (voltage or power) are available from typical units.

TYPES OF MICROPHONES

If we survey catalog listings, we find that many types of microphones are available, based on constructional principles. Yet if we analyze these types more closely we find that there are only two *basic* principles:

1. The microphone acts as a varying impedance—its resistance or reactance varies with the sound amplitude or pressure. This type requires a source of power (dc), and the current in the circuit will vary in accordance with the microphone impedance. In this category are found the *carbon microphone* and the *capacitor microphone*.
2. The microphone acts as a generator, and the voltage generated is proportional to the sound amplitude or pressure. This category includes the *dynamic microphone,* the *velocity microphone,* the *crystal microphone,* and the *ceramic microphone.*

CARBON MICROPHONES.   The carbon microphone was originally developed for use in telephony. Such a unit is shown in Fig. 17-2a. As a sound wave strikes the diaphragm, it moves in and out in accordance with the sound pressure. This varies the pressure on the carbon granules, changing the electrical resistance between the carbon disks. These disks, in turn, are connected to the external circuit, as shown in Fig. 17-2b, and the no-signal current is adjusted to the manufacturers' rated value (this may range from 5 to 50 mA). Then, when a sound wave strikes the diaphragm, as the sound pressure varies the current in the circuit will vary and an alternating voltage will be induced in the output winding. This electrical output should be a replica of the sound wave.

This microphone is known as a *single-button* carbon microphone, the "button" referring to the box containing the carbon granules. Because of

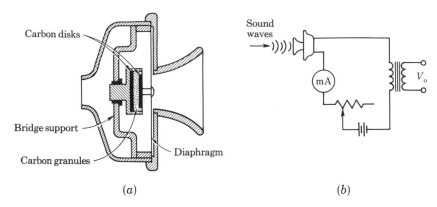

Carbon disks

Bridge support

Carbon granules

Diaphragm

Sound waves

mA

V_o

(a)   (b)

Figure 17-2   Carbon microphone.

the random changes in contact among the granules, a continuous hiss is heard with single-button microphones. To overcome this, a *double-button* unit was developed for use in radio applications. Its action is depicted diagrammatically in Fig. 17-3. The push-pull effect so created tends to balance out the hiss and any even-harmonic distortion.

Carbon microphones have impedance values that range from 50 to 200 Ω. Of course, since they are used with a transformer, this value can be modified as desired by using the proper turns ratio. These microphones have some drawbacks. Their frequency response, although adequate for speech, is not as good as that of the newer types. The distortion level is also higher, and even with the double-button construction the hiss is not completely eliminated. As a result, carbon microphones are no longer used for radio broadcasting. However, these units are appreciably more sensitive than any other type of microphone; they are rugged; and the initial cost is low. Consequently, they are still in use in portable equipment and wherever low cost and high output level are more important than fidelity considerations.

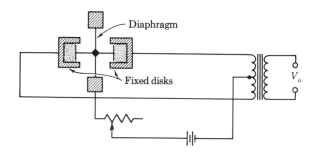

Diaphragm

Fixed disks

V_o

Figure 17-3   Action of a double-button microphone.

CAPACITOR MICROPHONES.   The capacitor microphone was at one time the standard for use in the radio broadcasting field. Its fidelity is superior to the carbon type, and it is free from the hiss common to the carbon units. Basically this microphone consists of two circular metal plates separated by a dielectric; or in other words, it is essentially a capacitor. One of the plates is the flexible diaphragm. Sound waves striking this diaphragm cause it to vibrate, thereby varying the distance between the plates. The change in capacitance value is proportional to the pressure changes due to the sound wave. The construction and basic circuit used with this type of microphone are shown in Fig. 17-4. To produce an electrical output, it is necessary to use a dc power source with this unit. In this respect, it is similar to the carbon microphone. This time, however, since the capacitance value is low, the reactance is high and a high-voltage source is needed. Supply voltages range from 200 to 600 V.

The operation of this microphone can be readily explained with reference to Fig. 17-4b. When the circuit is first energized, a charging current flows, charging the capacitor with the polarity as shown. The amount of charge Q depends on the capacitance value and the supply voltage ($Q = CE$). Once the capacitor is charged, the current drops to zero. This basic charging operation is therefore a transient effect. When the sound wave pressure causes the diaphragm to move inward, the capacitance increases. Now the capacitor can hold a higher charge. Again current flows, in the direction of the solid arrow. During the next half-cycle of the sound wave, the pressure decreases below normal; the diaphragm moves outward and the capacitance of the microphone decreases. Due to the reduced capacitance, it can no longer hold the previous charge. Electrons flow out of the lower plate and into the upper plate. This flow is shown by the dotted arrows. The reversal of current produces an alternating voltage across resistor R. This is the output voltage. Capacitor C couples this ac signal to the amplifier, at the same time blocking the dc component. The two components R and C act in the same manner as the collector (or drain) resistor and coupling capacitor of any R-C coupling network.

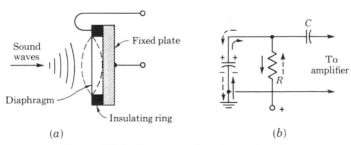

(a)   (b)

Figure 17-4   Capacitor microphone principle.

Unfortunately, this microphone has its drawbacks. We have already seen one — it requires a high-voltage source (usually referred to as the *polarizing voltage*). It also has extremely low output compared to any other type. In addition, it has a very high impedance. This high impedance creates two problems. First, the interconnecting leads are subject to stray electrostatic field pickup and, particularly because of the low signal level from the microphone, the stray field pickup may be stronger than the signal itself. The result is excessive noise and hum in the output. Second, even the small shunting capacitance of the connecting leads has a low reactance compared to the microphone impedance and will seriously impair the high-frequency response. Because these high-impedance effects are so drastic, it is necessary to mount the first stage of amplification directly within the microphone shell or stand. This microphone is seldom used in commercial applications because of these disadvantages, but it is still used in laboratory sound measurement tests.

CRYSTAL MICROPHONES.   This type of microphone belongs in the "generator" category; that is, it produces its own voltage and does not require an external power source. The voltage is produced by the *piezoelectric effect*. (The term *piezo* means pressure.) Certain crystalline materials such as Rochelle salts and quartz exhibit this effect. If they are subjected to a mechanical strain or pressure, a difference of potential will be developed across the opposite faces. Rochelle salt crystals are used for microphones because they produce a higher output voltage for a given mechanical strain. One type of construction uses two thin crystals cemented together in a differential arrangement termed a *bimorph* cell. When a sound wave strikes this composite plate, one crystal increases in length while the other decreases. This causes the bimorph cell to bend, much like the bending of two dissimilar metals when heated. The bending, in turn, produces the piezoelectric effect. Two such bimorph cells are used in a microphone as shown in Fig. 17-5*a*. Their voltages are series-additive to produce a still higher output.

If sensitivity is more important than wide-range frequency response (as for example in public address speech systems), the principle shown in Fig. 17-5*b* is used. The sound waves strike a diaphragm that is mechanically linked to the bimorph cell. Vibration of the diaphragm will cause mechanical distortion of the cell, producing an output voltage. The output from this diaphragm type of microphone is 10 to 15 dB higher than can be obtained from the direct action of the sound wave on the bimorph cell.

The crystal microphone has a high impedance and can be connected directly across the gate resistor of a FET amplifier.* This impedance is not

---

*With a BJT amplifier, an emitter-follower circuit and/or bootstrapping should be used to obtain a sufficiently high input impedance.

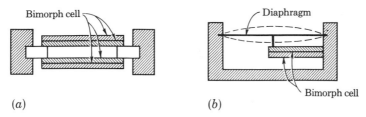

Figure 17-5   Crystal microphone construction.

nearly as high as for the capacitor microphone, so that the first amplifier stage can be remote from the microphone. However, because of the high impedance, the length of line should be kept reasonably short (less than 30 ft) to minimize shunt capacitance effects and the line should be shielded to reduce stray pickup. This microphone has some very desirable features. Electrically, it can have excellent frequency response (40 to 12,000 Hz); it has a relatively high output level; and it does not require an external power source. Mechanically, it is relatively rugged and lightweight. The microphone also has its drawbacks. The piezoelectric effect of the Rochelle salt crystal drops off rapidly as the temperature increases above 100°F (37.6°C), and the unit can be permanently damaged if exposed to temperatures above 125°F (46.1°C).

CERAMIC MICROPHONES.   It was discovered in 1946 that ceramic materials could be made piezoelectric. This led to the development of the *ceramic* microphone. In principle of operation it is similar to the crystal type. In fact it is almost exactly like the crystal microphone in all characteristics, with one added advantage: the ceramic unit will last indefinitely under adverse temperature and moisture conditions that would ruin the Rochelle salt crystal.

DYNAMIC MICROPHONES.   Figure 17-6 shows the basic constructional features of the *dynamic* microphone. When a sound wave strikes the diaphragm, it moves in and out in accordance with the sound pressure.

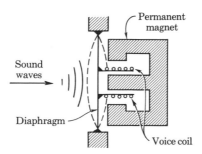

Figure 17-6   Dynamic microphone construction.

As the diaphragm vibrates, it causes the coil to move back and forth within the air gap of the permanent magnet structure. The coil cuts the magnetic field and a voltage is induced in the coil proportional to the sound pressure. Because of the motion of the coil, this microphone is also known as a *moving-coil* microphone.

Since the coil has relatively few turns, the microphone has a low output impedance (5 to 20 Ω). Usually an impedance-matching transformer is built into the microphone case itself. Then, either by selector switch or by attaching to the proper pair of secondary leads, various output impedances are available. Common values are: low—50, 150, 250 Ω; high—30,000 or 40,000 Ω. With proper design, excellent frequency response (40 to 20,000 Hz) can be obtained. In addition, sensitivity is good; the microphone is rugged and is not subject to temperature or moisture effects. It is therefore widely used—both as a general-purpose unit and as a high-quality microphone.

VELOCITY MICROPHONES.   The basic principle of the dynamic microphone is also applicable to the velocity microphone. In this case, however, there is no diaphragm, and instead of a coil of wire, a flat ribbon of aluminum alloy is used as the moving element. Because of this, the microphone is also known as a *ribbon* microphone. The ribbon is suspended between the pole pieces of a U-shaped permanent magnet, and as it vibrates within the magnetic field, it cuts the field and a voltage is induced. When a sound wave passes by the ribbon, the force acting on the ribbon is proportional to the difference in pressure on the front and back of the ribbon, which is in turn dependent on the velocity of the air particles. This action led to the name *velocity* microphone.

Since the moving element is merely a strip of conducting ribbon, the impedance of this microphone is extremely low. Therefore impedance step-up transformers are always built into the microphone housing to raise the output impedance. Again, since the moving element is only a single conductor, the induced voltage is low and the sensitivy of this type of microphone is lower than for other types. On the other hand, it is capable of excellent frequency response.

MICROPHONE COMPARISONS.   A comparison of the various types of microphones is shown in Table 17-1. It should be realized that the values given are average values, and better or poorer characteristics can be obtained from any individual unit in any classification.

The physical appearance of a variety of microphones can be seen in Fig. 17-7. Their descriptions are as follows: (*a*) floor-stand type—dynamic, cardioid pattern; (*b*) floor-stand type—dynamic, omnidirectional pattern; (*c*) desk-stand type—ceramic, cardiod pattern; (*d*) hand type—available as dynamic or carbon; (*e*) desk-stand type—velocity, bidirectional; (*f*) floor-stand type—available as dynamic or crystal; (*g*) contact microphone.

Table 17-1
MICROPHONE COMPARISONS

| TYPE | IMPEDANCE | OUTPUT LEVEL | |
		dBV at 1 dyne/ sq cm	dBm at 10 dynes/ sq cm
Capacitor	very high	−60	—
Carbon, double-button	100–200 Ω	−33*	−29
Carbon, single-button	50–100 Ω	−23*	−19
Ceramic (see Crystal)		—	—
Crystal, direct-cell	high (1 MΩ)	−54	—
Crystal, diphragm	high (1 MΩ)	−48	—
Dynamic	5–20 Ω	−58*	−55
Velocity	approx 1 Ω	−63*	−60

*This voltage level is based on $Z_o = 40{,}000 \ \Omega$.

(a)   (b)   (c)   (d)

(e)   (f)   (g)

Figure 17-7   Typical microphones (*Electro-Voice and Turner*).

RECORDED PROGRAM SOURCES

The microphones we have been discussing are used to pick up "live" programs—that is, programs which are occurring at that time. However, on many occasions music or speech is recorded for playing back at a later time. Two types of recordings are in common use—disc (phonograph records) and tape. Let us examine each of these media briefly.

PHONOGRAPH RECORDS.   In disc recording, a blank disc is placed on the turntable; then, as the turntable rotates, a *cutting head* is caused to move across the face of the disc, cutting a continuous spiral groove. If no signal is applied to the cutting head, its *stylus* (the cutting needle) does not vibrate and the groove is smooth, as shown in Fig. 17-8*b*. When a signal is fed to the head, the side-to-side vibrations of the stylus produce the irregular groove shown in Fig. 17-8*c*. The groove is "modulated" by the sound signal.

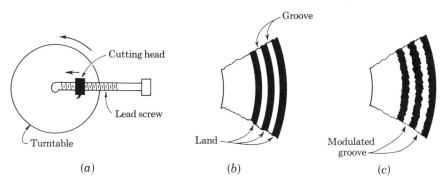

Figure 17-8 Recording turntable and record sections.

Two types of cutting heads are in common use—the magnetic and the crystal. In the magnetic head, the electrical output of the recording amplifier is fed to a coil mounted within a magnetic field. The interaction of the varying field (created by the signal) and the fixed magnetic field causes the coil to move, and the stylus, attached to the coil assembly, moves *sideways* in accordance to the signal-voltage variations. In the crystal cutting head, the piezoelectric effect converts the signal-voltage variations into stylus movements.

To play back a disc recording, a pickup arm with a *playback* cartridge is used. One such unit is shown in Fig. 17-9*a*. This unit uses a *turnover* cartridge with two needles, a 1-mil needle for use with microgroove records

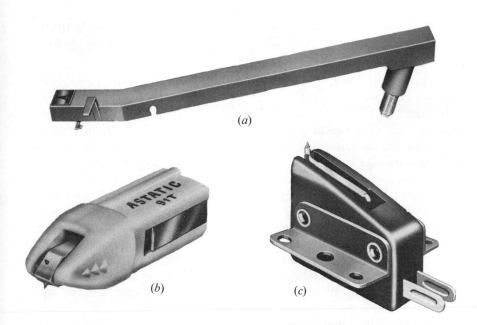

(a)

(b)   *(c)*

Figure 17-9   Phonograph pickup and playback heads (*Astatic Corporation*).

(178 to 300 lines per inch), and a 3-mil needle for use with standard groove records* (88 to 120 lines per inch). Figure 17-9*c* shows a cartridge with a single 1-mil needle.

Except for the needle itself, playback cartridges are very similar to cutting heads. They may be of the magnetic, crystal, or ceramic types. In each case, the side-to-side motion of the needle as it rides in the modulated groove generates an output voltage. In the magnetic type, the needle motion causes a coil to move within a magnetic field and a voltage is induced in the coil. Magnetic pickups have low impedance, approximately 100 to 300 Ω. The voltage output is relatively low—in the range of 10 to 30 mV. A popular modification of the magnetic type of pickup is the *variable reluctance* pickup. In this case, the coil is fixed and movement of the needle varies the reluctance of the magnetic path, changing the flux linking the coil. This pickup also has low impedance and low output voltage. With crystal and ceramic pickups, the piezoelectric effect converts the needle motion into an output voltage. These pickups have high impedances (approximately 1 MΩ) and appreciably higher output voltages, ranging from approximately 0.5 to 3.5 V. The frequency response for playback cartridges may range from a low of 50 to 4000 Hz to as high as 30 to 15,000 Hz. In general, the higher

*Standard groove is used in the older 78-rpm records.

quality magnetic units are considered to have the superior frequency response; however, ratings of 30 to 15,000 Hz have been given in manufacturers' literature for ceramic cartridges.

TAPE RECORDING.   Flexible tapes for recording are made by applying a magnetic iron oxide as a coating on either a cellulose acetate or a polyester base material. Recording tapes are available in various forms: the open reel (7-in reel containing 1200 ft, or 5-in reel containing 600 ft); the two-track cassette; and the eight-track cartridge. Both the cassette and the cartridge have the advantage of requiring no threading (they are continuous loops) and no rewinding. One track is played in one direction, and the next track in the opposite direction. In each case (reel, cassette, or eight-track), the magnetic tape is made to pass in front of a recording or playback head. The basic elements of a head are (1) a soft iron core with a small air gap and (2) a coil (see Fig. 17-10a). When recording, the amplified signal voltage is fed from the output of the recording amplifier to this coil. A varying flux is produced in the core and across the air gap. As the tape is run past this air gap, it is magnetized in accordance with the input signal.

If the input signal is removed and the magnetized tape is now rerun past the same air gap, the changing magnetic field of the moving tape links the soft iron core and cuts the coil. The voltage induced in the coil is a replica of the original signal voltage. The same head can therefore be used for recording and playback. A typical tape recorder head assembly is shown in Fig. 17-10b. This unit incorporates two separate heads: one is a combination record-playback head, the other is called the *erase head*. By feeding a supersonic frequency (25 to 50 kHz) to this head, the tape is demagnetized. Since the previous program material is erased, the tape can be reused. The voltage output from a playback head is very low, in the range of 1 to 10 mV. Consequently, high-gain amplifiers are needed for use in tape recorders, and

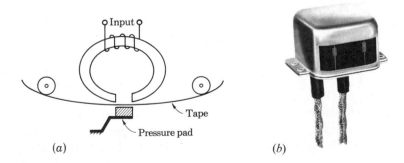

(a)   (b)

Figure 17-10 Tape recorder basic operation and a typical record playback and erase head *(Shure).*

extreme attention must be given to minimize hum and extraneous noise pickup.

In addition to the recording and playback of audio material, tape recorders are also used to record and play back video signals. A television field is recorded line by line on the tape, together with the line and frame synchronizing signals. For playback, the signals are fed through the video amplifier section of the TV receiver to the cathode-ray tube. Obviously, video tape recorders must have far better frequency response than sound recorders.

STEREO RECORDING.   The aim in stereophonic reproduction is to give sound a spatial dimension, or directivity. For example, at a concert hall string instruments may be heard at one side of the orchestra and the brass section on the other side. Also, when an actor walks across a stage as he speaks, his voice can be heard to move with him. Stereophonic sound can be reproduced electronically by using two speakers spaced at least 6 ft apart. Of course, the audio signals fed to the two speakers must differ in the same manner as the sound reaching the left and right ears of the audience "on location." This is readily done by using a "left" and a "right" microphone to pick up the original sounds. In monophonic reproduction, the outputs from the two (or more) microphones are combined. In stereophonic reproduction, each pickup must be processed as a separate message channel all the way through to its own loudspeaker.

In stereophonic records, two separate sound tracks are recorded. The two tracks are cut into opposite walls of the same groove at 45° angles. This makes possible the use of a pickup cartridge with only one stylus— but two transducer mechanisms. Because of the 90° space relationship of the sound tracks, this one needle excites each transducer independently, producing "left" and "right" outputs with negligible interaction (crosstalk). Simplified sketches of a ceramic and a magnetic stereo cartridge are shown in Fig. 17-11. The stereo pickup head has four output leads (or terminals),

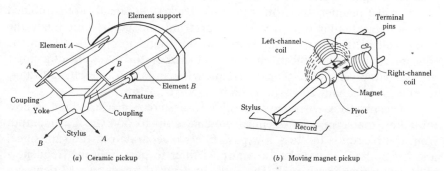

(a) Ceramic pickup   (b) Moving magnet pickup

Figure 17-11   Basic action in stereo cartridges (*Pickering*).

two for each channel. These outputs are fed to separate amplifiers, and their outputs in turn feed the "left" and the "right" speakers.

Stereo tapes are made by feeding the left and right microphone outputs to separate tracks—one for the left channel and one for the right. Obviously, a dual head is necessary both for recording and for playback. The same open-reel, cassette, and eight-track tapes that are used for monophonic recording are also used for stereo. However, since the tracks are now used two at a time the total playing time is cut in half.

QUADRAPHONIC SOUND.   For quadraphonic sound, four separate pickup areas are used—front left (FL) and front right (FR) as in stereo, plus back left (BL) and back right (BR). When recording on tape, each of these pickups is fed through a separate head and recorded on separate tracks of the same tape. For this, the eight-track cartridge is generally used. Four of the tracks are used simultaneously to record in one direction and the other four for recording in the other direction. Consequently, rewinding is unnecessary. Such a cartridge is known as a *Quad-8* or a *Q-8*.

When records are used for quadraphonic sound we have a problem: a record groove has only two walls. How do we get four channels of sound information on what are essentially two tracks? Two techniques are being used—*discrete four-channel* and *matrix four-channel*. Let us review each of these systems briefly.

In the discrete four-channel system, the sound signals from the front left and back left are combined in an *adder circuit* to produce a (FL + BL) output. This *sum signal* occupies a frequency spectrum of (roughly) from 20 Hz to 15 kHz. In addition, the back left signal is inverted and then combined again with the front left to produce a (FL − BL) output. This *difference signal* is then used to frequency-modulate a 30-kHz carrier. This raises the audio (difference) signal to a frequency spectrum of from 20 to 50 kHz. The total frequency span is shown in Fig. 17-12. Notice that there is no intermixing between the sum signal and the "raised" difference signal. This total signal is used to modulate the inner wall of the record groove.

In like manner, the front right and back right signals are combined to produce a similar total frequency spectrum and used to modulate the outer wall of the groove. Therefore, all the sound information from the four microphones has effectively been recorded on the two walls of a single groove. However, notice that the frequencies contained in each wall range from a low of about 20 Hz to as high as 50 Hz.

The playback cartridge is similar to a stereo cartridge—but it now must be capable of reproducing frequencies up to 50 kHz. Each output from the pickup is fed to a low-pass filter to separate out the sum signal and is simultaneously fed to a bandpass filter to pass only the frequency-modulated difference signal. This is in turn put through an FM demodulator, where the difference signal is converted back to the audio range.

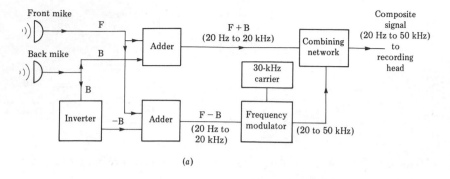

(a)

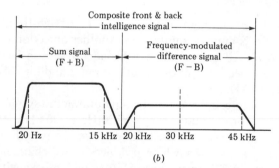

(b)

Figure 17-12   Signal treatment and frequency spectrum of the composite intelligence signal applied to one wall of record groove.

Now we combine the sum and the difference signals by adding and by subtracting them. From one of the cartridge outputs we get

$$(FL + BL) + (FL - BL) = 2 FL$$

and

$$(FL + BL) - (FL - BL) = 2 BL$$

Notice that we have recovered separately a front left and a back left audio signal. Using the same treatment, the signal from the other cartridge output yields 2 FR and 2 BR—or the separated front right and back right audio signals. Each of these signals is fed to its own amplifier and speaker to produce quadraphonic sound. Figure 17-13 shows this signal processing in block diagram form for one of the outputs of the cartridge. This method gives excellent separation and purity of the four individual microphone signals. It was developed by the Japan Victor Company (JVC) and is also used in the United States by RCA. The JVC records are called *CD-4 discs* (compatible discrete four-channel). RCA calls its records *Quadra-discs*.

The discrete four-channel system also has disadvantages. There are manufacturing problems in the records and in the cartridge—due to the high frequencies involved. (A special stylus, the Shibata stylus, is recom-

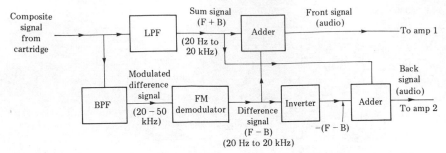

Figure 17-13 Separating the front and back audio signals from one output of the quadraphonic cartridge.

mended for better tracking of the high-frequency signals.) High-frequency noises and capacitance effects must be minimized. Another objection is that these records cannot be played on-the-air at a broadcast station because broadcast regulations (and equipment) do not permit handling of frequencies above (about) 15 kHz.

These problems do not occur in the matrix quadraphonic system. All frequencies are within the normal audio range of 20 Hz to 20 kHz. Many matrix systems have been proposed and tried. In all cases, the four discrete-channel pickups are fed to an *encoder* which produces a two-channel output, with the rear-channel intelligence phase shifted and mixed with its front-channel signal. This two-channel output is then processed by using standard stereo techniques. In playback, the stereo output is first fed to a *decoder* to theoretically recover the original four-channel sounds. Unfortunately, the separation achieved with matrix systems is not good. Appreciable crosstalk occurs. With one matrix system, it is possible to get good separation between front left and back right, but the separation between front left and back left is poor. Another system gives good front left-to-right separation, but the front-to-back crosstalk is high. In general, a sound originating in one pickup area will produce appreciable output from three speakers. (It was this weakness of the matrix systems that led to the development of the CD-4 records.)

It is claimed by the matrix-system proponents* that complete channel separation is not necessary. Psychologically, the listener is satisfied by the overall effect—particularly since the listener does not know what the original sound distribution was. However, to meet the competition from the discrete four-channel recordings, the two major matrix-system manufacturers have added special feedback control circuits to improve separation. Among such circuitry are the *vario-matrix circuit* used with the QS system and the *parametric logic circuit* used with the SQ system. A full discussion of these circuits is beyond the scope of this text.

*Two of the better matrix systems are the Sansui QS and the CBS-Sony SQ.

LOUDSPEAKERS

The ultimate goal of an audio system is to convert the amplified electrical signals back into sounds. For this purpose one or more loudspeakers are used as the "load" for the power amplifiers. The next few sections will cover a few of the more important points about loudspeaker characteristics.

THE DYNAMIC LOUDSPEAKER.   The most commonly used type of loudspeaker by far is the *dynamic* loudspeaker. Its basic construction and external appearance can be seen from Fig. 17-14. The signal voltage, from the output of the power amplifier, is fed to the *voice coil,* creating a flux that varies in accordance with the audio signal. The magnetic interaction between the fixed field of the permanent magnet and the varying field due

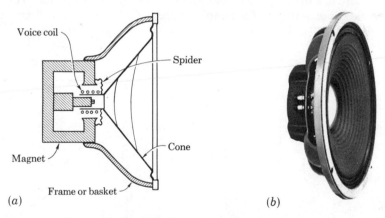

(a)   (b)

Figure 17-14   The dynamic loudspeaker (*Lansing*).

to the voice coil causes the coil to move axially with respect to the center leg of the magnetic structure. Since the voice coil is rigidly attached to the cone, the coil movement causes vibration of the speaker cone. When the cone moves forward, the air immediately adjacent to its *front* surface is compressed. Then, as the coil is pulled back, the air in this same region is rarefied. These pressure changes produce sound waves. To permit free motion of the cone, the outer rim of the cone is attached to the frame by means of a flexible suspension. At the inner end of the cone a *spider* is used to keep the voice coil centered within the air gap space. This spider must be flexible enough to allow full excursion of the voice coil and cone.

BAFFLES.   In most of the small ac/dc table-model radios, the loudspeakers are mounted directly on the chassis. When such a chassis is out of its cabinet, the sound quality seems flat or lacks body. Yet when the set and speaker are back in the cabinet the tone improves again. This is not an aural illusion. A loudspeaker that radiates in the open air—as is the case when a speaker is out of its cabinet—is very inefficient, and its efficiency is poorest at the lower frequencies. It was mentioned above that as the cone moves forward, the air pressure in front of the cone surface increases. At this same instant, the air pressure at the rear face of the cone must *decrease*. If the free-flow air path from the front surface to the back of the speaker is short *compared to the wavelength of the sound wave,* air flows around the edge of the cone and nullifies the pressure difference, preventing the formation of sound waves. This effect is more pronounced at the lower frequencies (longer wavelengths). When the speaker is replaced in the cabinet, the length of air path from front to back of the speaker is increased and the low-frequency response is improved. The cabinet acts as a *baffle* to prevent direct air flow from the front to the back surface and vice versa.

There are many types of baffles in common use. Each has its advantages and disadvantages. The simplest type is the open-back cabinet shown in Fig. 17-15. The increase in the length of air path is shown in the side

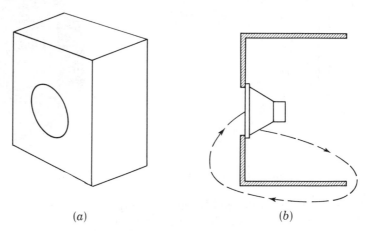

(a)   (b)

Figure 17-15 Open-back speaker cabinet.

view by the dotted line. To be effective, the cabinet should be large enough so that this air path is at least a half wavelength at the lowest frequency desired to be reproduced. (At 50 Hz this distance is approximately 22 ft.)* Obviously, the larger the cabinet, the better the low-frequency baffling.

Carrying this idea to its ultimate conclusion leads to *the infinite baffle.* Ideally this would imply mounting a loudspeaker at the center of a solid surface of infinite dimensions, so that sound waves from the front and rear

---

*Wavelength (λ) can be found from the relation $\lambda = v/f$, where v is the velocity of sound waves in air (approximately 1130 ft per second) and f is the desired frequency.

surfaces of the cone can never intermix. A more practical application of the infinite baffle principle would be to mount the speaker on a large unbroken wall between two rooms, so that the front cone surface radiates into one room and the rear surface into the other room. Although technically practical, such a solution may not be desirable for other reasons.

Returning to the cabinet idea, the back of the cabinet can be closed to prevent cancellation of the pressure differential. (This type of cabinet is sometimes also called an infinite baffle because the back wave can never come around to cancel the front wave.) However, with a totally enclosed box a back pressure is created in opposition to the motion of the cone. This back pressure would damp or reduce the cone excursion and impair the low-frequency response. The effect is minimized by using an enclosure of sufficient volume. The larger the box, the larger the volume and the lower the back pressure developed by the cone excursion. Similarly, since larger-diameter speakers move more air, they require larger cabinets. For a 12-in speaker a cabinet volume of over 9 cu ft is recommended. To prevent the back wave from bouncing and rebounding off the inside surfaces of the enclosure, the walls of the cabinet must be padded with a sound-absorbent material of sufficient thickness.

With the totally enclosed speaker baffle, the only sound reaching the listener is due to the sound waves radiated from the front surface of the cone. The sound waves generated by the rear surface are lost. This represents a waste of energy. A loudspeaker's response tends to fall off at the lower frequencies. It is possible—particularly with regard to these lower frequencies—to use this wasted sound energy to reinforce the front wave. However, the sound wave from the rear surface must be delayed just long enough (one half-cycle) so that by the time it reaches the front, the cone has reversed its direction of motion; now both waves (front and back) are in phase. This principle is utilized in the speaker enclosures shown in Fig. 17-16. The cabinet on the left is known as a *bass-reflex cabinet*. For optimum results, the volume of the enclosure and the area of the port must be matched to the particular loudspeaker characteristics. This cabinet has two advantages over the totally enclosed structure in that it provides somewhat better low-frequency response and for a given size of speaker a smaller enclosure can be used. (For a 12-in speaker, the bass-reflex cabinet requires a volume of only 6 cu ft.)

The enclosure shown in Fig. 17-16b uses an *acoustical labyrinth* to delay the sound wave from the rear of the speaker. Maximum sound reinforcement is produced at the frequency which makes the labyrinth path equal to a half wavelength. The path is lined with sound-absorbing material so as to absorb the energy of the higher-frequency waves (above 150 Hz). This prevents variation of the output level with frequency, due to some of the higher frequencies coming out in phase and others out of phase. In

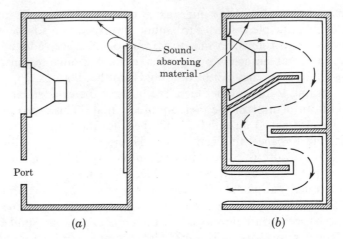

Figure 17-16   Bass-reflex and labyrinth-type speaker enclosures.

comparison with the previous enclosures, the labyrinth type requires less space for the same speaker and the same low-frequency response.

A type of enclosure that is ideally suited for placement in the corner of a room is the *folded-horn cabinet* shown in Fig. 17-17. With proper design, excellent low-frequency response—down to 30 or even 20 Hz—can be obtained. The efficiency of this design depends on the length and flare of the horn. Because of their ability to move large masses of air, horn-loaded speakers are more efficient than direct radiating cones.

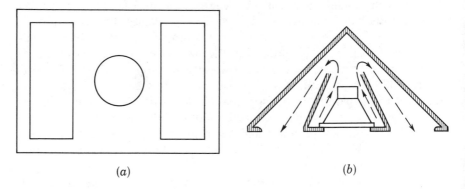

Figure 17-17   Folded-horn enclosure. **(a)** Front view. **(b)** Top view.

HIGH-FIDELITY SPEAKER SYSTEMS

We have seen that with proper baffling good low-frequency response is made possible. However, in order for a baffle to be effective the loudspeaker itself must be capable of producing an output at these low frequencies. This in itself is not too difficult to achieve. But for high-fidelity applications, a good output is also required at the high frequencies (15,000 Hz or even higher). Unfortunately, the design aspects that lead to good high-frequency response tend to impair the low-frequency response. By careful design *and compromise,* a good *extended-range* speaker can be produced with a flat frequency response from approximately 45 to 10,000 Hz. This is an average value.

DUAL-SPEAKER SYSTEMS.   If a wider frequency range is desired, it is better to use two speakers, one specifically designed for good low-frequency response and the other for good high-frequency response. In this way, no compromise need be made and the overall efficiency is higher. The low-frequency speaker is called a *woofer;* the high-frequency speaker, a *tweeter.* The woofer in most respects looks like any other dynamic loudspeaker; however, its high-frequency response falls off rapidly. The tweeter may be a much smaller cone-type speaker, or it may use a diaphragm-type driver and a horn. The latter type is more efficient. Tweeters have excellent high-frequency response, but their response falls off rapidly at the lower frequencies.

In such a dual-speaker system, the acoustical power output comes from the woofer at the low frequencies and from the tweeter at the high frequencies. However, at some intermediate frequency both speakers are delivering equal power. This is the *crossover frequency,* and can be considered as the point where the power shifts from one speaker to the other. In a well-designed system, this crossover occurs when each speaker is delivering half rated power. Obviously, to achieve this result the speakers must be matched. To simplify matching, woofer and tweeter specifications state the crossover frequency for the unit.

Figure 17-18 shows two tweeter units of the horn type. In addition these units also use some form of *acoustic lens* to help disperse the high-

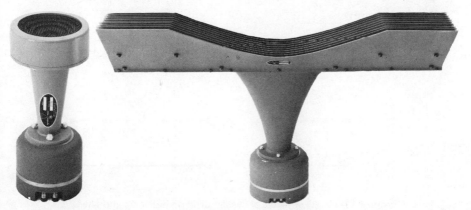

Figure 17-18   Horn-type tweeters with acoustic lenses (*Lansing*).

frequency sound waves over a wide angle. The unit in Fig. 17-18*a* has a crossover frequency of 1200 Hz and a 90° dispersion angle. The longer horn used with the tweeter of Fig. 17-18*b* results in a crossover frequency of 500 Hz, and its larger lens produces a dispersion angle of 130° by 65°.

DIVIDING NETWORKS.   To get optimum results from a dual-speaker system, it is important to feed to each unit only those frequencies that it can handle best. The low frequencies, up to the crossover frequency, should be fed to the woofer, and the high frequencies, from the crossover value and up, should be fed to the tweeter. This is accomplished by means of a *dividing or crossover network*. Any combination of high-pass and low-pass filter circuits can be used for this purpose. Some of the more commonly used circuits are shown in Fig. 17-19. A very simple technique is to connect the speakers in parallel but to insert a reactance (*L* or *C*) in series with each speaker as shown in Fig. 17-19*a*. The reactance of the series element should equal the loudspeaker voice-coil impedance at the crossover frequency. At higher frequencies, the reactance of the capacitor decreases, putting a higher voltage across the tweeter. Simultaneously, since the reactance of the inductance increases, the voltage applied to the woofer decreases. Conversely,

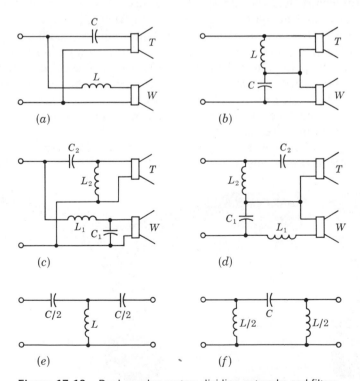

Figure 17-19   Dual-speaker system dividing networks and filters.

for frequencies below the crossover value, the voltage across the woofer increases while the signal level fed to the tweeter decreases.

Figure 17-19*b* shows a series arrangement for accomplishing the same result. Again, the reactance values (for *L* and *C*) should equal the voice-coil impedances at the crossover frequency. In this case, however, the inductor bypasses the low frequencies away from the tweeter while the shunting action of the capacitor keeps the high frequencies out of the woofer.

If sharper separation of the frequencies fed to each speaker is desired, sharper cutoff filters can be used. The L-type filter shown in Figs. 17-19*c* and *d* will achieve this result. For the parallel speaker connections the component values can be obtained from the following relationships:

For low pass:
$$L_1 = \frac{R_o}{\pi f_c} \tag{17-1}$$

$$C_1 = \frac{1}{\pi f_c R_o} \tag{17-2}$$

For high pass:
$$L_2 = \frac{R_o}{4\pi f_c} \tag{17-3}$$

$$C_2 = \frac{1}{4\pi f_c R_o} \tag{17-4}$$

where f_c is the crossover frequency and R_o is the speaker impedance. For the series-speaker combination, the constants are

$$L = \frac{R_o}{2\sqrt{2}\pi f_c} \tag{17-5}$$

and

$$C = \frac{1}{\sqrt{2}\pi f_o R_o} \tag{17-6}$$

Still sharper separation can be obtained by replacing each of the above L-filter sections by a T-filter section (Fig. 17-19*e*) or a π-filter section (Fig. 17-19*f*). The component values are found from the relationships shown above, except that where two capacitors or inductors are used, each is made half of the equivalent L-filter value.

COAXIAL AND MULTICONE SPEAKERS.   Instead of using two completely separate structures as the woofer and tweeter, a rather ingenious constructional technique permits mounting the tweeter coaxially within the framework of the woofer. Such a design is shown in Fig. 17-20*a*. Notice that a horn tweeter is used in this design. These units have two completely separate magnetic structures and voice coils, and require a dividing network

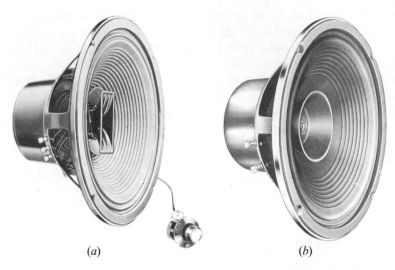

(a)   (b)

Figure 17-20   Coaxial and multicone speakers (*Electro-Voice*).

to feed the proper frequencies to each unit. In many designs, the network is built into the base and only one pair of output leads is necessary.

The loudspeaker shown in Fig. 17-20b is sometimes also referred to as a coaxial speaker. It does use a coaxial cone structure; however, it has only a single voice coil and a single magnetic structure. By careful design, a *mechanical* crossover is achieved, so that while the large cone handles the low frequencies, above this crossover frequency only the smaller high-frequency apex moves. Each cone surface can therefore be designed for optimum response within its frequency range. A frequency response of from 30 to 13,000 Hz is claimed for the speaker shown in Fig. 17-20b.

THREE-WAY SPEAKER SYSTEMS.   Still better results are claimed for a three-way speaker system. The audio frequency spectrum is divided into three limited ranges, and a separate speaker (or cone) is used for each range. The simplest technique involves the use of three individual speakers—the woofer, the mid-range, and the tweeter. An electrical network consisting of a low-pass, a band-pass, and a high-pass filter is then used to feed the proper frequency range to each speaker. One manufacturer ingeniously combines the three speakers into one concentric assembly (similar to the coaxial speaker) and calls this combined unit a "triaxial" speaker. It should be noted that in this combined assembly each unit has its own magnetic structure, its own voice coil, and its own cone (or horn).

Another technique for obtaining a three-way concentric assembly utilizes the dual cone with common voice coil and mechanical crossover for the low-range and mid-range cones and a coaxial tweeter with electrical crossover network for the higher frequencies. If you reexamine Fig. 17-20*a*, you will see that in addition to the coaxially mounted horn it also employs a mid-range cone. This unit is classified as an "integrated three-way speaker system" and has a rated frequency response of 30 to 19,000 Hz.

HORN-TYPE SPEAKERS.   Since a horn is more efficient than a baffle in converting electrical power into acoustic power, horn-loaded transducers are often used in public address and paging systems. This is particularly true in outdoor installations. However, instead of using a cone-type loud-speaker to feed the horn, a special driver unit is employed. These driver units have a voice coil and magnetic structure similar to the dynamic speaker. The large cone, however, is replaced with a much smaller diaphragm. The assembly is enclosed in a rugged weatherproof housing with a threaded neck for attachment to the horn. A driver and horn are shown in Fig. 17-21.

(*a*)   (*b*)

Figure 17-21   The reflex trumpet (*University Loudspeakers*).

The low-frequency response of these units is directly dependent on the length of the horn. However, a shorter horn has a wider angle of cover-age. Consequently, a compromise must be made between fidelity (at low frequency) and sound coverage. One manufacturer makes four trumpet lengths available. The longest model is $6\frac{1}{2}$ ft long for a low-frequency cut-off of 85 Hz. To reduce the overall physical size of the unit, the horn is often folded within itself to produce what is known as a *reflex* or reentrant trumpet. This construction can be seen in Fig. 17-21*b*.

AUDIO LINES

In a table-model radio, only a few inches of wire may be needed to inter-connect the power amplifier (through the matching transformer, if used) to the loudspeaker. However, in many audio systems the loudspeaker, or speakers, may be hundreds or even thousands of feet away from the amplifier chassis. Such a situation must obviously exist in an arena, in a hospital paging system, in a large auditorium, or in a hotel music system. When long distances are involved, several questions arise:

1. Do we use direct-coupled loads?
2. If we use transformer-coupled loads, do we
   (*a*) Mount the output transformer on the amplifier chassis?
   (*b*) Put it with the speaker?

First we must realize that in any long-distance system, high power levels are involved and efficiency becomes important. For maximum power transfer from amplifier to load, transformer coupling is necessary. Therefore, direct-coupled loads are not used.

Now let us examine the two alternatives when using transformer coupling. Figure 17-22 shows the conditions that exist if we mount the output transformer on the amplifier chassis and then run connecting wires to the loudspeaker. The line interconnecting these components is connected to the 8-Ω output impedance of the transformer secondary on one end and to the 8-Ω voice coil of the speaker at the other end. It is therefore called an *8-ohm* or a *low-impedance line*. Notice carefully that there is nothing special about these wires; they could be any suitable pair of wires. They have no particular impedance of themselves; the only reason they are called an 8-ohm line is because they are connected to an 8-Ω source and/or 8-Ω load.

This type of connection is satisfactory if the line is only a few feet in length; however, let us see what might happen if the distance from the speaker to the transformer is 3000 ft. Assume that the wire being used is No. 20. This wire has a resistance of 10 Ω per 1000 ft. The total resistance of the line is 60 Ω. Since the speaker or load resistance is only 8 Ω, most of the power will be wasted in the lines and less than 12 percent will reach the loudspeaker. Even using No. 12 wire at 1.62 Ω per 1000 ft the line resistance is 9.7 Ω, and less than half power would be delivered to the loudspeaker. To keep the power loss to within 10 percent would require a No. 0 wire. But the diameter of a No. 0 wire is almost $\frac{1}{3}$ in! Obviously, changing the wire size is not the answer.

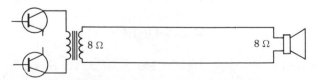

Figure 17-22   A low-impedance line.

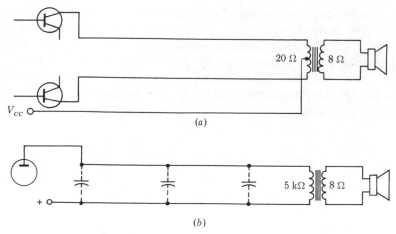

Figure 17-23   Effects with transformer at the speaker location.

The second alternative is to put the transformer at the speaker location. This is shown in Fig. 17-23. Diagram (*a*) represents a fairly low-impedance transistor output stage feeding into the audio line. True, the line impedance is not as low as the previous 8 Ω, but it is still quite low and the power loss on the line is still too high. Notice also that the dc collector supply voltage has to be brought out to the speaker location. We need a third wire, and this adds to the line losses.

With vacuum-tube amplifiers, we have the opposite condition. Tubes have relatively high output resistances. Let us consider a 5000-Ω value. With the output transformer at the speaker location, the line is connected to the 5-kΩ side of the transformer; at the amplifier end it is connected to the tube and power supply, which requires a 5-kΩ matching load. This line is now considered a 5-kΩ line, or a *high-impedance line*. Compared to the 5000 Ω of the load, even the 60 Ω of the No. 20 wire is negligible. So there is no problem of power loss. But now we have a new problem. Notice the capacitors shown with dotted lines. These represent the output capacitance of the tube, the distributed capacitance of the line, and the distributed capacitance of the transformer. The total capacitance can be high enough to make the shunting reactance low compared to the 5000 Ω of the load. The high-frequency response can be seriously affected.*

The solution is to use a line of some intermediate impedance value. Commonly used line impedances range from 50 to 600 Ω. In this way, the impedance can be made high enough that the losses due to ohmic resistance of the wires are not serious and low enough to make shunt capacitance effects negligible. To obtain such an intermediate line impedance value

*This line will also have to meet the electrical code requirements for high-voltage lines.

requires using one transformer, T_1, at the amplifier chassis to match the output stage to the desired line impedance value and then using a "line to voice-coil" transformer, T_2, at the speaker location to step down from the line value to the voice-coil value. Such an arrangement is shown in Fig. 17-24. Transformer T_1 may be either a step-up or a step-down unit, depending on the impedance required by the output stage as compared to the line impedance.

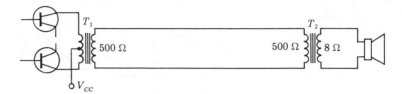

Figure 17-24   A 500-Ω line.

Audio lines of moderate impedance are also used to interconnect input sources to the amplifier when the distance between them is too long. As before, if a low-impedance line is used, the line loss is too great, whereas if a high-impedance line is used, the shunt capacitance effect would destroy the high-frequency response. In addition, a high-impedance line presents another problem. Since the signal level of a source may be very low (10 mV for some microphones), the line can pick up stray hum or noise fields of comparatively strong amplitude. To prevent such stray pickup, shielded wire is used for the line. This in turn aggravates the shunt capacitance losses. For long runs, lower impedance lines are necessary.

REVIEW QUESTIONS

1. Name four sources that could be used to feed audio signals to an audio amplifier.

2. Name six types of microphones.

3. Name four major characteristics that are used to describe or compare microphone performance.

4. Describe briefly what is meant by each of the following: **(a)** omnidirectional pattern; **(b)** bidirectional pattern; **(c)** cardioid pattern.

5. Refer to Fig. 17-1: **(a)** What is this type of graphic representation called? **(b)** What quantities are plotted in these graphs? **(c)** Which kind of directional characteristic is represented by diagram (*a*)? (*b*)? (*c*)? **(d)** Give the approximate decibel loss for each of the three for a sound incidence angle of 30°. **(e)** Repeat part (d) for a sound incidence angle of 90°. **(f)** Repeat part (d) for a sound incidence angle of 180°.

6. (a) How does the frequency response of microphones compare with the response of amplifiers? **(b)** What is implied by uniform or flat response rating?

7. What is meant by the nominal impedance of a microphone?

8. (a) Name two electrical quantities that are used as references for microphone sensitivity ratings. (b) What value or values of each quantity are used as the reference *level?* (c) What determines *which* quantity is used?

9. (a) What is the basic unit used to express microphone sound pressure levels? (b) Give three sound pressure levels that might be used in evaluating microphones. (c) What is meant by the term *microbar?* (d) If a sound level is listed as 10 bars, what is the true pressure?

10. Give the five common ways by which microphone ratings may be specified.

11. How does a rating expressed in dbV differ from one expressed in dBm?

12. (a) Explain briefly the basic principle of the carbon microphone. (b) What effect does an increase in sound pressure have on the electrical resistance? (c) What is meant by a *button?* (d) How many buttons are used in the microphones? (e) State two limitations of these units. (f) In what applications would these units be found? Why?

13. (a) Explain briefly the basic principle of the capacitor microphone. (b) What is meant by *polarizing voltage?* (c) What is the major disadvantage of this type of microphone?

14. (a) Explain briefly the basic principle of the crystal microphone. (b) In what respect is a ceramic microphone superior?

15. (a) Explain briefly the basic principle of the dynamic microphone. (b) In what way does the velocity microphone differ?

16. (a) Which types of microphones are essentially high-impedance units? (b) Which are low-impedance units? (c) Of the low-impedance types which has the highest output level and which has the lowest output level? (d) Of the high-impedance types, which has the highest output level?

17. Explain briefly how sound is recorded on a disc.

18. Explain briefly how sound is reproduced from a disc recording.

19. (a) Name two basic types of playback cartridges. (b) How do their impedance values compare? (c) How do their output voltages compare?

20. What basic property makes a tape suitable for recording purposes?

21. (a) Explain briefly the construction of a tape recording head. (b) How does a recording head compare with a playback head? (c) What is the range of output voltage from a playback head?

22. Explain briefly how sound is recorded on tape.

23. Explain briefly how sound is reproduced from a recorded tape.

24. (a) What is the function of an erase head? (b) What signal is fed to this head?

25. (a) Explain briefly how tape can be used for recording TV programs. (b) Can a standard sound tape recorder be used for TV program recording? Explain.

26. (a) Are two grooves used to record stereo sounds on a record? Explain. (b) Are two styli used to pick up the two stereo sound tracks from a record? Explain. (c) Can a single amplifier, but with two speakers, be used to reproduce stereo program material? Explain.

27. (a) When an eight-track tape cartridge is used to record a stereo program, what happens to its total playing time? (b) Explain why this is so.

28. For quadraphonic sound reproduction, how many sound pickup points are there? Where are they located?

29. (a) Are cassettes popular for recording quadraphonic sound on tape? Explain. **(b)** Are eight-track cartridges suitable? **(c)** What happens to their playing time?

30. Name two systems used for recording quadraphonic sounds on records.

31. Refer to Fig. 17-12: **(a)** Does this represent the recording or the playback technique? **(b)** Does it represent the right or the left side of the room? **(c)** Why is an "inverter" block used in the lower row of diagram (*a*)? **(d)** Why are a carrier and frequency modulation used in this lower row? **(e)** Does diagram (*b*) represent the frequencies applied to the left wall, the right wall, or both walls?

32. Refer to Fig. 17-13: **(a)** What does the block marked "LPF" represent? **(b)** Why is it needed? **(c)** What does the block marked "BPF" represent? **(d)** Why is it necessary? **(e)** Could a high-pass filter be used in its place? Explain. **(f)** Why is a demodulator needed after the BPF block? **(g)** Explain how a "front" signal is obtained from the output of the top row "adder" block.

33. Compare the quality of a hi-fi phono cartridge for stereo with one for quadraphonic reproduction.

34. Give two disadvantages of the discrete four-channel system.

35. Compare the separation obtainable with discrete four-channel and matrix systems.

36. (a) In a quadraphonic matrix system, how are the four microphone outputs recorded on the disk? **(b)** How are the four playback amplifier voltages obtained from the record?

37. Refer to Fig. 17-14: **(a)** What is connected to the voice coil? **(b)** What is the effect of this input? **(c)** What is the purpose of the spider?

38. Why is a baffle needed with cone-type loudspeakers?

39. Name five types of baffles that can be used with cone-type speakers.

40. (a) What is a dual-speaker system? **(b)** What are the individual units called? **(c)** What advantage does it have over a single speaker?

41. (a) Where is a dividing network used, and why? **(b)** What is another name for this device? **(c)** Draw the diagram for a parallel, L-type network. **(d)** How can sharper separation be achieved?

42. What is a coaxial speaker?

43. What is a multicone speaker, and how does it differ from a coaxial?

44. What is a three-way speaker system?

45. (a) How does a horn-type speaker differ from a cone-type speaker? **(b)** What relationship exists between length of horn and frequency response? **(c)** How is the physical size of these units reduced to practical values?

46. (a) What is the disadvantage of a low-impedance line? **(b)** What is the disadvantage of a high-impedance line? **(c)** How are intermediate line values obtained?

47. Refer to Fig. 17-23: **(a)** Is the line in diagram (*a*) a high-impedance or a low-impedance line? **(b)** In diagram (*b*) is this a high-impedance or a low-impedance line? **(c)** Explain the difference between these two cases.

48. Refer to Fig. 17-24: **(a)** Is this a high-impedance or a low-impedance line? Explain. **(b)** Is T_1 a step-up or a step-down transformer? Explain.

PROBLEMS

1. A high-impedance microphone has an output rating of 46 dB below 1 V per dyne per square centimeter. **(a)** What would its rating be for a sound reference pressure of 10 dynes per square centimeter? **(b)** What would its rating be for a reference level of 1 V per microbar? **(c)** Find the open-circuit output voltage corresponding to the original reference level.

2. If the microphone of Problem 1(c) were used in a condition where the average sound-pressure level is 20 dynes per square centimeter, what would be the average open-circuit output voltage?

3. A high-impedance microphone produces an open-circuit voltage of 24 mV at a sound pressure of 10 microbars. Specify the rating for this microphone, using **(a)** 10 microbars as the pressure reference level and **(b)** 1 microbar as the pressure reference level.

4. A dynamic microphone (10-Ω impedance) has a rating of -55 dB. The reference level is 1 mW per 10 dynes per square centimeter. Find: **(a)** the rated power output; **(b)** the voltage output across a 10-Ω load; **(c)** the voltage output when used with a step-up transformer to 500 Ω.

5. A dynamic microphone (impedance = 8 Ω) has a rating of 58 dB below 1 mW per 10 microbars. It is used at an average sound level of 20 dynes per square centimeter. Find: **(a)** its rated output (watts); **(b)** the actual power output; **(c)** the actual voltage output across an 8-Ω load.

18

Special Audio Circuits

Having discussed the various devices used to feed signals into an amplifier and the devices used to convert the amplified output back into sound, let us turn our attention once more to the amplifier itself. This time we will consider circuits which are of particular interest in audio applications. This chapter will cover volume controls, tone controls, multiple-speaker connections, and a complete audio amplifier.

VOLUME-CONTROL CIRCUITS

An amplifier used for a specific industrial application can very often be designed for some fixed value of gain. In fact, constancy of this gain value is sometimes of paramount importance. However, the gain required from an audio amplifier varies, depending on the source to be used (tuner, phono, or microphone) and even more on the specific type of source (such as crystal or dynamic microphone). Generally, the output required also varies. The output level determines the volume of the sound from the speaker. Volume requirements will vary with the type and size of the room, size and mood of audience, type of program, and so forth. These problems are readily solved by employing a *volume control* in the audio amplifier. The simplest and most common type of volume control is a potentiometer used to control the level of the signal voltage fed to the input of an amplifier stage. The circuit shown in Fig. 18-1 is suitable for use with FETs (and vacuum tubes). The control merely replaces the gate resistor of the standard R-C coupled amplifier.

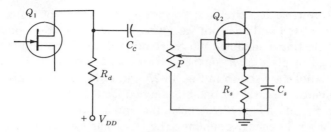

Figure 18-1   Simple volume control.

Since the ear responds logarithmically to changes in sound intensity, potentiometers for use as audio volume controls are made with a *logarithmic taper*. Starting with the potentiometer arm at the bottom (fully counterclockwise), the resistance is zero, and it increases gradually as the control is turned clockwise. At the halfway point, the resistance may be only 10 to 20 percent of maximum. However, the rate of change in resistance increases as the control is advanced, and the resistance rises rapidly to maximum value as the control is turned to the fully clockwise position.

When applying the simple potentiometer circuit of Fig. 18-1 to BJTs, care must be taken to see that the dc bias or operating point is not affected by variation of the gain control. Several suitable circuits are shown in Fig. 18-2. Capacitor C_1 of Fig. 18-2a breaks the dc path between the gain control and the collector circuit. Similarly, capacitor C_2 isolates the control from the base circuit. In this way, variation of the gain control cannot possibly affect the dc potentials of either the collector on the one side or the base of the next transistor.

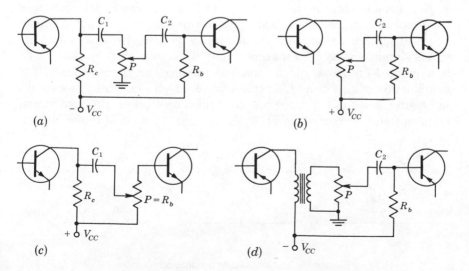

Figure 18-2   Methods for gain or volume control.

However, such a circuit is "overdesigned." The potentiometer *total* resistance is fixed and does not vary with the setting of the potentiometer arm. By selecting a suitable total resistance value, the potentiometer itself can be used in place of R_c of the previous stage or R_b of the next stage. These variations are shown in Fig. 18-2b and c. Notice that this technique also eliminates the need for one blocking capacitor.

The circuit of Fig. 18-2d shows the application of a volume control to transformer interstage coupling. The base-bias system in each of these circuits uses a simple series resistor R_b for fixed bias. Voltage-divider biasing and better stability could be had by adding another resistor from base to ground; or voltage feedback could be employed by returning R_b to the collector instead of to the power supply V_{CC}. Speaking of the power supply, notice that some are shown as positive, others negative. However, notice also that the positive supply source is used with an *NPN* transistor and the negative supply with a *PNP* unit. Of course, any circuit could be used with either transistor as long as the correct supply polarity is maintained.

Sometimes it is desirable to place a control in a circuit where impedance values are important and must be maintained constant. In such cases, the above potentiometer is undesirable. Notice that at low volume levels it tends to short-circuit one side. A typical situation where this effect is detrimental is in an audio line, where it is necessary to control the volume of one or more speakers independently at the speaker locations. The simple potentiometer would upset the line impedance, and the mismatch created would vary with the setting of the control. To prevent this mismatch, special attenuators known as *pads* are used.

There are three common types of pads, the *L pad,* the *T pad,* and the *H pad.* Each of these is shown in Fig. 18-3. Let us analyze the action of the simplest of these—the L pad as shown in Fig. 18-3a. First it should be noted that the controls R_1 and R_2 are ganged, so that both values are varied simultaneously. For maximum output (control fully clockwise), R_1 is zero and R_2 has a very high value compared to the line impedance. The combination of zero series resistance and very high shunt resistance results in no attenuation of the signal and no loading on the line. For zero output (control fully counterclockwise), R_2 is zero and R_1 is equal to the line im-

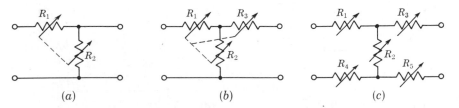

(a)   (b)   (c)

Figure 18-3   Constant-impedance attenuators.

pedance. Control R_2 short-circuits the load for zero output, while R_1 is the correct value to maintain impedance matching on the line. For in-between settings, the resistances of the controls are such that $R_1 + (R_2$ in parallel with the load) will always equal the line impedance. Obviously, then, these controls must be tailor-made for a specific line impedance. One disadvantage of this control is that proper impedance is not maintained on the output or load side. However, in many cases this matching is not necessary.

When proper impedance match is required on both input and output sides, the T pad of Fig. 18-3b can be used. The addition of a third variable resistor, R_3, makes this possible. This new resistor is made identical to R_1. The impedance on the output side is now maintained constant in the same manner as explained for the input side of the L pad. In addition, the value of R_2 is modified to correct for this added resistance. All three controls are ganged together for single control.

When a line is balanced with respect to ground, use of an L pad or T pad would upset the balance. To preserve balance, both lines must be treated alike and therefore variable resistors must be inserted in both lines. This gives rise to the H pad shown in Fig. 18-3c. The variable resistors R_1, R_3, R_4, and R_5 are all alike and all five resistors are ganged for single-control action.

TONE-CONTROL CIRCUITS

From our discussion of frequency distortion it would appear that the ideal audio amplifier should have a flat frequency response over the audio frequency range. (For high fidelity, this range can be considered in round figures as from 20 to 20,000 Hz.) Yet there are a number of reasons why a flat response is not the answer. In general, signal sources (microphones, pickups, records, and tapes) have poor response at the low- and high-frequency ends. The same defect is also found in the loudspeaker units used to convert electrical power back into sound. To compensate for these deficiencies, the amplifier's frequency response can be peaked at the high- and low-frequency ends, so that the *overall* response is flat.

A second reason for incorporating tone controls in an amplifier is to compensate for peculiarities of the human ear. The hearing sensitivity of the average human ear varies with the frequency of the sound and also with the intensity of the sound. Experiments by many scientists have set two limits of audibility:

1. *Threshold of audibility*—the power level at which a sound is just barely audible in the absence of any distracting sounds. Using 1000 Hz as a reference, this sound pressure for the normal ear is taken as 10^{-16} W per square centimeter.

2. *Threshold of feeling or pain*—the power level at which sounds reach a "near-pain" intensity. Using the threshold of audibility as a reference level (0 dB), this maximum sound level is considered as 120 dB higher.

The results of a series of experiments on audibility performed by H. Fletcher and W. A. Munson are shown in Fig. 18-4. These graphs are often referred to as the Fletcher-Munson curves. Each curve, representing an equal loudness level contour, shows the sound intensity required at any frequency to produce a loudness equal to the loudness at 1000 Hz. Several important conclusions can be made from this study:

1. The ear is most sensitive to sounds in the 3000 to 4000-Hz region.
2. The frequency response of the ear varies with the sound intensity. At very loud sound levels approaching the pain level, the ear frequency response is fairly flat. At low sound levels, the ear is very insensitive to low-frequency sounds and also has reduced sensitivity to high-frequency sounds.

This last point has great significance when applied to home music systems. If an orchestral selection is being reproduced at low volume levels, the low-frequency and high-frequency tones may not be heard at all, or at best they will sound too weak. Good tonal balance can be restored by using compensation circuits in the amplifier.

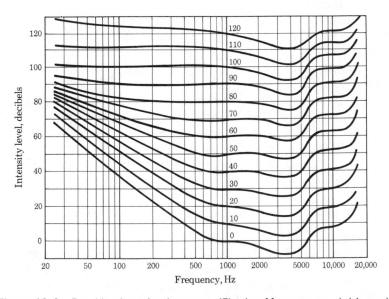

Figure 18-4 Equal loudness level contours (Fletcher-Munson curves) (*Jensen*).

Tone compensation techniques fall basically into three categories depending on the type of circuits used to achieve the compensation. These categories are the resonant-circuit type, the inverse-feedback type, and the frequency-discriminating network type. In addition, combinations of the basic circuitry are also used. Let us examine each of the basic types in more detail.

RESONANT-CIRCUIT COMPENSATION.   The impedance of a parallel circuit is a maximum at resonance and decreases toward zero for frequencies to either side of resonance. Furthermore, the impedance at resonance is resistive. By proper choice of components it is possible to produce resonance at any desired frequency and for any desired impedance value. Since, within limits, the gain of an amplifier depends on the impedance of the load in its output circuit,* consider what would happen to the gain of an amplifier if its load were to contain a parallel-resonant circuit. At the resonant frequency, the load impedance would be a maximum and the gain would be high; at frequencies off resonance, the gain would drop, approaching zero for frequencies further and further off resonance. This idea is utilized in tone-compensation circuits.

Figure 18-5 shows a typical tone-control circuit of the resonant-circuit type. Notice that in addition to a fixed resistor, two parallel-resonant circuits are used as the load impedance. Also, each parallel circuit is shunted by variable resistors (R_1 and R_2). These variable resistors are the tone controls. The circuit $L_1 C_1$ is resonant at some high frequency, such as 16,000 Hz, while $L_2 C_2$ is resonant at some low frequency, such as 40 Hz. At mid-frequency (around 1000 Hz) the impedance of each parallel circuit is practically zero and the total load impedance Z_e depends mainly on resistor R_d. At lower frequencies, the impedance of the $L_2 C_2$ circuit begins to rise until it is a maximum at 40 Hz. A similar rise in impedance occurs at 16,000 Hz due to $L_1 C_1$. The gain of the circuit varies with the impedance, providing bass and treble boost. If tone controls R_1 and R_2 are set at maximum resistance value, maximum boost is obtained. Depending on the circuit component values used, maximum boosts of up to 20 dB are readily possible. The amount of boost can then be adjusted by means of the tone controls. The frequency response curves obtained for various settings of these controls are shown in Fig. 18-5b. This circuit can be used directly with tubes, and—with component value and bias changes—it can also be used with BJTs. A high input impedance (such as obtained with emitter feedback or with bootstrapping) is desirable to prevent excessive loading of the tuned circuits.

One point should be made clear regarding this circuit *and all tone-compensation circuits:* the price paid for such compensation is loss of gain.

*$h_{fe}(R_e/r_i)$ for a BJT; $g_{fs}R_e$ for a FET; $g_m R_e$ for a tube.

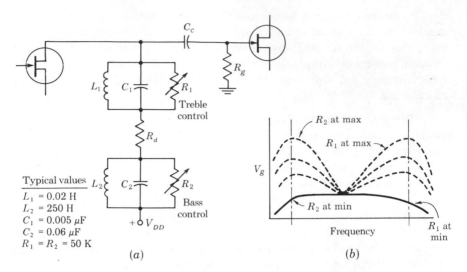

Typical values
$L_1 = 0.02$ H
$L_2 = 250$ H
$C_1 = 0.005$ μF
$C_2 = 0.06$ μF
$R_1 = R_2 = 50$ K

(a)

(b)

Figure 18-5   Resonant-circuit type of bass-treble control.

This can be explained as follows. If the FET used in the above circuit were capable of a maximum gain of 60 and a boost of 20 dB is desired, then *the circuit is deliberately designed for a mid-frequency gain of only 6.* The maximum gain of 60 is obtained only at the boosted frequencies. The 10-to-1 voltage gain (at the high and low end compared to mid-frequency) results in a 20-dB boost. In the circuit of Fig. 18-5, this means that the value of R_d is selected to produce a gain of 6, and the L, C, R_1, and R_2 values are selected to produce the full gain of 60. If the tone controls are then set for zero boost, the gain of this stage will be flat at 6. Of course, this loss of gain can be made up by adding another voltage amplifier stage.

INVERSE-FEEDBACK COMPENSATION.   In Chapter 16 it was seen that inverse feedback causes a loss of gain. Therefore, if a feedback network is added to a circuit so as to produce 20-dB inverse feedback, the output at all frequencies will be reduced by 20 dB. Now if by some technique we reduce the amount of feedback at 40 Hz, the gain at that frequency will increase. The control used to reduce the feedback becomes the bass boost control. In similar fashion, another control to reduce the 16,000-Hz feedback becomes the treble control.

An application of this principle is shown in Fig. 18-6. With both tone-control potentiometers set at their lowest point, the circuit reduces to a simple current-feedback circuit identical to Fig. 16-4b. If the parallel resistance value of P_1 and P_2, compared to the drain resistor R_d, is such that it will produce 20 dB of feedback, the gain for *all* frequencies is reduced by 20 dB.

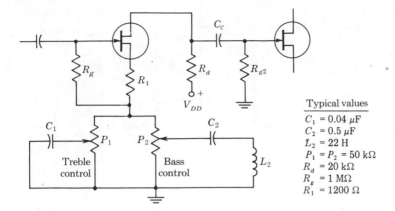

Typical values

C_1	= 0.04 μF
C_2	= 0.5 μF
L_2	= 22 H
$P_1 = P_2$	= 50 kΩ
R_d	= 20 kΩ
R_g	= 1 MΩ
R_1	= 1200 Ω

Figure 18-6   Inverse-feedback type of bass-treble control.

When the arm of potentiometer P_1 is at its uppermost point, capacitor C_1 shunts the resistance of the potentiometer. The reactance of the capacitor decreases at the higher frequencies, thereby decreasing the feedback voltage at these frequencies. Since the effectiveness of this bypassing action depends on the setting of the potentiometer arm, this control determines the level of the treble boost. Sharper treble boost action can be obtained by replacing capacitor C_1 by a capacitor and inductor that are series-resonant to a specific boost frequency.

In similar fashion, inductor L_2 serves to bypass the load frequencies around potentiometer P_2. The setting of P_2 therefore determines the level of the bass boost. Notice capacitor C_2 in this circuit. It could be series-resonant with L_2 at some specific boost frequency, but it serves an even more important function as a blocking capacitor to prevent the inductor from altering the dc operating conditions as the setting of P_2 is varied. This method of tone control can also be used with BJT circuits using emitter feedback.

COMPENSATION BY FREQUENCY-DISCRIMINATING NETWORKS.   Every R-C circuit is a frequency-discriminating circuit. This principle can be used in a number of variations to produce tone control. One application is shown in Fig. 18-7. The R-C network acts as a frequency-sensitive voltage divider between the output from the previous stage and the input to the next stage. The component values used will vary depending on the impedances of the stages being coupled. The values shown here apply to high-impedance circuits such as FETs, BJTs with bootstrapping, or BJTs with emitter feedback. P_1 and P_2 are the bass and treble controls, respectively. Maximum boost is obtained when the controls are at the top of their travel. Zero boost—or flat response—is obtained with the controls approximately

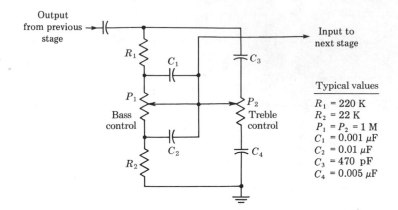

Figure 18-7   R-C network bass-treble control.

at center. Conversely, by setting either control below center that end of the frequency band can be attenuated below the mid-frequency level. Such a condition may at times be desirable. For example, if the noise level of the signal is high, it may be preferable to reduce the noise level by attenuating the high-frequency response even though full fidelity is impaired.

To simplify analysis, the circuit is redrawn in Fig. 18-8a as it would appear when both controls are set for maximum boost. Using the typical component values from Fig. 18-7, the impedance Z_A at 16,000 Hz is approximately 21,000 Ω (capacitive), while the impedance Z_B is approximately 22,000 Ω (resistive). The output voltage *at this frequency* is approximately 3 dB less than the input.* Because of the increase in reactance of C_3, Z_A rises to approximately 200,000 Ω at 1000 Hz and to 220,000 Ω (resistive) at 50 Hz. Simultaneously, Z_B, because of the effect of C_2, rises to approximately 28,000 Ω at 1000 Hz and to 300,000 Ω (capacitive) at 50 Hz. Treating these values vectorially, the voltage-divider action causes a loss of 18 dB at 1000 Hz and only 3 dB at 50 Hz. Compared to mid-frequency, the high- and low-frequency ends are boosted by approximately 15 dB.

Figure 18-8b shows the circuit redrawn to represent both controls set for maximum attenuation. Phasor analysis of the impedance relations shows that the output voltage is less than the input voltage by approximately 39 dB at 16,000 Hz, 22 dB at 1000 Hz, and 34 dB at 50 Hz. Compared to mid-frequency, the high- and low-frequency ends have a *loss* of 17 and 12 dB, respectively. Notice that in each case, boost or attenuation, the mid-frequency output voltage is approximately 20 dB below the input value. This type of tone-control circuit has a 20-dB insertion loss. An additional voltage amplifier stage is needed to make up for this loss. Obviously,

---

*Approximate values are used instead of exact calculations to focus attention on the action of the circuit rather than on the mathematics.

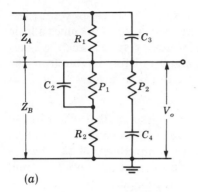

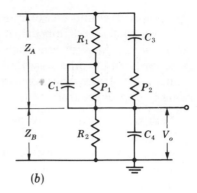

Figure 18-8   Circuit of Figure 18-7—redrawn for **(a)** maximum boost condition and **(b)** maximum attenuation condition.

tone-compensation circuits cannot be obtained without first deliberately attenuating the mid-frequencies by an amount equal (at least) to the boost desired and then providing for additional amplification.

MULTIPLE-SPEAKER CONNECTIONS

In high-power audio amplifier systems, the output of the power amplifier stage is fed to a load consisting of one or more loudspeakers. For maximum power transfer, the load impedance must be matched to the power amplifier. This can be done by using an output transformer having the proper turns ratio. When the load consists of a single speaker, the solution is very simple: we merely use an output transformer with the proper turns ratio to match the load to the amplifier output. The problem becomes slightly complicated by the fact that loudspeaker voice-coil impedances range from 3.2 to 16 Ω. Transformer manufacturers solve this by supplying suitable taps on the secondary winding or by using multiple-coil secondaries. By connecting these secondary coils singly, in series, or parallel, a variety of output impedances can be obtained. Sometimes the primary may also consist of a tapped or multiple-coil arrangement, so that a wide variety of primary impedances is also possible. Such a transformer can match literally any amplifier to any speaker and is known as a *universal output transformer*. Therefore, either by selecting the correct output transformer or by using the proper taps on a universal output transformer, a single loudspeaker can be matched to a power amplifier.

However, there are many applications where an amplifier must feed many speakers. Sometimes the loudspeakers used are identical and we want an equal distribution of power among the speakers. Such a situation would apply to a public address system for a large hall or open arena. In a paging system or intercommunication system we might need to deliver a high power level into one or more speakers in a noisy shop, yet far less power to a small office. Impedance matching in such cases becomes a little more involved. Also, in any such multiple-speaker system each speaker may be far from the others, and all of them are probably remote from the amplifier itself. Interconnections would therefore be made via audio lines ranging in impedance from 50 to 600 Ω.

SERIES-CONNECTED SPEAKERS.   Let us assume that we wish to supply equal power to each of six 8-Ω speakers. We could interconnect these loads in series. This would make the total impedance 48 Ω, which is close enough to 50 Ω so that the output transformer at the amplifier could have a secondary line impedance of 50 Ω. Unfortunately, there are serious disadvantages to series connections. If for any reason one speaker burns out or otherwise becomes open-circuited, none of the other five will function. Since there is no sound from any speaker, time-consuming tests must be made before the defective unit is located.

Another disadvantage arises if we desire unequal power distribution among the units. In a series system, equal current flows through each voice coil, and if the voice coils have equal impedances, equal power is delivered to each speaker. Using the above paging system illustration, either the volume level in the shop is too low or the office volume is too loud. Of course, separate volume controls could be incorporated into each unit. This wastes power at the speaker and in addition requires a higher power capability from the amplifier.

By using speakers with different voice-coil impedances, the power delivered to each speaker would differ. The unit with the highest voice-coil impedance would have the highest power level and vice versa. Such a solution is not practical since there are only three commonly used voice-coil impedances (3.2, 8, and 16 Ω) and the maximum variation is only 5:1. The desired power level variation is often much greater. Because of these disadvantages, particularly the total loss of output if one speaker fails, series-connected speakers are not recommended.

PARALLEL-CONNECTED SPEAKERS.   The above speakers could be connected in parallel. This would solve the worst objection to the series connection. Now, if one speaker fails (open-circuited), the others will continue to deliver sound output (actually the sound level from each speaker will increase) and the defective unit will therefore be self-evident. However, we still have a problem if we want a different sound level from each speaker. Again we need special, differing voice-coil impedances, but this time in reverse. Since the voltage across each voice coil is equal, *the unit with the lowest impedance will have the highest power level.*

The most serious objection to direct paralleling of the speakers is the resulting low value of line impedance. Using the six 8-Ω speakers in a parallel-connected system, the total line impedance reduces to 1.33 Ω. This is much too low; the resulting power loss in the lines could not be tolerated. Paralleling the speakers directly cannot be used, except for relatively short runs using only two speakers. It is not practical for paging or for large public address systems.

AUDIO LINE—IMPEDANCE-MATCHING TRANSFORMERS.   A parallel speaker system can be used if the total impedance is raised to a suitable line value—for example, 500 Ω. This can be done by using individual impedance-matching transformers at each speaker location. The technique is illustrated in the following example.

EXAMPLE 18-1

Six 8-Ω speakers are to be used in an auditorium sound system. The amplifier has a maximum power rating of 50 W; its output transformer has line impedance taps at 125, 250, and 500 Ω. Equal sound level is desired from each speaker. Specify the transformer impedance needed at each location.

Solution

1. Use the 500-Ω line tap, since that will give the lowest line loss.
2. For equal sound levels, each speaker should take one-sixth of the total power and the impedance for each speaker unit should be the same. Since six units are to be paralleled and their total equivalent impedance must be 500 Ω (to match the line), each unit impedance must be $6 \times 500 = 3000$ Ω.
3. At each location, use a line-to-speaker transformer having impedances of 3000 Ω primary and 8 Ω secondary.

The general technique should be obvious. For equal power distribution, use a transformer at each location whose secondary impedance matches the speaker voice coil and whose primary impedance is n times the line impedance, where n is the number of speakers to be used. This applies equally well even if the speaker voice-coil impedances are not alike. Remember that each transformer secondary matches its own speaker voice coil.

Before we consider a situation requiring different sound levels at the various locations, let us review some basic power–voltage relations. For any given line impedance Z_o and total power level P_T, the voltage across the line can be found from the general equation $P = V^2/R$. And since Z_o is a resistive load,

$$V^2 = P_T Z_o$$

This same value of voltage is applied across each speaker unit (parallel system), and the power taken by any one unit P_1 must be

$$P_1 = \frac{V^2}{Z_1}$$

Substituting for V^2 its equivalent $(P_T Z_o)$, we get

$$Z_1 = \frac{P_T}{P_1} Z_o$$

(18-1)

This equation indicates that the impedance at any one location, for a given power level, must be equal to the line impedance multiplied by the power level ratio (amplifier power to desired speaker power). Obviously the speaker unit impedance must be higher than the line impedance. This will produce the desired power level at each speaker. To obtain a proper impedance match, it is necessary to observe one more factor: *the sum of the power levels assigned to each speaker location must equal the total power level used at the main amplifier for calculating line voltage.* An example will illustrate this point.

EXAMPLE 18-2

A 50-W amplifier with a 500-Ω output line tap is to be used in a factory music-paging system to supply the following power levels at each of six locations: large shop (A), 20 W; medium shop (B), 13 W; shipping room (C), 10 W; main office (D), 5 W; two private offices (E and F), 1 W each. The speakers to be used are three 16-Ω 25-W speakers for locations A, B, and C; a 10-W 8-Ω speaker at D; and two small 3.2-Ω speakers at E and F. Select suitable matching transformers and draw a diagram for the interconnections necessary.

Solution

1. Calculate the total power that the amplifier must deliver:

$$20 + 13 + 10 + 5 + 1 + 1 = 50 \text{ W}$$

2. For a 50-W total and 500-Ω line the individual transformer primary impedances needed at each location are

$$Z_A = \frac{P_T}{P_A} Z_o = \frac{50}{20} \times 500 = 1250 \ \Omega$$

$$Z_B = \frac{P_T}{P_B} Z_o = \frac{50}{13} \times 500 = 1920 \ \Omega$$

$$Z_C = \frac{P_T}{P_C} Z_o = \frac{50}{10} \times 500 = 2500 \ \Omega$$

$$Z_D = \frac{P_T}{P_D} Z_o = \frac{50}{5} \times 500 = 5000 \ \Omega$$

$$Z_E = Z_F = \frac{P_T}{P_E} Z_o = \frac{50 \times 500}{1} = 25,000 \ \Omega$$

3. The transformers required are: (A) 1250-Ω primary to 16-Ω secondary; (B) 1920-Ω primary to 16-Ω secondary; (C) 2500-Ω primary to 16-Ω secondary;

(*D*) 5000-Ω primary to 8-Ω secondary; (*E* and *F*) 25,000-Ω primary to 3.2-Ω secondary.

4. The interconnections are shown in Fig. 18-9.

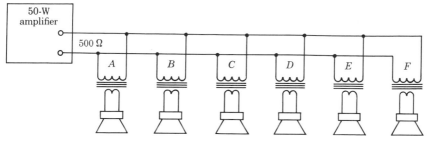

Figure 18-9   Multiple-speaker distribution system.

In Example 18-2, the sum of the power levels at each speaker location conveniently match the full power rating of the amplifier. But even if the main amplifier is changed to one of higher or lower rated power output, there will be no mismatch. Notice that the theory discussion terminated in an impedance value proportional to the power *ratio,* and not on the amplifier power rating itself. If the amplifier power output is decreased (new amplifier or just reduced volume level), the output from each speaker is reduced *proportionately* but the power ratio and impedance match remain correct.

IMPEDANCE MATCHING—70-V LINE.   Transformer manufacturers have come out with an additional system for rating the transformers, so that impedance matching, at any power level, should be obvious. This system is referred to as the *constant-voltage method,* and the commonly used voltage is 70 V. Unfortunately, the title of this method is a misnomer because the output voltage will vary with the sound level or power output.

The first step in this system is to provide an extra tap on the amplifier output transformer, in addition to the previous typical line impedance taps (such as 500, 333, 250, 125, and 50 Ω). This new tap is brought out and marked 70 V. The impedance of this new tap is such that *when the amplifier delivers its rated output, the line voltage will be 70 V.* The actual line impedance value of this new tap can be found from the basic power relation. For example, if the amplifier rating is 100 W, the line impedance corresponding to the 70-V tap will be

$$Z_o = \frac{V^2}{P} = \frac{70.7 \times 70.7}{100} = \frac{5000}{100} = 50 \ \Omega$$

(Notice that 70.7 V is used in place of 70. This is the exact voltage at the tap.)

In other words, if an output transformer *for use with a 100-W amplifier* had a 50-Ω tap, it would now also be labeled as the 70-V tap. Similarly, an output transformer for use with a 10-W amplifier would have its 500-Ω tap marked 70 V ($Z_o = V^2/P = 5000/10 = 500\ \Omega$).

Therefore, if we compare several amplifiers *of different power ratings*, each using the 70-V tap, the line impedance will vary from unit to unit, but in all cases the line voltage is 70.7 V. This, then, is the meaning of the term *constant-voltage method*. Regardless of the amplifier rating, if operated at its rated power level the line voltage is always the same—70.7 V.

The power taken by any load connected to this constant-voltage line varies inversely with its resistance ($P = V^2/R$). Conversely, for a given power level the load should have a resistance equal to $R = V^2/P$. (Using the loudspeaker as the load, resistance is replaced by impedance.) Also, since the line voltage is always 70.7 V, V^2 can be replaced by the fixed value 5000. The final step should now be obvious. To deliver a given power level P_1 to a loudspeaker at location 1, its impedance Z_1 should be

$$Z_1 = \frac{5000}{P_1} \tag{18-2}$$

Equation 18-2 satisfies the power aspects, and, as before, the impedance will be matched if the sum of the power levels assigned to each individual speaker location equals the total power rating of the amplifier. If the amplifier is operated below its power rating, the line voltage will drop to less than 70.7 V and the power delivered to each speaker will in turn be reduced, but the impedance match will be maintained. Let us try this technique on the public address system of Example 18-2.

EXAMPLE 18-3

The amplifier of Example 18-2 is to be used on its 70-V line tap to distribute power to the six speakers as before. Find the primary impedance needed for each transformer.

Solution

$$Z_A = \frac{5000}{P_A} = \frac{5000}{20} = 250\ \Omega$$

$$Z_B = \frac{5000}{P_B} = \frac{5000}{13} = 384\ \Omega$$

$$Z_C = \frac{5000}{P_C} = \frac{5000}{10} = 500\ \Omega$$

$$Z_D = \frac{5000}{P_D} = \frac{5000}{5} = 1000 \ \Omega$$

$$Z_E = Z_F = \frac{5000}{P_E} = \frac{5000}{1} = 5000 \ \Omega$$

Compare the primary impedances found in Example 18-3 with the corresponding values found in Example 18-2 when using a 500-Ω line. The values in this problem are exactly one-fifth of the corresponding previous values. Obviously, the 70-V line impedance must also be one-fifth of the previous value, or 100 Ω, and it is. (For a 50-W amplifier and a 70-V line, $Z = V^2/P = 5000/50 = 100 \ \Omega$.)

Transformer manufacturers then added the final simplification. With a constant-voltage system, the power taken by any speaker load is inversely proportional to its impedance. Therefore the *speaker transformer* primary taps can be marked directly in watts instead of in ohms. With such markings on a universal transformer no calculations are needed. The installer merely connects the line to the tap marked with the power he wishes delivered at that location. Such a transformer, for low power levels, is shown in Fig. 18-10.

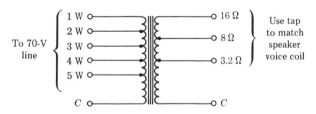

Figure 18-10   A universal speaker-to-line transformer for use with 70-V line.

A COMPLETE AUDIO AMPLIFIER

Figure 18-11 shows a fairly simple audio amplifier containing many of the circuit aspects that we have discussed in this and earlier chapters. This circuit is actually one channel of the audio section of an AM/FM stereo receiver. The schematic as a whole can look imposing. Analysis is simplified by studying the circuit in sections. We know that all amplifiers must have a power supply, small-signal amplifiers, and power amplifiers. Let us approach this schematic in that order.

POWER SUPPLY.   The main power components are shown in the lower right corner of Fig. 18-11. The power transformer has two secondary

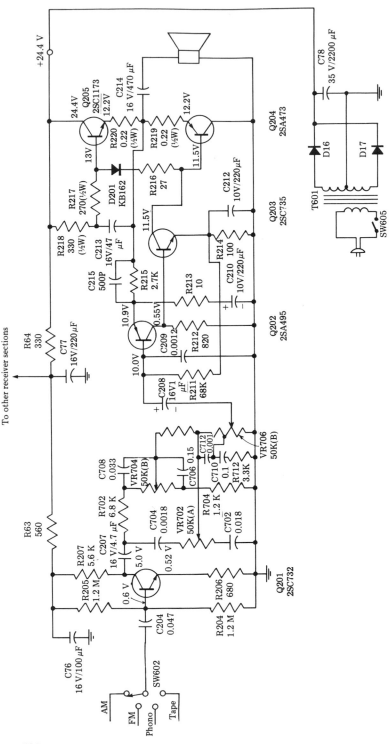

Figure 18-11 A complete audio amplifier (*Telex*).

windings—a low-voltage winding for the pilot light circuits* and a center-tapped main winding. The rectifier circuit uses this winding in the popular single-phase center-tap full-wave circuit. The filtering circuit is a little harder to identify because it is spread out. Actually, it uses a capacitor filter with several additional stages of R-C filtering. The input capacitor is C78 (35 V/2200 μF). At this point we take off the dc feed for the power output transistors. A high degree of filtering is not required for this stage since the signal level is high and the ripple content will be negligible in comparison. Also, since this is a push-pull stage any ripple introduced via the power supply should cancel. On the other hand, at the input stage the signal level is much lower and therefore the filtering action must be improved. Notice how this is accomplished:

1. An R-C filter (R64 and C77) is used to supply all the other receiver stages.
2. A second R-C filter (R63 and C76) is used for additional filtering for the input stage.

These R-C filters serve another purpose as *decoupling filters*. Whenever a multistage amplifier is fed from a common power supply, there is always some danger of positive feedback causing oscillations. For example, without these filters the ac component of the collector currents of each stage is returned directly to the common power supply filter, C78. The reactance of the filter capacitor may not be negligible at low audio frequencies. For C78, $X_C \cong 8 \; \Omega$ at 10 Hz, which may seem negligible. However, with the high collector currents in the output stages, a sufficient ac voltage can be developed across C78. This voltage is applied (along with the dc) to the collector of the input stage, Q201, creating an "output signal" at that frequency. This unwanted signal is amplified by Q202 and Q203 and is applied to the output stages in proper phase to reinforce or increase their ac components of collector current—at that frequency. The process is regenerative (positive feedback), and the amplifier breaks into oscillations. No actual input signal is needed to create this effect. It can be started by any irregularity in electron flow. Since the power supply impedance becomes significant only at low frequencies, oscillations are sustained only at very low frequencies, producing a distinctive "put-put-put" sound in the loudspeaker. From its sound, this malfunction is known as *motorboating*.

Now let us consider the effect of the decoupling filters. This time, the ac component of collector current for Q201, flowing through R207, can take either of two paths—through R63 (560 Ω), through R64 (330 Ω), and into the common power supply impedance; or direct to ground via the decoupling bypass capacitor (C76). Obviously, even at 10 Hz the choice is

---

*The pilot light circuitry is not shown.

approximately 4:1 through the capacitor. Additional decoupling (another 4:1) is then provided by the second filter (R64, C77), thereby preventing feedback from the common power supply.

SMALL-SIGNAL AMPLIFIERS.   The first stage of this amplifier is Q201. Notice that it is a standard *NPN* R-C coupled amplifier, with voltage-divider base bias (R205, R204) and emitter stabilization bias (R206). Since the emitter resistor is not bypassed, this also introduces ac inverse feedback. The collector load resistor is R207. An input signal (from the AM or FM output of the tuner or from a separate phono or tape player) is fed through the selector switch SW602 to the base of Q201. Capacitor C204 prevents upsetting of the dc base bias by the input circuitry. The output from Q201 is capacitively coupled to the next stage, Q202, through the tone and volume control circuits. VR702 is the treble control; VR704 is the bass control. These controls will either boost or attenuate the high-frequency and low-frequency response. This circuitry and action are quite similar to the R-C network base-treble control of Figs. 18-7 and 18-8, and no further discussion is now necessary.

The output from the tone controls is applied across VR706. This is the *loudness* or volume control. This circuitry is similar to that described in Figs. 18-1 and 18-2. However, notice that this control has a fixed tap —in addition to the rotating potentiometer arm—and an additional R-C circuit is connected to this tap. This additional circuitry produces a *compensated loudness control;* that is, as the general loudness (or volume) is lowered, the high and low frequencies are not attenuated to the same extent. In this way, we compensate for the reduced sensitivity of the ear at low loudness levels. (See the Fletcher-Munson curves of Fig. 18-4.)

The next three stages are direct-coupled. To reduce the power supply voltage requirements, Q202 and Q203 are of different types.* Q202 is a *PNP* unit. Its collector is connected to ground through R212, while the emitter is fed from the positive dc supply. Notice that the low collector potential becomes the proper base bias for Q203, the *NPN* transistor. Notice also that the dc supply for this transistor is not taken directly from the power supply, but rather from the center point between the two output transistors. Since capacitor C214 charges to one-half of the supply voltage, this feeds 12.2 V to Q202, eliminating the need for a voltage-dropping resistor. More important, the center tap between Q204 and Q205 is also the output point. Therefore, in addition to dc we are also getting inverse feedback to the emitter of Q202, stabilizing operating point and gain and minimizing distortion.

Q203 is an *NPN* unit with emitter bias stabilization. The collector voltage is dropped from the supply bus by resistors R218, R217, R216, and the diode D201. Its output is direct-coupled to both Q204 and Q205.

*This technique was shown in Chapter 14 (see Fig. 14-7).

POWER OUTPUT STAGE.   The output stage uses Q204 and Q205 in complementary push-pull. Notice that one is an *NPN* and the other a *PNP* unit. Therefore, a paraphase drive is not needed. The single-ended output from Q203 will drive these transistors in push-pull. The no-signal operating point for Q204 and Q205 is at "projected" cutoff to prevent crossover distortion. The voltage drops across diode D201 and resistor R216 provide this bias. Emitter resistors R219 and R220 are used for bias stabilization to prevent thermal runaway. Notice that they are very low resistance values. Since the two transistors are in series across a single power supply, their junction—the common-emitter connection—is at half the power supply voltage. Capacitor C214 is therefore needed to keep direct currents out of the speaker load. The capacitor also has another function—as a power supply source. When Q205 conducts, C214 charges (to half the supply voltage) with the polarity as shown. This not only serves as the dc supply for Q202, as mentioned earlier, but it is also the dc supply for Q204 while Q205 is in cutoff. Obviously, a large capacitor is needed to store sufficient charge.

Two features of this schematic are of great help to the technician who may be called upon to test or service this unit:

1. Notice the potentials marked at the terminals of each transistor. These are the normal voltages that should be measured from these terminals to ground. Discrepancies between these and actually measured values can give a clue to the cause of any malfunction.
2. All component values (and ratings) for capacitors and resistors are given. This data, in combination with the voltage readings mentioned above, should simplify troubleshooting.

REVIEW QUESTIONS

1. Why are volume controls needed in audio amplifiers?

2. Explain the action of potentiometer P in Fig. 18-1.

3. Refer to Fig. 18-2: **(a)** In diagram (*a*), what is the function of capacitor C_1? **(b)** In diagram (*a*), explain why the potentiometer arm is not connected directly to the base of the transistor, thereby eliminating C_2 and R_B. **(c)** In diagram (*a*), what is the function of capacitor C_2? **(d)** Explain why the circuit variation shown in diagram (*b*) is possible. **(e)** Explain how the change from diagram (*a*) to (*c*) is possible.

4. **(a)** Under what condition is the potentiometer volume control unsuitable? **(b)** Name three other types of controls that might be used in such cases.

5. The control shown in Fig. 18-3*a* is to be used with a 500-Ω line. **(a)** What is the maximum value of R_2? **(b)** What is the value of R_1 at this setting? **(c)** What is the minimum value of R_2? **(d)** What is the value of R_1 at this setting?

6. **(a)** How does the value of R_2 in Fig. 18-3b compare with R_2 in Fig. 18-3a? **(b)** Compare the R_1 values in these units. **(c)** What is the purpose of the added component R_3? **(d)** How does its resistance variation compare with R_1 or R_2?

7. When is it necessary to use the unit of Fig. 18-3c?

8. Give two fundamental reasons that make the use of tone-compensation circuits very desirable.

9. **(a)** What is meant by the *threshold of audibility*? **(b)** What is meant by the *threshold of feeling*? **(c)** What is the sound intensity difference in decibels between these two levels?

10. Refer to Fig. 18-4: **(a)** To what frequency is the ear most sensitive? **(b)** What intensity level (above the threshold) is needed to make a 50-Hz sound just audible? **(c)** Repeat part (b) for a 10,000-Hz sound. **(d)** How much boost of treble or bass would be needed at a sound intensity of 100 dB above the threshold of audibility to compensate for the ear? Explain.

11. Name three basic types of tone-compensation circuits.

12. Refer to Fig. 18-5: **(a)** State the two basic principles on which the action of this circuit is based. **(b)** When the controls R_1 and R_2 are set at maximum, why does the response in the center region drop so low? **(c)** What would be the effect (if any) of removing R_d? Explain. **(d)** Why does reducing the resistance of R_1 lower the output at the high frequencies? **(e)** Explain why it does not also affect the mid-frequencies. **(f)** Compared to a "normal" R-C amplifier, does this circuit really boost the high and low frequencies? Explain.

13. Refer to Fig. 18-6: **(a)** How does the gain of this stage compare with the gain of a "normal" R-C amplifier? Explain. **(b)** What kind of feedback circuit is used? **(c)** How does variation of P_1 affect the mid-frequency gain? Explain. **(d)** What is the effect of this control? **(e)** Explain the effect of setting P_2 at maximum (1) on the high-frequency response and (2) on the low-frequency response. **(f)** What is the function of C_2? **(g)** Which components determine the mid-frequency feedback? **(h)** From *examination* of these component values, what is the approximate feedback and maximum gain obtainable at mid-frequency?

14. Refer to Fig. 18-7: **(a)** What is the position of the controls for maximum boost? **(b)** At what positions is flat response obtained? **(c)** What additional feature not found in the previous circuits does this circuit provide?

15. Refer to Fig. 18-8a: **(a)** How are the tone controls set for this circuit condition to apply? **(b)** At 16 kHz, why is Z_A approximately 21,000 Ω and capacitive? **(c)** Why is Z_B approximately 22,000 Ω at this frequency? **(d)** Why is the output *down* 3 dB at this frequency? **(e)** Why is this a 15-dB *boost* compared to mid-frequency?

16. **(a)** What is a universal output transformer? **(b)** How is the effect obtained? **(c)** What is the advantage of such a unit?

17. State two objections to series-connected multiple-speaker systems.

18. **(a)** What is the objection to direct paralleling of speakers in a multiple-speaker system? **(b)** Under what condition might direct paralleling be used? **(c)** How is the objection of part (a) eliminated?

19. Refer to Fig. 18-9: **(a)** If it is desired to deliver equal power at each speaker, must all voice coils have identical impedances? **(b)** Must all transformers have identical secondary impedances? **(c)** Must all transformers have identical primary impedances?

Explain. **(d)** In order to match the amplifier, what value of primary impedance is necessary? **(e)** If the amplifier is operated so as to produce a 32-W output, how much power is delivered to each speaker?

20. Refer to Fig. 18-9: **(a)** If different power levels were desired at each location, could the speakers be identical? **(b)** What components would be affected? **(c)** If location D should have the highest power level, how would the components differ? **(d)** What relation must exist between the amplifier output impedance and the primary impedance at the unit locations?

21. **(a)** In the constant-voltage multiple-speaker system, what voltage value is generally used? **(b)** In a given amplifier, is this voltage constant? Explain. **(c)** Under what condition will this voltage be obtained? **(d)** What is constant about this value?

22. If Fig. 18-9 represented a 70-V line, what simple relation would be used to determine **(a)** the transformer primary impedance at each location and **(b)** the transformer secondary impedance at each location?

23. Refer to Fig. 18-10: **(a)** What determines the choice of secondary tap? **(b)** What determines the choice of primary tap? **(c)** If it is desired to deliver 2 W using a speaker with a 3.2-Ω voice coil, what connections are made? **(d)** If the line terminal is changed to the "5W" position, why does this deliver more power to the speaker? **(e)** How does the impedance seen on the primary side compare when delivering 2 W to a 3.2-Ω speaker or to an 8-Ω speaker? Explain.

24. Refer to Fig. 18-11: **(a)** What type of rectifier circuit is this? **(b)** What type of filter circuit does it use? **(c)** State two functions for R63, C76. **(d)** What is the function of SW602 at the input? **(e)** How is the base bias for Q201 obtained? **(f)** What is the function of R206? **(g)** Of R207? **(h)** What type of coupling is used between Q201 and Q202? **(i)** What is the function of VR702? **(j)** How does it affect the response when at the top position? **(k)** When at the bottom position? **(l)** Which component is the bass control? **(m)** What is the function of VR706? **(n)** Why is there also a tap connection to this control? **(o)** What type of coupling is used between Q202 and Q203? **(p)** Why is the collector of Q202 returned to the ground side of the circuit instead of to the positive supply (as for Q201)? **(q)** How is the +10.9 V at the emitter of Q202 obtained? **(r)** Is there any inverse feedback applied to this transistor? Explain. **(s)** How is the base bias for Q203 obtained? **(t)** Is any bias stabilization used? Explain. **(u)** How is the 11.5 V on the collector of Q203 obtained? **(v)** What type of output circuit does this amplifier use? **(w)** How is the base bias for Q204 and Q205 obtained? **(x)** What is the purpose of this bias? **(y)** What is the function of R220? **(z)** State two functions for capacitor C214.

25. State two servicing features included in the schematic of Fig. 18-11.

PROBLEMS AND DIAGRAMS

1. Draw a schematic diagram showing a one-stage P-channel FET amplifier with a simple volume control.

2. Draw a schematic diagram showing a volume control incorporated between two NPN transistors with R-C interstage coupling, voltage-divider bias, and voltage-feedback stabilization.

3. Draw the diagram for **(a)** an L pad; **(b)** a T pad; **(c)** an H pad.

4. The circuit of Fig. 18-5 uses a FET with $g_{os} = 3.33\ \mu$mho and $g_{fs} = 6000\ \mu$mho; $R_d = 15\ \text{k}\Omega$; $R_g = 0.5\ \text{M}\Omega$; $C_c = 0.05\ \mu$F. All other values are as shown on the diagram. Calculate: **(a)** the frequencies of maximum bass and treble boost; **(b)** the gain at these boost frequencies when the tone controls are set at minimum; **(c)** the gain at these frequencies when the controls are set for maximum boost (assume that the impedance of the resonant circuits is very high compared to R_1 and R_2); **(d)** the decibel boost at each frequency.

5. In Fig. 18-6, using the typical given values and the FET of Problem 4, calculate **(a)** the percent feedback with both tone controls set at minimum; **(b)** the maximum possible gain at these settings; **(c)** the maximum possible gain at 50 Hz when P_2 is at maximum boost; **(d)** the maximum possible gain at 16,000 Hz with P_1 at maximum boost.

6. Draw the circuit diagram for an *NPN* transistor amplifier using emitter-feedback tone control and voltage-divider bias.

7. In Fig. 18-7, the controls are set for maximum boost. Using the given typical values, and a 1-V input signal, calculate **(a)** Z_A, Z_B, and V_o at 16 kHz; **(b)** Z_A, Z_B, and V_o at 1000 Hz; **(c)** Z_A, Z_B, and V_o at 50 Hz. **(d)** Using the output voltage at 1000 Hz as reference, calculate the decibel gain or loss at the other two frequencies. **(e), (f), (g),** and **(h)** Repeat (a), (b), (c), and (d) with the controls set for maximum attenuation. (*Note:* Phase relations must be considered.)

8. A 25-W amplifier is to be used with a 200-Ω line to feed equal power to four 6-Ω speakers. Calculate: **(a)** the impedance ratings for the transformers at each speaker location; **(b)** the line voltage when the amplifier delivers its rated power output; **(c)** the line voltage when the amplifier power output is reduced to 10 W.

9. The amplifier of Problem 8 (25 W, 200-Ω output impedance) is to supply the following speaker loads: 10 W to an 8-Ω speaker; 5 W each to a 16-Ω speaker and a 6-Ω speaker; 3 W to one 3.2-Ω speaker and 2 W to another 3.2-Ω speaker. Find: **(a)** the transformer impedance ratings needed at each location; **(b)** the line voltage when the amplifier delivers its rated output; **(c)** the line voltage for an amplifier output of 10 W; **(d)** the power delivered to each speaker under condition (c).

10. A 100-W amplifier is used with a 500-Ω line to deliver power as follows: 20 W each to three 16-Ω horns; 5 W each to five 8-Ω speakers; and 1 W each to five 3.2-Ω speakers. Find the transformer impedance ratings needed at each location.

11. The speaker system of Problem 9 is to be fed from a 25-W amplifier with a 70-V line tap. Specify the transformer impedance ratings needed at each location. Compare with Problem 9, and account for any differences or similarities.

12. The speaker system of Problem 10 is to be fed from a 100-W amplifier with a 70-V tap. Specify the transformer impedance ratings needed at each location.

19

Linear Integrated Circuits

The integrated-circuit industry got its start early in the 1960s with support from NASA and the military. Their space and missile systems used hundreds of thousands of individual electronic components. The need for miniaturization and for improved reliability of these systems led to the development of integrated circuits (ICs). In the early days, the price of an IC unit was far in excess of its discrete-component counterpart, and so the government was the only consumer. With improved manufacturing techniques, costs came down. Further reduction in unit cost was then achieved by volume production.

For industrial applications—where cost is a major consideration—the advantages of the integrated circuit are realized only when a number of identical basic circuits are used repeatedly throughout the product. High-volume production then makes the ICs cost-competitive with discrete-component circuits. Such a field is digital data processing. However, since the late 1960s, further manufacturing improvements and cost reductions have made ICs practical in all fields of electronics.

MONOLITHIC ICs.   Based on construction, one type of integrated circuit is known as a *monolithic IC*.* These circuits are made from a single crystal of material (usually silicon). Conductors, resistors, capacitors, diodes, and transistors are all formed on the same chip by using manufacturing processes common in the semiconductor industry (diffusion, epi-

---

*This name is derived from the Greek words *monos* meaning single and *lithos* meaning stone.

taxial growth, masking, and etching). The monolithic process results in the smallest possible structure and the lowest cost. Also, since there are no "connecting wires" between components, integrated circuitry has far better reliability than discrete-component circuits.

A cross-sectional sketch of a simple monolithic integrated circuit is shown in Fig. 19-1, together with its schematic. The fabrication process starts with a P-type substrate onto which a thin N-type epitaxial layer is grown. Then, by exposing this N layer to oxygen gas at high temperature (1000°C), an insulating layer of silicon dioxide (SiO_2) is formed over the entire surface. This protects the wafer from possible contamination. The process is called *passivation*. Next, by masking and etching, the protective SiO_2 layer is removed in four places and P-type impurity is diffused into the unprotected areas through the epitaxial N layer and reaching into the P-type substrate. At this point the wafer looks as shown in Fig. 19-1a.

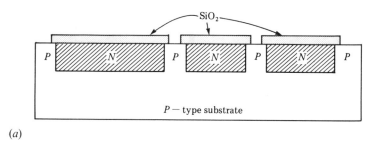

(a)

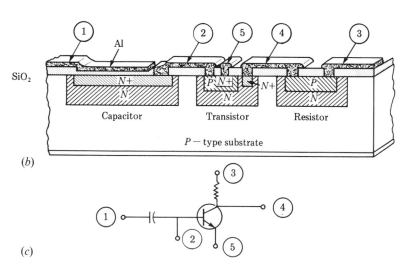

(b)

(c)

Figure 19-1   A monolithic silicon integrated circuit.

Notice the three islands of N-type material, separated from each other by back-to-back P-N junctions. If the substrate is maintained negative in polarity with respect to the N-type islands, the resistance of these reverse-biased junctions approaches infinity and the N islands are electrically insulated from each other.

By continuing this selective process of passivation, masking, etching, and additional doping, the end product shown in diagram (*b*) is formed. Notice that the aluminum deposition is used to make interconnection between the IC components and to the external terminals (① to ⑤). Now compare Fig. 19-1*b* with its schematic, Fig. 19-1*c*. The capacitor is formed by the outer aluminum layer and the heavily doped $N+$ layer separated by the SiO_2 as the dielectric. The transistor is obviously an NPN type. Notice that the aluminum layer marked ② interconnects the $N+$ side of the capacitor to the P material, or base of the transistor, and also serves as an external terminal point. Terminal ⑤ connects directly to the emitter ($N+$ region) of the transistor, and terminal ④ connects to the larger $N-$ region, the collector. Notice that terminals ④ and ③ connect to opposite ends of the P region of the third island. This P layer is used as the resistor, and the resistance value depends on the resistivity, length, and thickness of the layer.

COMPONENT DENSITIES.   The integrated circuit of Fig. 19-1 is obviously a very simple one. There are only three components on the one wafer. On such a scale, the process would be too costly to be industrially practical. In actual practice, a wafer (of about 1×1 in) is divided into about 400 chips, each 50×50 mil, or 2500 mil^2. Many components— such as transistors, diodes, or resistors—can be fabricated on one such chip. Depending on the component density, the processing is classified as *large-scale integration* (LSI), *medium-scale integration* (MSI), or *small-scale integration* (SSI). Since ICs were first used in digital computer circuits, the criterion for classification is based on the number of *logic gates** contained within the one chip. Large-scale integration is defined as the interconnection of 100 or more circuits of logic-gate complexity by silicon processing. Since a single gate may use one or more transistors plus associated resistors, LSI implies the interconnection of hundreds of components in one chip. Consequently, instead of a simple circuit, entire electronics systems or subassemblies may be contained in one LSI package. Again using the logic-gate criterion, an MSI chip is defined as an integrated circuit with greater than 15-gate complexity.

At first thought, one might think that an LSI chip with its hundreds of components would be far more expensive than a simple SSI unit. This is

---

*A logic gate (such as an AND or OR circuit) is a basic building block in data processing systems.

not so. The fabrication steps are the same in each case. The original design and development costs for the LSI are more expensive, because the tolerances for masking, diffusing, and etching are much tighter when the component density is increased. Yet with high-volume production these costs are spread over many units, and the per-unit cost of the LSI chip is not much higher. This could lead to another false impression—that all integrated circuits should have high component density (LSI). Again this is not so. As we increase the complexity of the integrated circuit, its versatility decreases. For example, whereas a transistor and resistor could be used in myriads of circuit applications, a complete LSI subsystem could probably be used only in the application for which it was originally designed. Remember that we need high-volume application to warrant the first cost of LSI processing. So unless the application has high-volume demand, LSI methods would not be practical and small-scale integration would be preferable.

Another technique that brings costs down is batch fabrication, with many chips processed simultaneously. For example, starting with a wafer which is subsequently cut into 400 chips, we obviously produce 400 chips at the same time. In addition, we process many wafers together as a batch. This further increases the total number of chips produced in one set of operations.

PROS AND CONS FOR MONOLITHIC ICs.   Some of the advantages resulting from the use of monolithic integrated circuitry should be readily apparent:

1. The small size of the chip, compared to its discrete-component counterpart.
2. The low cost of the chip, compared to the cost of the individual components and the labor in interconnecting them into a circuit.
3. The high reliability of an IC. Once the circuit is properly designed and the processing is fixed, all ICs are identical. There will be no misconnection of the individual components, no poor solder joints, no wiring board, no wires to break.

In addition there are other advantages that may not be so obvious:

1. *Improved performance.*   Since the cost of a complex IC is not materially greater than that of a simple circuit, it becomes practical to use sophisticated circuitry to obtain improved operational features. With discrete components, such added circuitry would be too expensive.
2. *Better-matched components.*   Because transistors are made simultaneously—on the same chip—and using the same processing, their characteristics are more closely matched. This is an ideal advantage for differential amplifiers.

3. *Better temperature stability.* Again, because the components are on a common substrate, temperature differences are minimized and compensation is more effective.

4. *Serviceability.* This is obtained in complex systems through the use of replaceable or plug-in boards. An example of this is the modular Quasar television receiver.

Unfortunately, monolithic integrated circuitry also has its drawbacks:

1. IC transistors have a lower high-frequency limit than discrete units. (This is due mainly to the parasitic capacitances inherent in the isolation process.)

2. In Fig. 19-1 we saw the construction of an *NPN* transistor. IC transistors are generally of this type. *PNP* units are more difficult to produce, and such IC *PNP* transistors have appreciably lower current gain than an *NPN* IC unit or a discrete *PNP* unit.

3. In Fig. 19-1 we also saw the construction of an IC resistor. The range of values obtainable by this technique is from about 20 to 25,000 Ω. Higher resistance would require too much space. A further limitation is that accurate resistance values cannot be obtained. Tolerances of ± 10 percent (at 25°C) are about the best that can be expected. What is worse is that the resistance values are very temperature-sensitive. A redeeming feature is that the change with temperature is uniform, so that resistance ratios remain constant.

4. Similarly, capacitance values obtainable by IC processing are limited to about 200 pF. Larger values would require excessive surface area.

5. No practical inductor values are obtainable in IC form.

Hybrid ICs. Many of the limitations of the monolithic IC can be overcome by using a *hybrid* form of integrated circuit. Such a circuit may consist of any combination of thin-film or thick-film elements, discrete elements, and monolithic ICs—all within a single package. The advantage of this type of circuitry is that optimum design techniques can be used for each type of component. Hybrid ICs are especially useful when large (or accurate) resistance values, large capacitance values, or inductors are required. For example, although the range of values using thin-film techniques is about the same as with monolithic ICs, these resistors can be adjusted (by scribing, wet anodizing, or laser-beam cutting) to very accurate values. On the other hand, by using thick-film techniques resistor values in the megohm range are possible. Hybrid technology is also useful in small production runs since the initial costs are lower than for monolithic construction. Finally, since monolithic devices have low power-dissipating capability, hybrid technology is necessary when high power levels are involved. Figure 19-2 shows the constructional details of a hybrid IC that can deliver up to

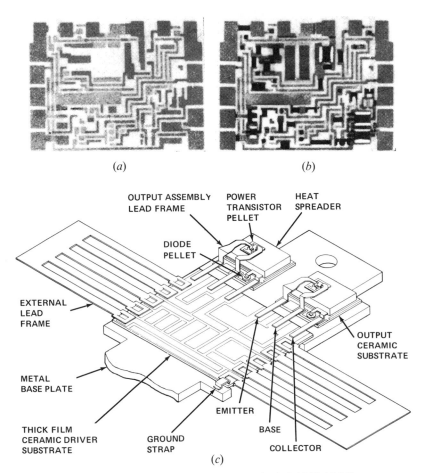

(a)   (b)

OUTPUT ASSEMBLY   POWER   HEAT
LEAD FRAME   TRANSISTOR   SPREADER
                   PELLET

DIODE
PELLET

EXTERNAL
LEAD
FRAME

OUTPUT
CERAMIC
SUBSTRATE

METAL
BASE PLATE

EMITTER

THICK FILM   BASE
CERAMIC DRIVER   GROUND
SUBSTRATE   STRAP   COLLECTOR

(c)

Figure 19-2   Constructional details of a hybrid IC (*RCA*).

100 W of output power. This IC is constructed in two sections: a thick-film signal-input and driver section, and a power transistor output section. Figure 19-2*a* shows the conductor pattern and a crossover dielectric printed and fired on a 1.2 × 1.6 in ceramic substrate. Figure 19-2*b* shows the input and driver substrate complete with resistor, semiconductor, and capacitor chips. In all, it consists of 23 resistors, 7 capacitors, 6 diodes, and 9 transistors. This substrate (after testing) is then bonded to the base plate of the pretested output section, as shown in Fig. 19-2*c*. The output section consists of two output transistors mounted on copper heat-spreader blocks. Two diodes shunt the emitter-collector terminals of these power transistors to protect them from reversal potentials that may occur during switching when used with transformer-coupled loads. An external view of this hybrid IC is shown in Fig. 19-3*d*.

PACKAGING.   Integrated circuit units are available in a variety of packages. Some look like a simple transistor using the standard TO case. Of course they have more leads. Such a unit is the CA3028B shown in Fig. 19-3a. This unit uses a TO-5 case (approx $\frac{1}{3}$ in diameter and $\frac{1}{4}$ in high). A second very common package is the *dual in line*. They are available as plastic or ceramic packages. A plastic unit (the CA3070) is shown in Fig. 19-3b. Another type is the *flat pack*. This is also available as a plastic or ceramic package. Figure 19-3c shows a ceramic flat pack (CD4016). All these units are monolithic ICs. On the other hand, Fig. 19-3d shows a hybrid flat-pack unit on a molded-epoxy package. (The details of this IC were discussed in the previous section.)

(a)

(b)

(c)

(d)

Figure 19-3   Typical IC packaging (*RCA*).

IMPACT OF ICs ON INDUSTRY.   The widespread use of integrated circuits has already begun to affect everyone engaged in the electronics industry. Let us first consider the technicians. When servicing discrete-component circuits, most of their time is spent in locating the defective component. To aid them, they refer to circuit schematics which also show the key voltages (and resistance values) at the leads of a transistor (or at the pins of a vacuum tube). Once the defective component has been located, the replacement time and cost are generally only a small fraction of the total time and cost. The major part of any service charge is the troubleshooting time. With ICs, technicians will not be able to measure voltages at the leads of a transistor. Nor will they be able (in general) to replace an individual resistor or capacitor. Except for outboard components, they must replace a complete IC unit. If the technicians are knowledgeable, the defective IC should be readily apparent. Troubleshooting time and cost are greatly reduced, and since ICs are rather inexpensive, total service costs are reduced. In analyzing any malfunction, technicians must be systems conscious. They must understand the functions and interrelations of the various blocks in the system. This will require a higher level of knowledge.

Engineers who design ICs will need reorientation for this field. With discrete components, resistors and capacitors are cheap compared to transistors (or tubes). With ICs the reverse is true, so the engineers' designs should minimize the use of passive components in favor of active devices. For example, we know that an unbypassed emitter resistor—in a CE circuit—will reduce the stage gain due to inverse feedback. With ICs it is preferable to make up this loss by including another stage within the IC rather than adding a bypass capacitor. Using a three-transistor differential amplifier is even better. Not only is the bypass capacitor not needed, but the constant-current third transistor replaces the emitter resistor. In some cases, transistors are also used in place of collector load resistors, with the bias set to give the desired equivalent resistance value. When resistors are used, it is important that the circuit action should not be critically dependent on the individual resistor values. (Remember that resistor tolerances can be as wide as ± 25 percent.) Instead, resistor voltage-divider actions are preferable, since resistor ratios are under better contol. Finally, the engineers who *use* ICs need to be more systems designers than circuit designers. Their work involves interconnection of blocks and subsystems instead of individual components.

As the use of ICs increases, manufacturers of passive components (resistors, capacitors, and inductors) will be faced with a falloff in the demand for low-power units. Their production will have to concentrate on high-power and/or special discrete units that are compatible in size with integrated circuits and can be used outside the IC package. The need for component manufacturing will decrease.

Linear integrated circuits (LICs) are now used in every field of electronics—by the military, in industrial applications, and in consumer (household) products—wherever amplification and/or control is needed. The automotive industry is one of the latest penetrations. ICs are used to control fuel injection, ignition, voltage regulation, and in a number of safety devices such as antiskid control. Consumer applications include radio, audio, TV, and household appliances such as washers, driers, and ovens. In the field of music, ICs have revolutionized the organ industry. They not only produce the basic frequencies (for the keyboard), but by means of frequency dividers and amplifiers they also produce harmonics to duplicate the sounds of other instruments—and even the human voice. Some organ models feature pizzicato, tremolo, vibrato, sustain, and wa-wa rock effects. They also have automatic rhythms (for example, beguine, waltz, samba, rock), a glissando maker, and follow-the-player rhythms (bass drum, cymbal, brush, snare drum). Such organs are now virtually one-man bands—and all due to integrated circuitry. Obviously, ICs range from very simple general-purpose units to complex specialized devices. Manufacturers' catalogs describing these devices often run to over 500 pages. Let us examine some of these ICs.

OPERATIONAL AMPLIFIERS (OP AMPS)

The operational amplifier circuit, or op amp as it is colloquially known, was originally developed for use in analog computers. The very name of this circuit stems from the fact that it can be connected so as to perform a variety of mathematical operations. Among these are inversion, addition, subtraction, differentiation, and integration. Yet, basically, any operational amplifier is just a high-gain dc-coupled amplifier and can therefore be used in any application where amplification is desired. As a discrete-component circuit, the operational amplifier was never popular. Its cost was too high, and it had stability problems due to poor matching of the transistors. Integrated circuitry solved both problems. Monolithic IC operational amplifiers are quite inexpensive, and since they are fabricated on one chip, matching and temperature stability are good. (Furthermore, additional stabilizing circuitry can be added at minimal extra cost.) In recent years, the growth of the integrated-circuit industry has zoomed, with sales exceeding millions of dollars a month.

Functionally, the internal circuitry of an operational amplifier can be divided into three parts:

1. The input section is a differential amplifier with high input impedance. This also provides for inverting and/or noninverting input. With BJTs in the input stage, the input resistance may range from 10 to 100 kΩ. For higher impedance (up to a megohm) Darlington pairs can be used for each side of the differential amplifier. By using FETs, input impedances of up to 100 MΩ are possible.
2. The second section is designed to provide the full gain of the op amp. Quite often it is also a differential amplifier. Terminals are usually brought from this section to allow for external control of

the frequency and phase response of the unit to suit the specific application.

3. The output section is generally a low-impedance power output stage. It may be an emitter follower—single-ended or push-pull—or it may be a class B power stage. These circuits reduce the output loading and tend to make the gain and frequency response of the unit independent of the load. In addition, this stage provides level (voltage) translation and may include output short-circuit protection. Level translation is necessary in a dc amplifier to remove the dc component in the output, translating the output voltage to a ground reference. Obviously, the schematic of a monolithic IC op amp can be quite complex. This is illustrated in Fig. 19-4 by the schematic of the LM308, a 1973 vintage operational amplifier. No attempt will be made to discuss this circuit here. Our interest is in how to use these devices, rather than in how to design them.

SYMBOL.   The internal components of an IC cannot be replaced or altered. Consequently, there is nothing to be gained in showing the

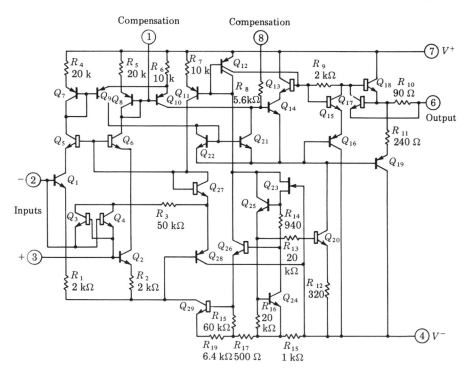

Figure 19-4   Schematic of the LM308 operational amplifier (*National Semiconductor Corp.*).

internal circuitry of an IC when drawing schematics that include ICs. Instead, these devices are represented symbolically by triangles—with, at times, suitably numbered terminals.* The basic symbol for an operational amplifier is shown in Fig. 19-5. This could be called an "ac equivalent" symbol, since only the signal input and output connections are shown. The (−) terminal is the *inverting input* terminal. A signal applied between this input terminal and ground will appear 180° out of phase at the output (with respect to ground). The (+) terminal is the *noninverting input* connection, and it is used if phase reversal is not wanted. The location of these terminals in the IC symbol is standard—except that the (+) and (−) inputs may be reversed if the overall diagram is simplified. The other terminals —power supply points, ground, and compensating circuit connections— may be shown anywhere along the upper or lower sides of the triangle, as convenient.

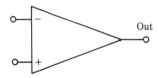

Figure 19-5   IC operational amplifier symbol.

CHARACTERISTICS AND SPECIFICATIONS.   Manufacturers' data sheets for operational amplifiers give the ratings and characteristics of their various models. Before we can select or use one properly, we must understand the meanings of these parameters. Let us examine some of these terms:

1. *Input offset voltage.*   An ideal operational amplifier would have zero output voltage when the voltage applied between the two input terminals is zero. (This also applies for single-ended input, with the other input terminal grounded.) Such a condition requires perfect matching of the differential amplifier pairs. This is approached with integrated circuitry, but not perfectly. If exact zero output is required, it may be necessary to apply a small dc bias voltage between the input terminals (through two equal resistors) to compensate for imbalance. This voltage is called the *input offset voltage*. In well-designed units the offset voltage is quite low—of the order of 1 mV.

---

*Whether the numbered terminals are shown or not depends on the specific application. Sometimes, only some of the terminals being used are shown.

2. *Input bias current.*   Both inputs of the operational amplifier require bias currents, which are the base currents of the differential pair. The average of these two input currents is called the *input bias current.*

3. *Input offset current.*   This refers to the difference in the bias currents into the two input terminals when the output is zero. It is again an indication of mismatch in the differential amplifier stage. This becomes important with high-resistance input sources, because there will be a large unbalance in the input bias voltages.

4. *Slew rate (SR).*   Because of capacitance effects within the IC and because of externally connected capacitors, time constants are created that prevent the output voltage from changing as rapidly as the changes in the input signal. This is both a frequency and an amplitude* limitation and is most evident with step-voltage inputs. (The rate of rise of a step voltage is theoretically infinite.) This limitation is specified as the *slew rate,* defined as the maximum rate of change in output voltage with a large amplitude step function applied at the input. It is expressed in volts per microsecond ($V/\mu s$). A high slew rate is desirable if high-frequency and/or high-amplitude operation is contemplated. (A common value is around $1 \ V/\mu s$.)

   With sine-wave input signals, operating beyond the bandwidth limit (beyond the cutoff frequency) will reduce the output level—but the output is still a sine wave. On the other hand, operation beyond the slew rate limit will cause distortion of the signal. From the equation of a sine wave ($v = V_m \sin \omega t$, and remembering that $\omega = 2\pi f t$), differentiating with respect to time and setting this to zero, we get the rate of change in voltage—which for a sine wave is $2\pi f V_m$. For undistorted output, the slew rate (SR) should not exceed this value:

$$SR \geq 2\pi f V_m \tag{19-1}$$

Since slew rate is in volts per microsecond, then if frequency is in megahertz the voltage amplitude will be in volts. This equation can be used to determine the maximum undistorted output available from an op amp of given slew rate—or vice versa, to select an op amp of suitable slew rate for a given frequency and amplitude of signal.

---

*It should be obvious that for any given frequency (or timing) of the input, the higher the amplitude of the signal, the faster the voltage must rise in the given time (time of one quarter-cycle).

EXAMPLE 19-1

It is desired to obtain a peak-to-peak sine-wave output of 6.0 V. The signal frequency is 50 kHz. What is the minimum slew rate specification for the IC?

Solution

1. For a peak-to-peak of 6.0 V, the output amplitude (V_m) is 3.0 V.
2. $SR = 2\pi f V_m = 6.28 \times 0.05 \times 3.0 = 0.94$ V/μsec

In this design, the op amp selected should have a slew rate of at least 0.94 V/μsec in order to avoid distortion of the input-signal waveform.

5. *Power supply (voltage) rejection ratio (PSRR).* This is the ratio of the change in input offset voltage to the change in the supply voltage producing it. Since the change is small, it is expressed in microvolts per volt (μV/V).
6. *Power supply sensitivity. (S).* This is sometimes used in place of PSRR. It is a general term that specifies the ratio of change in some specified parameter to the change in supply voltage causing it. Again, it is expressed in μV/V units.

Obviously, these are not all the specified parameters. Although the others are not discussed here, they should be understood either directly from their titles or from previous discussions.

GAIN AND FEEDBACK.   Depending on the number of stages and other design factors, the gain of an operational amplifier ranges from a low of (about) 1000 up to about 1 million. However, the gain for any one type —from the same manufacturer—may differ by a fairly wide margin. For example, the data sheet for the Raytheon RC4739 gives the minimum gain as 20,000 and the typical gain as 300,000. This is a variation of 15 to 1! Furthermore, the gain of any given unit will also vary widely with frequency of operation and with supply voltage. Such gain variations cannot be tolerated in any serious application. In Chapter 16, we saw that inverse feedback is used with discrete-component amplifiers to stabilize gain. The same technique is used with ICs. As before, the penalty is reduced gain, but there is gain to spare in these integrated circuits, so this presents no problem. The gains of 1000 to 1 million mentioned above are the gains without feedback, or open-loop gains A_{VOL}. This gain is sometimes given in decibels (that is, 20 log of the voltage gain) or as the volts output for an input of 1 mV (V/mV).

The IC is obtained from the manufacturer without feedback, and inverse feedback is then applied by the user to fit his specific requirements. Remember, as was shown in Chapter 16, that the higher the feedback level,

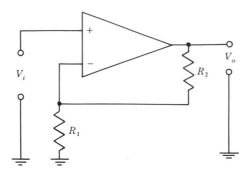

Figure 19-6   Application of inverse feedback to a noninverting op amp.

the better the gain stability but the lower the closed-loop gain. Let us see how this feedback is applied to an operation amplifier. Figure 19-6 shows an op amp connected as a noninverting amplifier.* With the input signal fed to the $(+)$ terminal, the output is in phase with the input. Now notice that part of this output voltage is fed from the voltage divider (R_2-R_1) to the inverting input terminal. This is *series-fed* inverse *voltage* feedback, and as explained in Chapter 16, the feedback factor β is $R_1/(R_1 + R_2)$. Also, the gain with feedback (see Equation 16-4) is

$$A_{\text{fb}} = \frac{A}{1 + A\beta}$$

Furthermore, since the open-loop gain of an op amp is very high, we can use the approximate equation (16-5) with excellent accuracy:

$$A_{\text{fb}} = \frac{1}{\beta}$$

And now, substituting for β from the voltage-divider values, we get for a noninverting op amp:

$$A_{\text{fb}} \text{ (noninv)} = \frac{R_1 + R_2}{R_1} = 1 + \frac{R_2}{R_1} \tag{19-2}$$

Notice that the gain of the operational amplifier is now dependent only on the external feedback resistor values.

EXAMPLE 19-2

The data sheet for the RC4739 operational amplifier shows an open-loop gain range of 20,000 to 300,000. One such unit is used in the feedback circuit of Fig. 19-6, with $R_1 = 1\ \text{k}\Omega$ and $R_2 = 39\ \text{k}\Omega$.

*It should be noted that this is not an actual working diagram but a simplified diagram intended merely to show the feedback connection.

1. Using the minimum value of A_{VOL} and Equation 16-4, calculate the closed-loop gain.
2. Repeat, using Equation 19-2.
3. Repeat, using Equation 19-2 and the maximum value of A_{VOL}.

Solution

1. $\beta = \dfrac{R_1}{R_1 + R_2} = \dfrac{1}{1 + 39} = 0.025$

$$A_{fb} = \dfrac{A_{VOL}}{1 + A_{VOL}\beta} = \dfrac{20{,}000}{1 + 20{,}000(0.025)} = 40$$

2. A_{fb} (noninv) $= 1 + \dfrac{R_2}{R_1} = 1 + \dfrac{39}{1} = 40$

3. A_{fb} (noninv) $= 1 + \dfrac{R_2}{R_1} = 1 + \dfrac{39}{1} = 40$

From this example, notice that the "approximate" equation is just as accurate as the basic equation. Note also that in spite of the 15-to-1 variation in open-loop gain of this device type, the gain with feedback is stabilized.

In the schematic of Fig. 19-7, the input signal is applied between the $(-)$ terminal and ground. The circuit is therefore an *inverting* amplifier, and the output voltage is 180° out of phase with the input signal at a. Notice that resistor R_2 provides a feedback path from the output to point a. Since the feedback is in phase opposition to the input, we have inverse feedback. Resistor R_1 is in series with the input signal source. This resistor is necessary to prevent short-circuiting of the feedback voltage if the source resistance is zero (or small compared to the desired value for R_1). This effectively places R_1 in parallel with the input resistance of the op amp, and we have a shunt-fed voltage-feedback circuit.

Since the gain of an operational amplifier is very high (approaching infinity in the ideal case), the feedback voltage (V_f) is almost equal to the

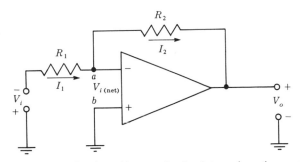

Figure 19-7   Application of inverse feedback to an inverting op amp.

signal input voltage (V_i) and the net input voltage $V_{i(net)}$ approaches zero. This places point a at virtual ground level. Therefore the input current $I_1 = V_i/R_1$. Also, the feedback current $I_2 = V_o/R_2$. Now, remembering that the input resistance of an open-loop op amp is very high (approaching infinity in the ideal case), negligible current flows into the amplifier. Consequently we can say that $I_2 = I_1$, or

$$\frac{V_o}{R_2} = \frac{V_i}{R_1}$$

Rearranging terms,

$$\frac{V_o}{V_i} = \frac{R_2}{R_1}$$

And since V_o/V_i is the closed-loop gain,

$$A_{fb} \text{ (inv)} = \frac{V_o}{V_i} = \frac{R_2}{R_1} \tag{19-3}$$

INPUT AND OUTPUT IMPEDANCE.   As we saw in Chapter 16, inverse feedback alters the effective input and output impedance of the circuit. Since in both cases (inverting or noninverting amplifier) we use voltage— or shunt-derived—feedback, the output impedance will decrease, approaching closer to the ideal value of zero. The output impedance (or resistance) with feedback (Z_{of}) becomes

$$Z_{of} = \frac{Z_o}{1 + A_{VOL}\beta} \tag{19-4}$$

where $\beta = R_1/R_2$ in the inverting amplifier

$\beta = R_1/(R_1 + R_2)$ in the noninverting amplifier

With series-fed feedback (i.e., the noninverting circuit), the input impedance (or resistance) increases with feedback and, as shown in Chapter 16, becomes

$$Z_{if} \text{ (noninv)} = Z_i(1 + A_{VOL}Z_i) \tag{19-5}$$

In the inverting amplifier circuit, the feedback is shunt-fed, and in Chapter 16 we saw that this reduces the input impedance (or resistance). The virtual grounding of point a in Fig. 19-7 puts resistor R_1 across the input of the op amp. This resistor value is generally much lower than the open-loop input resistance, and so for all practical purposes the input impedance (Z_{if}) with feedback becomes

$$Z_{if} \text{ (inv)} \cong R_1 \tag{19-6}$$

EXAMPLE 19-3

Each section of the Motorola MC1537L dual operational amplifier has $A_{VOL} = 45{,}000$, $Z_o = 30\ \Omega$, and $Z_i = 400\ k\Omega$. The unit is used as an inverting amplifier with $R_1 = 5100\ \Omega$ and $R_2 = 620\ \Omega$. Find the gain, input impedance, and output impedance with this feedback.

Solution

1. $A_{fb} = \dfrac{R_2}{R_1} = \dfrac{620{,}000}{5100} = 121$

2. $Z_{if} \cong R_1 = 5100\ \Omega$

3. $\beta = \dfrac{R_1}{R_2} = \dfrac{5100}{620{,}000} = 0.00823$

$$Z_{of} = \frac{Z_o}{1 + A_{VOL}\beta} = \frac{30}{1 + (45{,}000 \times 0.00823)} = 0.079\ \Omega$$

TYPICAL OP AMP WORKING CIRCUIT.   None of the circuits we have seen so far has been operational. Figure 19-8 shows a basic working circuit for an inverting amplifier. Notice that there are two power supply input points: $+V$ and $-V$. This requires a dual power supply with the common—the negative of the $+V$ supply and the positive of the $-V$ supply —grounded. A dual or split supply is desirable because the op amp incorporates one or more differential amplifiers. From the discussion of differential amplifiers in Chapter 14, you will recall that one power source was fed to the collectors of the differential pairs and the other power source fed the emitters. The supplies were therefore labeled V_{CC} and V_{EE}, respectively. Even though the internal connections may now differ, some manufacturers still use the V_{CC} designation for the $+V$ terminal and the V_{EE} designation for the $-V$ terminal. Notice also that each power supply feed

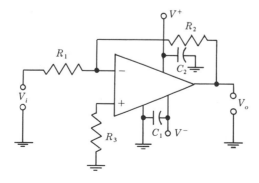

Figure 19-8   Op amp working circuit.

point is bypassed to ground. C_1 and C_2 are decoupling capacitors. Their use minimizes the possibility of circuit interaction through the power supply. (A typical value for these capacitors is 0.1 μF.)

If you compare Fig. 19-8 with the earlier simplified diagram (Fig. 19-7), you will notice that a resistor, R_3, has been added. Each input terminal, + and −, must have a dc return to ground—directly or through the input source. To reduce offset voltage imbalance, the resistance of these two input paths should be equal. Therefore, R_3 is added to balance the feedback network resistor R_2 if offset imbalance is objectional.

MULTIPLE OP AMP UNITS.   To reduce overall circuitry costs when constructing ICs, high-density chips are desirable. Yet there is a limit to what can be included in an operational amplifier. High-quality units incorporate such features as short-circuit protection, voltage regulation, and internal compensation. Even so, component density is still not very high. It is therefore practical to put two or more completely independent operational amplifiers on the same chip. For example, the RCA CA3747T, the Motorola MC1435F, the Fairchild μA747, and the National LM1558 are all dual operational amplifiers; the Fairchild μA7351 is a triple operational amplifier; and the RCA CA324E, the Raytheon RM4136, the Motorola MC3401P, and the National LM2900 are examples of quad (four-unit) packages.

OTHER GENERAL LICs

We mentioned earlier that monolithic ICs are economically practical only when they are produced in large quantities. One way of ensuring high-volume utilization is to provide versatility. This is what made use of the operational amplifier so widespread. Of course, a simple transistor is even more versatile—but to use only one transistor on a chip is hardly economical. The obvious answer, then, is to combine several independent transistors on the same chip and let the user interconnect them as he wishes. Such an IC package is called a transistor array. The RCA CA3086 combines three transistors and a differentially connected pair on one chip. The National LM3026 consists of two independent differential pairs, each with its own constant-current transistors. The Fairchild 3036 is a package of four independent transistors. General Electric's D40C is an internally-connected Darlington pair with a gain factor (h_{fe}) of 10,000 to 40,000. Its external appearance is similar to the plastic versawatt single transistor shown in Fig. 3-2d. Similar combinations (and many others) are available from all IC manufacturers.

LICs FOR THE CONSUMER MARKET

It was mentioned in Chapter 14 that an operational amplifier can be operated from a single power supply by grounding the negative power-supply terminal.* The disadvantage in such operation is that some performance characteristics—particularly CMRR—will be impaired. However, in consumer electronics applications the cost of a dual supply is generally not warranted, while the loss in performance is not serious. Consequently, when op amps are employed in consumer circuits the single-supply mode is used. Figure 19-9 shows one section of a dual operational amplifier (LM381) used for one channel of a stereo phono amplifier. Resistors R_1 and R_2 form the feedback voltage divider that sets the gain of the amplifier, as described earlier. R_4, C_2, and C_3 form a frequency-selective feedback network to provide proper RIAA equalization. The other half of the IC is used in an identical circuit to amplify the second channel of the stereo recording. This IC also has individual internal power supply decoupling regulator circuits for each of the op amps, and it can operate from a supply voltage of from 9 to 40 V.

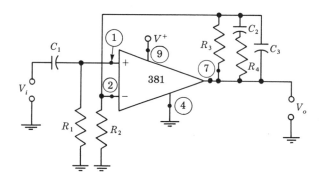

Figure 19-9 Typical IC phono preamp.

SPECIAL-PURPOSE CIRCUITS.   The operational amplifier has wide application because it can be used in a variety of circuits. On the other hand, high utilization can also be achieved by tailoring an IC for a specific function (or combinations of functions) in a consumer product with high sales volume. One such product is the television receiver. Millions of color

*Caution: If the IC has a separate ground terminal, it must not be grounded when the IC is used with a single power supply.

TV receivers are sold each year. Sections of these receivers can be replaced by integrated circuitry. If the receiver manufacturers agree as to what combinations of stages should be grouped into IC units, high-volume production runs can be made. This brings the cost per unit down, below the cost of the discrete components they replace. Such has been the case, and ICs are now available to replace most of the discrete-component circuitry in the TV receiver. ICs are available for the sound system, the video system, the chroma system, and the deflection system. Even the automatic fine tuning circuitry can be obtained as an IC chip. This is a relatively simple unit with only 13 transistors, 4 diodes, 11 resistors, and 2 capacitors. At the other extreme is the chroma demodulator with 51 transistors, 5 diodes, 62 resistors, and 4 capacitors—in a standard 16-lead dual-in-line package.

Let us examine one unit in some detail—the 3065.* Figure 19-10 shows a block diagram of this IC unit together with a few outboard com-

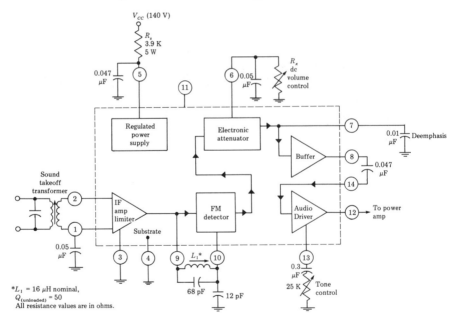

Figure 19-10 An IC TV sound system (*RCA*).

ponents, forming a complete TV receiver sound system. It combines a multistage IF amplifier-limiter, an FM detector, an electronic attenuator, a zener diode power-supply regulator, and an audio amplifier-driver—all in a 14-lead dual-in-line plastic package. The output from this unit can drive a discrete or an IC push-pull transistor power amplifier. Of special

*It is available from several manufacturers by the same number.

interest is the electronic attenuator, which together with the outboard variable resistor, R_x, performs the function of a conventional volume control. However, this system has a distinct advantage. When using the conventional potentiometer volume-control circuit, the audio signal must travel from the amplifying unit to the volume control and back. Obviously, the potentiometer itself must be conveniently located on the front panel of the TV receiver cabinet. The long leads to and from the control can pick up hum or noise voltages. Shielded leads must be used to minimize such pickup. This problem is particularly serious with remote-control units. In this circuit, the electronic attenuator (which itself contains several transistors) is physically located right next to the rest of the signal circuitry. Control of volume is achieved by changing the dc bias levels of the transistors within the attenuator, by means of the variable resistor R_x. Only a single unshielded lead is required between the attenuator and the control resistor, and there is no possibility of hum or noise pickup since this lead carries only dc. Whereas such an involved volume-control system would be too expensive using discrete components, as part of an IC it is now cheaper than the cost of installing shielded leads.

Similar advances have been made in both the audio amplifier field and in AM/FM receiver circuits. Among the ICs available for such use are RF-IF amplifiers, IF amplifier-limiters, detectors, audio preamplifiers, FM stereo multiplex decoders, phase-locked loop stereo decoders, dual audio stereo preamplifiers, four-channel SQ decoders, class A and class B audio drivers, and dual-channel power amplifiers.

A complete AM radio receiver can be obtained by combining an AM radio system IC* and an audio power amplifier IC.† Figure 19-11 shows such a system in block diagram form, together with the necessary outboard components. It contains a converter stage, IF amplifier stages, a detector, an audio preamplifier, and a zener-diode voltage regulator. It provides automatic gain control (AGC) for the IF stages and a buffer amplifier to drive a tuning meter. A brief signal-flow analysis is as follows: ① starting at the antenna, the signal is selected by the first tuned circuit and coupled to the input terminal of the converter ②. A local oscillator signal is also applied to this terminal. The IF signal from the converter output ③ is coupled through the IF transformer to the input of the first IF amplifier ④. The output from this stage ⑥ is coupled through a second IF transformer to terminal ⑧, the input of the second IF amplifier. (Terminal ⑦ provides for neutralization to prevent regeneration.) The output from this stage is internally connected to the detector, and the detector output appears at terminal ⑨. Volume and/or tone controls can be inserted between ter-

*Such as the Fairchild μA720, the National LM1820, or the RCA CA3088E.
†Such as the Fairchild μA706 or the National LM380.

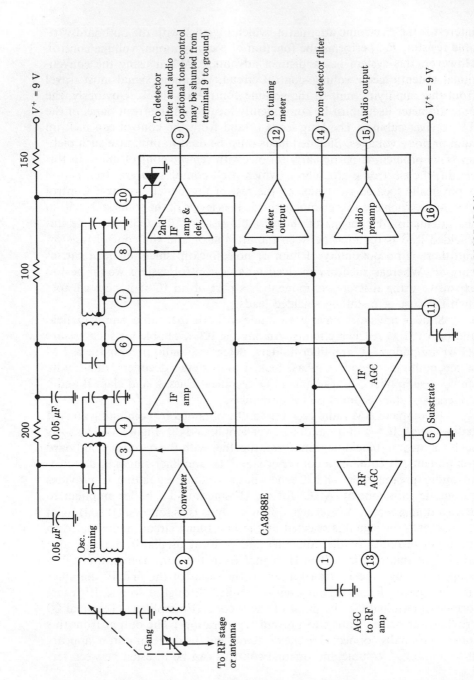

Figure 19-11 Functional block diagram of the CA3088E (*RCA*).

minals ⑨ and ⑭, as the audio is fed to the preamplifier. The audio signal is amplified by the preamplifier and fed from terminal ⑮ to the power amplifier (not shown).

Another example of a complete systems IC is the LP2000 microtransmitter made by Lithic Systems, Inc. It is effectively a wrist-radio transmitter dream come true. In a single chip, it contains an oscillator, a buffer amplifier, a preamplifier, a modulated output stage with a 50-mW RF power output, and a voltage regulator. With the addition of a microphone, an antenna, and a 6-V battery, we have the complete transmitter function for a 27-MHz walkie-talkie.

A COMPLETE IC AUDIO AMPLIFIER

With the advances made in integrated circuitry, it is possible to obtain a complete audio amplifier in a minimum of space and with a minimum of external components. If the power levels are moderate, the entire amplifier is contained in a monolithic package. An example of this is the Motorola MFC9000, rated at 4 W continuous sine-wave power output into an 8-Ω load and housed in a dual-in-line package approximately 0.75×0.30 in area. It requires only a 15-mV input signal level for full power output. It also incorporates short-term short-circuit protection. Only six resistors and eight capacitors are needed externally to complete the amplifier circuit. Another unit of this type is the RCA CA810Q, with a power output of 7 W into a 4-Ω load. Both these units are ideally suited as the audio sections in many television, radio, and phonograph systems. Two and four such units can be used for stereo and quadraphonic applications.

For higher power outputs—as in quality high-fidelity systems—it is necessary to use hybrid ICs. Earlier in this chapter we examined one such unit (see Figs. 19-2 and 19-3d). Hybrid units are also available from other manufacturers. For example, the Bendix BHA-0002 is a class B quasi-complementary amplifier that delivers an output of 15 W with an input of 350 mV. The Sanken SI-1050H hybrid IC has a power output of 50 W but requires an input of 700 mV. Obviously, both these units require a preamplifier. They could also be used ideally when the preamplifier is incorporated in the tuner section IC, such as the 3065 unit of Fig. 19-10.

Figure 19-12 shows the schematic of a complete IC audio amplifier using one section of a quad operational amplifier (Motorola MC3301P) as the preamplifier and a hybrid IC (Sanken SI-1050H) as the main amplifier. The gain of the preamplifier is set by selecting the value of R_1 (in comparison with R_4). R_2 is then made the same as R_1 for balance. The output from the preamplifier is fed to the volume control and then to the tone controls. These circuits are similar to those in Figs. 18-7 and 18-11. The

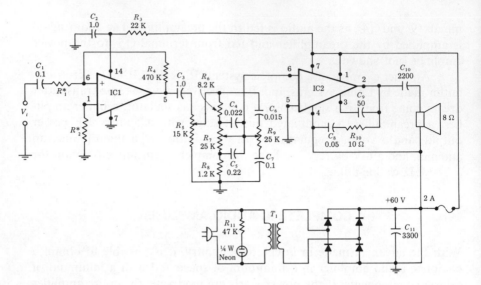

Figure 19-12 An IC audio amplifier (*IC–2 section and power supply courtesy Sanken*).

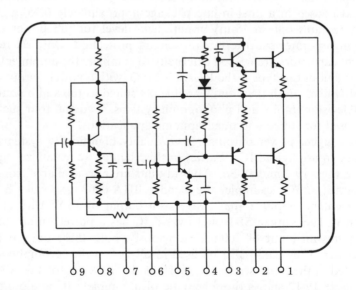

Figure 19-13 Internal connections of the hybrid IC SI-1050A (*Sanken*) :

1. V_{CC}
2. Output (to a capacitor)
3. Feedback
4. Ground for output
5. Ground for input
6. Input
7. V_{CC}
8. Spare
9. Spare

output from the tone-control circuits is, in turn, applied as the input to the main amplifier, IC-2. If the gain of the preamplifier is properly set, the input to pin 6 should be (approximately) the 700 mV required for full power output—with the volume control set at maximum and the tone controls at center position for flat response. The power supply for this amplifier is a full-wave bridge circuit with capacitor filter.

The circuit details for the main amplifier are shown in Fig. 19-13. The input signal is applied to pin 6 and is capacitively coupled to the base of an *NPN* transistor. This first stage is a standard common-emitter amplifier with voltage-divider base bias, emitter dc stabilization, and inverse feedback coupled from the final output (pin 2) through the 50-μF external capacitor, to pin 3, and through a series resistor to the emitter leg. The output from this first stage is capacitively coupled to a second *NPN* transistor, also with voltage-divider base bias and dc emitter stabilization. This stage is direct-coupled to an inverted Darlington quasi-complementary output circuit, as described in Chapter 15. (See Fig. 15-24.) In total, this IC contains 17 resistors, 8 capacitors, 1 diode, and 6 transistors.

REVIEW QUESTIONS

1. **(a)** What is the meaning of the term *monolithic IC*? **(b)** Name four components that can be formed on such a chip.

2. Refer to Fig. 19-1: **(a)** What is the function of the silicon diode layer? **(b)** What is the process of adding this layer called? **(c)** What is the desired (approximate) electrical resistance between the *N* islands in diagram (*a*)? **(d)** How is this effect obtained? **(e)** What material is used to make electrical connections to (or between) the sections of the unit in diagram (*b*)? **(f)** What is the meaning of the *N*+ designation?

3. Refer to Fig. 19-1*b*: **(a)** What constitutes the capacitor? **(b)** What type of transistor does this construction produce? **(c)** How is the resistor effect obtained?

4. **(a)** What does LSI stand for? **(b)** What does this term mean?

5. **(a)** How do the fabrication steps for an LSI chip compare with those for one of low component density? Explain. **(b)** On a one-chip basis, which would be more expensive to produce—the LSI or the SSI? **(c)** Why is this so? **(d)** Under what condition can the per-chip cost be similar?

6. Give six advantages of monolithic ICs over discrete-component circuitry.

7. Why is better performance practical with ICs and not with discrete circuitry?

8. Give five shortcomings of monolithic integrated circuitry.

9. **(a)** What is a *hybrid* IC? **(b)** Give three advantages of hybrid construction over monolithic.

10. Why is the unit in Fig. 19-2 called a hybrid IC?

11. Name three commonly used IC packaging methods.

12. What is the effect of ICs with regard to **(a)** servicing time, **(b)** service costs, **(c)** technician training requirements, **(d)** circuit design engineering, **(e)** component manufacturing?

13. (a) How did the operational amplifier get such a name? **(b)** What makes this IC unit so popular? **(c)** Name three functional sections of any operational amplifier.

14. Figures 19-4 and 19-5 are representations of the same IC unit. Which would you use **(a)** in the fabrication of the unit? Why? **(b)** In the overall circuit of a system using this unit? Why?

15. Refer to Fig. 19-5: **(a)** What does the $(-)$ terminal signify? **(b)** What does the $(+)$ terminal signify? **(c)** Can a positive pulse input signal be applied to the $(-)$ terminal? Explain. **(d)** What will the output be?

16. Explain each of the following terms briefly: **(a)** input offset voltage; **(b)** input bias current; **(c)** input offset current; **(d)** slew rate; **(e)** power supply rejection ratio; **(f)** power supply sensitivity.

17. (a) Why does the slew rate of an operation amplifier limit the frequency of the input signal it can handle without distortion? **(b)** Why does it also limit the amplitude of the input signal? **(c)** For high-frequency and/or high-amplitude inputs, which is preferable—a high slew rate or a low one?

18. (a) What is meant by the open-loop gain of an operational amplifier? **(b)** What determines this value? **(c)** What is (approximately) the range of values obtainable? **(d)** For a given IC type, is the open-loop gain the same? Explain. **(e)** For a given unit, is the open-loop gain constant? **(f)** How can the gain of the unit be stabilized?

19. Refer to Fig. 19-6: **(a)** Is this an actual working diagram? Explain. **(b)** Why is this called a noninverting op amp? **(c)** How is feedback obtained? **(d)** Why is this *inverse* feedback? **(e)** Why is this *voltage* feedback? **(f)** Why is this *series-fed* feedback? **(g)** How do we evaluate the feedback factor for this circuit?

20. Refer to Example 19-2: **(a)** Why do we get the same answer for the closed-loop gain with Equation 16-4 as with Equation 19-2? **(b)** Why do we get the same answer with an open-loop gain of 20,000 and with an open-loop gain of 300,000?

21. Refer to Fig. 19-7: **(a)** Why is this an inverting amplifier? **(b)** Is this a shunt-fed or a series-fed feedback system? **(c)** Is the equation for the closed-loop gain in this circuit the same as for series-fed circuits? Explain any differences.

22. (a) Will inverse feedback affect the input impedance of the circuit? **(b)** The output impedance? **(c)** Does voltage feedback increase or decrease the output impedance? **(d)** Give the equation for this effect. **(e)** Does series-fed feedback increase or decrease the input impedance? **(f)** Give the equation. **(g)** Does shunt-fed feedback increase or decrease the input impedance? **(h)** Give the equation.

23. Refer to Fig. 19-8: **(a)** Compared to Fig. 19-7, what makes this a working diagram? **(b)** What type of power supply does this circuit use? **(c)** What is the function of capacitors C_1 and C_2? **(d)** Which resistors provide the feedback? **(e)** How do we evaluate the closed-loop gain of this circuit? **(f)** What is the function of resistor R_3? **(g)** Is its use always necessary? Explain.

24. (a) What is a *quad* operational amplifier? **(b)** Does it cost approximately four times as much as a single unit? **(c)** What is the advantage of a quad?

25. (a) What is a transistor *array*? **(b)** Give some examples. **(c)** Why are array units used?

26. Refer to Fig. 19-8: **(a)** Can this circuit be used with a single power supply instead of with the dual or split supply shown? **(b)** What is the disadvantage of such a change? **(c)** In a house radio application, which operation would be preferred—single supply or dual supply? Explain. **(d)** To convert to single-supply operation, what must be done with the negative V terminal? **(e)** What other change must be made?

27. Refer to Fig. 19-9: **(a)** Is this an inverting or a noninverting amplifier? **(b)** What resistors essentially determine the feedback factor? **(c)** What is the function of the C_3, C_2, and R_4 network?

28. **(a)** Under what condition is a special-purpose integrated circuit economically practical? **(b)** In what broad field would the use of special-purpose ICs be warranted?

29. Refer to Fig. 19-10: **(a)** Approximately what portion of a complete monochrome TV receiver would you estimate this to be? **(b)** What is the function of the block marked "electronic attenuator"? **(c)** How is the effect of this circuit varied? **(d)** What advantage does this system have over a "conventional" control? **(e)** Would this system be used in a discrete-component circuit? Explain.

30. Refer to Fig. 19-11: **(a)** What is the name generally given to this IC? **(b)** Name two items needed to make this a fully operating device. **(c)** Explain why the components connected to terminals ②, ③, ④, ⑥, ⑦, and ⑧ are not included within the IC itself. **(d)** How does the audio output signal from the detector get to the audio amplifier? **(e)** Explain why this connection is not made internally.

31. **(a)** Is it possible to construct a complete audio amplifier using only one IC and some external passive components? **(b)** Up to what power output level are monolithic ICs available for such use? **(c)** Give an example of such a unit.

32. **(a)** Can ICs be used for higher-power audio amplifiers? **(b)** What type of ICs? **(c)** Give an example.

33. Refer to Fig. 19-12: **(a)** What type of unit is IC-1? **(b)** What is its function in this circuit? **(c)** What determines its gain? **(d)** What is the probable function of R_3, C_2? **(e)** What is the function of R_5? R_7? R_9? **(f)** What type of unit is used for IC-2? **(g)** What is its rated power output? **(h)** What input signal level is required to obtain this power output? **(i)** What is the probable function of capacitor C_{10}? **(j)** What is the purpose of the four diodes shown between T_1 and C_{11}? **(k)** What is the function of capacitor C_{11}?

34. Refer to Fig. 19-13: **(a)** What is the circuit used with the first transistor called? **(b)** How is its base bias obtained? **(c)** Is there any dc bias stabilization? Explain. **(d)** Is there any inverse feedback? Explain. **(e)** What type of coupling is used between the first two transistors? **(f)** What type of coupling is used between the second and the next stage? **(g)** Name the circuit connection used for the two transistors shown in the upper right of the diagram. **(h)** Repeat for the two BJTs in the lower right of the diagram. **(i)** Repeat for all four of these transistors.

PROBLEMS AND DIAGRAMS

1. Draw the symbol used to represent an operational amplifier.

2. A sine-wave output of 10 V peak to peak is required. The operating frequency is 100 kHz. What slew rate specification must an operational amplifier have to deliver this output without distortion?

3. An operational amplifier circuit is set for a gain of 50. The slew rate is 1.5 V/μs. The input signal is a 500-kHz sine wave with a 20-mV peak-to-peak swing. Will the output be distorted?

4. The operational amplifier of Problem 3 is used with an input signal level of ± 50 mV peak. **(a)** What is the highest frequency it can handle without distortion? **(b)** If the circuit gain is increased to 100, what change, if any, will this have on the frequency limitation?

5. Draw a circuit showing an operational amplifier used as a noninverting amplifier with feedback.

6. The Fairchild μA709 has an open-loop gain range of 25,000 to 70,000. It is used as a noninverting amplifier (see Fig. 19-6) with $R_1 = 5.1$ kΩ and $R_2 = 270$ kΩ. Calculate the closed-loop gain **(a)** using Equation 16-4 and both A_{VOL} limits and **(b)** using Equation 19-2.

7. Each section of a National LM3900 quad operational amplifier has a minimum open-loop gain of 1200. It is used in the circuit of Problem 6. Calculate the closed loop gain **(a)** using Equation 16-4 and **(b)** using Equation 19-2. **(c)** Why is there a difference between these two answers, whereas in Problem 6 the two equations gave the same answers?

8. The RCA CA3033 has a typical open-loop gain of 90 dB. It is used as a noninverting amplifier (Fig. 19-6) with $R_1 = 3.3$ kΩ and $R_2 = 300$ kΩ. **(a)** What is the open-loop voltage gain? **(b)** What is the gain with feedback?

9. Draw a circuit showing an operational amplifier used as an inverting amplifier with feedback.

10. The Motorola MC1741 has a minimum open-loop gain of 150 μV/V. It is used as in Fig. 19-7, with $R_1 = 2.2$ kΩ and $R_2 = 220$ kΩ. **(a)** What is the open-loop voltage gain? **(b)** What is the gain with feedback?

11. In Problem 10, what value of resistor R_2 will produce a closed-loop gain of 160?

12. The op amp of Example 19-2 is reconnected as in Fig. 19-7, using the same resistor values ($R_2 = 39$ kΩ and $R_1 = 1$ kΩ). **(a)** What is the closed-loop gain now? **(b)** Using the answer obtained in Example 19-2 as reference, what is the percent change in gain?

13. Draw a working diagram for an operational amplifier connected as a noninverting amplifier with feedback.

14. Repeat Problem 13, but using a single power supply.

Appendix A

The Cathode-Ray Tube

The modern cathode-ray tube is undoubtedly the most versatile indicating and measuring device ever developed. The "moving element" in this indicating device is a beam of electrons. Since electrons have practically no mass, this moving element is not limited by the inertia effects common to all mechanical indicators. As a result, the cathode-ray beam can be deflected at a very rapid rate and can therefore respond to frequencies in the megahertz range.

There are numerous industrial applications of the cathode-ray tube. One of the most common is the cathode-ray oscilloscope used widely in schools, laboratories, production plants, and service stations to observe and analyze the waveshapes fed to or obtained from electronic equipment. Its application in television receivers should be familiar to everyone. Navigational devices such as radar and racon use cathode-ray tubes as indicating devices. Aircraft landing systems such as Ground Control Approach (GCA) use several of these tubes to guide a plane in a blind landing. In addition, since sound, light, and heat can be converted into electrical impulses, cathode-ray tubes are being used extensively in fields other than electronics—in medicine, for research study, and in production control.

CATHODE-RAY TUBE CONSTRUCTION

The constructional details of cathode-ray tubes vary somewhat with the size of the tubes, the purpose for which they are intended, and the manufacturer of the units. These tubes can be divided into two main categories:

electrostatic and electromagnetic. Each type has special advantages that make it more adaptable in certain applications. Most cathode-ray oscilloscopes use electrostatic tubes. Consequently, this is the only type that is discussed here. (Television receivers are now designed with electromagnetic tubes. Radar units employ the electromagnetic type exclusively.)

A typical cathode-ray tube of the electrostatic type is shown in Fig. A-1. The tube consists essentially of five major components: the envelope, the tube base, the electron gun assembly, the deflection plate assembly, and the fluorescent screen. Let us consider each of these components in detail.

ENVELOPE.   For proper operation and control of the electron beam, the tube must be evacuated. Most cathode-ray tubes are enclosed in a glass envelope to maintain this vacuum. In addition, the envelope also serves as a housing and support for the other components of the tube. The large surface area of the glass and the vacuum inside the envelope both contribute to develop terrific air pressure strains. As a result, tubes (particularly the larger television tubes) must be handled with care. Jarring and banging must be avoided. Manufacturers often recommend that gloves and protective goggles be worn when handling these tubes.

The inner walls of the tube, between the neck and screen, are usually coated with a conducting material known as *aquadag*. This coating is electrically connected to the accelerating anode, so that electrons striking the walls of the tube are returned to the circuit. In addition this coating acts as a shield against external electrostatic disturbances. The fluorescent coating is applied to the inner face of the tube to form the screen. Some tubes have a second aquadag coating on the outer surface of the glass envelope. Now we have two conducting layers (inner and outer coatings) separated by a dielectric (the glass envelope). This forms an appreciable capacitance. (In many television receivers this capacitance is used as part of the high-voltage filtering system. When a double-coated tube is used, care must be taken to replace it with a similar tube—otherwise it will be necessary to add a capacitor to take the place of the tube aquadag-to-glass capacitance.)

TUBE BASE.   The internal elements of the tube are terminated in prongs at the tube base. This simplifies the problem of making connections to the external circuit and at the same time makes replacement of the tube comparatively simple.

ELECTRON GUN ASSEMBLY.   The electron gun assembly performs several functions. First, it provides the source of electrons for the indicating beam. Second, it focuses these electrons into a narrow beam. At the same time it also accelerates the beam and directs it toward the face of the tube so that it strikes the screen with sufficient energy to cause fluorescence. In addition it also controls the quantity of electrons in the beam, thereby controlling the intensity of the visible glow. These functions are accomplished by the various components of the gun assembly. Referring to Fig. A-1, the electron gun assembly consists of the elements marked K, G_1, G_2, A_1, and A_2. Let us discuss each element in turn.

1. *Cathode* (K).   Cathode-ray tubes use indirectly heated cathodes, which consist of a cylinder of nickel or nickel alloy approximately $\frac{1}{8}$ in. in diameter and $\frac{1}{2}$ in. long. The end of this cylinder facing the screen is capped, the cap being coated with oxide. This ensures a copious supply of electrons aimed, in general, in the desired direction. The cathode is raised to emission temperature by a heater wire inserted through the opposite end of the cathode cylinder. To reduce magnetic effects of the heater current, the heater coil is wound in a double spiral so that the magnetic field of half the winding is canceled by the equal and opposite field of the other half of the winding. The heater wire is made of tungsten or tungsten-alloy wire coated with an insulating material for protection.

2. *Control grid* (G_1).   The control grid in cathode-ray tubes is a metal cylinder which completely encloses the cathode. The end of this cylinder facing the screen is capped by a disk with a small aperture in its center. Electrons emitted by the cathode must pass through this opening to reach the screen. The "effective" size of this aperture is controlled by variation of the grid bias.

3. *Preaccelerator grid* (G_2).   The purpose of the preaccelerator grid is to supply a strong fixed positive potential to pull the electrons emitted by the cathode through the grid aperture at high velocity.

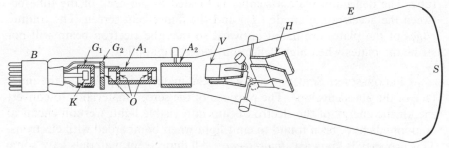

Figure A-1   Schematic arrangement of electrodes in a cathode-ray tube:

B :	tube base	A_2 : high-voltage electrode
K :	cathode	(anode 2)
G_1 :	control electrode	V : vertical deflection plates
G_2 :	accelerating electrode	H : horizontal deflection plates
A_1 :	focusing electrode	E : envelope
	(anode 1)	S : fluorescent screen

(*Note:* Electrodes K, G_1, G_2, A_1, and A_2 constitute an electron gun.)

In general, this electrode is also a metal cylinder with a small aperture. The size and shape of this structure vary with the tube model and manufacturer's design. In fact, in many of the simpler tubes this structure is eliminated and the function of acceleration is taken over by the next electrode.

4. *Focusing anode* (A_1). This structure is another metal cylinder. Its length varies in design depending on whether or not a preaccelerator grid is used and on the size of the preaccelerator grid, if it is used.

5. *Accelerating anode* (A_2). The purpose of this anode is to supply additional accelerating force to the electron beam. It is therefore operated at a high positive potential compared to the cathode. This structure is another metal cylinder with a disk and aperture.

DEFLECTION PLATE ASSEMBLY. The beam produced by the electron gun must now pass through the deflection system before it hits the screen. This deflection plate assembly provides a means for moving the point at which the electron beam strikes the screen. It consists of two pairs of plates, the plane of one pair being at right angles to the plane of the other pair. The *vertical deflection plates* are mounted horizontally in the tube. By applying proper potentials to these plates the electron beam can be made to move up and down, or vertically on the face of the tube. Notice that the nomenclature of the plates applies not to their physical mounting position but rather to the direction of the deflection they will produce. The *horizontal deflection plates* are mounted in a vertical plane—but they cause the electron beam to be deflected left and right, or horizontally across the face of the tube. The deflection plate assembly is located in the neck of the tube between the accelerating anode (A_2) and the fluorescent screen. The trailing edges of the plates are flared outward so that the electron beam will not strike the plates when high deflection forces are applied.

FLUORESCENT SCREEN. The screen material is applied to the inner face of the glass envelope. The purpose of the screen material is to convert the kinetic energy of the electron beam into visible light. Certain chemical compounds have been found to emit light when bombarded with electrons. This property is known as *fluorescence*. All fluorescent materials have some afterglow; that is, they continue to emit light for some time after the original excitation by bombardment has passed. This second property is distinguished from fluorescence and is called *phosphorescence*. The duration of phosphorescence varies with the type of material used for the screen. Commercially, the phosphorescence or *persistence* characteristics of a screen material are classified as short, medium, or long. Each type has its special advantages. For example, if a tube is to be used to observe a waveshape of short duration and that is not repeated (such as a transient), a long-persistence screen will cause the image to linger. The waveshape can be studied even after the original impulse has passed on. Again, when studying waveshapes of very low frequency, a long-persistence screen will reduce flicker of the image. On the other hand, when the image viewed changes rapidly in character, a short-persistence screen is necessary. Otherwise, the afterglow of the previous image will interfere with the new image.

The characteristics of various types of screen materials can be obtained from the manufacturers of cathode-ray tubes. Standard cathode-ray tubes are usually classified with a phosphor code designation. Some of the common types are:

Type P1 screen: green trace, medium persistence, suitable for general oscilloscope use.

Type P2 screen: bluish-green trace, long persistence, suitable for observation of transient signals or low-frequency reccurrent signals.

Type P4 screen: white trace, suitable for television application. Its persistence is well balanced to minimize flicker and still give clear pictures of rapidly changing images.

Type P5 screen: blue trace of high photographic actinity, short persistence, particularly suitable when photographic records of waveshapes are desired.

EFFECT OF CONTROL GRID POTENTIAL

The intensity or brilliance of the spot produced on the screen depends (among other factors) on the number of electrons in the beam. Since the grid structure is closest to the cathode, variation of the grid potential is the most effective way of controlling the strength of the beam. This action of the grid is similar to the controlling effect of a conventional grid on the plate current of a vacuum tube. Cathode-ray tubes are normally operated with the grid structure at some negative potential compared to cathode. A potentiometer, connected to vary this grid bias, serves as *intensity control.*

In addition, the electrostatic field between the cylindrical grid and the cathode accomplishes another important function. The field acts like an optical lens in that it concentrates the electrons into a small beam. This action is illustrated by Fig. A-2.

An electron traveling along a horizontal axis is not affected by the electrostatic field and will continue to travel in a straight line. However, consider electrons that are emitted at some vertical angle, such as along paths *OA* or *OC*. As these electrons cross the electrostatic field, they are repelled in the direction of each line of force that they cross. The resultant

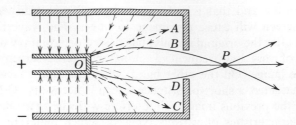

Figure A-2 Electron lens action of grid-cathode field.

forces cause these electrons to follow a curved path, such as *OBP* and *ODP*. Notice that the paths of these electrons cross over at point *P*. This is true for all electrons whose paths allow them to pass through the aperture. Point *P* can therefore be considered as the focal point or *crossover point* for the grid-cathode "electronic lens." Any electrons whose original paths are too divergent will hit the walls of the grid cylinder and will be removed automatically from the beam.

Thus we see that the action of the grid is twofold: it concentrates the electrons into a point source for good focusing, and it controls the number of electrons in the beam, thereby controlling the intensity of the spot on the tube face. A precaution concerning the setting of the intensity control should be emphasized at this point. When electrons strike the screen, most of the kinetic energy of these electrons is dissipated as heat at the point of impact. *If a small, very bright spot is kept stationary too long in one place a hole will be burned in the screen.* Therefore, keep the intensity as low as permissible for good observation, particularly if the spot is stationary.

ELECTROSTATIC FOCUSING

We have seen that the grid action focuses the beam at the crossover point. However, this point is just outside the grid cylinder itself. Beyond this point the beam diverges again. Additional focusing is needed to make the beam focus at the screen. This effect is produced by the electrostatic field between the focusing anode (A_1) and the accelerating anode (A_2). Both these anodes are operated at positive potentials with respect to cathode. However, the accelerating anode is operated at a higher positive potential, so that it is positive compared to the focusing anode. Figure A-3 shows the electrostatic field between these two anodes.

The focusing action of this electrostatic field can be explained by reference to Fig. A-3. Electrons traveling along the radial axis of the tube do not cross the lines of force and are not affected by the field. They continue to travel along the axis and hit the screen at point *O*. However, starting from the crossover point (*P*), some electrons tend to take diverging paths,

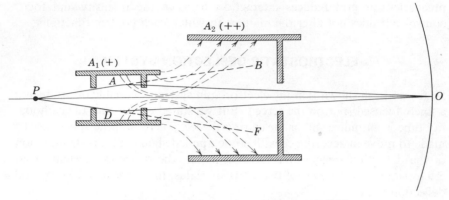

Figure A-3   Focusing action of electrostatic field.

such as *PAB* or *PDF*. As these electrons cross the electrostatic field in the region of the focusing anode, they are repelled by the field in the direction of the lines of force, or *toward the central axis of the tube*. The resultant motion is along the curved lines *PAO* and *PDO*. But the lines of force toward the end of the accelerating anode are outward! Why doesn't the beam diverge again? The explanation is simple. First, the field strength is getting much weaker. (Notice how far apart the lines are.) Second, the beam is now moving at such high velocity (due to the acceleration caused by the positive anodes) that the pull of the field at this point has very little effect. Furthermore any electrons that are too divergent to pass through the apertures are intercepted and removed from the beam by the disks in the focusing and accelerating anodes. By varying the potential of the focusing anode, the shape of the electrostatic field can be altered so as to converge the beam until the smallest possible spot is produced on the screen. This action is accomplished by connecting the focusing anode to another potentiometer, which acts as the *focus control*.

From the above discussion it should be obvious that there is some interaction between the intensity and focusing controls. For example, variation of the intensity control changes the field between grid and cathode and causes the crossover point to shift slightly. But this shift affects the focus. It is necessary to alter the focus control with change in intensity. In addition, since the potential of the focusing anode is the first positive potential near the cathode, any change in this voltage varies the attractive force pulling electrons from the cathode through the grid aperture. Therefore resetting the focus control also varies the intensity of the beam to some extent. This situation was avoided by the addition of the preaccelerator grid between the grid structure and the focusing anode. The preaccelerator grid is electrically connected to the accelerating anode and both these structures are operated at a high *fixed* positive potential. The insertion of this preaccelerator grid reduces interaction between the intensity and focus control but does not alter the manner in which each control functions.

ELECTROSTATIC DEFLECTION SYSTEM

So far we have shown that the electron gun system is capable of producing a finely focused spot on the screen. But we stated earlier that the cathode-ray tube is an indicating or measuring device. Therefore, the spot must be made to move in accordance with some applied signal. This is the function of the deflection system. We have seen that the deflection system in an electrostatic tube consists of two pairs of plates, the vertical and horizontal deflection plates.

Let us consider the action of the vertical deflection plates, shown in Fig. A-4. Remember that the vertical plates are mounted horizontally in the tube but produce an up-and-down deflection of the beam.

For the first example, we will assume that there is no difference in potential between plates A and B and that the beam travels along the axis of the tube, striking the screen at point O. Now let us apply a voltage across the deflection plates, so as to make plate A positive compared to plate B. An electrostatic field will be set up between the plates. The direction of the lines of force will be upward, from B to A. Under this condition, as the beam passes through the deflection plate area it will be deflected upward by the force of the field, striking the screen at some point C. The voltage applied across the deflection plates has caused the spot on the screen to move from O to C.

If the voltage applied across the plates is doubled, the strength of the electrostatic field is also doubled. The angle (β) through which the beam is deflected is doubled and the beam strikes the screen at D. The distance that the spot moves on the face of the beam is proportional to the deflec-

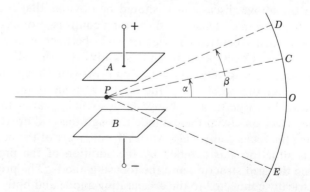

Figure A-4 Action of vertical deflection plates.

tion angle of the beam, which in turn is proportional to the voltage applied across the deflection plates. Now if this same voltage is applied to the deflection plates with reversed polarity (plate *A* negative and plate *B* positive), the direction of the electrostatic field will be downward. But the strength of the field will be the same as before. The beam will be deflected downward by an equal amount and will strike the screen at *E*. The direction of motion of the spot will therefore indicate the polarity of the applied signal voltage.

What would happen if a sine wave of voltage were applied across the deflection plates? The spot would move up and down above and below point *O*, an amount proportional to the *peak* value of the sine wave. Due to the persistence of the screen material, you would see a vertical line.

In exactly the same manner, dc voltages applied to the horizontal deflection plates would produce motion of the spot in a horizontal direction. If voltages were applied to both sets of plates simultaneously, the electron beam would be acted upon by both the horizontal and the vertical fields of force. The spot would move at some angle across the face of the tube, depending on the relative field strengths. For example, if the voltages applied to each set of plates were equal, the spot would move along a 45° line. This effect is shown in Fig. A-5*a*. If the voltage applied to the vertical

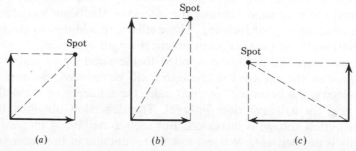

Figure A-5 Deflection of spot by combined vertical and horizontal deflection plate fields.

plates is stronger than on the horizontal plates, the spot will move as shown in Fig. A-5*b*. On the other hand, if we reverse the polarity of the signal on the horizontal plates and also make it stronger than the voltage applied to the vertical plates, the spot will move as shown in Fig. A-5*c*. It should be obvious from this discussion that by application of suitable voltages of correct polarities, the spot can be made to move to any location on the face of the screen. Remember that the amount of horizontal motion depends only on the horizontal deflection voltage and is independent of the vertical deflection voltage. The same applies to the amount of vertical motion.

DEFLECTION SENSITIVITY

The cathode-ray tube, by itself, is relatively insensitive. It may be necessary to apply several hundred volts across the deflection plates to produce full-scale deflection of the spot. *Deflection sensitivity* is the term used to compare the merits of various types of tubes. It is a measure of the deflection voltage required to produce a given spot deflection on the screen.

Several factors affect the sensitivity of the deflection system. A little thought should make these factors obvious. In the preceding section, we saw that a given voltage applied to the plates produced a definite *angle* of deflection. Therefore, the further we place the screen from the deflection plates, the greater will be the spot deflection produced by a given voltage. In construction of cathode-ray tubes, it is common practice to put the vertical plates next to the accelerating anode and follow with the horizontal deflection plates. Obviously the vertical deflection system will have a higher deflection sensitivity than the horizontal deflection system. To increase sensitivity by increasing the distance from plates to screen requires a longer tube structure. On the other hand, with a longer tube the electrons would tend to slow down, reducing the intensity of the spot. Also, due to the mutual repelling action among the electrons, the beam would tend to diverge, resulting in poor focusing. These effects, in addition to mechanical considerations, limit the maximum practical length of a tube.

If we lower the potentials applied to the anodes (A_1 and A_2), the acceleration of the electrons in the beam will be reduced. Because of the lower velocity, the beam will remain under the influence of the deflecting plates' field for a longer time interval. The deflection obtained from a given deflection voltage is increased. But such a method of increasing the sensitivity is not desirable. We just saw that reduction of the beam velocity results in loss of intensity and definition (i.e., poor focus). When selecting the operating anode potentials, a compromise must be made between higher potentials for strong intensity and good focus on the one hand and lower potentials for higher deflection sensitivity on the other.

For any given tube, the first of these two factors—distance between deflection plates and screen—is fixed by the manufacturer. On the other hand, the potential applied to the accelerating anode may vary over reasonably wide limits as chosen by the user of the tube. But this affects the amount of spot deflection for a given deflection voltage, even for the same tube, and should not be charged against the merits of the tube. To eliminate this variable, manufacturers specify deflection sensitivities in terms of 1 kV on the accelerating anode. Typical units used are *volts per inch per kilovolt, volts per centimeter per kilovolt,* or *volts per millimeter per kilovolt.* Each of these units specifies the amount of deflection voltage needed for a given spot deflection when the anode voltage is 1 kV. Sometimes the sensitivity is expressed as the amount of deflection produced when 1 V is applied to the deflection plates—i.e., *millimeters per volt per kilovolt.* Both methods give the same information. This second form should be recognized as the reciprocal of the first. An average sensitivity value for commonly used tubes is approximately 30 V per inch per kilovolt. An example will illustrate the use of this deflection sensitivity factor.

EXAMPLE A-1

What value of deflection voltage is needed to produce a 3-in deflection on a cathode-ray tube having a sensitivity of 30 V/in/kV if the anode voltage is 5000 V?

Solution

1. At 30 V/in, 3 in would require 90 V.
2. Since the anode voltage is greater than 1 kV, the deflection would be reduced. For 5000 V, $5 \times 90 = $ **450 V** would be needed across the deflection plates.

INTENSIFIER BANDS

The need for compromise between low potentials for high sensitivity or high potentials for good definition and intensity is remedied to a great extent by use of *postdeflection acceleration.* The beam travels at moderate velocities until it passes the deflection plate system. Then it is further accelerated to give high intensity. This method requires an additional electrode known as an intensifier band. This electrode is actually a ring of conducting material placed between the aquadag coating and the screen. Connection to the band is made by a contact directly on the body of the tube. The potential applied to the intensifier electrode may be several times that of the normal accelerating anode (A_2). By this means deflection sensitivities from three to five times higher than in previous models can be obtained.

In certain applications such as observation or photography of transients, or when enlargement by a projection lens system is desired, very high spot intensities are necessary. To produce such high intensities without materially affecting the spot size or deflection sensitivity, tubes using several intensifier bands have been developed. One typical model (Fig. A-6) uses three intensifier bands located near the screen end of the cathode-ray tube. The potentials for these high-voltage electrodes are obtained from a high-voltage power supply. These bands apply increased accelerating volt-

Figure A-6   High-voltage cathode-ray tube with intensifier bands.

ages to the electron beam in three equal steps. An additional band, operated at ground potential, is located between the normal accelerating anode (A_2) and the three intensifier rings. The purpose of this band is to shield the deflection plates from the intensifier field and from external electrostatic fields.

POSITIONING CONTROLS

In the discussion of the deflection system, we assumed that the beam would travel along the axis of the tube and strike the screen at the center when there was no difference in potential across the deflection plates. This may or may not be true, depending upon constructional tolerances in the manufacturing and assembling processes. In addition, it is sometimes desirable that the beam start from some point other than the center—such as top, bottom, left, or right edge of the screen. To adjust the starting point of the beam, it is necessary to use a vertical and a horizontal positioning control.

To illustrate the action of these controls let us examine the schematic diagram of a typical cathode-ray tube connected to suitable operating voltages. (See Fig. A-7.) Two separate positioning controls are used, one for the vertical and another for the horizontal plates (see Fig. A-7b).

In studying this diagram, you may be puzzled by the voltage-divider connections. Why are the accelerating anode and one plate of each deflection pair operated at ground potential? Why is the ground connection made near the positive end of the divider? The reason for this is merely as a safety measure. In the use of equipment containing a cathode-ray tube, the operator is most likely to come in contact with the high voltage applied to the screen, the deflection plates, and the positioning controls. It is therefore of definite advantage to operate these components at or near ground potential.

Now let us check the action of the positioning controls. You will notice from Fig. A-7b that the positioning control potentiometer is con-

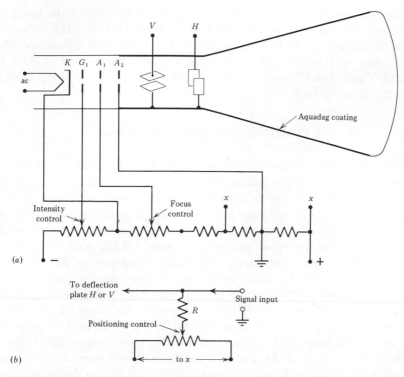

Figure A-7 Schematic diagram of cathode-ray tube connected to operating potentials.

nected across the voltage divider so that while one end is *slightly* positive compared to ground the other end is negative by an equal amount. What is the deflection plate potential when the potentiometer is set at midpoint? It is at ground or zero potential. But this is also the potential of the opposite deflection plate. Since there is no difference in potential between the plates, the electron beam is not affected and should strike the screen at the center.

If the potentiometer is moved to the right, the upper vertical deflection plate (or right horizontal deflection plate, facing the screen) is made positive. The electron beam will be attracted by this positive plate and will be deflected upward (or to the right). The magnitude of this displacement will depend on how far the potentiometer is advanced to the right. In a similar manner, movement of the potentiometer to the left will make the upper (or right) plate negative, causing the beam to be displaced in the opposite direction.

What is the purpose of resistor R in Fig. A-7b? Will it reduce the positioning voltage applied to the deflection plates? Since the deflection plates draw no current, the answer is no! Then what is it there for? Remember that we have been discussing only the positioning of the beam on the screen. The signal to be examined on the screen is fed to the deflection plates through the right-hand input leads. If resistor R is eliminated, this signal may very well be short-circuited through the voltage divider to ground. This resistor is used to isolate or *decouple* the positioning voltage from the signal voltage.

BALANCED DEFLECTION SYSTEM

The circuit described above for positioning and for deflection by the signal voltage is called an unbalanced deflection system, since voltages are applied only to one plate, the opposite plate being at ground potential. Such a system causes a slight defocusing action, known as deflection distortion. The distortion is negligible with small-screen oscilloscopes since the anode voltage is low and the deflection voltage needed for full-scale deflection is also low. But the distortion increases with the magnitude of the deflection voltage and may be objectionable on large-screen tubes or higher anode voltage tubes, such as projection tubes. The cause for this defocusing action can be understood from the following discussion. Earlier in this appendix, we saw how the electrostatic field created by the difference in potential between the focusing anode and the accelerating anode caused the electrons in the beam to focus at the screen. The spot was well defined—small and round—because this field was symmetrical with respect to the axial length of the tube. (See Fig. A-3.) Naturally if the beam is to focus at the screen, it still has appreciable width while passing through the deflection plate assembly. We can also see from Fig. A-7 that one deflection plate of an unbalanced deflection system is always at the same potential as the accelerating anode (A_2), that is, ground potential. Now if a high deflection voltage is applied to the other deflection plate, a new electrostatic field will be created between the accelerating anode and only this deflection plate. The field is definitely unsymmetrical around the tube axis, and since the beam has appreciable width at this section, the amount of deflection for each electron will vary depending on its location with respect to the tube axis. So—although the spot is deflected, it has lost definition and now appears larger in diameter. If a sine-wave voltage is applied to the vertical plate, the beam will move up and down on the face of the screen, producing a vertical line. But if the focusing is adjusted for smallest spot size at the center of the screen, the spot will widen at the top and bottom of the beam travel. This type of distortion is shown in Fig. A-8.

To overcome this defect, the better circuits use a balanced deflection system, employing a *paraphase amplifier* (sometimes called push-pull amplifier). If a deflection voltage of 200 V is to be viewed, it is applied as

Figure A-8   Defocusing action due to unbalanced deflection system.

+ 100 to one plate and − 100 on the opposite plate. Obviously, in this system neither plate of the horizontal or vertical deflection pair is grounded. An electrostatic field is now created between the accelerating anode and *each* plate of the deflection system. This field is symmetrical. In fact the field potential at the central axis is zero—just as if the deflection voltage were zero. By this means, defocusing action is almost completely eliminated.

ELECTRON-BEAM PATH

If you refer to Fig. A-7 for a moment, you will see that the beam starts at the cathode and by the action of the gun is shot toward the screen. In its path, the beam is affected by the action of the various controls and by the deflection system and finally hits the screen. What happens now to the electrons in the beam? Do they stay on the screen, or how do they complete a path? If they accumulated on the screen, we would have trouble. The accumulated electron charge on the screen would soon repel the beam.

When the beam hits the screen, it produces secondary-emission electrons. These bounce back to the aquadag coating surrounding the walls of the tube near the screen. Since the aquadag coating is electrically conductive, the electrons return through this coating to the second anode and back to the power supply voltage divider. The electron circuit is therefore completed through the action of the secondary-emission electrons and the aquadag coating.

Appendix B

The Cathode-Ray Oscilloscope

Probably the most familiar application of the cathode-ray tube is in the cathode-ray oscilloscope. This versatile instrument has gained wide popularity in laboratories, production lines, schools, and even on the service bench. The oscilloscope is used primarily for the observation of the shape of the voltage waveforms in various types of electrical circuits. Its use has spread to many other industrial fields. Pressure changes, vibrations, light, and sound can be converted to electrical voltages which can be studied by use of the oscilloscope. The applications of this instrument to industry are limitless. In the field of medicine, the oscilloscope has become a familiar diagnostic and supervisory tool. It is the very heart of all intensive care and cardiac care units.

There are two other uses of the cathode-ray oscilloscope that are not so well known. These are to measure the frequency of an unknown voltage and the phase shift caused by electrical components. Before discussing the uses of the oscilloscope in detail, let us study the component sections of this unit. A block diagram (Fig. B-1) will simplify this analysis.

SIMPLE OSCILLOSCOPE—BLOCK DIAGRAM

Oscilloscopes are available with various degrees of sophistication—and of course with commensurate price tags. For observation of waveshapes, a relatively simple model will suffice. Since such a unit is easier to understand, we will use a basic model for our block-diagram analysis. The added features found in more sophisticated models will be discussed later.

ATTENUATOR.   When we wish to study the shape of a given voltage waveform, it is fed to the VERT INPUT terminals. Frequently the signal to be observed has a high amplitude and may cause damage to the oscilloscope. An attenuator is therefore required to reduce the amplitude of the signal to a value that can be handled safely. Several factors must be carefully considered in the design of suitable attenuators:

1. The input impedance must be as high as possible to prevent or reduce loading of the circuit under test.
2. The amount of attenuation should be variable so as to handle signals of various amplitudes.
3. For any setting of the attenuation control, the amount of attenuation should be constant regardless of the frequency of the voltage under test.
4. Stray capacitances should be avoided; otherwise high-frequency signals may be bypassed to ground.

AMPLIFIERS.   In an earlier section, we saw that a common value of deflection sensitivity is approximately 30 V per inch per kilovolt. Using an oscilloscope with an anode voltage of only 1 kV, how much deflection would we get if the signal under observation had an amplitude of 0.1 V? Now do you see why amplifiers are needed? With suitable amplifiers, an oscilloscope can be used to investigate waveshapes of very weak amplitudes. The deflection sensitivity with amplifiers may be as high as 2 mV per centimeter, depending on the number of stages used in the amplifier.* The amplifier must include a variable gain control (VERT GAIN) for adjusting the amplitude of the deflections on the screen to some convenient level. This gain control is in addition to a step attenuator.

The use of amplifiers solves one problem—making the oscilloscope suitable for observation of low signal levels—but it creates a new one. Unless these amplifiers are properly designed for wideband frequency response and minimum phase shift, they introduce serious distortion in the screen pattern for signals of low or high frequencies, thereby limiting the usefulness of the oscilloscope. Because of price limitations, the amplifiers in the lower-priced oscilloscopes are seldom suitable for frequencies below 20 Hz or above 200 kHz. In the expensive models, frequency ranges as low as 5 Hz and as high as 50 MHz can be obtained. In addition, oscilloscopes with dc amplifiers are available for studying very-low-frequency effects and complex waves with dc components or when a dc reference must be indicated.

*The unit shown in Fig. B-1 has a vertical amplifier sensitivity of 0.03 V per inch (rms).

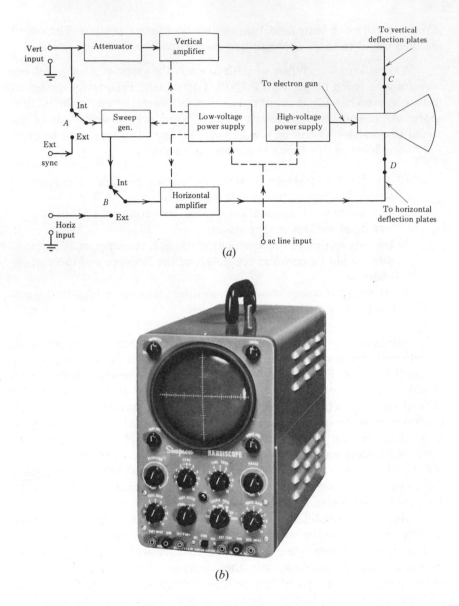

(a)

(b)

Figure B-1   Block diagram and external view of a basic oscilloscope *(Simpson)*.

On the other hand, the cathode-ray tube, without amplifiers, gives true reproduction of waveforms even at very high frequencies (approximately 200 MHz maximum). This requires that the signal be fed directly to the deflection plates. Such a connection can be made in Fig. B-1 by removing link C and applying the signal to the vertical deflection plates.

Many commercial units have this provision. You must remember, however, that direct connection can be used only when the signal to be observed has sufficient amplitude.

TIME BASE—SWEEP GENERATOR.   So far we have been discussing the application of a signal voltage to the vertical deflection plates. This causes the electron beam to move up and down at the same rate as the frequency of the applied signal. But under these conditions all we will see on the screen is a vertical line! While the beam is moving up and down, if we could also cause the beam to move horizontally from left to right *at a uniform rate of speed,* the pattern on the screen would show the full variation of signal voltage with time. As soon as a full cycle of the signal voltage has been traced, the beam should be returned *quickly* back to the left-hand side of the screen so that it can start to trace out a second cycle.

How can this timing action be effected? Since timing is a horizontal motion, a suitable voltage should be applied to the horizontal plates. But what constitutes a suitable voltage? If we apply a voltage across the horizontal deflection plates, so as to make the right-hand plate positive with respect to the grounded left-hand plate, the beam will be deflected to the right. If we increase the potential of this plate at a steady rate, the beam will be pulled more and more to the right at a uniform speed. Once the beam reaches the right-hand side of the screen, we will want it to return quickly back to the left. This can be done by rapidly reducing the potential of the right-hand plate. But this is an exact description of the "voltage versus time" variation of a sawtooth wave! Our problem is solved. By applying a sawtooth voltage to the horizontal plates, we can view the exact pattern of the signal on the vertical plates. In the block diagram (Fig. B-1) this sawtooth voltage is produced by the sweep generator, amplified by the horizontal amplifier, and applied to the horizontal deflection plates. (Switch *B* must be in the position marked "INT.")

Let us see how the application of a sine wave to the vertical plates and a sawtooth to the horizontal plates gives us the desired screen pattern. Referring to Fig. B-2, at time = 0, the sine-wave voltage amplitude is zero and the sawtooth voltage amplitude is also zero. The beam strikes the screen at point *A*. At time = 1, both the sine-wave voltage and the sawtooth voltage have increased. The beam is deflected upward by the sine-wave voltage and to the right by the sawtooth voltage, striking the screen at *B*. At time = 2, the sine wave has reached its maximum value; the sawtooth voltage has increased slightly. The beam is deflected to its highest point and slightly more to the right, causing the spot to appear at *C* on the screen. At time = 3, the sine-wave voltage has decreased, but the sawtooth voltage is still increasing. The beam drops down but still moves to the right, hitting the screen at *D*. Finally, at time = 8, the sawtooth voltage is at its maximum, causing the spot to appear on the screen at point *I*. The time interval

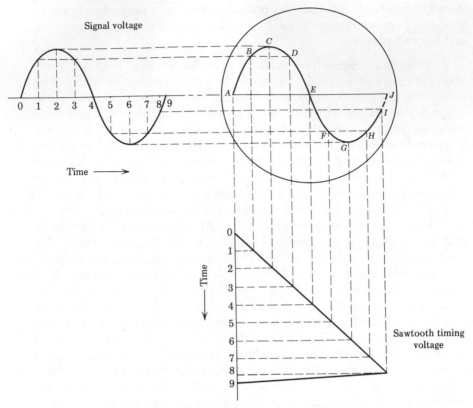

Signal voltage

Time ⟶

Time

Sawtooth timing voltage

Figure B-2   Production of screen pattern.

0 to 8 represents the forward sweep time. During the time interval 8 to 9 the beam returns back to the starting point. This time interval 8 to 9 is often called the *retrace time* or *flyback time*. Since the retrace interferes with the pattern to be observed, most circuits cut off the electron beam current by biasing the cathode-ray tube grid and "blanking out" the screen. Obviously the section of the sine wave from I to J is lost. For this reason, it is important that the retrace time be as short as possible—in other words the sawtooth voltage should drop to zero as rapidly as possible.

If the frequency of the sine-wave signal were doubled, two cycles of sine wave would correspond, in time, to one cycle of sawtooth. Obviously, using the same analysis as above, two cycles would appear on the screen. Also, if the frequency of the sawtooth wave were reduced so that the time for one sawtooth cycle corresponded to the time for three sine-wave cycles, three cycles of the signal would be seen on the screen. If a one-cycle pattern is desired on the screen, the sawtooth frequency and signal frequency must be the same.

The frequency of the sweep voltage is controlled by the dials marked COARSE FREQ and FINE FREQ. The course control changes the frequency in large steps or bands; the fine control gives full variation within each band. For example, as shown in Fig. B-1, the coarse control is set for a frequency band of 500 to 3000 Hz. With the fine control at zero, the sweep frequency would be at the low end of the band, or approximately 500 Hz. Setting the fine-frequency dial fully clockwise (at 10) would produce a sawtooth frequency of approximately 3 kHz. At a setting of 6, as shown, the sweep frequency should be somewhere around 2000 kHz. The exact value cannot be determined by the dial setting. In practice the fine-frequency dial is adjusted until the pattern stands still.

SYNCHRONIZATION.   When the sweep frequency is equal to (or a submultiple of) the signal frequency, we have seen that the desired pattern is traced on the screen. Also, with each cycle of the sawtooth voltage the pattern traced on the screen is identical with the previous one, and the resulting image, due to the retentivity of the eye and the persistence of the screen, is stationary. Figure B-3 shows what happens if the sawtooth frequency is slightly faster than the signal frequency. Obviously, each cycle of the sawtooth wave corresponds to less than one cycle of the signal. Successive patterns on the screen appear to have been shifted to the right. As a result the image is not stationary but appears to drift to the right. The speed with which the pattern moves will increase if the difference between the two frequencies is increased. On the other hand, if the sweep frequency is slightly too slow, the pattern will drift to the left.

Although sweep circuits are designed with fine control of frequency, no simple sweep circuit will hold its frequency exactly constant. The resulting drift of the screen pattern is annoying. To remedy this, a synchronizing voltage is fed into the internal sweep generator circuit to lock the sweep

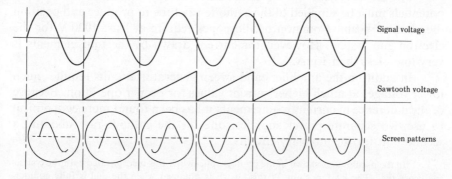

Figure B-3 Apparent motion of screen pattern owing to slightly higher sweep frequency.

frequency to the desired conditions. The synchronizing voltage can be taken internally from the signal being fed to the vertical amplifier or from any desired external source. When internal synchronization is used, each signal viewed on the screen is synchronized (in phase relation) with respect to itself. Therefore all patterns will appear to have identical phasing. When each signal is of individual interest only, this apparent phasing is of no objection. But quite often we are examining two waves and we are particularly interested in the phase relation of one compared to the other. In this case *external* synchronization is necessary, using the same sync signal as the reference for both observations. This reference signal could be some third voltage wave—or, better yet, one of the two signals under observation The sync voltage is applied between the EXT SYNC input terminal and ground, and switch *A* (block diagram Fig. B-1) is set to EXT.

EXTERNAL HORIZONTAL INPUT.   When using the sweep voltage to view waveforms on the screen, switch *B* on the block diagram is set so as to connect the sawtooth voltage through the horizontal amplifier to the scope plates. The gain of the horizontal amplifier is set so as to give the desired horizontal spread to the pattern on the screen. For example, if a section of the full pattern needs careful analysis, increasing the gain of the horizontal amplifier will spread the wave out.

But you will recall that the oscilloscope can also be used for frequency-comparison measurements or phase-shift measurements. Under these conditions, the internal sweep circuit is not desired. Switch *B* (block diagram Fig. B-1) is set to EXT so as to connect the horizontal amplifier to the horizontal input terminals.* Now the two signals to be compared for frequency or phase shift are connected—one to the vertical input and the second to the horizontal input. The details of how these measurements are made are given later in this appendix.

POWER SUPPLIES.   We saw earlier in this appendix that proper potentials must be supplied to the cathode-ray tube to produce and position the electron beam. Common oscilloscopes require at least 1000 V for the electron gun system. However, the current drawn by the tube elements is very low—less than 1 mA.

In addition the amplifiers and sweep generator circuits require much lower voltages at much higher current values for proper operation. Because of the difference in current requirements it has been found more economical to use two separate power supplies, one for the cathode-ray tube and a second unit for the auxiliary circuits.

*In the oscilloscope shown in the photograph, switch *B* is incorporated with the coarse-frequency dial. The EXT position of this switch is obtained when the dial is fully counter-clockwise.

LABORATORY MODELS

A typical laboratory model oscilloscope is shown in Fig. B-4. Essentially, the operation of this unit is the same as that described for the basic model. However, a glance at the front panel controls confirms that this is a more complex unit. One reason for the added complexity is that this is a dual-channel oscilloscope. Two signals can be observed simultaneously on the screen. This requires duplication of all vertical input signal controls. (Note the two sets of similarly labeled controls below the screen face and also the two input terminals.)*

In circuitry, the major differences between this oscilloscope and the basic unit described earlier are that in this unit:

1. The amplifier gains are highly stabilized. (The gain will not change with time, temperature, or supply voltage, variations.) This requires a much more complex and expensive amplifier circuitry.
2. The vertical gain is varied in fixed calibrated steps, using precision resistors instead of a continuously variable potentiometer.
3. The sweep timing is also varied in fixed calibrated steps.

Now let us study the action of the major controls in more detail.

CH 1 VOLTS/DIV.   This is the vertical gain control for channel 1. It can be set (in steps) from a low-gain position of 10 V/div (fully counterclockwise) to a maximum gain of 5 mV/div (fully clockwise). If the signal being analyzed is weak, higher gain is needed, and this switch is turned clockwise to obtain a greater screen deflection. Obviously, the deflection will increase in "jumps" as the VOLTS/DIV switch is turned clockwise. Now notice that there are two concentric knobs on this control. It is a dual control. The above step variations are obtained by rotating the larger (outer) knob. In addition, if the inner knob is now rotated, continuously variable gain can be obtained between the settings of the main (outer) knob. However, when using the "continuously variable" knob, the gain calibration is no longer valid. As a warning to the user, the red indicator (UNCAL) will light. Calibration is restored by turning the inner knob fully counterclockwise. (The screwdriver controls marked GAIN and STEP ATTEN BAL are used to readjust the calibration if it becomes necessary.)

AC GND DC.   This lever switch is used to select the method of coupling the input signal to the oscilloscope amplifiers. In the AC position,

*A cable is shown plugged into the channel 1 input terminal.

any dc component of the input signal is blocked and only the ac component will be viewed on the screen. This action is necessary when the dc component is high compared to the ac signal level (for example, 100 V dc and 0.5 V ac) and we wish to examine the detail in the ac component. With the dc removed, the VERT GAIN setting can be increased until the ac waveshape fills the screen. The low-frequency limit of this oscilloscope on ac setting is about 2.0 Hz. On the other hand, when it is desired to view the complex waveform, ac + dc components, then the DC position of this switch must be used. In the GND setting, the input circuit is grounded (without shorting the input signal) and no signal reaches the amplifiers. (This is useful for adjusting the zero position of the trace to the center of the screen.)

Notice that identical controls are also used for the channel 2 vertical system. In addition, because of this second channel three other controls are provided.

INVERT PULL.   When pulled out, this control inverts the channel 2 signal 180° from its original phasing.

MODE.   The outer (or larger) knob on this dual control selects the vertical system operating mode. In the CH 1 or CH 2 position only that channel's signal will be viewed on the screen. In position ALT both signals will be seen. Since this cathode-ray tube has only one electron beam, the display is switched at the end of one sweep from the channel 1 signal to the channel 2 signal and back and forth. Due to the persistence of the screen phosphors and that of the eye, both signal waveforms are seen simultaneously. When the mode switch is set on CHOP, again a dual-trace display is seen. This time the switching back and forth is at a rate of about 500 kHz, so that each waveform is broken up into approximately 1-μs segments. In the ADD position, the channel 1 and channel 2 signals are added and their phasor sum is displayed on the screen.

TRIGGER.   This switch affects the relative phasing of the horizontal sweep voltage and the vertical input signals. In the NORMAL position, the sweep circuit is triggered from the channel being displayed. On the other hand, in the CH 1 ONLY position the sweep is always triggered by the channel 1 signal regardless of which vertical signal is being displayed. This is of advantage if we wish to determine the phase of the channel 2 signal relative to the channel 1 signal. This point is dealt with in detail later in this appendix.

So far we have been discussing the controls affecting the vertical deflection system. Now let us concentrate on those that set up the horizontal deflections. These controls (on the right half of the front panel) are much more complex than the sweep-circuit controls of the basic oscilloscope.

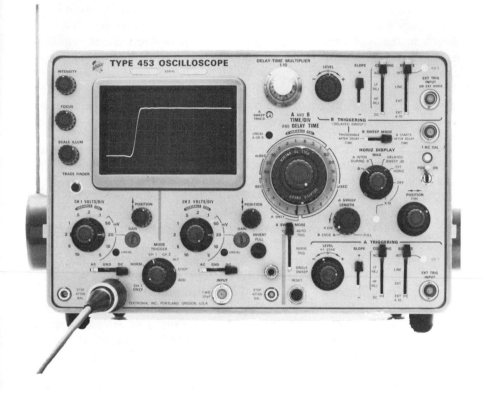

Figure B-4   Laboratory model oscilloscope (*Tektronix*).

This model has two sweep systems. The more commonly used one (the *A* sweep) is essentially similar to that of the basic oscilloscope, but it is calibrated. The other (the *B* sweep) is a delayed sweep.

  HORIZ DISPLAY—MAG.   This is a dual control. The outer knob selects the sweep system. Position *A* (as shown) is for the main or *A* sweep system. The next position clockwise (A INTEN DURING B) produces one form of the delayed or *B* sweep. The third position (DELAYED SWEEP B) is for another form of the delayed sweep. The last position (EXT HORIZ) disconnects the internal sweep generator and provides a means for applying an external signal (through the horizontal amplifiers) to the horizontal deflection plates. Notice that the external signal is applied at the upper right terminal marked EXT TRIG INPUT or EXT HORIZ.

  The inner knob of this dual control is for the MAG switch, which has two positions: OFF or X10. The normal position is OFF. In the X10 position it increases or magnifies the sweep speed to 10 times the rate set by the main sweep speed dial. This is useful when we wish to examine in detail only a small portion of the waveform seen on the screen. With this switch in the X10 position, the center division of the normal display is expanded to fill the full screen width. Any portion of the normal display can be examined by resetting the POSITION control until the desired portion of the waveshape is moved to the center of the screen, and then switching to X10. An indicator light will be energized to remind the operator that he is using the X10 magnification.

A TRIGGERING.   A group of interrelated switches and controls are involved in maintaining good stability of the screen display. They are located in the lower right quadrant of the front panel; their functions follow:

SOURCE.   This lever switch selects the source of the trigger signal. On INT, the trigger signal is obtained from the vertical deflection system (either from the channel 1 signal or alternately from the channel 1 and channel 2 signals, depending on the setting of the trigger switch in the vertical deflection system). When the SOURCE switch is in the LINE position, the trigger signal is obtained from the line voltage. In the EXT position, an outside signal can be used for the triggering action. This external signal is applied by using the EXT TRIG INPUT connector. If this external signal is too strong (and causes distortion of the observed waveshape), then the EXT ÷ 10 position can be used. This attenuates the external signal to (approximately) one-tenth its amplitude before applying it to the trigger circuit.

COUPLING.   This switch determines whether the trigger circuit will respond to the positive-going or to the negative-going slope of the trigger signal (depending on whether the switch is in the up or the down position).

LEVEL.   The outer knob of this dual control selects the amplitude point on the trigger signal at which the sweep is triggered. By setting this level control—in combination with the SLOPE switch, the displayed waveshape can be made to start at any point in its cycle. This can be seen from Fig. B-5. In Fig. B-5a, the slope switch is set for a positive slope (rising waveshape) and the level is high and positive. In Fig. B-5b, the slope switch is set for triggering on the positive slope of the waveform, but this time the level is low and negative. In Fig. B-5c, the slope is negative and the level is high and negative.

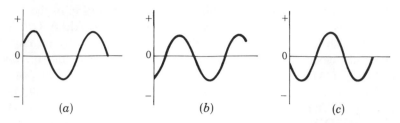

Figure B-5 Effect of trigger slope and level on the displayed waveform.

HF STAB. This section of this dual control is used to remove jitter in the displayed waveform. It is effective only with high-frequency signals.

A SWEEP MODE. This lever switch determines the operating mode for the *A* sweep. In the AUTO TRIG position, the sweep is initiated by the applied trigger signal as set by the group of *A*-triggering controls. If there is no trigger signal, the sweep runs free at the sweep rate selected by the A TIME/DIV switch. The NORM TRIG position differs from the above only in that no trace is displayed when there is no trigger signal. The SINGLE SWEEP mode is used when the signal to be displayed is not repetitive (or varies in amplitude, shape, or time). With such signals, the conventional repetitive sweep may produce an unstable display. To obtain another single-sweep display, the RESET button must be pressed. The RESET light (located inside the reset button) will light when the *A* sweep generator circuit has been reset and is ready to produce a sweep.

A AND *B* TIME/DIV AND DELAY TIME. This is essentially a triple control. The larger (or outer) knob selects the timing of the *A* or *B* sweep. The *A* sweep setting is read between the two black lines on the clear plastic flange. (As shown in Fig. B-4, it is set at 0.1 μs per division.) The setting for the *B* sweep is indicated by the white dot on the rim of this knob. (Notice that the *B*, or delayed, sweep is also set at 0.1 μs per division.) Whenever the two sweeps are in coincidence—the white dot of this outer knob set to the same position as the black lines on the plastic ring—the two controls are locked together, and the sweep rate of both sweep generators is changed simultaneously by this single knob. This is the normal setting for changing the *A* sweep rate. To change the *B* sweep rate independently, the knob is pulled out. This disengages the clear plastic flange, and only the *B* sweep generator is now changed. Actually, then, this knob is used to set either the *A* and/or *B* sweep generator timing.

The smaller (or inner) knob is the A VARIABLE control. This provides continuously variable sweep rates between the settings of the TIME/

DIV switch. When this control is being used, the calibration of the TIME/ DIV switch is no longer valid. A warning light (UNCAL A OR B) is turned on. Calibration of the sweep time base is restored by turning this control fully clockwise to its CAL position. (A similar control, the B VARIABLE–CAL, is found on the right side of the instrument case.)

The instant at which the sweep starts can be delayed by using the B sweep system. The amount of delay is set by a combination (product) of the A TIME/DIV and the DELAY–TIME MULTIPLIER. Use of this delayed sweep provides for a more accurate (detailed) view of a small portion of the full display. In this respect it is similar to the MAG setting of the horizontal display switch. However, whereas the MAG setting gives a magnification of 10, the delayed sweep can offer magnifications to 1000. Since use of the delay sweep is not common, further discussion of this feature is omitted.

The rest of the switches and controls on this panel affect the B (or delayed) sweep. Again, since this is not the commonly used condition, description of these control actions is left to the manufacturers' instructional manuals.

APPLICATIONS OF THE CATHODE-RAY OSCILLOSCOPE

We have already seen one application of the oscilloscope—examination of waveshapes. This is probably the most important use of this piece of equipment. For this purpose, the signal voltage is connected to the vertical input terminals of the oscilloscope and the gain of the vertical amplifier is adjusted to give sufficient vertical deflection on the screen. The sweep circuit is set for internal sweep and adjusted for a stationary pattern. The horizontal amplifier gain is adjusted so that the pattern sweeps the full width of the screen.

In addition there are four other valuable applications of the oscilloscope which are not too widely used, probably because they are not too well understood. Let us consider each of these applications in turn.

VOLTAGE MEASUREMENT.   Since the oscilloscope has a very high input impedance, it can readily be used in place of a vacuum-tube voltmeter for measurement of voltage amplitudes, within the frequency limitation of the vertical amplifier. When using a basic (uncalibrated) oscilloscope it is preferable to turn the sweep circuit off. The signal to be evaluated is fed to the VERT INPUT and GND terminals and the VERT GAIN is adjusted for a suitably large deflection. The pattern on the screen will be a vertical line, with the height of this line proportional to the peak-to-peak value of the applied voltage. Now the "unknown" signal is removed and a known voltage is applied to the VERT INPUT. The VERT GAIN control must not be reset. By comparing the two vertical deflections, the "unknown" amplitude is evaluated. For convenience in "calibrating," the oscilloscope in Fig. B-1 has a 1-V peak-to-peak signal* that can be fed to the vertical amplifier by jumping from the 1-V *P-P* terminal to the VERT INPUT terminal. (The ground connection is made internally.)

If dc voltages are to be measured (using the oscilloscope of Fig. B-1), the amplifiers in this unit cannot be used. However, if the voltage is high enough, direct connection can be made to the vertical deflection plates. Naturally the oscilloscope must be calibrated again for such use. Direct connection to the plates can also be used for frequencies above the range of the amplifiers.

Voltage measurements using a laboratory model (calibrated) oscilloscope are quite simple. In the unit shown in Fig. B-4, the "unknown" signal is applied to either the channel 1 or the channel 2 input, with the MODE switch set to display the channel used. The VOLTS/DIV switch is set for a maximum viewable height, and the sweep and triggering controls are set for a stable display of several cycles. The peak-to-peak value of the wave is evaluated by multiplying the peak-to-peak deflection in divisions by the VOLTS/DIV setting of the VERT GAIN dial. For low-frequency signals below about 20 Hz (or for dc voltages), the DC position of the INPUT coupling switch must be used.

PHASE-SHIFT MEASUREMENTS USING LISSAJOUS FIGURES.   One of the most important applications of the oscilloscope, next to observation of waveshapes, lies in its ability to indicate various degrees of phase shift. You will recall from your studies of ac circuits† that phase shift may occur when a signal is passed through an R-C or R-L circuit. Since electronic equipment contains such components, it is often desirable to know how much phase shift, if any, they introduce. For example, in television amplifiers and in oscilloscope amplifiers, phase shift should be reduced to a negligible value or compensating circuits must be included to make the effects of phase shift constant for signals of all frequencies. It should be obvious that any discussion of phase shift is limited to only one frequency at a time. If a signal of 1000 Hz is fed into a unit, the output frequency is naturally also 1000 Hz—but what is the phase relation of the output voltage compared to the input voltage? The phase shift of this same unit (between input and output) can also be checked at any other frequency. Discussion

---

*Other manufacturers supply other values of calibrating voltages (or more than one value).

†J. J. DeFrance, *Electrical Fundamentals,* Englewood Cliffs, N.J.: Prentice-Hall, Inc., 1969, part 2, chap. 9.

of phase relation between two signals not of the same frequency has no meaning, since the phase relation is continuously changing.

To measure the phase shift caused by any piece of electronic equipment is a simple matter. Merely connect one input of the oscilloscope (horizontal or vertical) to the input voltage being fed to the unit under test and connect the other input of the oscilloscope to the output of the unit under test. Obviously, the internal sawtooth sweep circuit should be disconnected, and the horizontal deflection plates and amplifier should be connected to the external signal source. This is done by setting switch B of the Fig. B-1 block diagram to the EXT position or the HORIZ DISPLAY switch of Fig. B-4 to EXT HOR. The pattern obtained on the screen will not be a reproduction of the waveshape of the input (or output) voltage signal, but rather a peculiar-shaped figure. This pattern is called a *Lissajous figure* (named after the French scientist who developed this type of measurement). The pattern obtained will vary depending on the phase relation and relative amplitudes of the two signals being compared.

It is desirable, when making these tests, that the signals reaching the horizontal and vertical plates be of equal amplitude. This does not mean that the original signals being compared must have the same amplitude. Any difference in their signal strengths can be made up by using compensatingly greater oscilloscope amplifier gain for the weaker signal. A legitimate question at this time is how do we adjust the H and V gain for equal voltages on the horizontal and vertical deflection plates? The method is very simple. Set the horizontal gain to zero. With no signal on the horizontal deflection plate, the pattern on the screen is a vertical line. Adjust the vertical gain for a good-sized deflection. After noting the vertical amplifier gain (or the VOLTS/DIV) setting, turn this control to zero (or the INPUT switch to GND) and increase the gain of horizontal amplifier until the resulting horizontal line is equal in length to the vertical line previously obtained. Now merely adjust the gain of the vertical amplifier to the setting recorded above. The signals reaching the horizontal and vertical deflection plates are equal in amplitude.

Figure B-6 shows a point-by-point development of the Lissajous figures that are obtained for various angles of phase shift between 0 and 180°. In each of these diagrams, the signal fed to the vertical input terminals of the oscilloscope is leading the signal fed to the horizontal input terminals. Using similar point-by-point developments it can be shown that the patterns obtained for vertical input signals lagging by the same angle of phase shift are identical. In other words, a 30° (or any other angle) phase shift between the two signals—regardless of which signal is leading (or lagging)—produces identical screen patterns. Since this is so, is it necessary to develop Lissajous figures for phase shifts greater than 180°? Obviously not—one signal leading by some angle θ greater than 180° is the same as that signal

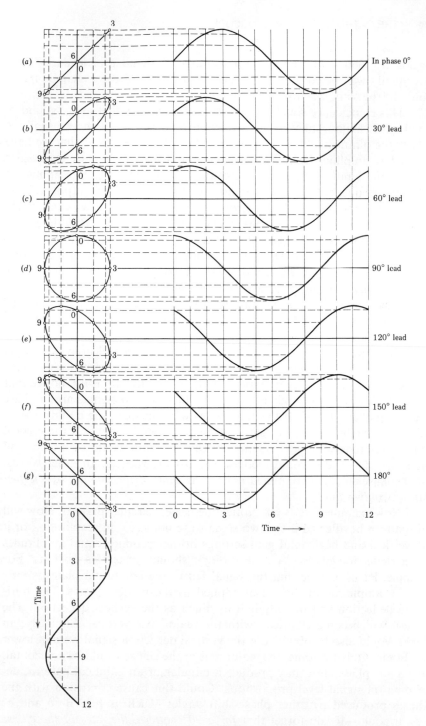

Figure B-6 Effect of phase shift on Lissajous figures (1 : 1 frequency ratio).

lagging by $(360 - \theta)$ degrees. For example, a signal that leads by 300° can equally well be considered as lagging by 60°, and the Lissajous figure would be the same as shown in Fig. B-6c.

This coincidence simplifies the pattern structures that must be understood when making phase-shift measurements. But—and there usually is a fly in the ointment—it adds a complication. Let us suppose that we have checked an amplifier for phase shift and found that a 15° phase shift exists between the input voltage (E_i) and output voltage (E_o) of the amplifier at a certain frequency. Now we would like to know if the output voltage leads or lags the input voltage. From the pattern analysis, we cannot tell. It would seem that our measurements were of no avail. But there is a solution. Let us insert an R-C circuit between the E_o signal and the oscilloscope, with the constants (R and C) deliberately selected to add approximately 15° *lead* to the E_o signal. If the Lissajous elliptical pattern now widens, indicating an *increase* in phase toward 30°, the original output voltage (E_o) must have been leading! On the other hand, if the pattern collapses toward a slant line, the original output voltage must have been lagging.

Now that this ambiguity of lead or lag has been cleared, let us get back to Fig. B-6. Remember that these patterns are obtained only when the two signals are of equal amplitude. Referring to Fig. B-6a, notice that a straight line at a 45° angle is obtained when the two signals are in phase and of equal amplitude. If the amplitudes were unequal, the pattern would still be a straight line but the slope of the line would change. If the vertical signal amplitude were greater, the vertical deflection would be greater and the slope of the line would increase toward the vertical. In other words, the 0° phase shift is determined *not by the slant of the line but by the fact that it is only a line!*

Now examine Fig. B-6d. Notice the perfect circular pattern. How will this pattern be affected if the two signals are not equal in amplitude—or if the vertical and horizontal gain settings do not produce equal amplitudes at the deflection plates? A little thought should make this obvious. For example, let us assume that the signal fed to the horizontal plates is only half the amplitude of the signal applied to the vertical plates. The horizontal deflection will be only half as much as the vertical deflection. The pattern will become elliptical, with the major axis vertical. Similarly, an ellipse would also be obtained if the vertical deflection signal were of lower amplitude. Only this time, the major axis of the ellipse would be horizontal. So—a 90° phase shift may produce a circular or an elliptical pattern, depending on signal level proportions. Could this cause confusion with the ellipses produced by other phase-shift angles—such as Fig. B-6b and c? Yes, it could—if you forget that *for a 90° pattern, the axes of the ellipse must be vertical and horizontal.*

Evaluating the degree of phase shift for intermediate angles is more confusing—if the two signal levels are not equal. The elliptical patterns shown in Fig. B-6 will vary in thickness and also in the slant of the major axis of the ellipse.

It would seem that except for 0° and 90° phase shifts, evaluation of phase shift by this method is only a good guess. However, the oscilloscope is capable of much better accuracy. The system explained below will give a high degree of accuracy. In addition, when using this second method, it is not necessary to have both signals of equal amplitude. Reference to Fig. B-7 will make the procedure clear. Merely measure the maximum vertical deflection of the pattern (y maximum) and the vertical distance intercepted by the pattern as it crosses the y axis (y intercept). The position of the y axis can be determined either by estimating the center line of the pattern or by temporarily turning the horizontal gain to zero. The vertical line left on the screen will coincide with the y axis. (For convenience the location of the y axis can be made to coincide exactly with the center of the screen by using the horizontal centering control.) When this axis is determined, the horizontal gain should be adjusted to a suitable value.

The phase-shift angle (θ) between the two signals is obtained from the following relation:

$$\sin \theta = \frac{y_{\text{intercept}}}{y_{\text{max}}}$$

Once the sine of the phase-shift angle is calculated, the angle itself can be found from trigonometric tables or slide rule scales. From Fig. B-7, we can readily see that when the phase shift is zero, as in (a), the y intercept is zero; as the phase shift increases, the y intercept increases and $\sin \theta$ from our formula also increases; in the extreme case of 90° phase shift, as in (d), the y intercept equals y maximum and $\sin \theta = $ unity.

A more rigorous proof of the above formula can be made from Fig. B-8. Here the signal applied to the vertical plates leads the horizontal

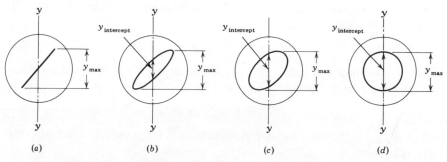

Figure B-7   Evaluation of phase-shift angle from y maximum and y intercept.

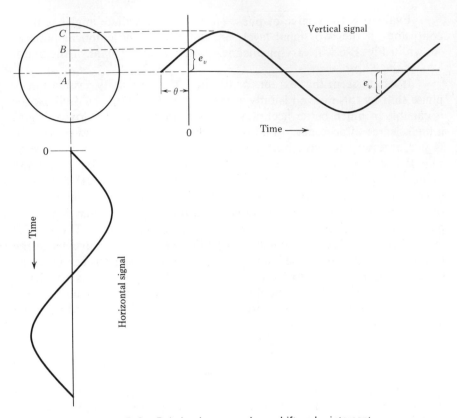

Figure B-8   Relation between phase shift and y intercept.

voltage by an angle θ. If both signals were in phase, the projection on the oscilloscope axis for time = 0 would be at the center of the coordinate axis, A. However, due to the lead of the vertical signal voltage, the projection of this point is at B. Since the amplitude of the horizontal voltage is zero at time = 0, the y intercept A-B on the screen is caused by the instantaneous voltage value of the signal on the vertical plates at time = 0. Expressed mathematically,

$$y_{\text{intercept}} = e_v$$

But the instantaneous value of this voltage depends on its own maximum value and the degree of phase shift θ, or

$$e_v = E_{v(\text{max})} \sin \theta$$

where $E_{v(\text{max})}$ is the maximum vertical deflection (y_{max} or A-C). Substituting this value, we get

$$e_v = y_{\text{intercept}} = y_{\text{max}} \sin \theta$$

Solving for sin θ,

$$\sin \theta = \frac{y_{\text{intercept}}}{y_{\text{max}}}$$

So far we have discussed the y intercept and y maximum values obtained from the positive half-cycle of the signal applied to the vertical deflection plates. If we repeat the analysis for the negative part of the cycle, we would get equal but negative values for the y intercept and y maximum. Therefore in calculating the phase shift between the signal applied to the vertical and horizontal deflection plates we can use either the half-values (positive or negative) as proved from Fig. B-8 or the total values as shown in Fig. B-7; doubling both the y intercept and y maximum gives the same result for sin θ. In practice it is usually easier to measure the total values.

A word of caution must be added at this point. Remember that the amplifiers in the oscilloscope (horizontal and vertical) may not be identical. At certain frequencies, these amplifiers may introduce phase shifts of their own. If the phase shift is not equal for the horizontal and vertical amplifiers, then the phase shift as measured from the scope pattern may be due not only to phase difference between the original signals but also to phase shift introduced by the H and V amplifiers.

The solution to this seeming vicious circle is quite simple. First check the H and V amplifiers by feeding the *same* signal to both the H and V inputs. If the amplifiers are identical, the pattern seen will be a straight line. If any other pattern is seen, calculate the phase difference between the H and V amplifiers and whether the horizontal or vertical amplifier output is leading (or lagging). Now feed the signals to be tested one each to the H and V inputs. Again calculate the phase difference from the scope pattern. Correct this value by the phase difference between the horizontal and vertical scope amplifiers found previously. The remaining phase difference must be due to shift between the test signals.

Since the phase-shift difference between the vertical and horizontal oscilloscope amplifiers may also vary with the input signal frequency, this precaution must be made separately for all frequencies above and below the limitations of these amplifiers.

PHASE-SHIFT MEASUREMENTS USING EXTERNAL SYNCHRONIZATION. The preceding method for measurement of phase shift seems rather involved. Actually the difficulty lies in the verbal explanation of the procedure rather than in the technique itself. It is harder to explain than it is to show. There is another method sometimes used which avoids the necessity of checking the oscilloscope amplifiers and is somewhat simpler to understand and to apply. The accuracy of this second method is not as good, but it is often sufficient. Let us assume that we wish to determine the phase shift (if any) introduced between input and output of some electronic unit. Using the basic oscilloscope of Fig. B-1:

1. Feed the input voltage (E_i) of the unit under test to the vertical input terminals of the oscilloscope. Adjust the vertical gain of the oscilloscope amplifier to give maximum usable deflection on the screen.
2. Using the internal sawtooth generator, adjust the coarse and fine frequency control to produce a one-cycle pattern as stationary as possible. Also adjust the horizontal gain to give a large but convenient horizontal sweep across the screen face. (Since one cycle is 360°, convenient deflections would be 12, 18, or 36 spaces; i.e., one space equals 30° , 20°, or 10°, respectively.)
3. Using *external synchronization,* apply the minimum synchronizing voltage needed to lock the screen pattern. As a source of this external synchronizing voltage, we can use the same input voltage (E_i) that is being fed to the unit under test (and at present, also to the vertical input terminals of the oscilloscope).
4. Draw the pattern seen on the screen (preferably trace), showing the exact starting point and zero axis of the wave.
5. Now disconnect the signal voltage (E_i) from the oscilloscope and reconnect the vertical input terminals of the oscilloscope to the output of the unit under test. Adjust the vertical gain for the same vertical deflection as in step 1.
6. Since the frequency of this output voltage (E_o) is the same as the input voltage (E_i), no change need be made to any other control.
7. Repeat step 4 above. Since the synchronizing signal is the same for step 4 and step 7, the two waveshapes (E_i and E_o) are drawn with respect to a common reference. Their phase relation can now be identified as follows:
   (*a*) If the two waves are in coincidence (they both start at zero and rise in a positive direction), they are in phase.
   (*b*) If the E_o wave starts at some positive value while E_i starts at zero and rises in a positive direction, then E_o leads E_i. The amount of lead can be calculated by noting the spacing between some identical point on each wave. For example, with the horizontal gain adjusted for 36 spaces equal to 360°, if the zero (and going positive) point for E_o occurs two spaces to the left of the zero (and going positive) point of E_i, E_o leads by 20°.
   (*c*) If the E_o wave appears shifted to the right of the E_i wave, then E_o is lagging and the amount of lag can be calculated from the number of spaces difference, as above.

The dual-channel laboratory oscilloscope shown in Fig. B-4 is ideally suited for such phase-shift measurements. The two signals to be evaluated are fed one each to the two vertical input terminals; the VOLTS/DIV switches are set for (approximately) equal screen deflections; the MODE switch is set for ALT so that both patterns are displayed simultaneously; and the trigger is set for CH 1 ONLY, so that both waveforms will be triggered at the same timing instant. The rest of the procedure should be obvious from our earlier discussion.

FREQUENCY MEASUREMENT USING A FREQUENCY STANDARD. Another basic application of the oscilloscope is in the measurement of frequency. In addition to an oscilloscope, it is necessary to have a source of known frequencies as a standard for comparison. (A calibrated audio or RF oscillator is suitable as the standard.) The signal of unknown frequency is fed to the vertical input terminals of the oscilloscope; the standard frequency source is connected to the horizontal input terminals. Obviously the internal sawtooth sweep circuit is not used. (Switch *B* in the block diagram of Fig. B-1 is set to EXT or the HOR DISPLAY switch of Fig. B-4 is set to EXT HOR. This connects the horizontal input terminals through the horizontal amplifier to the horizontal deflection plates.) Again, as for phase-shift measurement, the patterns developed on the screen will be Lissajous figures. Analysis of these patterns will give the frequency ratio between the two signals.

Let us analyze the pattern that is formed when the unknown frequency is twice the standard frequency. In other words, two cycles of the unknown frequency will correspond in time to one cycle of the standard frequency. Figure B-9 shows a point-by-point development of the pattern. The method used is the same as in Fig. B-2.

In stating frequency ratios, it is common practice to state the horizontal frequency first and then the vertical frequency. In Fig. B-9, the frequency ratio would be 1:2; that is, one horizontal to two vertical. Let us add two tangent lines to the oscilloscope pattern of Fig. B-9. Notice that the *horizontal motion* of the electron beam causes the beam to touch the *vertical tangent line AB* only once, at point 6. On the other hand, the *vertical motion* of the beam causes it to touch the *horizontal tangent line* twice, at points 5 and 1. By this method, the frequency ratio of any Lissajous figure can be determined. Merely count the number of times the beam sweeps horizontally to vertically. If you prefer formulas, this relation can be given by

$$\frac{\text{Frequency on horizontal axis}}{\text{Frequency on vertical axis}} = \frac{\text{number of loops touching the vertical line}}{\text{number of loops touching the horizontal line}}$$

The pattern shown here (Fig. B-9) applies only if the two signals are starting with an in-phase relationship—starting at zero and increasing in a positive direction. It should be obvious from our previous discussion on phase shift that with any other starting phase relation, the pattern would

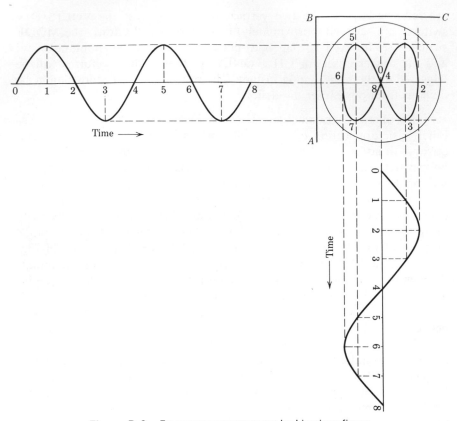

Figure B-9 Frequency measurement by Lissajous figure.

be different. However, regardless of the exact shape of the pattern, the beam would still make one horizontal sweep to two vertical sweeps. In terms of the formula, the pattern would approach the vertical tangent line once and the horizontal tangent line twice. The frequency ratio can still be readily identified as 1:2.

If the two frequencies are not exactly in a 1:2 ratio, the starting phase relation will vary continuously. If they start in phase, the angle will gradually increase to 360° (or back to 0°). As a result, the pattern seen on the screen will change continuously from the horizontal figure eight shown in Fig. B-9 to an upright (or inverted) U and back again. In any case, the pattern will always retain the 1:2 relationship.

Let us use an example to illustrate frequency measurement by Lissajous figures.

EXAMPLE B-1

What is the frequency of each of the unknown signals (Fig. B-10) if the frequency applied to the horizontal plates is 1000 Hz?

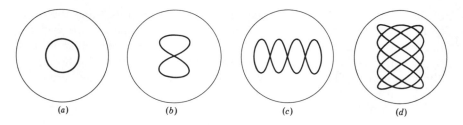

Figure B-10

Solution

1. Since the beam sweeps vertically once and horizontally once, the ratio is 1:1. Therefore the unknown frequency is also 1000 Hz.
2. The beam sweeps horizontally twice but only once in the vertical direction. The ratio is 2:1. This time the unknown frequency is only half the standard frequency, or 500 Hz.
3. The frequency ratio is 1:4. The unknown frequency must be 4000 Hz.
4. Five horizontal sweeps are made while the vertical motion makes three sweeps. The vertical frequency is at a slower rate (three-fifths of 1000, or 600 Hz).

When the standard frequency is fixed in value, the size of the oscillo-scope screen limits the frequency range of the unknown signals that can be measured. For example, if the difference between known and unknown frequencies is too great (10:1 or 1:10), the number of tangent loops may be too difficult to evaluate. Also, if the frequency difference is too small, such as 10:9 or vice versa, again the number of tangent loops may be too many to count.

The 1:1 ratio (since it has the least number of loops) is the easiest to identify and the most accurate in its indication. Therefore, the simplest method of measuring the frequency of an unknown signal is to compare it to the output from a calibrated variable-frequency oscillator. Vary the frequency of this standard oscillator until a 1:1 pattern is obtained. (Although Fig. B-10a shows a circle for the 1:1 pattern, remember that this pattern may also vary from a straight line to an ellipse or circle—depending on phase shift.) Then merely read the frequency indicated on the dial of the variable-frequency oscillator. The accuracy of the determination is obviously limited to the accuracy of the oscillator used as the standard.

FREQUENCY MEASUREMENT USING CALIBRATED TIME BASE.   Obvi-ously, this technique can only be used with an oscilloscope that has a high-stability sweep and a calibrated TIME/DIV control. The "unknown" signal is fed to the vertical input, and the vertical gain is adjusted for a convenient display height. The A-sweep TIME/DIV control is set to display one cycle of the waveform in just under eight major scale divisions;* and the wave-form is centered on the screen. Then the period (or time for one cycle) of the unknown signal is equal to the horizontal span of one cycle multiplied by the TIME/DIV setting. For example, if one cycle spans 7.4 horizontal divisions and the TIME/DIV switch is set at 0.1 ms, then the period (T) is $7.3 \times 0.1 = 0.73$ ms and the frequency is

$$f = \frac{1}{T} = \frac{1}{0.73} \times 10^3 = 1370 \text{ Hz}$$

FREQUENCY RESPONSE BY SQUARE-WAVE TESTING.   You will recall from your studies of complex waves† that a square wave has many harmonics and that the amplitude of these harmonics is fairly strong. For example, the fifteenth harmonic has an amplitude equal to one-fifteenth of the fundamental frequency amplitude. Also, from an analysis of this wave you will recall that the steepness of the wave front is due to the high-order harmonics. These characteristics of a square wave give us a very quick method of checking the frequency response of an amplifier by using a square-wave generator and an oscilloscope. A square wave is fed into the amplifier and the output wave is examined on the oscilloscope for distortion of the square-wave pattern. Since we are examining the waveshape, the signal is fed to the vertical input and the internal sawtooth generator is used for horizontal sweep.

For checking the low-frequency response, the square-wave frequency should be as low as the lowest frequency limit desired. If the amplifier has good response at this low frequency, the screen pattern will be undistorted (Fig. B-11a). If the gain of the amplifier falls off at this low frequency, the pattern will show a dip in the center (Fig. B-11b). Usually low gain at low frequencies is accompanied with a high degree of phase shift. The combined effect of low gain and phase shift is seen in Fig. B-11c and d. Figure B-11d is obtained when the response drops off to 10 percent and the phase shift is 80 percent. Notice that this waveshape looks like the output from a differentiating circuit. This is exactly what happens. R-C circuits in a poorly designed amplifier will cause differentiation of the square-wave input.

To test the high-frequency response of an amplifier, the square-wave signal should have a frequency of approximately one-tenth of the maximum frequency desired. If the output as seen on the screen is a good square

---

*The region between the first and ninth major divisions provides the most linear time measurement. Therefore, for accurate timing or frequency measurements the first and last divisions of the display should not be used.

†J. J. DeFrance, *Electrical Fundamentals,* Englewood Cliffs, N.J.: Prentice-Hall, Inc., 1969, part 2, chap. 4.

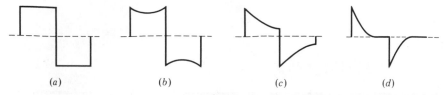

(a)   (b)   (c)   (d)

Figure B-11   Effect of amplifier low-frequency response on square-wave pattern.

wave, the amplifier has good response up to at least the fifteenth harmonic of the square-wave frequency. Figure B-12 shows the effects of poor high-frequency response. The rounding of both edges of the square wave represented in (a) is due to poor high-frequency response without phase shift. This pattern represents a loss of gain of approximately 30 percent at 10 times the square-wave frequency. Figure B-12b represents a similar loss of

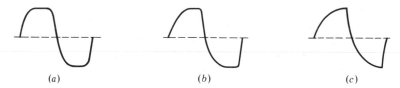

(a)   (b)   (c)

Figure B-12   Effect of poor amplifier high-frequency response on square-wave pattern.

gain but with attending phase shift. The curve of Fig. B-12c results from very poor gain and high phase shift. It represents a loss of 30 percent gain at a frequency of approximately twice the square-wave frequency.

So far we have discussed four basic uses of the oscilloscope. To summarize, they are:

1. Viewing of waveshapes—for analysis
2. Measurement of voltage
3. Measurement of phase shift
4. Measurement of frequency

In addition we included a fifth use—checking of frequency response by square-wave testing. Actually this last application is not a basic use, but rather a *specific* application of viewing a waveshape for analysis. Similarly, the oscilloscope can be used for many other specific applications of these four basic functions. For example, the oscilloscope is used for calibration of signal generators; alignment of AM, FM, and TV receivers; troubleshooting and servicing of receivers; and modulation checks of percentage and linearity in transmitters. Again, these are not new uses but applications of the four basic techniques.

Appendix C

Current and Voltage Letter Symbols

Current and voltage values are often represented by letter symbols. This is especially true in equations and diagrams. Much confusion can arise if these symbols are interpreted incorrectly. The symbols used in this text are in accordance with IEEE standards. An explanation follows.

GENERAL

1. All *fixed* values—such as dc (or average), rms, and maximum values—are represented by capital letters (E or V for voltages, I for currents).
2. All *instantaneous* values are represented by lowercase letters (e or v, and i).

To further distinguish current and voltage values, suitable subscripts are added. In general, it is necessary to distinguish between dc components, ac (or varying) components, and *total* values (dc + ac). These subscripts also indicate the element or elements to which the currents or voltages apply.

APPLICATION TO TRANSISTORS

1. Supply voltages (dc)—use *doubled capital-letter subscript* for the element being supplied. Examples:
   (*a*) V_{EE} emitter supply voltage
   (*b*) V_{DD} drain supply voltage

2. All other dc values—use appropriate *capital-letter subscript(s)* for the elements involved. Examples:
   (a) I_B base current
   (b) V_{BE} voltage between base and emitter
3. Values of the ac (varying) component only—use appropriate *lower-case subscript(s)*. Examples:
   (a) I_e rms value of emitter current
   (b) I_{em} maximum value of emitter current
   (c) i_b instantaneous value of base current
   (d) v_{ce} instantaneous value of collector-to-emitter voltage
4. Total values (ac + dc)—use appropriate *capital-letter subscripts*. Examples:
   (a) I_D instantaneous value of the *total* drain current
   (b) I_{BM} maximum value of the *total* base current
   (c) v_{BE} instantaneous value of the total base-to-emitter voltage

It should be noted that total values (ac + dc) are measured with a zero (absolute) value as reference, whereas ac components are measured with their own average values as references. The above designations are illustrated in Fig. C-1 for the drain current in a field-effect transistor.

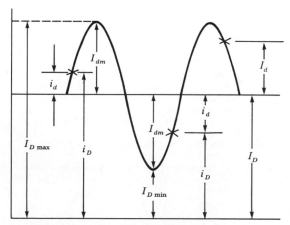

I_{Dmax} — maximum value of the total current.
i_d — instantaneous value of the ac component.
i_D — instantaneous value of the total current.
I_{dm} — maximum value of the ac component.
I_{Dmin} — minimum value of the total current.
I_D — average value, or dc component.
I_d — rms value of the ac component.

Figure C-1   Letter symbol for FET drain currents.

APPLICATION TO ELECTRON TUBES

Letter symbols are used in much the same manner with electron tubes. However, instead of using lowercase and capital-letter subscripts to distinguish between ac and total or dc components, we now use *different* letters, and we always use lowercase subscripts. In the grid circuit, g denotes an ac component, while c denotes a total or a dc component. In the plate circuit, p is used for the ac component, and b is used for the dc or for the total value. This nomenclature is illustrated in Fig. C-2 to distinguish voltage values in the plate circuit of a vacuum tube.

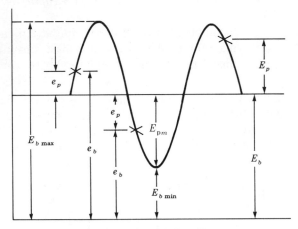

$E_{b\,max}$ — maximum value of the total plate voltage.
e_p — instantaneous value of the ac component.
e_b — instantaneous value of the total plate voltage.
E_{pm} — maximum value of the ac component.
$E_{b\,min}$ — minimum value of the total plate voltage.
E_p — rms value of the ac component.
E_b — average value, or dc component.

Figure C-2   Letter symbols for vacuum-tube plate voltages.

Answers to Odd-Numbered Numerical Problems

Chapter 1

1. (a) 34
   (b) 4
   (c) 2, 8, 18, 6
   (d) 6
   (e) no
3. (a) 179 Ω
   (b) 90.9 Ω
5. (a) 3000 Ω
   (b) 600 Ω
7. 1300; 13; 4.33

Chapter 2

5. 0.41 mA
7. 1.51 mA; 49.5 mA
9. 65.7
11. 4.17 mA
13. 0.99
15. 0.88
17. 29

Chapter 3

5.

V_{GS}	I_D(mA)	
	(a)	(b)
0	9.9	9.7
−1	6.3	6.2
−2	3.5	3.4
−3	1.25	1.2
−4	0.2	0.2

7. 2500 μmho
9. 25 μmho; 40,000 Ω
11.

V_{GS}	I_D
0	2.5 mA
−0.5	1.30
−1.0	0.49
−1.5	0.07
−1.8	0

13. 3.97×10^{-4}
15. (a) 578 μmho; 3200 μmho
    (b) 1170 μmho; 2066 μmho
17. (a) 1633 μmho
    (b) 1746 μmho

Chapter 4

3. (a) 0.5 mA
   (b) 50 kΩ
   (c) 33 kΩ

Chapter 4 (*cont.*)

5. (a) 4.4 mA
   (b) 22.7 kΩ
   (c) 14.3 kΩ
7. (a) 20
   (b) 11.4 kΩ
   (c) 1750 μmho
   (d) 1754 μmho
9. (a) 0.2 V
   (b) −0.8 V
   (c) −1.2 V
   (d) 3.18 mA
   (e) 2.30 mA
   (f) 0.88 mA
11. (a) 0.2 V
    (b) −2.3 V
    (c) −2.7 V
    (d) 0.5 mA
    (e) 0.2 mA
    (f) 0.3 mA

Chapter 5

1. 450 V
3. 2.49 V
5. 960 turns
7. 0.815 A
9. 0.3 A
11. (a) 333 V
    (b) 2.5 V
    (c) 5.0 V
    (d) 6.33 V
13. (a) 135 V
    (b) 212 V; 60 Hz
    (c) 89.9 V; 120 Hz
15. (a) 424 V
    (b) 42.4 V
17. 1:4.63 step-up
19. 270 V
21. (a) 200-0-200 V
    (b) 566 V
23. (a) 432 V
    (b) 288 V; 120 Hz
    (c) 679 V
25. 1:2.78 step-up
27. 2336 V
29. 2431 V

Chapter 6

1. 12.8%

Chapter 6 (*cont.*)

3. (a) 155 V
   (b) 45 V
   (c) 13 V
   (d) 9.62%
5. (a) 32 ms
   (b) 16.7 ms
   (c) 484 V
7. 905 V
9. 38.4°
11. 720 mA
13. (a) 4.0%
    (b) 53 V
15. (a) 64.2 μF
    (b) 227 V
17. (a) 259 V
    (b) 1.89%
19. 252 V
21. (a) 889 V
    (b) 1 : 7.41 step-up
23. 830 μF
25. (a) 1.38%
    (b) 0.345%
    (c) 0.04%
27. 36 mA
29. 17.2%
33. (a) 113.6 Ω
    (b) 1.76 W
    (c) yes
35. (a) 45.5 Ω
    (b) 2.16 W
    (c) no

Chapter 8

1. (a) 27 W
   (b) 9 times
3. (a) 5
   (b) 0
   (c) 1
   (d) 2
   (e) 5
   (f) 1
5. (a) 1.629
   (b) 3.571
   (c) 0.0969
   (d) 2.861
   (e) 5.615
   (f) 1.571
7. 10.8 dB
9. 0.158 W
11. 1.89 V

Chapter 8 (*cont.*)

13. (a) 10 W
    (b) 0.00631 μV
    (c) 40.8 dB
    (d) −27 dB
15. +114.8 dB
17. 3.56 dB; 1.51 mV
19. (a) 10 W
    (b) 8.94 W
    (c) 70.7 μW
    (d) 102 dB

Chapter 10

3. 1330 Ω
5. 22.1 μF
7. (a) 47 kΩ
   (b) 40 kΩ
   (c) 41.7 kΩ
9. (a) 22 V
   (b) 15 V
   (c) 7 V
13. 37.7
15. (a) 1.28 mA
    (b) 0.71 mA
    (c) 0.22 mA
17. (a) 0.89 mA
    (b) −2.4 V
19. (a) 0.62 mA; 10.4 V
    (b) 1290 Ω
    (d) 8.1 V
    (e) 20.3
21. (a) 1370 Ω
    (b) 12.4 V
    (c) 10.3
23. (a) 15,400 Ω
    (b) 9.7 V
    (c) 24.2
25. (a) +3 V
    (b) 9 : 1
    (c) 1714 Ω
27. (b) 8.6 mA, 162 V

Chapter 11

3. (b) 2.9 mA; 6.0 V
   (c) 1360 Ω
5. (a) 1.98 mA; 49.5
   (b) 3.68 mA; 46
7. (a) 5.0 mA
   (b) 1.77 mA
   (c) 5.3 V; 1.87 V
   (d) 25
   (e) 17.7

Chapter 11 (*cont.*)

11. (a) 120 μA
   (b) 6.12 mA
   (c) R_b = 157 kΩ
       R_c = 1500 Ω
       R_e = 163 Ω
15. (a) 1.02 mA
   (b) 0.954 mA
   (c) 0.950 mA
23. 3.95 μA; 4.74 mA; 18.6 V
25. 0.67 mA; 3.04 V; 8.38 μA
27. 17.3 μA; 2.60 mA; 6.32 V
29. (a) 0.981 mA; 2.18 μA; 3.27 V
   (b) 11.9 kΩ
   (c) 117
   (d) 2.34 V; 117
31. (a) 2.02 mA; 20.2 μA; 7.0 V
   (b) 1290 Ω
   (c) 79.6
   (d) 3.96 V; 39.6

Chapter 12

1. (a) 1662 Ω; 240 kΩ
   (b) 36.3; 175; 6335
3. −0.974
5. 18.7 V
7. (a) 613 Ω; 16.7 kΩ
   (b) 42.4; 692; 29,340
   (c) 6.00 V
9. (a) 5934
   (b) 47,500
   (c) 48,000
11. (a) 27.9
   (b) 3000
   (c) 4960

Chapter 13

1.

I_B	I_C
100 μA	1.7 mA
200	2.9
300	4.0
400	4.8
500	5.45
600	6.1

3. (a) −50 mA
   (b) −60 mA
   (c)

I_B	I_C
0.1 mA	10 mA
0.2	18.5

Chapter 13 (*cont.*)

3. (c)

I_B	I_C	(*cont.*)
0.3	26.1	
0.4	33.0	
0.5	39.0	
0.6	44.0	
0.7	49.0	
0.8	54.0	
0.9	56.2	

5. (a) 0.45 mA
   (b) 11.5%
7. (a) 0.20 mA
   (b) 6.56%
9. (a) 4.0 mA; 10.0 V
   (b) −4.1 V; +6.6 V
   (c) 5.45 & 1.70 mA
   (d) 0.42 mA
   (e) 22.4%
11. C_i = 172 pF; C_o = 4.0 pF
13. (a) 82; 8.2 V
   (b) 58.3 kHz
   (c) 60; 6.0 V
15. (a) 121
   (b) 2.01 MHz; 85.5
17. (a) 45°
   (b) 3.86°
   (c) 39.8 kHz
19. (a) 121
   (b) 51.8 Hz
21. (a) 82
   (b) 0.75 Hz
23. −11.5 dB @ 20 Hz
   −8.24 dB @ 2 MHz
25. (b) −9 dB @ 40 Hz; −9.5 dB @ 30 kHz
   (d) −5 dB @ 80 Hz; −7 dB @ 20 kHz
29. (a) 76 kHz
   (b) 133 kHz
   (c) 89.7; 52.2; −4.7 dB

Chapter 14

3. 64.5:1 & 32.4:1
7. 42.9 kΩ each
13. (a) 116
   (b) 0.08
   (c) 63.2 dB
15. 2000 (66 dB)

Chapter 15

1. 64 W
3. 48.4°C
5. 0.938°C/W

Chapter 15 (*cont.*)

7. (b) 1.75 A; 17.5 V
9. (a) 27.2%
   (b) 13.1 W
17. (a) 20 W
    (b) 20 W
    (c) 40 W
19. (a) 9.72 W
    (b) 6.22 W
    (c) 9.72 W
    (d) 9.72 W
21. (a) 3.1 W
    (b) 22.5%
    (c) 59%
23. (a) 12.5 W
    (b) 6.25 W
25. (a) 2.2 A; 18 V
    (b) 0.04–1.8 A; 3.0–17.5 V
    (c) 1.06 A & 9.0 V
    (d) 18 V

Chapter 16

1. (a) 20
   (b) 14.3
3. 24.6 dB
5. 24.1 dB
7. 0.47%
13. 2.44%
19. (a) 114
    (b) 6.25%
    (c) 14.0
23. (a) 2025–14,400
    (b) 21.7–22.0
    (c) 0.16–0.023%
25. 35.7 kΩ
27. (a) 4.54%
    (b) 350 kΩ
33. $R_1 = 462$ kΩ
    $R_2 = 8.39$ kΩ
39. (a) 250 Ω
    (b) 235 Ω
41. 775 kΩ; 26.7 Ω
45. (a) 22.8 kΩ
    (b) 582 kΩ
55. −5.37 dB

Chapter 17

1. (a) −26 dBV
   (b) −46 dBV
   (c) 5 mV
3. (a) −32.4 dBV
   (b) −52.4 dBV

Chapter 17 (*cont.*)

5. (a) 1.58×10^{-9} W
   (b) 3.17×10^{-9} W
   (c) 0.159 mV

Chapter 18

5. (a) 54.1%
   (b) 1.85
   (c) 16.7
   (d) 16.7
7. (a) $Z_A = 21$ kΩ $\underline{/-84.5°}$
   (b) $Z_B = 22$ kΩ $\underline{/-2.6°}$
      $V_o = 0.677$ V
   (b) $Z_A = 185$ kΩ $\underline{/-33°}$
      $Z_B = 27.1$ kΩ $\underline{/-36°}$
      $V_o = 0.128$ V
   (c) $Z_A = 220$ kΩ $\underline{/0°}$
      $Z_B = 254$ kΩ $\underline{/-61.3°}$
      $V_o = 0.623$ V
   (d) +14.5 dB @ 16 kHz
      +13.7 dB @ 50 Hz
   (e) $Z_A = 180$ kΩ $\underline{/-2.3°}$
      $Z_B = 1.99$ kΩ $\underline{/-85°}$
      $V_o = 0.011$ V
   (f) $Z_A = 235$ kΩ $\underline{/-35.4°}$
      $Z_B = 700$ kΩ $\underline{/-55.3°}$
      $V_o = 0.757$ V
   (g) $Z_A = 3.37$ MΩ $\underline{/-69.6°}$
      $Z_B = 636$ kΩ $\underline{/-1.98°}$
      $V_o = 0.099$ V
   (h) −36 dB @ 16 Hz
      −17.7 dB @ 50 Hz
9. (a) for Z_A, 500:8
      Z_B, 1000:16
      Z_C, 1000:16
      Z_D, 1667:3.2
      Z_E, 2500:3.2
   (b) 70.7 V
   (c) 44.7 V
   (d) 4W, 2W, 2W, 1.2 W, 0.8 W
11. $Z_A = 500:8$
    $Z_B = 1000:16$
    $Z_C = 1000:16$
    $Z_D = 1667:3.2$
    $Z_E = 2500:3.2$

Chapter 19

3. Yes, (just)
7. (a) 51.6
   (b) 53.9
11. 352 kΩ

Index